35th European Peptide Symposium 2018

Dublin, Ireland
26-31 August 2018

Editors:

Patrick Timmons
Michal Lebl
Chandralal Hewage

ISBN: 978-1-5108-8036-8

Printed from e-media with permission by:

Curran Associates, Inc.
57 Morehouse Lane
Red Hook, NY 12571

CURRAN ASSOCIATES INC. proceedings.com

Some format issues inherent in the e-media version may also appear in this print version.

Printed by Curran Associates, Inc. (2019)

For permission requests, please contact European Peptide Society
at the address below.

European Peptide Society
c/o Professor Paula Gomes
Rua do Campo Alegre, s/n
PT-4169-007 Porto
Portugal

Phone: +351 220 402 563

pgomes@fc.up.pt

Additional copies of this publication are available from:

Curran Associates, Inc.
57 Morehouse Lane
Red Hook, NY 12571 USA
Phone: 845-758-0400
Fax: 845-758-2633
Email: curran@proceedings.com
Web: www.proceedings.com

Proceedings of the 35th European Peptide Symposium

26th-31st August 2018 | Dublin City University, Ireland

Edited by:

Patrick Timmons
University College Dublin,

Chandralal Hewage

University College Dublin,

Michal Lebl
Prompt Scientific Publishing

European Peptide Society

Welcome to the 35th EPS in Dublin

Dear friends, partners and fellow colleagues

On behalf of the organising committee, it is our pleasure to invite you to join in the 35th European Peptide Symposium, 35EPS, on 26-31 August 2018 at the Dublin City University, Dublin, Ireland. Ireland has been known as the land of Saints and Scholars. The capital city of Dublin lies at the mouth of the river Liffey embracing the Irish Sea. There is evidence of a community living in Dublin since the fifth century. It is a city rich in historical and cultural attractions including the unique Book of Kells, the famous Guinness brewery, the neighbouring hills of Tara and the monastic village of Glendalough.

The European Peptide Society is a non-profit association established for the public benefit to promote in the advancement of education and scientific study of the chemistry, biochemistry, biology and pharmacology of peptides in Europe and in neighbouring countries. In co-operation with John Wiley & Sons, the EPS publishes the Journal of Peptide Science and the EPS Newsletter. The EPS supports the organisation of the European Peptide Symposium, sponsors local and specialised meetings and workshops, and supports and promotes young scientists with travel grants and dedicated seminars.

The European Peptide Symposium is a well-established biennial peptide conference that has grown substantially. It has been well attended by Nobel Laureates, world-class scientists, leading scientific investigators, world-renowned commercial partners, publishers and leading companies in the area of peptide research and its applications. This symposium includes sessions covering every aspect of the research, education and innovation related to the peptide science.

The 35EPS will take place at the Helix, a purpose built conference venue associated with the Dublin City University. The Helix is located approximately 5 km north of Dublin city centre short distance away from the Dublin InternationalAirport. The scientific sessions, poster sessions and exhibition will take place in the Helix and its adjacent rooms. Dublin City University has excellent on-site accommodation, restaurants and cafes which will cater for symposium delegates.

The 35EPS symposium will start on 26th of August with award and prize giving lectures and a welcome reception. In addition to plenary and invited speaker presentations, there will be poster presentations and a mini symposium for young scientists for covering every aspect of peptide science. The symposium gala dinner will take place on the 30th of August at the Crowne Plaza Hotel and the excursions are planned on Friday for sampling the history and culture of Dublin and its surroundings towns.

The executive council, the scientific advisory board, and the program committee hope to make the 35EPS symposium an exciting and memorable scientific event. You will also have the opportunity to visit the hospitable and attractive city of Dublin with many cultural attractions. It is our pleasure to welcome you to Dublin, and we hope that you enjoy the symposium and your visit.

35 EPS Organising Committee
Chandralal Hewage, University College Dublin (Chairman)

Dear colleagues and friends in the worldwide peptide community

The 35th edition of the European Peptide Society symposium, 35EPS, is well on the move! We are happy that this symposium signals the expansion of our society's sphere of influence into a so far unvisited EPS member nation. With a rich cultural heritage, a forward-looking society and, more to the point, a dynamic biotech sector, Ireland is an ideal location for talking shop on peptides in the 21st century. In continuation of the highly successful 34EPS in Leipzig, to be followed by equally enticing calls in Barcelona (2020) and Florence (2022), this upcoming 35EPS promises to be a unique opportunity to learn about the ever-growing impact of peptide research in a multitude of scientific areas, to share your own results and interests, and to build productive professional ties with like-minded scientists on a worldwide scale. Assisted by capable scientific and program committees, Prof. Chandralal Hewage, the symposium chair, is planning not only a first-rate scientific event but has also arranged for top-notch conference facilities, including affordable accommodation, and for an attractive social program.

On behalf of the European Peptide Society, I look forward to greeting you personally in Dublin, August 26-31, 2018!

With best personal regards,
David Andreu
President, European Peptide Society

Organisation

Organising Committee

Chairman

Chandralal Hewage, University College Dublin

Co-Chairs

Marc Devocelle, Royal College of Surgeons Ireland
Tia Keyes, Dublin City University
Finbarr O'Harte, University of Ulster

Committee Members

Fintan Kelleher, Tallaght Institute of Technology
Cliona O'Farrelly, Trinity College Dublin
Brian Green, Queens University Belfast
Denis Shields, University College Dublin
Eoin Fleming, University College Cork
Patrick Timmons, University College Dublin

Scientific Advisory Board

Alethea Tabor, University College London
Chandralal Hewage, University College Dublin
David Andreu, Pompeu Fabra University
Ernest Giralt, University of Barcelona
Florine Cavelier, Universitè de Montpellier
John Howl, University of Wolverhampton
John Wade, University of Melbourne
Knud Jensen, University of Copenhagen
Marc Devocelle, Royal College of Surgeons Ireland
Marcey Waters, University of North Carolina
Maria Luisa Mangoni, Sapienza Università di Roma
Maurice Manning, University of Toledo
Meritxell Teixidó, IRB Barcelona
Norbert Sewald, Bielefeld University
William Lubell, Université de Montréal

EPS Council

EPS National Representatives

Christian Gruber, Austria
Steven Ballet, Belgium
Tamara Pajpanova, Bulgaria
Andreja Jakas, Croatia
Vaclav Cerovsky, Czech Republic
Knud Jensen, Denmark
Ale Närvänen, Finland
Florine Cavelier, France
Oliver Seitz, Germany
George Spyroulias, Greece
Gábor Mező, Hungary
Chandralal Hewage, Ireland
Assaf Friedler, Israel
Paolo Rovero, Italy
Dirk Rijkers, The Netherlands
Bengt Erik Haug, Norway
Zbigniew Szewczuk, Poland
Miguel Castanho, Portugal
Vladislav Deigin, Russia
Katarina Krskova, Slovakia
Mojka Lunder, Slovenia
Meritxell Teixido, Spain
Ulf Göransson, Sweden
Helma Wennemers, Switzerland
John Howl, United Kingdom

Awards

Awards

Miklós Bodanszky Award

The Miklós Bodanszky Award is presented in commemoration of his outstanding contributions to peptide science. The award is given to the scientist who made significant contributions to peptide-based drug research in the period of ten years after having obtained the PhD degree. The Award is presented during the EPS Symposium and is sponsored by BCN PEPTIDES. The 2018 award will be announced at the 35th EPS opening ceremony.

Awardee: **Norman Metanis**; The Hebrew University of Jerusalem, Israel

Leonidas Zervas Award

The Leonidas Zervas Award is presented in commemoration of his outstanding contributions to peptide science. The award is given to the scientist who has made the most outstanding contributions to the chemistry, biochemistry and/or biology of peptides in the five years preceding the date of selection, and is sponsored by the donation of Dr. Rao Makineni. The 2018 award will be announced at the 35th EPS opening ceremony.

Awardee: **Christian Hackenberger**; FMP Berlin, Germany

Josef Rudinger Memorial Award

The Josef Rudinger Memorial Lecture Award is presented "in commemoration of Josef Rudinger's role in the foundation of the European Peptide Symposia and of diverse contributions he made to peptide chemistry". The Award is presented during the EPS Symposium and is sponsored by PolyPeptide Group. The 2018 award will be announced at the 35th EPS opening ceremony.

Awardee: **James P. Tam**; Nanyang Technological University, Singapore

ESCOM Science Foundation (ESF) Dr. Bert L. Schram Symposium and Awards

On behalf of the ESCOM Science Foundation (ESF), 35EPS is pleased to announce the Dr. Bert L. Schram Young Investigators' Mini Symposium (YIMS) for speakers under 35 years of age. Additionally, there will be oral presentation and poster presentation awards and a networking mixer.

The Dr. Bert L. Schram Oral Presentation Awards

There will be two oral presentation awards, €750 for the winner and of €500 for the runner-up. They will also receive a signed certificate in the joint names of EPS and ESF. Awards will be presented by ESF representatives at the symposium banquet.

Awardees:

1st place: **Tyler Lalonde**; University of Western Ontario, Canada
2nd place: **Sabine Schuster**; Eötvös Loránd University, Hungary

The Dr. Bert L. Schram Poster Presentation Awards

There will be six poster presentation awards. The top two winners will receive each a cash amount of €500 and the next four winners a cash amount of €250 each. They will also receive a signed certificate in the joint names of EPS and ESF. Awards will be presented by ESF representatives at the symposium banquet.

Awardees:

1st place: **Jasmine Egli**; ETH Zürich, Switzerland
2nd place: **Mario Felicio**; Instituto de Medicina Molecular, Portugal
3rd place: **Steven Verlinden**; Vrije Univeriteit Brussel, Belgium
4th place: **Reem Mousa**; The Hebrew University of Jerusalem, Israel
5th place: **Thomas Schmitz**; University of Bonn, Germany
6th place: **Vera Neves**; Instituto de Medicina Molecular, Portugal

Wiley Poster Presentation Awards:

Awardees

1st place: **Shunliang Wu**; University of Copenhagen, Denmark
2nd place: **Rumit Maini**; University of Tokyo, Japan
3rd place: **Lena Steinacker**; Technical University of Darmstadt, Germany

The Dr. Bert L. Schram Networking Mixer

The Networking Mixer is a buffet lunch bringing together EPS young scientists and international senior scientists from academia and corporate research. This is an exclusive event and the attendees will gain invaluable insights on current research trends, and orientation on career planning and professional opportunities. Attendees will have ample opportunity to network and meet and greet EPS officers and ESF representatives.

Symposium Report

As Finbarr O'Harte, co-chair of the 35th European Peptide Symposium (EPS), highlighted during his concluding remarks, this meeting was hosted in a venue consistently named for a peptide event. Held at the Helix on the campus of Dublin City University (DCU), the EPS biennial international symposium happened for the first time since 1989 in Ireland, one of the recent member states which joined the EPS in 2008.

The Symposium was opened on the 26th of August by Chandralal Hewage (University College Dublin), Chair of the 35th EPS, who introduced Brian MacCraith, President of DCU, and David Andreu, Chairman of the EPS. They welcomed the participants on behalf of DCU and the EPS Executive Committee, respectively. Following the traditional gong banging, the Symposium started with the presentation of the Miklós Bodanszky Award to Norman Metanis (The Hebrew University of Jerusalem, Israel) for his outstanding contributions to peptide science using selenium chemistry. Next, Norbert Sewald, Scientific Affairs Officer of the EPS, announced Christian Hackenberger (Leibniz-Forschungsinstitut für Molekulare Pharmakologie, Berlin, Germany) as the winner of the Leonidas Zervas Award, in commemoration of his outstanding contributions to peptide science involving functional peptide- and protein-conjugates for extra and intracellular targeting. The opening ceremony concluded with the presentation of the Josef Rudinger Memorial Award to James P. Tam (Nanyang Technological University, Singapore). This Award commemorates Josef Rudinger's role in the foundation of the European Peptide Symposia and the diverse contributions he made to peptide chemistry. Its current edition acknowledges James Tam's life-long dedication to peptide science and to the peptide scientific community.

The main Symposium took place from Monday the 27th to Thursday the 30th of August, consisting in 9 Invited Lectures and 70 Oral Presentations, delivered over 14 sessions. Their topics included: Biologically Active

Participants of the 35th EPS in one of the Helix Foyers, where the coffee/tea breaks and exhibitions took place (left) and in the Helix Theatre, where the oral presentations were delivered (right).

Recipients of the Dr. Bert L. Schram Poster Presentation Awards, with David Andreu, Chairman of the EPS, Elizabeth Schram, President of the ESF, Johan Elgerma, Secretary/Treasurer of the ESF (left), Chandralal Hewage, Chair of the 35th EPS, Norbert Sewald, Scientific Affairs Officer of the EPS (right).

Peptides; Glycopeptides, Lipopeptides and Macromolecular Assemblies; Amino Acids, Proteins and Synthesis; Prodrugs, Conjugates, Targeting and Uptake; Peptide Mimetics; Proteomics, Bioinformatics, Structural and Conformational Studies; Peptides in Chemical Biology and Therapeutics; Nanotechnology, Imaging and Analytical Techniques. In addition to the large number of local participants, over 525 international participants from more than 27 different countries could share discussions and data on international research in the field of peptide science during these sessions, the coffee/tea and lunch breaks, as well as the closing reception and the conference dinner. As for the 34th EPS, German delegates constituted the largest group of participants, followed by representatives from the USA, UK, France, Switzerland and Japan. Other Asian and European countries, as well as Australia and Canada were also represented among the participants. 33 exhibitors also took part in the Symposium.

The programme of the 35th EPS also included a Young Investigator Mini Symposium (YIMS) on Monday afternoon, the ESCOM Science Foundation (ESF) Dr. Bert L. Schram Networking Mixer during a buffet lunch on Tuesday and three poster sessions during which 260 posters were presented. Eleven investigators under 35 years of age presented at the YIMS. Tyler Lalonde (University of Western Ontario, Canada) and Sabine Schuster (Eötvös Loránd University, Budapest, Hungary) received the Bert L. Schram Awards for best oral presentations, presented by ESF representatives Elizabeth Schram and Johan Elgerma. They also presented the six Dr. Bert L. Schram Poster Presentation Awards, selected among 94 posters running in the competition.

In addition, three poster awards from Wiley were presented by Nuno Santos (Universidade de Lisboa, Portugal). Also sponsored by the ESF, the Networking Mixer lunch is an exclusive event bringing together EPS young scientists, ESF representatives, EPS officers and international senior scientists from academia and industry. It provided a unique opportunity to PhD and postdoctoral researchers to network and discuss current research topics, as well as career options and professional opportunities.

A special issue of the *Journal of Peptide Science* containing the abstracts of the Invited, Oral and Poster presentations, the details of the sponsors and exhibitors, as well as the proceedings of the 35th Symposium will be published in the coming months.

Three hundred participants joined the conference dinner on Thursday night. On Friday, participants could choose between the tours of the National Stud and Japanese Gardens, Dublin City and Guinness Storehouse, Malahide Castle and Gardens, Trinity College and Book of Kells, to discover some of Ireland's most popular tourist attractions.

EPS Executive Committee, ESF representatives and 35th EPS Chair. Pictured are (back row, l-r) Krzysztof Rolka, Communication Officer of the EPS, Johan Elgerma, Secretary/Treasurer of the ESF, Norbert Sewald, Scientific Affairs Officer of the EPS, David Andreu, Chairman of the EPS; (front row, l-r) Elizabeth Schram, President of the ESF, Anna Maria Papini, Treasurer of the EPS, Paula Gomes, Secretary of the EPS, Chandralal Hewage, Chair of the 35th EPS.

The Scientific Advisory Board, Alethea Tabor (University College London), David Andreu (Pompeu Fabra University), Ernest Giralt (University of Barcelona), Florine Cavelier (Université de Montpellier), John Howl (University of Wolverhampton), John Wade (University of Melbourne), Knud Jensen (University of Copenhagen), Maria Luisa Mangoni (Sapienza Università di Roma), Maurice Manning (University of Toledo), Meritxell Teixidó (Institute for Research in Biomedicine, Barcelona), Norbert Sewald (Bielefeld University) and William Lubell (Université de Montréal), was substantially involved in the running of the Symposium. Tia Keyes (Dublin City University), co-chair of the 35th EPS, was also instrumental in the operations of the event on the campus of DCU, including the scientific and social events and the accommodation. A very special thanks to the sponsors of the 35th EPS as well; without their generous support, the Symposium could not have taken place. During the closing ceremony, Patrick Timmons (University College Dublin) was also presented with a token of appreciation, to acknowledge his significant contribution to the organisation of the 35th EPS. At the end of this ceremony, David Andreu handed over the EPS gong to Meritxell Teixidó for the 36th EPS, which will be held in Barcelona, Spain on August 30 - September 4, 2020.

Written by Marc Devocelle (Royal College of Surgeons in Ireland), National Representative for Ireland at the EPS Council.

David Andreu, Chairman of the EPS, handing over the EPS gong to Meritxell Teixidó, Chair of the 36th EPS.

Sponsors, Exhibitors and Supporters

Silver Sponsors

CEM
BACHEM
SCIENCE FOUNDATION IRELAND

Bronze Sponsors

Merck
CBL Peptide Technology and Manufacturing
Fáilte Ireland

Award and other Sponsors

BCN PEPTIDES
EPS - Dr. Rao Makineni
PolyPeptide Group
ESCOM Science Foundation - Dr. Bert L. Schram
WILEY

Supporters

ChemPep
MDPI - molecules
ORPEGEN

Exhibitors

aapptec
Advanced ChemTech / peptides&elephants
Almac Group
BCN PEPTIDES SA
Belyntic GmbH – HPLC-free Peptide Purification
Biametrics
Biopeptek Pharmaceuticals
Biotage AB, from Sweden
C.A.T. GmbH & Co KG
Chinese Peptide Company
CSBio Co.
GL BIOCHEM (SHANGHAI) LTD
Gyros Protein Technologies
HECHENG
INTAVIS
Iris Biotech GmbH
Luxembourg Bio Technologies
ONTORES
PeptART Bioscience GmbH
Peptides International / Peptide Institute
PolyPeptide Group – The Ultimate Peptide Partner
Polypure
Purolite Life Sciences
Rapp Polymere GmbH
Senn Chemicals AG
Sunresin New Materials Co., Ltd
Teledyne ISCO
YMC Europe GmbH – The Selectivity Company
ZEOCHEM AG
Zhejiang Peptites Biotech Co., Ltd

Sponsors

Programme

	August 26 Sunday PM
14.00-17.00	**REGISTRATION**
16.30-19.00	**OPENING CEREMONY**
	AWARDS
	Miklós Bodanszky Award
	AL1 - Norman Metanis, The Hebrew University of Jerusalem, Israel Lessons from Selenium Chemistry to Study Protein Science
	Leonidas Zervas Award
	AL2 - Christian Hackenberger, FMP Berlin, Germany The power of chemoselectivity: Functional peptide- and protein-conjugates for extra- and intracellular targeting
	Josef Rudinger Memorial Award
	AL3 - James P Tam, Nanyang Technological University, Singapore Advances in Site-specific and Linkage-specific Ligation

	August 27 Monday AM
	Chairs: David Andreu
08.30-09.10	**IL1 - David R Spring**
	Adventures in Drugging Undruggable Targets
	Chairs: Tia Keyes, Glenn King
	Biologically Active Peptides 1
09.10-09.25	**OP1 - Nermina Malanovic**
	Lipid domain formation induced by antimicrobial peptides OP-145 and SAAP-148 results in efficient killing of Escherichia coli
09.25-09.40	**OP2 - Steven Ballet**
	Engineering Hybrid Peptidomimetics for Improved Pain Treatments
09.40-09.55	**OP3 - Norelle Daly**
	An engineered cyclic peptide alleviates symptoms of inflammation in a murine model of inflammatory bowel disease
09.55-10.10	**OP4 - Sarah Jones**
	Penetrating the Impenetrable: Progress towards a Non-hormonal Male Contraceptive
10.10-10.25	**OP5 - Marta De Zotti**
	Peptide-Based Biopesticides
	Chair: Marc Devocelle
11.00-11.30	**IL2 - Finbarr O'Harte**
	The design and assessment of the therapeutic potential of apelin-13 peptide mimetics for alleviating metabolic dysfunction in diabetes and obesity
	Chairs: Ekaterina Kolesanova, David Spring
	Glycopeptides, Lipopeptides and Macromolecular Assemblies
11.30-11.45	**OP6 - Andy Wilson**
	Development and Exploitation of Photo-Crosslinking Methodology to Study Protein-Protein Interactions
11.45-12.00	**OP7 - Claudia Bello**
	Toward homogeneous glycoproteins via auxiliary-assisted sequential glycosylation and ligation of peptides
12.00-12.15	**OP8 - Oliver Zerbe**
	Peptide-Guided Assembly of Armadillo Repeat Protein Fragments
12.15-12.30	**OP9 - Veronica Dodero**
	Are gluten-related disorders a new protein aggregation disease?
12.30-12.45	**OP10 - Eoin Scanlan**
	Chemical Synthesis of Glycopeptides and Glycoproteins Using Acyl-thiol-ene Mediated 'Click' Ligation

	August 27 Monday PM
	Chairs: Jane Aldrich, David Craik
	Young Investigator Mini Symposium
16.15-16.30	**MS1 - Jakob Gaar** Generation and characterisation of novel antibodies from selectively Advanced Glycation Endproduct (AGE) modified Collagen Model Peptides (CMPs)
16.30-16.45	**MS2 - Lea Albert** Light-controlled inhibition of MLL1 methyltransferase by azo-containing peptides: towards optoepigenetic leukemia regulation
16.45-17.00	**MS3 - Chun Yuen Chow** Development of NaV-Selective Agonists with Potential for Treatment of Dravet Syndrome Epilepsy
17.00-17.15	**MS4 - Varsha Thombare** Exploiting thioamide reactivity in peptide synthesis
17.15-17.30	**MS5 - Karlijn Hollanders** Zn-catalyzed tert-butyl nicotinate-directed amide cleavage for applications in peptide synthesis and peptidomimetic design
17.30-17.45	**MS6 - Daria Roshchupkina** Synthesis of peptides glycated at Lys residues
17.45-18.00	**MS7 - Tyler Lalonde** Targeting Ghrelin Receptor Homodimer: Bivalent Ligands with Exceptional Binding Affinity and Potency
18.00-18.15	**MS8 - Elise Naudin** De novo designed proteins catalyzing amide bond forming reactions
18.15-18.30	**MS9 - Sabine Schuster** Synthesis and biochemical evaluation of GnRH-III-drug conjugates
18.30-18.45	**MS10 - Esben Matzen Bech** Directing PYY3-36 internalization through half-life extenders
18.45-19.00	**MS11 - Clara Pérez-Peinado** Mechanisms of bacterial membrane permeabilization of snake venom-derived peptides crotalicidin (Ctn) and Ctn[15-34]

	August 28 Tuesday AM
	Chair: Paula Gomes
08.30-09.10	**IL3 - Marion de Jong** Theranostic radiopeptides
	Chairs: Zbigniew Szewczuk, John Wade
	Amino Acids, Proteins and Synthesis 1
09.10-09.25	**OP11 - Chuanliu Wu** Multicyclic peptides constrained through noncanonical disulfide bonds and thioether crosslinks
09.25-09.40	**OP12 - Tsubasa Inokuma** Novel methodology for the synthesis of α-indolyl-glycine containing peptide via direct asymmetric Friedel–Crafts reaction to peptidyl imine
09.40-09.55	**OP13 - Muhammad Jbara** Palladium Prompted On-Demand Cysteine Chemistry for the Synthesis of Challenging Proteins
09.55-10.10	**OP14 - William Lubell** Synthesis and Biomedical Applications of Substituted Amino-Lactam and Amino-Imididazolone Constraints
10.10-10.25	**OP15 - Kirtikumar B Jadhav** Recombinant synthesis and purification of hTFF2 protein in S. Cerevisiae
	Chair: Anna Maria Papini
11.00-11.30	**IL4 - Meritxell Teixidó** Gate2Brain Shuttle Peptides, From Discovery to Applications
	Chairs: Marion de Jong, Finbarr O'Harte
	Biologically Active Peptides 2
11.30-11.45	**OP16 - Evelien Wynendaele** Do Quorum Sensing Peptides Play a Role in Psychiatric Disorders?
11.45-12.00	**OP17 - Hidehito Mukai** Mitocryptides: First Demonstration of Pathophysiological Involvements in Inflammatory Diseases
12.00-12.15	**OP18 - Thibaut Thery** Antifungal activity and immunomodulation of a de novo synthetic peptide
12.15-12.30	**OP19 - Shai Rahimipour** Targeting Amyloidogenic Proteins with Self-assembled Cyclic D,L-alpha-Peptides
12.30-12.45	**OP20 - Marina Rautenbach** Tyrocidines and Gramicidin S: Glorified cyclodecapeptide detergents or precise chemical scalpels

	August 28 Tuesday PM
	Chairs: Florine Cavelier, Fintan Kelleher
	Amino Acids, Proteins and Synthesis 2
16.15-16.30	**OP21 - Robert Zitterbart** Enabling Parallel Peptide Purification by a Novel Traceless Purification Linker
16.30-16.45	**OP22 - Christian Becker** Effects of non-enzymatic posttranslational modifications on protein function
16.45-17.00	**OP23 - Yoshio Hayashi** A New Aspect of Npys-based Solid Phase Disulfide Peptide Synthesis
17.00-17.15	**OP24 - Vincent Aucagne** N-2-hydroxybenzyl-cysteine peptide crypto-thioesters for native chemical ligation
17.15-17.30	**OP25 - Sira Defaus** Expanding the potential and multivalency of the B2T synthetic peptide vaccine against foot-and-mouth disease virus
17.30-17.45	**OP26 - Gerbrand van der Heden van Noort** A general method towards ADPr ribosylated peptides and proteins
	Chairs: Meritxell Teixidó, John Howl
	Prodrugs, Conjugates, Targeting and Uptake
17.45-18.00	**OP27 - Ming-Hsin Li** A Novel Evans Blue Organic Compound-appended PSMA-617 Peptide as SPECT Molecule Imaging Agent for Human Prostate LNCaP Animal Model
18.00-18.15	**OP28 - Gilles Subra** Sol-gel and peptides: an attractive route to unprecedented biomaterials
18.15-18.30	**OP29 - Ikuhiko Nakase** Arginine-rich cell-penetrating peptide-modified exosomes for macropinocytosis induction and effective cellular uptake
18.30-18.45	**OP30 - Adina Borbély** Novel Cryptophycin-based Conjugates for Tumor Targeting
18.45-19.00	**OP31 - Conan Wang** A Journey Alongside Cyclosporin on the Road to Peptide Oral Bioavailability

	August 29 Wednesday AM
	Chair: Paul Alewood
08.30-09.10	**IL5 - Knud Jensen**
	Selective N-terminal acylation of peptides and proteins with an optimized His-sequence
	Chairs: Tamara Paypanova, William Lubell
	Peptide Mimetics
09.10-09.25	**OP32 - Jung-Mo Ahn**
	Structure-Based Design of Alpha-Helix Mimetics to Target and Disrupt Estrogen Receptor-Coregulator Interactions in Breast Cancer
09.25-09.40	**OP33 - Jean-Alain Fehrentz**
	Ghrelin receptor ligands: from the bench to the drug on the market
09.40-09.55	**OP34 - Stefan Roesner**
	Macrocyclization of Small Peptides Enabled by Oxetane Incorporation
09.55-10.10	**OP35 - Carles Mas Moruno**
	The use of RGD peptidomimetics on biomaterials: new advances via selective integrin-subtype targeting
10.10-10.25	**OP36 - Steven L Cobb**
	Stabilising Peptoid Helices Using Non-Chiral Fluoro-Alkyl Monomers
	Chair: Norbert Sewald
11.00-11.30	**IL6 - Beate Koksch**
	Fluorine in Peptide and Protein Engineering
	Chairs: Ernest Giralt, Dick FitzGerald
	Proteomics, Bioinformatics, Structural and Conformational Studies
11.30-11.45	**OP37 - Jose Martins**
	Concerted biophysical and biological evaluation of Pseudomonas lipopeptides as a premise to unlock their application potential
11.45-12.00	**OP38 - Andrew Jamieson**
	Rationally Designed Peptidomimetics
12.00-12.15	**OP39 - Daniela Marasco**
	Amyloidogenicity of regions of Nucleophosmin 1: a direct link between protein misfolding and Acute Myeloid Leukemia
12.15-12.30	**OP40 - Quentin Kaas**
	Inhibition of nicotinic acetylcholine receptor subtypes by the ribbon isomers of α-conotoxins.
12.30-12.45	**OP41 - Zbigniew Szewczuk**
	Quaternary Ammonium Isobaric Labeling for a Relative and Absolute Quantification of Peptides

	August 28 Wednesday PM
	Chairs: Maria Luisa Mangoni, Wilfred van der Donk
	Peptides in Chemical Biology and Therapeutics 1
16.15-16.30	**Andrew White** Development of serine protease inhibitors using the 1,2,3-triazole motif as a disulfide mimetic in the cyclic peptide sunflower trypsin inhibitor-1
16.30-16.45	**Dorien Van Lysebetten** Peptide-Bile acid Cyclisation as a Tool for the Development of a Universal Vaccine Against Influenza A
16.45-17.00	**Piotr A Mroz** Stereochemically modified glucagon with improved biophysical parameters
17.00-17.15	**Benjamin Liet** Design, synthesis and study of multimeric peptidic conjugates for a new approach of anti-tumoral immunotherapy
17.15-17.30	**Nir Qvit** Engineered protein-protein interaction regulators for therapeutic applications
17.30-17.45	**Victor J Hruby** Utilizing Combinations of New Approaches to Peptide and Peptidomimetic Design for G-Protein Coupled Receptors
	Chairs: Nuno Santos, Knud Jensen
	Nanotechnology, Imaging and Analytical Techniques
17.45-18.00	**Friedrich Bialas** Immobilising integral membrane proteins on silica
18.00-18.15	**Konstantin Kuhne** Cathepsin B-Activatable Cell-Penetrating Peptides for Imaging Cancer-related Cathepsin B
18.15-18.30	**Neil O'Brien-Simpson** Designing antibiotic peptide polymer adjuvants for multidrug resistant bacteria
18.30-18.45	**Dorian J Mikolajczak** Tuning the Catalytic Activity and Substrate Specificity of Peptide-Nanoparticle Conjugates
18.45-19.00	**Ramon Subiros-Funosas** Lighting up programmed cell death in real-time: apoTRACKER as an in vivo compatible apoptosis diagnostic tool

	August 30 Thursday AM
	Chair: Luis Morodor
08.30-09.10	**IL7 - Dek Woolfson**
	Peptide design and assembly: from the test tube to inside cells
	Chairs: Alethea Tabor, Gábor Mező
	Peptides in Chemical Biology and Therapeutics 2
09.10-09.25	**OP53 - Anne Conibear**
	Synthetic integrin-binding immune stimulators target cancer cells and prevent tumor formation.
09.25-09.40	**OP54 - Ana Salomé Veiga**
	Activity of the viral-derived peptide pepR against Staphylococcus aureus biofilms and insights into the mechanism of action
09.40-09.55	**OP55 - Timothy Reichart**
	Development of Mirror-Image Monobodies for Targeted Cancer Therapies
09.55-10.10	**OP56 - Paul Alewood**
	Subtle modifications to oxytocin produce ligands that retain potency and improved selectivity across species
10.10-10.25	**OP57 - Roland Hellinger**
	Cyclotides as novel inhibitors of the human prolyl oligopeptidase
	Chair: Norbert Sewald
11.00-11.30	**IL8 - Wilfred van der Donk**
	Posttranslational Modifications during Cyclic Peptide Biosynthesis
	Chairs: Rosario Gonzalez-Muñiz, Christian Gruber
	Peptides in Chemical Biology and Therapeutics 3
11.30-11.45	**OP58 - Youhei Sohma**
	Near-infrared photoactivatable oxygenation catalyst of amyloid peptide
11.45-12.00	**OP59 - Minying Cai**
	Development of the MC1R Selective Ligands for the Melanoma Prevention
12.00-12.15	**OP60 - Sónia Henriques**
	Is the mirror image a true reflection? Lipid chirality in the activity of the prototypic cyclotide kalata B1
12.15-12.30	**OP61 - Jane Aldrich**
	Substitution of Aromatic Residues in the Macrocyclic Opioid Peptide [D-Trp]CJ-15,208 Alters the Opioid Activity Profile in vivo
12.30-12.45	**OP62 - Maayan Gal**
	Developing novel bi-specific peptide as an immunosuppressant using directed-evolution optimization

	August 30 Thursday PM
	Chair: Maurice Manning
14.15-14.45	**IL9 - Glenn F King** From pain to pain-killers: using spider-venom peptides to help us understand and treat chronic pain
	Chairs: Beate Koksch, Ale Närvänen
	Peptides in Chemical Biology and Therapeutics 4
14.45-15.00	**OP63 - Christina Schroeder** The use of peptide-membrane interactions in the design of selective and potent sodium channel inhibitors
15.00-15.15	**OP64 - Lydia Behring** Dipeptide-derived Alkynes as Novel Irreversible Inhibitors of Cathepsin B
15.15-15.30	**OP65 - Tamis Darbre** Potent Antimicrobial Peptide Dendrimers against Multi-Drug Resistant Pseudomonas aeruginosa and Acinobacter baumanii
15.30-15.45	**OP66 - Garry Laverty** Modifying short phenylalanine-phenylalanine peptide sequences to create multifunctional nanomaterials with biomaterial and drug delivery applications
	Chairs: Norman Metanis, Steven Ballet
	Amino Acids, Proteins and Synthesis 3
16.15-16.30	**OP67 - Abhishek Iyer** Reviving Old Protecting Group Chemistry for Site-Selective Peptide-Protein Conjugation
16.30-16.45	**OP68 - Hironobu Hojo** Use of selenoester for the efficient synthesis of protein
16.45-17.00	**OP69 - Hannes Ludewig** Eukaryotic macrocyclases enhance biocatalytic production of cyclic peptides
17.00-17.15	**OP70 - Juergen Machielse** Cost Effective peptide Purification via ZEOsphere DRP Mixed-Mode Chromatography
17.15-18.00	**CLOSING CEREMONY**
18.00-20.00	**BACHEM RECEPTION**
20.00-22.00	**CONFERENCE BANQUET**

Articles

The power of chemoselectivity: Functional peptide- and protein-conjugates for extra- and intracellular targeting

Christian Hackenberger

FMP and HU Berlin, Germany

https://doi.org/10.17952/35EPS.2018.001

Chemoselective reactions are essential synthetic tools employed in all areas of chemistry. In addition to their application in the total synthesis of various natural products, they have advanced research in the material and life sciences by allowing the conjugation of various biological molecules to polymers, nanoparticles and other materials as well as the straightforward chemical modification of proteins for biological and pharmacological purposes. Moreover, the development of chemoselective protein ligation strategies, spearheaded by the engineering of native chemical ligation[1] and expressed protein ligation[2] for the conjugation of two unprotected peptide or protein fragments from either synthetic or biological origin, have advanced the field of protein synthesis culminating in the impressive synthesis of naturally occurring posttranslationally modified proteins.[3] Another remarkable advancement in this area was the development of bioorthogonal reactions, which introduced the concept of performing chemical reactions in living organisms by modifying unnatural functional groups, so-called bioorthogonal reporters, present inside and outside of cells leading to the development of new probes to study extra- and intracellular pathways.[4]

In the Zervas Award lecture an overview is given on our recent investigations on the identification of new chemos-elective and bioorthogonal reactions for the synthesis and modification of functional peptides and proteins.[5] We apply these reactions to study functional consequences of naturally occurring posttranslational protein modi-fications (PTMs), in particular to study phosphorylation and O-GlcNAcylation events on the Alzheimer relevant Tau-protein.[6] Moreover, by utilizing chemoselective Staudinger reactions of azides with phosphites [7a,b,f] and phosphonites [7c,d,e] as well as the reaction of phosphites with electrophilic disulfides,[7g] we obtained phosphorylated Lys- and Cystein-peptides for to advance the proteomic analysis of these labile underexplored PTMs.[7f-h]

Another big emphasis of our group, which is highlighted in this presentation, is the generation of novel peptide- and protein-conjugates for pharmaceutical and medicinal applications as well as the cellular delivery of functional proteins and antibodies. For the latter study we employed cyclic cell penetrating peptides (cCPPs) to transport a functional full length protein to the cytosol of living cells as recently demonstrated by the direct delivery of GFP-conjugates.[8] For protein modification we use a combined approach of intein expression as well as recently developed bioorthogonal reactions and enzymatic ligations, for instance the so-called Tub-tag labeling.[9] This concept is finally applied to generate new antibody-drug conjugates as well as cell-permeable nanobodies, i.e. small antigen binding proteins that remain active within the reductive milieu inside living cells, to interfere with intracellular targets.[10]

References

1. P. E. Dawson, T.W. Muir, I. Clark-Lewis, S. B. Kent, Science, 1994, 266, 776–779.

2. T. W. Muir, D. Sondhi, P. A. Cole, Proc. Natl. Acad. Sci. USA, 1998, 95, 6705–6710.

3. For reviews see: a) C. P. R. Hackenberger, D. Schwarzer, Angew. Chem. Int. Ed. 2008, 47, 10030 - 10074; b) P. Siman, A. Brik, Org. Biomol. Chem. 2012, 10, 5684-5697.

4. H.C. Hang, C. Yu, D.L. Kato, C.R. Bertozzi, Proc. Natl. Acad. Sci. U.S.A. 2003, 100, 14846–14851; For reviews see: a) E. M. Sletten, C. R. Bertozzi, Acc. Chem. Res. 2011, 44, 666-676; b) E. M. Sletten, C. R. Bertozzi, Angew. Chem. Int. Ed. 2009, 48, 6974–6998.

5. D. Schumacher, C.P.R. Hackenberger, Curr. Opin. Chem. Biol. 2014, 22, 62-69.

6. a) C. Smet-Nocca, M. Broncel, J.-M. Wieruszeski, C. Tokarski, X. Hanoulle, A. Leroy, I. Landieu, C. Rolando, G. Lippens, C.P.R. Hackenberger, Mol. Biosyst. 2011, 7, 1420-1429; b) M. Broncel, E. Krause, D. Schwarzer, C.P.R. Hackenberger, Chem. Eur. J. 2012, 18, 2488-2492; c) O. Reimann, C. Smet-Nocca, C.P.R. Hackenberger,

Angew. Chem. Int. Ed. 2015, 54, 306-310; d) S. Schwagerus, O. Reimann, C. Despres, C. Smet-Nocca, C.P.R. Hackenberger, J. Pept. Sci. 2016, 22(5), 327-333.

7. a) R. Serwa, I. Wilkening, G. del Signore, M. Mühlberg, I. Claußnitzer, C. Weise, M. Gerrits, C. P. R. Hackenberger, Angew. Chem. Int. Ed. 2009, 48, 8234-8239; b) V. Böhrsch, R. Serwa, P. Majkut, E. Krause, C. P. R. Hackenberger, Chem. Commun. 2010, 46, 3176-3178; c) M.R.J. Vallée, P. Majkut, I. Wilkening, C. Weise, G. Müller, C.P.R. Hackenberger, Org. Lett. 2011, 13, 5440-5443; d) M.R.J. Vallée, L.M. Artner, J. Dernedde, C.P.R. Hackenberger, Angew. Chem. Int. Ed. 2013, 52, 9504-9508; e) M.R.J. Vallée, P. Majkut, D. Krause, M. Gerrits, C.P.R. Hackenberger, Chem. Eur. J. 2015, 21(3), 970-974. For the chemoselective synthesis of pLys-peptides see f) J. Bertran-Vicente, R.A. Serwa, M. Schümann, P. Schmieder, E. Krause, C.P.R. Hackenberger, J. Am Chem. Soc. 2014, 136(39), 13622-13628; g) J. Bertran-Vicente, M. Penkert, O. Nieto, M. Schümann, P. Schmieder, E. Krause, C.P.R. Hackenberger, Nature Comm. 2016, 7,12703. For a review see: h) A. Hauser, M. Penkert, C.P.R. Hackenberger, Acc. Chem. Res. 2017, 50(8), 1883 - 1893.

8. N. Nischan, H.D. Herce, F. Natale, N. Bohlke, N. Budisa, M.C. Cardoso, C.P.R. Hackenberger, Angew. Chem. Int. Ed. 2015, 54(6), 1950-1953.

9. a) D. Schumacher, J. Helma, F.A. Mann, G. Pichler, F. Natale, E. Krause, M.C. Cardoso, C.P.R. Hackenberger, H. Leonhardt, Angew. Chem. Int. Ed. 2015, 54(46), 13787-13791, b) D. Schumacher, O. Lemke, J. Helma, L. Gerszonowicz, V. Waller, T. Stoschek, P.M. Durkin, N. Budisa, H. Leonhardt, B.G. Keller, C.P.R. Hackenberger, Chem. Sci. 2017, 8, 3471-3478.

10. a) H. Herce, D. Schumacher, F.A. Mann, A. Ludwig, A. Schneider, M. Filies, S. Reinke, C. Cardoso, C.P.R. Hackenberger, Nature Chem. 2017, 9, 762-771; b) D. Schumacher, J. Helma, A.F.L. Schneider, H. Leonhardt, C.P.R. Hackenberger, Angew. Chem. Int. Ed. 2018, 57, 2314-2333.

Advances in site-specific and linkage-specific ligation

James Tam, Giang Kien Truc Nguyen, Antony Kam, Shining Loo

School of Biological Sciences, Synzymes and Natural Products Center, Nanyang Technological University, Singapore

https://doi.org/10.17952/35EPS.2018.003

Introduction

Enzymatic peptide synthesis began, nearly 80 years ago, with the modest goal of making small peptides by reverse proteolysis using the protease chymotrypsin. Recent advances in recombinant DNA methodologies, which have made proteins and biopharmaceuticals readily available, now demand a mild and aqueous-compatible enzymatic ligation method by a true ligase for site-specific modification and precision biomanufacturing of peptide/protein-based biomaterials and antibody-drug conjugates.

Currently, three different groups of ligase-like enzymes or ligases with demonstrated applications have been identified. The first two groups are ligase-like enzymes which include the transpeptidase sortase A and the modified subtilisin. The third, perhaps the largest group, belongs to the naturally-occurring ligases discovered in the biosynthesis of ribosomally-synthesized and post-translationally modified peptides (RiPP) which often display cyclic structures. They include TraF, PatG, PCY1, butelase 1, and POPB [1-5]. These ATP-independent and stand-alone ligases or cyclases have the advantage to perform both *in vitro* and *in vivo* ligation reaction unhampered by an enzyme complex. With the exception of butelase 1, most RiPP-derived ligases display low ligase efficiency and will need improvements to be fully exploited.

Butelase 1, discovered in the cyclic-peptide-producing plant *Clitoria ternatea*, is an Asn/Asp-specific ligase. It belongs to the asparaginyl endoprotease (AEP) family, but it is virtually devoid of protease activity. Butelase 1 recognizes a C-terminal Asx-containing tripeptide motif, Asn/Asp-HisVal, to form an Asx-Xaa peptide bond where Xaa is any amino acid, either intramolecularly or intermolecularly, yielding, respectively, head-to-tail cyclic peptides or site-specific modified proteins [5-14].

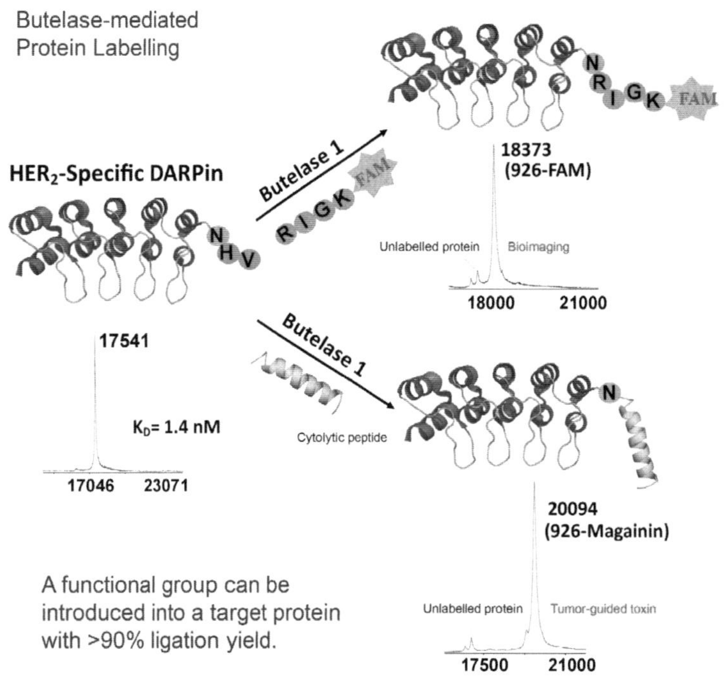

Figure 1: Butelase-mediated labeling of Her2-specific darpin 926 for bioimaging (RIGK-Fam) and targeted drug conjugates linking to a cytolytic peptide (magainin, see Figure 2).

Butelase is highly suitable for total synthesis of natural products because it does not produce any residual recognition signal sequence in the ligation product except for Asx. This advantage has been exploited for the successful synthesis of a family of high molecular weight (60-70 amin acods), hydrophobic circular peptide antimicrobials known as bacteriocins. Butelase has been shown to be compatible with orthogonal chemical ligation, or another ligase such as sortase A or subtiligase to provide variations of sequential chemoezymatic ligation or tandem ligation under one-pot conditions for modification of proteins, and antibody-drug conjugates.

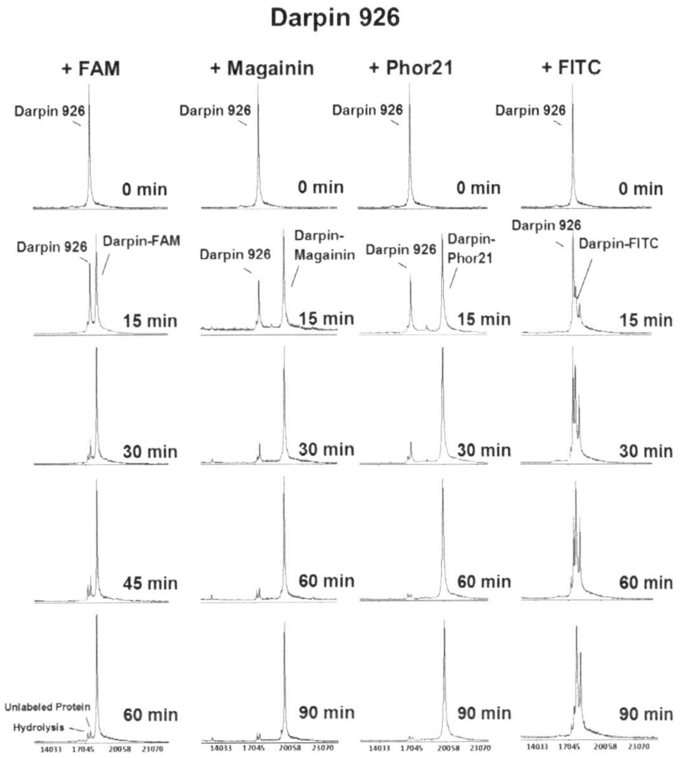

Figure 2: Time-course of butelase-mediated protein labeling of darpin 926 with FAM, Magainin, Phor21 and FITC. All reactions were performed with 100 nM butelase 1, 25 μM darpin-NHV, and 1mM peptide conjugates in 20 mM sodium phosphate, 1 mM EDTA, 40 μM β-mercaptoethanol at 42 °C.

Breast cancer is the most common cancer among women, with 1.6 million new cases and a mortality rate of > 0.5 million each year. About 20% of breast cancer are HER2-positive and treatment by monoclonal antibodies such as herceptin targeting HER2 receptor are effective. DARPins (designed ankyrin repeat proteins) represent a promising and tunable group of synthetic small non-immunoglobulin proteins. They have attracted increasing attentions as a low-cost replacement of antibodies in drug development because of their favorable molecular properties, small molecular sizes (6k vs 180k), high stability and ease of production by bacterial systems. DARPin 926, a re-engineered form of darpin, was designed to achieve high binding affinity specificity towards Her2 receptor.

Here we describe the use of butelase 1 for precise prepared DARPin-conjugates targeting the breast cancer marker epidermal growth factor receptor HER2 with various payloads.

Results and Discussion

In this study, we recombinantly expressed DARPin 926 using *E. coli* BL21 pLysS strain with a C-terminal Asn-His-Val motif for butelase 1 recognition. Following His-tag purification and TEV cleavage, the purified protein was subjected to butelase-mediated ligation.

Cell-type-specific targeting of the butelase-mediated darpin-conjugates were shown using live-cell confocal microscopy and flow cytometry. Figures 3 and 4 demonstrated that darpin-FITC preferentially binds to Her2-positive BT474 cells, and not to Her2-negative MCF-7 cells. These results suggested that we have successfully functionalized darpin 926 to achieve Her2-positive-cell-specific targeting using butelase-mediated ligation.

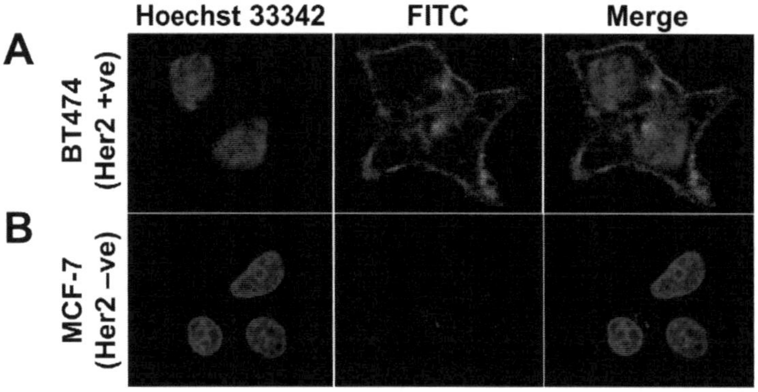

Figure 3: Confocal microscopy showed cell-type specific targeting of Darpin-FITC (green) clearly visible on the cell surface, using (A) Her2-positive (BT474), but not visible on (B) Her2-negative (MCF-7) breast cancer cells.

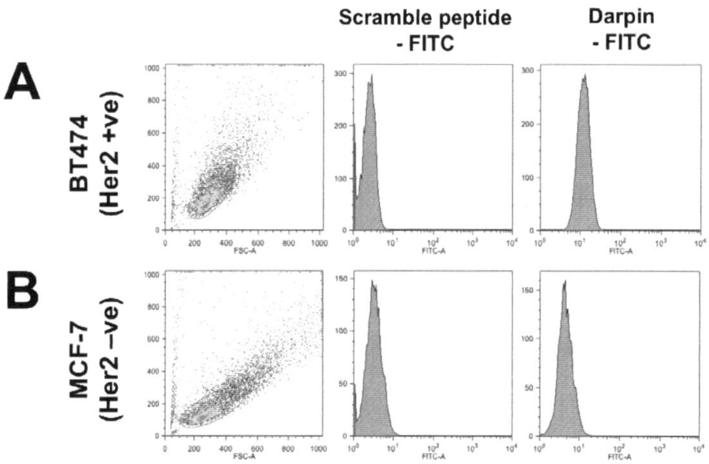

Figure 4: Flow cytometric analyses showed cell-type specific targeting of Darpin-FITC increased fluorescence intensity in (A) Her2-positive (BT474), but not in (B) Her2-negative (MCF-7) breast cancer cells.

Conclusions

In conclusion, we showed that butelase can be used for precision manufacturing of biological therapeutics with various payloads and a versatile tool to meet the demands of the coming age of precision medicine.

Acknowledgments

This research was supported by a Nanyang Technological University internal funding – 522 Synzyme and Natural Products Center (SYNC) and the AcRF Tier 3 funding (MOE2016-T3-523 1-003).

References

1. Nguyen, GKT, Wang, S, Qiu, Y, Hemu, X, Lian, Y, Tam, J P (2014). Nat Chem Biol 10: 732–8 doi: 10.1038/nchembio.1586.

2. Nguyen, GKT, Cao Y, Wang W, Liu CF, Tam JP (2015). Angew Chem Int Ed 54:15694-8 doi: 10.1002/anie.201506810.

3. Nguyen, GKT, Kam, A, Loo, S, Jansson, AE, Pan, LX, Tam, JP (2015). J Am Chem Soc 137: 15398–401 doi: 10.1021/jacs.5b11014.

4. Hemu, X, Qiu, Y, Nguyen, GKT, Tam, JP (2016). J Am Chem Soc 138: 6968–71 doi: 10.1021/jacs.6b04310

5. Nguyen, GKT, Hemu, X, Quek, JP, Tam, JP (2016). Angew Chem Int Ed 55: 12802–6 doi: 10.1002/anie.201607188.

6. Cao, Y, Nguyen, GKT, Chuah, S, Tam, JP, Liu, CF (2016). Bioconjug Chem 27: 2592–6 doi: 10.1021/acs.bioconjchem.6b00538

7. Yang, R, Wong, YH, Nguyen, GKT, Tam, JP, Lescar, J, Wu, B (2017) J Am Chem Soc 139: 5351–8 doi: 10.1021/jacs.6b12637

8. Bi, X, Yin, J, Nguyen, GKT, Rao, C, Halim, NBA, Hemu, X, Tam, JP, Liu, CF (2017). Angew Chem Int Ed 56:7822-5 doi: 10.1002/anie.201703317.

9. Bi, X, Yin, J, Hemu, X, Rao, C, Tam, JP, Liu, CF (2018). Bioconjug Chem 29: 2170–5 doi: 10.1021/acs.bioconjchem.8b00244.

10. Harmand, T, Bousbaine, D, Chan, AI, Zhang, X, Liu, DR, Tam, JP, Ploegh, HL (2018). Bioconjug Chem doi: 10.1021/acs.bioconjchem.8b00563

Adventures in drugging undruggable targets

David Spring

University of Cambridge, United Kingdom

https://doi.org/10.17952/35EPS.2018.008

Pharmaceutical chemistry research has traditionally focused upon a limited set of biological targets. Many other human disease-related targets have been termed 'undruggable' as they have proved largely impervious to modulation by small molecules. However, it is becoming increasingly evident that such targets can indeed be modulated; they are simply being challenged with the wrong types of molecules. Research using conformationally constrained peptides should provide hits against a broad range of targets with high frequency, including 'undruggable' targets.

Protein–protein interactions (PPIs) have emerged as attractive drug discovery targets in recent years due to their key roles in mediating various cellular functions. PPIs are however notoriously challenging to target; therefore, they have been termed 'undruggable'. The interfaces that characterize PPIs are often large, shallow, and highly flexible, making rational drug design difficult. Nevertheless, there is growing interest in exploring these, particularly through harnessing peptides as PPI-targeting drugs due to their potency, high specificity, and low toxicity. Peptides are also viewed as a bridging class of drugs that could potentially combine the desirable properties of small molecules with those of biologics. Despite numerous successful examples of peptide-based drugs that have already reached the market, peptides are often considered poor drug candidates because of their low bioavailability, rapid elimination, poor *in vivo* stability, and administration. These limitations are now gradually being mitigated by advances in peptide delivery and synthetic methodologies. An important advance in the development of peptide-based drugs is the introduction of the "stapling" strategy.

Incorporation of staples (conformational constraints) within the peptide has been extensively used to improve the bioavailability of these molecules; consequently, it is not surprising that a plethora of stapling techniques has been developed and has had a significant impact on the development of peptide therapeutics. Among the numerous stapling techniques known, two-component methodologies allow facile and divergent functionalization of peptides. The author's laboratory has pioneered a stapling technique that makes use of the double Cu-catalyzed azide–alkyne cycloaddition between di-azido peptides and functionalized di-alkynyl staples. In recent years, the author's laboratory has created biologically active, conformationally constrained peptides functionalized with cell-penetrating peptides, fluorescent tags, and photo cross-linking moieties, demonstrating the wide applicability of this methodology. The impact, advantages, limitations, and future applications of this technology and other two-component peptide stapling techniques on the development of clinically relevant peptides were highlighted.

Examples in the area of two-component peptide stapling methodology taken from the author's laboratory included:

- p53–MDM2 Interaction[1,2,3,4,5]
- Targeting the Genome-Stability Hub Ctf4[6]
- HNF1β – Importin α PPI[7]
- Tankyrase – Axin PPI[8]

Peptide therapeutics represent an emerging area in the pharmaceutical field. Considering their selectivity, efficacy, low toxicity, and cost of goods, peptides are the ideal molecules to target shallow areas of proteins, such as the interface of PPIs, thus unlocking a large number of extra and intracellular pathways that can be disrupted. The main drawbacks of this class of molecules are their inherent instability *in vivo* and their weak bioavailability. Significant progress has been made in increasing the resistance of the peptide to proteases and improving their cell-permeability and hence, tissue penetration. Macrocyclization of linear peptides is certainly one of the most successful methodologies to overcome many of the limitations of peptides, and a plethora of different techniques have been developed to this end. In this lecture, we have highlighted the recent advances made in peptide stapling, the advantage of the two-component peptide stapling techniques over one-component and, in particular, the achievements obtained with the rational design of two-component Cu-catalyzed azide–alkyne cycloaddition stapled peptides. The two-component peptide stapling methodologies involve an intramolecular reaction between the two amino acids on the linear peptide and a staple or linker that can bear a functional handle for further functionalization. After merging the two components together, the peptide may be constrained in the bioactive conformation minimizing the entropic penalty upon binding of the peptide to the targeted protein and increasing its stability to enzymatic degradation. This methodology has been successfully applied not only to constrain α-helical peptides, but also to random coils, therefore expanding its scope. Using the Cu-catalyzed azide–alkyne cycloaddition two-component peptide stapling technique developed in our group, we have created a toolbox of staples not only suitable for constraining different secondary structures of selected sequences, but also stapling at different positions and conditions while also bearing different functionalities. In addition, our two-component peptide stapling methodologies have been proven to be compatible with all the biogenic amino acids that may be present. The toolbox includes staples carrying fluorescent tags, cell-permeable motifs, biotin tags, nuclear localization sequences, and photoaffinity labelling moieties, showcasing the wide applicability of this system. Moreover, stapled peptides obtained were often found to be more active against the set target than the unstapled variants and wild-type peptides. Most importantly, they showed improved cell-permeability and *in vitro* stability. Encouraged by the positive results obtained so far, we envisioned that the scope of the two-component peptide stapling can be expanded further by adding more functionalized stapled peptides to the toolbox.

References

1. Y. H. Lau, P. de Andrade, S.-T. Quah, M. Rossmann, L. Laraia, N. Sköld, T. J. Sum, P. J. E. Rowling, T. L. Joseph, C. Verma, M. Hyvönen, L. S. Itzhaki, A. R. Venkitaraman, C. J. Brown, D. P. Lane, D. R. Spring. Functionalised staple linkages for modulating the cellular activity of stapled peptides. Chem. Sci. 2014, 5, 1804-1809. DOI: 10.1039/C4SC00045E

2. Y. H. Lau, P. de Andrade, N. Sköld, G. J. McKenzie, A. R. Venkitaraman, C. Verma, D. P. Lane, D. R. Spring. Investigating peptide sequence variations for 'double-click' stapled p53 peptides. Org. Biomol. Chem. 2014, 12, 4074-4077. DOI:10.1039/C4OB00742E

3. Y. H. Lau, P. de Andrade, G. J. McKenzie, A. R. Venkitaraman, D. R. Spring. Linear aliphatic dialkynes as alternative linkers for double-click stapling of p53-derived peptides. Chembiochem 2014, 15, 2680-2683. DOI: 10.1002/cbic.201402374

4. Y. H. Lau, Y. Wu, P. de Andrade, W. R. J. D. Galloway, D. R. Spring. A two-component 'double-click' approach to peptide stapling. Nature Protoc. 2015, 10, 585-594. DOI: 10.1038/nprot.2015.033

5. Y. H. Lau, Y. Wu, M. Rossmann, B. X. Tan, P. de Andrade, Y. S. Tan, C. Verma, G. J. McKenzie, A. R. Venkitaraman, M. Hyvönen, D. R. Spring. Double Strain-Promoted Macrocyclization for the Rapid Selection of Cell-Active Stapled Peptides. Angew. Chem. Int. Ed. 2015, 54, 15410-15413. DOI: 10.1002/anie.201508416

6. Y. Wu, F. Villa, J. Maman, Y. H. Lau, L. Dobnikar, A. Simon, K. Labib, D. R. Spring, L. Pellegrini. Targeting the genome stability hub Ctf4 by stapled-peptide design. Angew. Chem. Int. Ed. 2017, 56, 12866-12872. DOI: 10.1002/anie. 201705611

7. Target a Key Protein-Protein Interaction in Ovarian Cancer. Angew. Chem. Int. Ed. 2017, 56, 524-529. DOI: 10.1002/anie.201609427

8. W. Xu, Y. H. Lau, G. Fischer, Y. S. Tan, A. Chattopadhyay, M. de la Roche, M. Hyvönen, C. Verma, D. R. Spring, L. S. Itzhaki. Macrocyclised extended peptides: Inhibiting the substrate-recognition domain of tankyrase. J. Am. Chem. Soc. 2017, 139, 2245-2256. DOI: 10.1021/jacs.6b10234

The design and assessment of the therapeutic potential of apelin-13 peptide mimetics for alleviating metabolic dysfunction in diabetes and obesity

Finbarr O'Harte

Ulster University, United Kingdom

https://doi.org/10.17952/35EPS.2018.010

Introduction

To combat Type 2 diabetes mellitus (T2DM) there is an urgent need for development of new multi-faceted pharmaceutical agents, which induce weight loss and decrease both hyperglycaemia and associated complications, without causing adverse effects. Apelin is a circulating adipokine produced and secreted mainly by adipocytes and endothelial cells [1]. Several bioactive isoforms of apelin exist, including apelin-12, -13, -16, -17, -19 and -36 [2]. The apelin receptor, APJ is present in most tissues and the apelin/APJ (apelinergic system) has been shown to be involved in multiple metabolic processes including control of glucose homeostasis [3]. Rapid degradation and short half-life of native apelin isoforms (4-7 min) severely hinders the pharmacological exploitation of apelin peptides. To overcome the short half-life of apelin-13 related peptides, we have developed a range of enzyme resistant analogues. Two of these acylated analogues include (Lys^{8}GluPAL)apelin-13 amide and pGlu(Lys^{8}GluPAL)apelin-13 amide [4]. These stable apelin analogues stimulated insulin secretion from clonal pancreatic beta cells, primary culture of isolated mouse islets cells and proved to be the most potent of a series of analogues studied [5]. In the present study, metabolic and weight reducing effects of chronic (28 day) once daily administration of (Lys^{8}GluPAL)apelin-13 amide and pGlu(Lys^{8}GluPAL)apelin-13 amide were directly compared to the well-established acylated GLP-1 receptor mimetic liraglutide, using a high-fat fed mouse model diet-induced obesity-diabetes (DIO).

Methods

Apelin analogues were synthesised by EZ Biolabs (Carmel, IN, USA) at >95% purity. A gamma-glutamyl spacer (also found in liraglutide) with palmitate adjunct (GluPAL) was added to the side-chain of apelin Lys^{8} to promote binding to plasma proteins and reduce renal clearance, thus extending its *in vivo* bioactivity. Male NIH Swiss mice (Harlan UK Ltd., Blackthorne, UK) were maintained on a high fat diet (45% fat, Dietex International Ltd., Witham, UK) from 8 weeks old for a total of 150 days to evoke insulin resistance and diet-induced obesity-diabetes (DIO). Control mice was maintained on standard rodent diet (10% fat, Trouw Nutrition, Cheshire, UK). Groups of normal control and high-fat fed (HFF) mice (n=8) received once daily ip injections of either 0.9% saline vehicle (lean and high fat controls) or (Lys^{8}GluPAL)apelin-13 amide, pGlu(Lys^{8}GluPAL)apelin-13 amide or liraglutide (each at 25 nmol/kg bw) over a 28 day treatment period. The various once daily peptide treatments were extended beyond this period (up to 40 days) to allow for additional post-treatment investigations to be performed. Food intake, bodyweight, non-fasting blood glucose and plasma insulin concentrations were measured every 2-3 days up to day 28. Following the 28 day period, 16 h fasted mice were tested for ip and oral glucose tolerance, as well as an insulin sensitivity (25 U/kg insulin challenge). Blood samples were collected from tail snips of mice and blood glucose (Ascencia Contour meter) and HbA1c (PTS Diagnostic, IN, USA) were measured. Blood was taken from fasted mice for measurement of lipid profiles including HDL-cholesterol, LDL-cholesterol and triglyceride levels by an ILab 650 Clinical Analyser.

Results

Both acylated apelin-13 amide analogues significantly decreased non-fasted blood glucose ($P<0.05$ and $P<0.01$ Anova; Fig 1) and increased non-fasting plasma insulin (data not shown). Cumulative energy intake was also significantly decreased in apelin and liraglutide treated mice (27% decrease, $P<0.01$; Fig. 1). A significant decrease in bodyweight was noted over 28 days (3-7% bodyweight reduction), with all treatment groups compared to lean and high-fat fed saline treated mice ($p<0.01$ and $p<0.001$; Fig. 1). Apelin analogues and liraglutide showed an improved ip glucose tolerance (Fig. 1), reduced glycated haemoglobin (HbA1c Fig. 1) and improved insulin sensitivity (Fig. 1). All peptides showed a reduction in circulating triglycerides (Fig. 1G) and

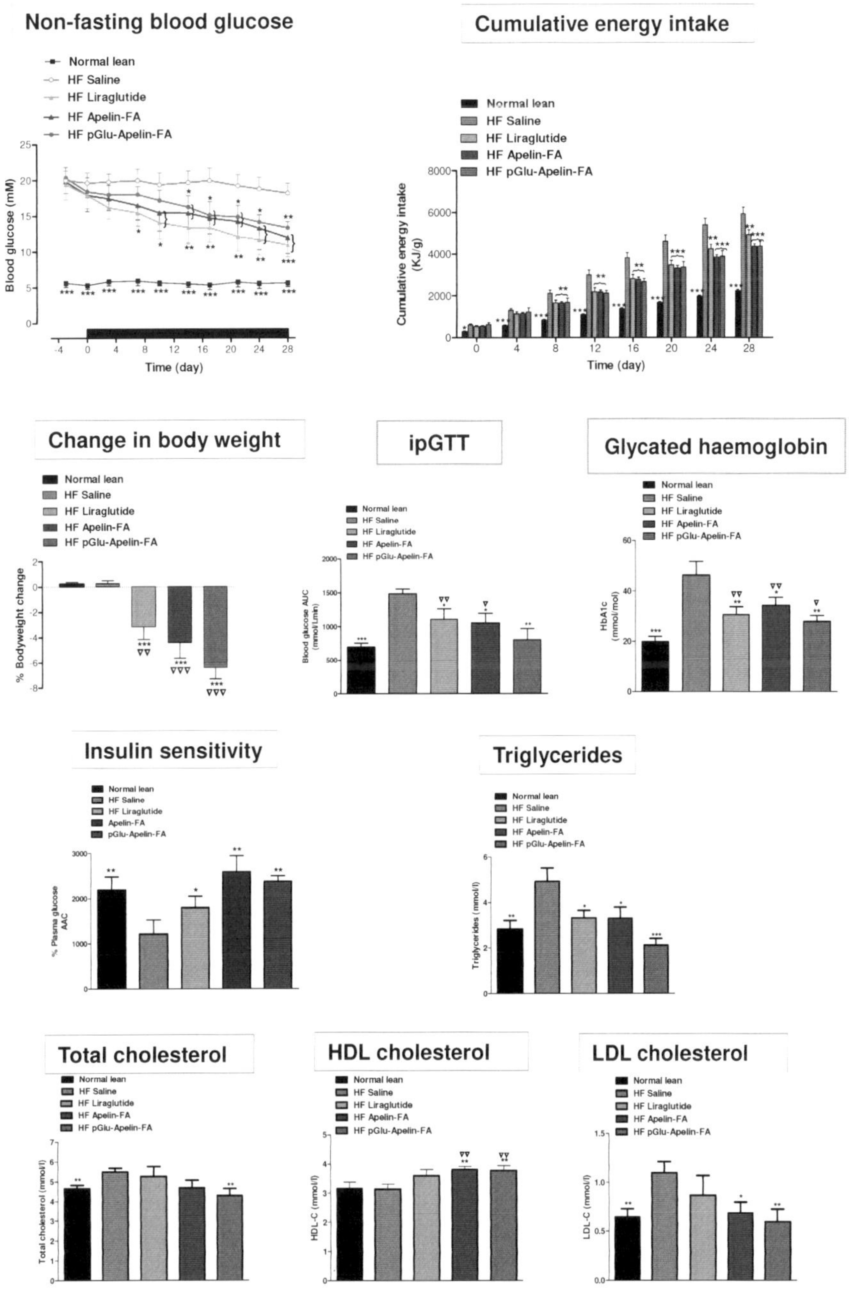
Non-fasting blood glucose
Normal lean
HF Saline
HF Liraglutide
HF Apelin-FA
HF pGlu-Apelin-FA
Blood glucose (mM)
Time (day)
Cumulative energy intake
Cumulative energy intake (KJ/g)
Time (day)
Change in body weight
% Bodyweight change
ipGTT
Blood glucose AUC (mmol/l.min)
Glycated haemoglobin
HbA1c (mmol/mol)
Insulin sensitivity
Apelin-FA
pGlu-Apelin-FA
% Plasma glucose AAC
Triglycerides
Triglycerides (mmol/l)
Total cholesterol
Total cholesterol (mmol/l)
HDL cholesterol
HDL-C (mmol/l)
LDL cholesterol
LDL-C (mmol/l)

Figure 1

both acylated apelin analogues demonstrated an improved lipid profile for HDL and LDL cholesterol compared to liraglutide (Fig. 1).

Discussion

Recent studies indicate an emerging involvement of apelin in energy metabolism and the pathophysiology of obesity [6]. In the present study, once daily (Lys^{8}gluPAL)apelin-13 amide and pGlu(Lys^{8}gluPAL)apelin-13 amide were compared to liraglutide in DIO mice in a 28 day study. Acylated apelin analogues resulted in marked reductions in energy intake as well body weight, improvements in non-fasting glucose, reduced glucose excursion after an ip glucose tolerance test and significantly reduced levels of glycated haemoglobin (HbA1c). The improvements may also reflect improvement in insulin action as evidenced by enhanced hypoglycaemic action of exogenous insulin. Apelin and its APJ receptors have been detected in the arcuate and paraventricular nuclei of hypothalamus, known to be key sites in central control of feeding behaviour and energy expenditure [7,8]. Apelin could also alter body adiposity independent of food intake by increasing energy expenditure [9]. Interestingly, apelin knockout (KO) mice exhibit reduced insulin sensitivity, glucose intolerance and hyperinsulinaemia [10]. In the present study with high fat fed DIO mice, pGlu(Lys^{8}gluPAL)apelin-13 amide reduced total cholesterol and both acylated apelin analogues reduced circulating triglycerides, LDL-cholesterol as well as increasing HDL-cholesterol. Cardiovascular benefits of apelin, including a reduction of blood pressure are already established [11]. Once daily administration of the GLP-1 mimetic, liraglutide replicated all of the benefits of apelin-13 analogues but it failed to completely improve lipid profile in DIO mice as shown previously in our laboratory [12]. In conclusion, the present study has shown that once daily administration of (Lys^{8}gluPAL)apelin-13 amide or pGlu(Lys^{8}gluPAL)apelin-13 amide ameliorated diabetes, evoked weight loss and decreased circulating lipids in DIO mice, with effects equivalent to or better than liraglutide. Overall the pGlu(Lys^{8}gluPAL)apelin-13 amide was the most effective analogue and was better than it equivalent non-acylated analogue tested previously [7].

References

1. Boucher J, Masri B, Daviaud D, Gesta S, Guigne C, Mazzucotelli A *et al.* Endocrinology 2005; 146:1764-1771.

2. Kawamata Y, Habata Y, Fukusumi S, Hosoya M, Fujii R, Hinuma S, *et al.* Biophysica Acta 2001; 23:162-171.

3. O'Carroll AM, Lolait SJ, Harris LE, Pope GR. J Endocrinol. 2013; 219:R13-R35.

4. O'Harte FPM, Parthsarathy V, Hogg C, Flatt PR. Biochem Pharmacol. 2017;146: 165-173.

5. O'Harte FPM, Parthsarathy V, Hogg C, Flatt PR. Peptides. 2018;100: 219-228.

6. Bertrand C, Valet P, Castan-Laurell I. Front Physiol. 2015; 6:115.

7. Parthsarathy V, Hogg C, O'Harte FPM, Flatt PR. Diabetes Obes Metabolism. 2018; 20:319-327.

8. Valle A, Hoggard N, Adams AC, Roca P, Speakman JR. J Neuroendocrinol. 2008; 20:79-84.

9. Higuchi K, Masaki T, Gotoh K, Chiba S, Katsuragi I, Tanaka K, *et al.* Endocrinology. 2007; 148:2690-2697.

10. Yue P, Jin H, Aillaud M, Deng AC, Azuma J, Asagami T, *et al.* Am J Physiol Endocrinol Metab. 2010; 298:59-67.

11. Mughal A, O'Rourke ST. Pharmacol Ther. 2018;pii: S0163-7258(18):30093-30097.

12. Millar P, Pathak N, Parthsarathy V, Bjourson AJ, O'Kane M, Pathak V *et al.* J Endocrinol. 2017; 234:255-267.

Acknowledgements

We would like to thank the funders DfE for awarding a PhD studentship to CH. Also, we thank Invest NI for their Proof of Concept PoC518 grant funding.

Are gluten-related disorders a new protein aggregation disease?

Veronica Dodero

Universität Bielefeld, Germany

https://doi.org/10.17952/35EPS.2018.013

Abstract

A new hypothesis about the relation of gliadin and its immunodominant 33-mer fragment aggregates as triggers of gluten-related disorders was presented.

Gluten-related disorders (1) are a complex group of diseases which involve the activation of the immune system triggered by the ingestion of gluten. These disorders have a high prevalence in western societies around 1-7% worldwide. The high prevalence is because gluten is present in wheat, rye, barley and some varieties of oats. It is accepted that the incomplete proteolysis of the gluten proteins is responsible for disease. It is assumed that there must be a particular susceptibility in an individual that could trigger this kind of diseases, but parameters have not yet been well clarified. Except for celiac disease (CD), where HLA-DQ genes are of invaluable importance in the diagnosis. Virtually all CD patients (>97%) carry the coding variants for HLA-DQ2 and /or HLA-DQ8 molecules. Although the genetic predispose is an essential factor in the pathogenesis of CD, it has low predictive value. This is illustrated by the fact that while 30-40% of the general population carries this genotype, the prevalence of CD worldwide is 1%.

Up to now, only a strict gluten-free diet (GFD) improves the clinical symptoms, diminishes the titers of the CD-specific antibodies and finally heals the intestinal mucosa and restores health. CD is an autoimmune enteropathy that may be present not only as intestinal manifestations but also as systemic ones, such as anemia, loss of weight, short stature, osteoporosis, and peripheral neuropathy. Other specific conditions associated with CD are, for example, Addison's disease, non-Hodgkin lymphoma, neurological disorders and type I diabetes mellitus. Interestingly, it is unknown which are the molecular events that start and persist tissue morphological changes, but the key role of gluten peptides is accepted. In the last years, a high percentage of the population worldwide is opting for a gluten-free diet because they feel "healthy." There is an estimation of self-assessment by adults affecting up to 13% of the population. In this regard, the Journal Science has recently published a feature article entitled "The war on gluten" (2). It has been shown that gastroenterologists worldwide are confused about the phenomenon of gluten avoiders. Up to know, it is unclear if this is a new diet trend, a style of life or a real public health problem.

Gliadins (3) are immunogenic proteins present in wheat gluten which are classified as α (25-35 kDa), β (30-35 kDa), γ (35-40 kDa) and ω (55-75 kDa) depending on their electrophoretic mobility. These proteins share a similar primary structure consisting of an N-terminal domain, a hydrophobic central domain which is rich in proline, glutamine, and phenylalanine, and, a non-repeating region including cysteine. It is important to mention that the amount and composition of gliadins depend strongly on the wheat variety and the specific cultivar. Also, the structure of the gliadin proteins is itself already complicated due to the heterogeneity of the individual proteins and the wide range of their molecular weights. Humans do not fully degrade these proteins, and after the normal gastric and pancreatic digestion, many peptides remains. The most resistant immunodominant peptide is a 33-mer amino acid fragment, which triggers an autoimmune response related to CD (4).

From a structural point of view, up to now, the research efforts to the understanding of the pathogenic related molecules were directed towards the identification, quantification, and separation of the components of gluten and connecting them with their pathological role *in vitro* and *in vivo* assays. Interestingly, the primary structure of the molecules involved in disease is known, but there is a lack of information about intrinsic behavior, like their folding and molecular organization under physiologically relevant conditions.

Gliadin is self-organized under aqueous physiological conditions: It was demonstrated that when commercial gliadin is dissolved and homogenized in water at pH 3.0 under low ionic strength, mainly α-gliadin remains in the solution self-organized. The self-assembly process is spontaneous and nano-spherical structures, micelle-type, are stabilized in the transparent water solution (5). Recently, it was confirmed that even at a dilute concentration

(0.1 mg/ml), gliadin is already a colloidal polydisperse system with an average hydrodynamic radius of 30± 10 nm. By cryo-transmission electron microscopy (cryo-TEM), it was detected mainly large clusters, but also prolate oligomers were visualized for the first time. Small angle X-ray scattering (SAXS) experiments supported the existence of gliadin prolate/ rod-like structures, suggesting that gliadin dimers are the minimal oligomeric unit organized longitudinally. To understand the basis of the oligomerization, an α2-gliadin model was built, obtained by ab initio prediction and analyzed by molecular dynamics (MD). Combining this model with aggregation predictors a 3D α2-gliadin model was created showing that there is a major expose surface prone to aggregation at the C-terminus that can be involved in the dimerization process of α2-gliadin (6). The presence of soluble superstructures of α-gliadin could explain the high proteolytical resistance, not only due to the high proline content in gliadin primary sequence but also because of inaccessibility of the enzymes due to the aggregation process.

The 33-mer gliadin peptide oligomerizes activating an innate immune response under accumulation conditions. The immunodominant33-mer gliadin peptide remains unprocessed, and it is found in human urine (7). Importantly, this 33 amino acid fragment is directed involved in celiac disease and probably in other immune pathologies associated with gliadin. Although, the immunological response observed in the chronic phase of the celiac disease is well characterized; the cause and the initiation of the inflammatory events are still obscure. From the molecular point of view, the 33-mer behaves as an amphiphile forming nano-aggregates, protofilaments, and lineal plated structures (8). The morphology of the nano-aggregates depends strongly on concentration,and the secondary structure ranges from random,and PPII to parallel beta structure (9). Recently, we reported that only large structures of 33-mer (>200 nm) induce an innate immune response in macrophages which is mediated by Toll-like receptor (TLR) 4 activation in humans (10). Our investigations showed that if the 33-mer peptide aggregates under accumulation conditions, can trigger a specific innate immune response.

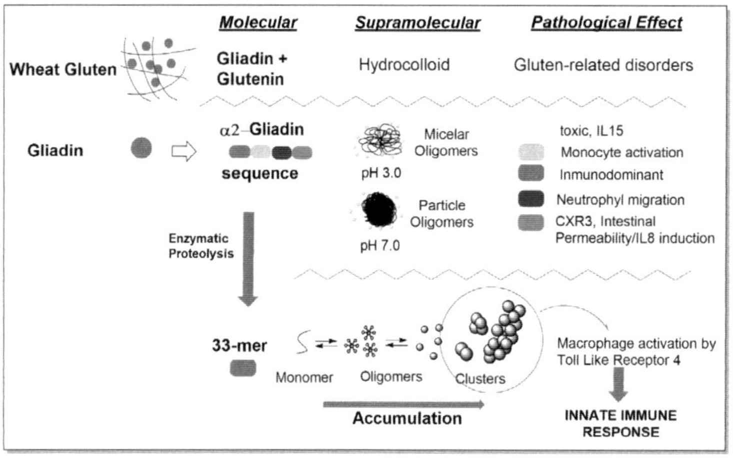

Scheme 1: This cartoon summarizes the relation between molecular structure, supramolecular organization and the pathological effect of gliadin sequence and 33-mer gliadin superstructures.

These findings are the proof of concept that the accumulation and aggregation of this peptide alone can be a hitherto unknown trigger of inflammation. Protein oligomerization and structural change with activation of the innate immune system are hallmarks of other human protein aggregation diseases.

References

(1) L. M. Lammers, M G. Herrera, V. I. Dodero, Translational Chemistry meets gluten-related disorders, ChemistryOpen 7 (2018) 217.

(2) K. Servick, The war on gluten, Science 360 (2018) 848.

(3) A. V. Balakireva,, A. A. Zamyatnin, Jr., Nutrients 8 (2016) 644.

(4) L. Shan, O. Molberg, I. Parrot, et al., Structural Basis for Gluten Intolerance in Celiac Sprue, Science 297 (2002) 2275.

(5) M. G. Herrera, T. V. Veuthey, V. I. Dodero, Self-organization of gliadin in aqueous media under physiological digestive pHs, Colloids Surf B Biointerfaces 141 (2016) 565.

(6) M. G. Herrera, D. S Vazquez, R. Sreij, M, Dreschler, Y. Hertle, T Hellweg, V. I. Dodero, Insights into Gliadin oligomerization behavior at digestive pH 3.0, Colloids Surf B Biointerfaces 165 (2018) 363.

(7) M. L. Moreno, A. Cebolla, A. Munoz-Suano et al., Detection of gluten immunogenic peptides in the urine of patients with coeliac disease reveals transgressions in the gluten-free diet and incomplete mucosal healing, Gut (2015) 1.

(8) M. G. Herrera, F. Zamarreno, M. Costabel, H. Ritacco, A. Hütten, N. Sewald, V. I. Dodero, Circular Dichroism and Electron Microscopy Studies *in vitro* of 33-mer Gliadin Peptide revealed Secondary Structure Transition and Supramolecular Organization, Biopolymers 101 (2014) 96.

(9) M. G. Herrera, L. A. Benedini, C. Lonez, P. L. Schilardi, T. Hellweg, J.-M. Ruysschaert, V. I. Dodero, Self-assembly of 33-mer gliadin peptide oligomers, Soft Matter 11 (2015) 8648.

(10) M. G. Herrera, M. Pizzuto. C. Lonez, K. Rott, A. Hütten, N. Sewald, J-M Ruysschaert, V. I. Dodero, Large Supramolecular Structures of 33-mer Gliadin Peptide Activate Toll-like Receptors in Macrophages, Nanomedicine: NBM 14 (2018) 1417.

Proceedings of the 35th European Peptide Symposium
Patrick B. Timmons, Chandralal M. Hewage, Michal Lebl (Editors)
16 *European Peptide Society & PSP, 2018*

Novel methodology for the synthesis of α-indolyl-glycine containing peptide via direct asymmetric Friedel–Crafts reaction to peptidyl imine

Tsubasa Inokuma, Kodai Nishida, Akira Shigenaga, Ken-Ichi Yamada, Akira Otaka

Tokushima University, Japan

https://doi.org/10.17952/35EPS.2018.016

Peptides which contain unnatural amino acids are attractive novel drug candidates due to their biological activity superior to those of the peptides consisting of only proteinogenic amino acids. [1] These peptides are conventionally prepared by two steps; (1) asymmetric synthesis of unnatural amino acid unit, and (2) installation of the unit by Fmoc or Boc Solid-phase peptide synthesis (Fig. 1-A). However, the present procedure requires multi-step of reaction, including removal of the protective groups of amino and carboxy groups and protection of the amine moiety by Fmoc group, to prepare diverse of peptide. To develop new methodology which addresses this problem, we envisioned an asymmetric construction of unnatural amino acid units in a growing peptide chain (Fig. 1-B). [3] If direct asymmetric 1,2-addition of α-iminopeptideis realized, changing of the nucleophiles enables easy access to a diverse range of unnatural amino acid-containing peptides.

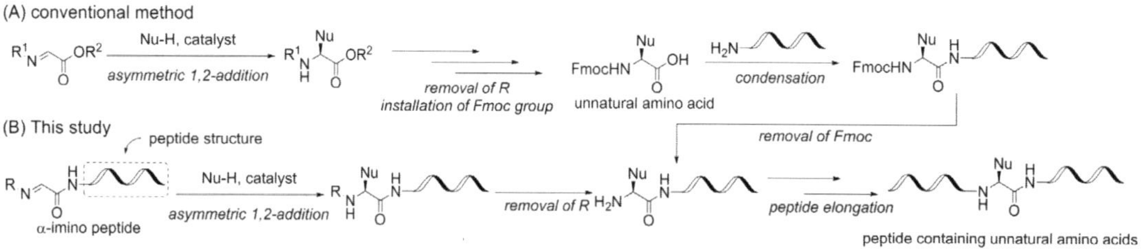

Figure 1: Procedures of the synthesis of peptides which contain unnatural amino acids.

We planned to use a hydrophobic anchor that makes the peptidic substrate soluble in less-polar organic solvents [2] because peptides are hardly soluble in less-polar organic solvents which are often most suitable for catalytic asymmetric reactions. As an unnatural unit of the target peptide, α-indolyl glycine is chosen because this unit is a promising analog to phenylalanine, tyrosine or tryptophan. An asymmetric construction of this unit was accomplished *via* a Friedel–Crafts reaction of N-2-nitrophenylsulfenyl (Nps)-protected α-imino ester and indole. [4] In this study, we planned to incorporate this reaction into our synthetic concept to synthesize an α-indolyl glycine-containing peptide. The outline of this study is shown in Fig. 2. At first, peptide elongation is performed on the amine moiety of a hydrophobic anchor using liquid-phase peptide synthesis (step 1). Next, condensation of N-Nps glycine on the growing peptide chain followed by oxidation of the N-terminal residue provides requisite imino peptide (step 2). Then, asymmetric Friedel-Crafts reaction of indole on the resulting imine is performed (step 3). Subsequent removal of the Nps group followed by usual manipulations for peptide elongation, cleavage of anchor, and global deprotection by acidic treatment would afford the desired peptides containing α-indolyl glycine (step 4).

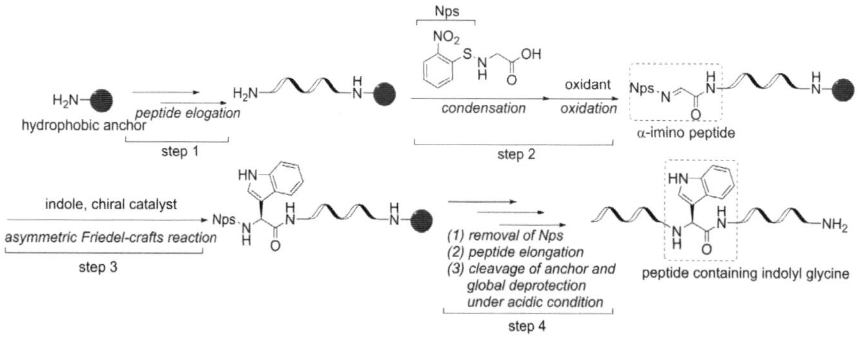

Figure 2: Outline of this study: Synthesis of a peptide bearing α-indolyl glycine.

We used bis(4-docosyloxyphenyl) methyl (Dpm (C22)) amine 1 which was reported by Takahashi as the hydrophobic anchor of our synthesis (Fig. 3). [2a] First, three glycine residues were installed into anchor 1 and the N-terminal amino group was protected by an Nps group. Next, MnO2-mediated oxidation of 2 was performed under our previously established conditions to give α-imino peptide 3. [5] In the presence of BINOL-derived chiral phosphoric acid 4 [6], an asymmetric Friedel–Crafts reaction of 3 and indole successfully produced α-indolyl glycine-containing adduct 5. Indole of this adduct 5 was then protected by an Alloc group by Alloc-imidazole [7]. After that 2-mercaptopyridine-mediated removal of the Nps group, condensation of Fmoc-Tyr(Ot-Bu)-OH, and removal of Fmoc group by treatment of piperidine afforded the protected tetrapeptide 6. Two-step deprotection involving acidic treatment and Pd-mediated removal of the Alloc group afforded unprotected tetrapeptide 7. Although the chemical yield and stereoselectivity of this process remain to be improved, we demonstrated the first example of direct asymmetric reaction to imino peptide.

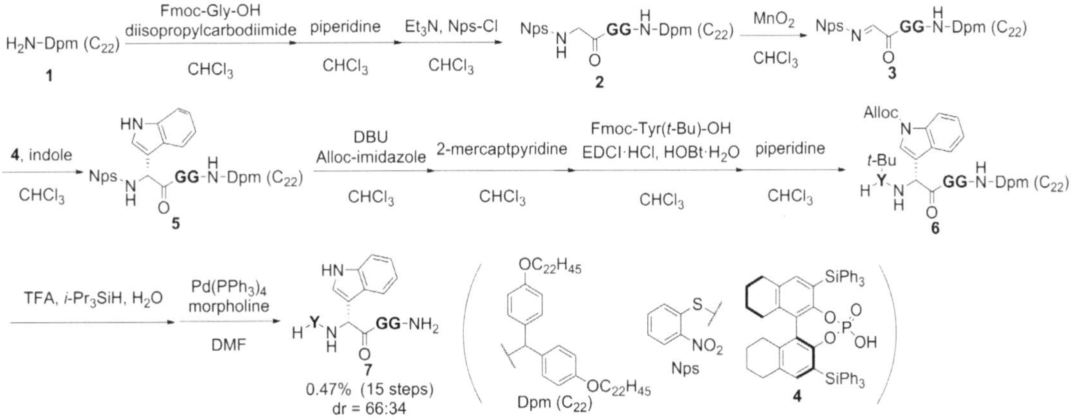

Figure 3: Direct asymmetric Friedel-Crafts reaction to peptidyl imine.

In summary, we developed an unprecedented methodology featuring a direct asymmetric reaction to peptidyl imine. By using this methodology, the preparation of α-indolyl glycine-containing peptide was achieved. Optimization of the condition to develop a truly practical methodology for synthesis of unnatural amino acid-containing peptide is currently underway.

References

1. W. Kang, H. Liu, L. Ma, M. Wang, S. Wei, P. Sun, M. Jiang, M. Guo, C. Zhou, J. Dou, Eur. J. Pharm. Sci., 2017, 105, 169.

2. (a) D. Takahashi, T. Yano, T. Fukui, Org. Lett., 2012, 14, 4514. (b) G. Tana, S. Kitada, S. Fujita, Y. Okada, S. Kim, K. Chiba, Chem. Commun., 2010, 46, 8219.

3. T. Inokuma, K. Nishida, A. Shigenaga, K. Yamada, A. Otaka, Heterocycles, 2018, in press. (DOI: 10.3987/COM-18-S(T)86)

4. M. J. Wanner, P. Hauwert, H. E. Schoemaker, R. de Gelder, J. H. van Maarseveen, H. Hiemstra, Eur. J. Org. Chem., 2008, 180.

5. T. Inokuma, T. Jichu, K. Nishida, A. Shigenaga, A. Otaka, Chem. Pharm. Bull., 2017, 65, 573.

6. N. V. Sewgobind, M. J. Wanner, S. Ingemann, R. Gelder, J. H. van Maarseveen, H. Hiemstra, J. Org. Chem., 2008, 73, 6405.

7. S. T. Heller, E. E. Schultz, R. Sarpong, Angew. Chem. Int. Ed., 2012, 51, 8304.

Stereoselective synthesis of a β-methylthio α-amino α-lactam dipeptide, a S-methyl-Cys-Val mimic

Azade Geranurimi, William Lubell

Departmentof Chemistry, Universit de Montral, Canada

https://doi.org/10.17952/35EPS.2018.018

Abstract

Pursuing β-substituted α-N-(Fmoc)amino-γ-lactam (Agl) dipeptides by displacements of the alcohol corresponding β-hydroxy-α-amino-γ-lactam (Hgl) residue, nucleophilic ring opening of cyclic sulfamidate 3 with thiocyanate has provided entry to trans-β-methylthio-Agl derivative 5 for use as a constrained S-methyl Cys derivative.

Introduction

Peptides are typically flexible and may adopt multiple conformers in solution. Among methods to restrict peptide movement and improve receptor binding affinity, cyclization has proven successful since the team at Merck led by Drs. Freidinger and Veber introduced α-amino-γ-lactam (Agl) residues into LH-RH.[1] Since their pioneering research, many have used Agl residues to favor peptide turn conformations by constraining ω, φ and ψ backbone dihedral angles.[2] Adding β-substituents to Agl residues has since been perceived as a means to enhance mimicry potential by simultaneously constraining backbone and side (χ) dihedral angles.[3-5]

Recently, a set of β-substituted α-amino-γ-lactam residues were synthesized using β-hydroxy-α-amino-γ-lactam (Hgl) residue 1 as a convenient precursor (Figure 1).[5,6] Mitsunobu reactions on the trans isomer of Hgl trans-1 with pro-nucleophiles having pKa values between 3 and 7 provided a variety of cis-β-substituted Agl analogs 2a-g in 49-80% yields.[6] Moreover, treatment of trans-Hgl 1 with diphenylphosphoryl azide (DPPA) as a hydrazoic acid equivalent under Mitsunobu conditions gave in 60% yield the corresponding β-azido-Agl residue, which was further converted to its phenyltriazole counterpart in 80% yield by copper-catalyzed azide alkyne cycloadditions.[6]

Attempts to employ Mitsunobu reaction on cis-Hgl cis-1 gave however only dehydro-lactam.6 Cyclic sulfamidate 3 was thus synthesized from cis-Hgl cis-1 to disfavor β-elimination to dehydro-lactam.[6] Ring opening of sulfamidate 3 has already provided β-azido-Agl 4a and constrained protected cysteine 4b (Figure 1). Exploring the chemistry of sulfamidate 3 further, we now report a method to provide constrained S-alkyl Cys residues (e.g., 5).

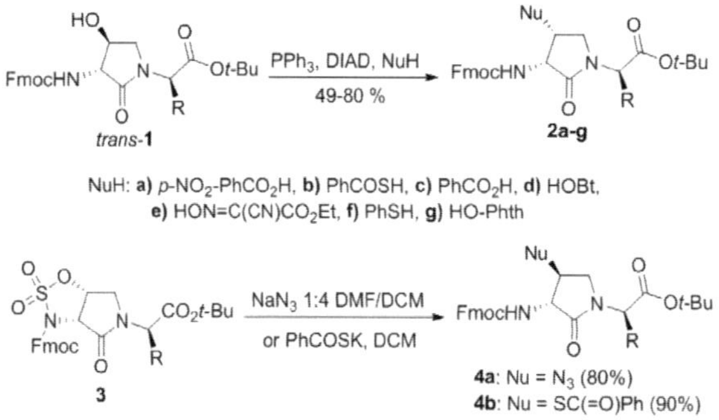

Figure 1: Synthesis of 4-substituted Agl residues by way of the Mitsunobu reaction and cyclic sulfamidate ring opening.

Results and Discussion

β-Thiocyano-Agl 4c was synthesized by reacting sulfamidate 3 with potassium thiocyanate in 1:1 DMF/DCM in 92% yield. Reduction of thiocyanide 4c with $NaBH_4$ and alkylation with iodomethane by way of a constrained Cys thiolate intermediate provided thioether 5 in 65% yield from 4c (Scheme 2).[7]

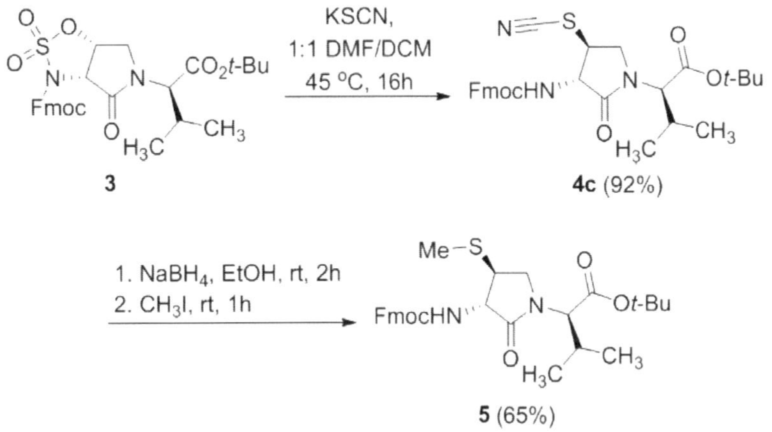

Scheme 2: Synthesis of constrained S-methyl Cys derivative [5]

Conclusions

Methylthio-Agl 5 may be considered a constrained isoleucine analog. Moreover, S-alkylation with other electrophiles may further expand the diversity of trans-β-substituted Agl residues. The synthesis of such amino acid residues with constrained backbone and side chain geometry, and their subsequent insertion into peptides is now under investigation and will be reported in due time.

Experimental

tert-Butyl (3S, 4S, 2'R)-2-[3-(Fmoc)amino-4-thiocyanato-2-oxopyrrolidin-1-yl]-3-methylbutanoate

A solution of sulfamidate 3 (1 eq., , prepared according to ref 6) in a mixture of DCM (2 mL) and DMF (2 mL) was treated with potassium thiocyanate (3 eq.,), stirred at 45 °C for 16 h, poured into 1 M NaH2PO4, and extracted with DCM. The combined organic phase was washed with brine, dried, filtered, evaporated to a residue, that was purified by column chromatography using a step gradient of 10-30% EtOAc in hexane. Evaporation of the collected fractions gave thiocyanide 4c as a white solid (): Rf = 0.3 (30% EtOAc in hexane); mp 56-60 °C; $[\alpha]_D^{25}$ 24° (c 1, $CHCl_3$); FT-IR (neat) νmax 3321, 2973, 2155, 1707, 1523, 1449, 1391, 1370, 1239, 1150, 1106, 1043, 781, 759, 738 cm^{-1}; ^{1}H NMR (500 MHz, $CDCl_3$) δ 7.76 (d, J = 7.6 Hz, 2H), 7.59 (ddd, J = 7.4, 3.8, 0.6 Hz, 2H), 7.41 (t, J = 7.4 Hz, 2H), 7.32 (td, J = 7.5, 1.0 Hz, 2H), 5.48 (s, 1H), 4.46 (m, 3H), 4.31 (s, 2H), 4.23 (t, J = 6.8 Hz, 1H), 3.83 (dd, J = 18.1, 9.1 Hz, 1H), 3.49 (t, J = 9.4 Hz, 1H), 2.23 (m, 1H), 1.48 (s, 9H), 1.02 (d, J = 6.7 Hz, 3H), 0.97 (d, J = 6.8 Hz, 3H); ^{13}C NMR (126 MHz, $CDCl_3$) δ 169.0, 156.5, 143.5, 141.3, 127.8, 127.1, 125.1, 120.1, 109.7, 82.7, 67.6, 60.9, 57.1, 48.1, 47.0, 45.9, 29.0, 28.0, 22.0, 19.5, 19.3; HRMS (ESI-TOF) m/z $[M+Na]^+$ calcd for $C_{29}H_{33}N_3O_5S$ 558.2033 found, 558.2017.

tert-Butyl (3S, 4S, 2'R)-2-[3-(Fmoc)amino-4-methylthio-2-oxopyrrolidin-1-yl]-3-methylbutanoate

A solution of thiocyanide 4c (1 eq.,) in EtOH (95%, 1 mL) was added drop-wise to a stirred solution of $NaBH_4$ (5 eq.,) in EtOH (95%, 1 mL) at rt. The mixture was stirred for 2 h, treated with iodomethane (2 eq.,), stirried for 1h, and treated with a solution of 1N HCl (2 mL). The mixture was extracted with EtOAc (3×5 mL). The combined organic phase was washed with brine, dried, filtered, evaporated to a residue, that was purified by column chromatography using a step gradient of 10-30% EtOAc in hexane. Evaporation of the collected fractions gave thioether 5 as a light yellow oil (): Rf = 0.5 (30% EtOAc in hexane); $[\alpha]_D^{25}$ 45° (c 0.2, $CHCl_3$); ^{1}H NMR (500 MHz, $CDCl_3$) δ 7.76 (d, J = 7.5 Hz, 2H), 7.60 (d, J = 7.4 Hz, 2H), 7.39 (t, J = 7.5 Hz, 2H), 7.31 (td, J = 7.5, 1.0 Hz, 2H), 5.26 (s, 1H), 4.45 (d, J = 9.1 Hz, 1H), 4.43 (d, J = 7.0 Hz, 2H), 4.34 (t, J = 8.9 Hz, 1H), 4.24 (t, J = 7.0 Hz, 1H), 4.06 (dd, J = 9.4, 7.5 Hz, 1H), 3.32 – 3.24 (m, 1H), 3.21 (t, J = 9.2 Hz, 1H), 2.24 – 2.14 (m, 4H), 1.47 (s, 9H), 1.00 (d, J = 6.7 Hz, 3H), 0.94 (d, J = 6.8 Hz, 3H); ^{13}C NMR (126 MHz, $CDCl_3$) δ 171.1, 169.5, 156.4, 143.8, 141.3, 127.7, 127.1, 125.2, 120.0, 82.3, 67.3, 60.7, 57.6, 47.8, 47.1, 45.7, 28.8, 28.1, 22.0, 19.4, 13.6; HRMS (ESI-TOF) m/z $[M+Na]^+$ calcd for $C_{29}H_{36}N_2O_5S$ 547.2237 found, 547.2228.

Acknowledgments

The authors would like to thank the Natural Sciences and Engineering Research Council (NSERC) of Canada for financial support of this research program.

References

1. Freidinger, R. M.; Veber, D. F.; Perlow, D. S.; Saperstein, R., Science 1980, 210, 656-658.

2. Freidinger, R. M., J. Med. Chem. 2003, 46, 5553-5566.

3. St-Cyr, D. J.; García-Ramos, Y.; Doan, N.-D.; Lubell, W. D., Aminolactam, N-Aminoimidazolone, and N-AminoimdazolidinonePeptide Mimics. In Peptidomimetics I, Springer2017; pp 125-175.

4. Wolf, J. P.; Rapoport, H., J. Org. Chem. 1989, 54, 3164-3173.

5. St.-Cyr, D.; Jamieson, A. G.; Lubell, W.D. Org. Lett. 2010, 12, 1652-1655.

6. Geranurimi, A.; Lubell, W. D., Org. Lett. 2018, 20, 6126-6129.

7. Gulea, M.; Hammerschmidt, F.; Marchand, P.; Masson, S.; Pisljagic, V.; Wuggenig, F., Tetrahedron: Asymmetry 2003, 14, 1829-1836.

Enabling parallel peptide purification by a novel traceless purification linker

Oliver Reimann[1], Dominik Sarma[1], Gavin Noble[2], Oliver Seitz[3], Robert Zitterbart[1]

[1]Belyntic GmbH, Germany
[2]Bachm (UK) Ltd., United Kingdom
[3]Bioorganic Synthesis, Department of Chemistry, Humboldt-Universitat zu Berlin, Germany

https://doi.org/10.17952/35EPS.2018.021

In contrast to peptide synthesis, peptide purification by high performance liquid chromatography (HPLC) cannot be easily varied in number and scale, which results in production restraints. Therefore, purification of peptides *via* preparative HPLC is a bottleneck, especially regarding higher throughput missions. Furthermore, purification by preparative HPLC is a formidable challenge when difficult peptides have low solubility, a tendency to aggregate or when truncations co-elute with the full-length peptide. Herein, we report the facilitated purification by catch (of crude peptides) and release (of pure peptides) that allows purification of standard and difficult peptides in parallel formats.

Our method is based on purification linkers that are coupled to the N-terminal end of the peptide in the last step of solid-phase synthesis. We have presented two linker systems, a base-labile system [1] and a reductively cleavable linker (scheme 1). The linker molecule must have three important functionalities. A traceless connector to the peptide, a chemical hock for the purification media and a cleavable core for the release of the peptide. The linker is connected to the peptide by activation of the carbonate with oxyma and formation of a carbamate. Owing to capping, truncations do not carry the linker. TFA cleavage liberates the peptide and the chemical hock, the amino-oxy moiety. Thus, the peptide is ready for purification. Only the full-length peptide obtained after solid-phase synthesis is immobilized onto an aldehyde-modified agarose resin by means of oxime ligation. This bio-orthogonal reaction is complete in less than one hour. Immobilization can be performed directly from the cleavage cocktail, avoiding laborious ether precipitation prior to purification and enabling the purification of strongly hydrophobic peptide. If precipitation is desired, denaturing agents such 8 M urea/guanidium chloride, TFA or organic solvents can be used to facilitate solubilization of aggregation-prone peptides. After immobilization, the truncated, acetylated peptide sequences are removed by simple washing.

The reductively cleavable linker 1 (scheme 1) is designed to allow liberation of the native peptide by a reduction-triggered 1,6-elimination, which yields the desired native peptide in pure form without occurrence of side reactions. A Staudinger reduction is used to reduce the azide to an aza-ylid with triphenylphosphine by liberation of nitrogen. The Aza-ylid is stable under the given conditions, allowing the washout of triphenylphosphine excess, followed by hydrolysis of the aza-ylid under acidic conditions, releasing the peptide. Finally, the beads are rinsed with TFA/H2O (9:1) to ensure complete dissolution and elution of any given peptide. The peptide can be directly ether precipitated to additionally remove the remaining one equivalent of triphenylphosphineoxid and gain the purified peptide as a white solid without lyophilization.

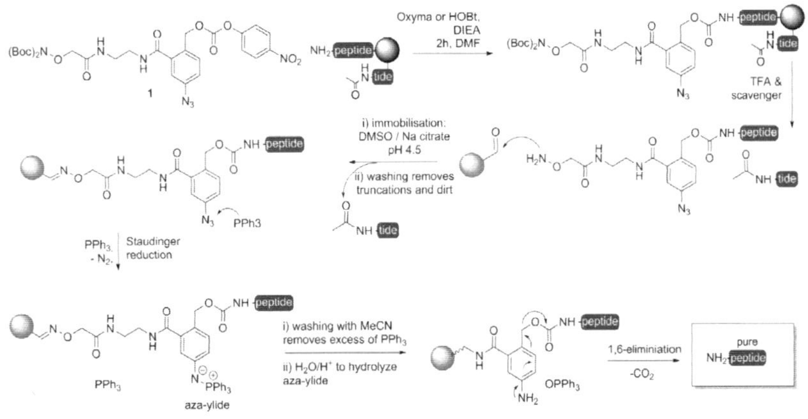

Scheme 1: Reaction overview of the peptide easy clean (PEC) process to purify peptides by reductively cleavable linker 1.

Some examples we have gained with peptides that were provided by BACHEM UK are shown in Figure 1. In the left panel the top trace shows a crude chromatogram of a histone sequence, that showed at least four capped co-eluting truncations in the main peak. The PEC-technology was able to remove those and rose purity up to 95%, by usage of the base-labile linker. Two other peptides that have been purified by the reductively cleavable linker are shown in the middle and right panel of Figure 1. Of note is the purification of the strongly aggregating [2] amyloid β-fragment 1-20 that was purified by the usage of urea for dissolution with a recovery of 50% and purity of 90%. By running flash chromatography of this peptide, a purity of 86% was achieved with a recovery of 38%.

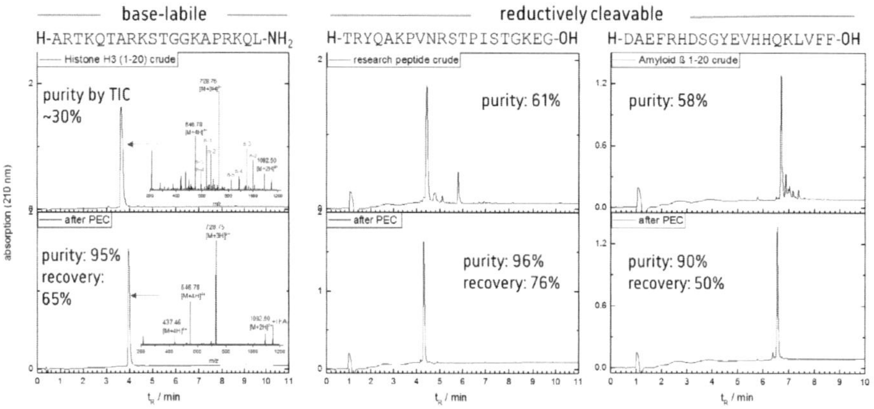

Figure 1: Examples of PEC-purified peptides before purification (crude) and after PEC-purification.

Together with Bachem UK we conducted a case study investigating the purification of a second H3 histone sequence. Dr. Gavin Noble in St. Helens England synthesized the peptide and it was equally split into two 200 μmol batches. We PEC-purified one batch by usage of the reductively cleavable linker and the second half was purified by use of a 2-dimensional approach HPLC, consisting of a first run with hexafluoro butyric acid and a second run with TFA. Before purification the peptide showed a crude purity of 34%. The 2-dimensional HPLC approach gained an apparent purity of 94% with a recovery of 50% (Figure 2). The PEC-purified peptide showed an 80% purity with a higher recovery of 80%. The needed time was about the same for both purifications, but with PEC four peptides were purified in parallel, while the HPLC purified one peptide. The PEC-process consumed 90% less organic solvents and 95% less acetonitrile in direct comparison.

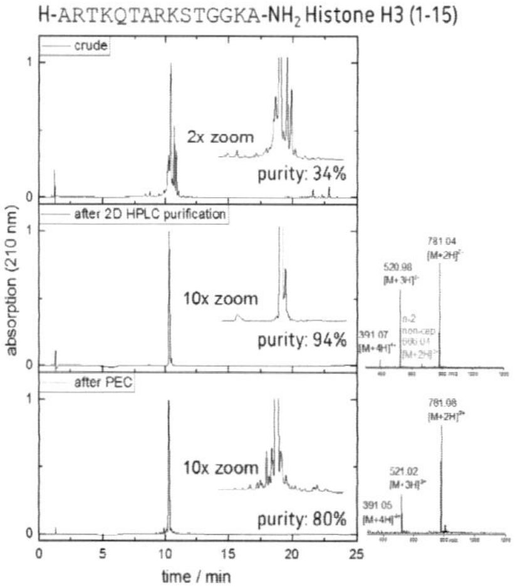

Figure 2: Case study of the purification of Histone H3 (1-15). 2-dimensional HPLC purification (middle trace) versus PEC-purification (lower trace).

By using catch & release based purification peptides can be efficiently isolated from complex mixtures and especially potentially co-eluting truncations. The feasibility of the method has been tested with many peptides, including very hydrophobic amyloid ß (1-20) rationally designed peptides with co-eluting truncations and base-sensitive sequences. Peptides were purified in parallel set-ups using vacuum manifolds to prove practicability. The tolerance of our catch & release process to denaturing agents and the ability to liberate peptides of a high purity by traceless reductive cleavage renders the method universal, paving the way for a true alternative to HPLC while reducing the consumption of organic solvents by 75-90%.

References:

[1] O. Reimann, D. Sarma, O. Seitz, R. Zitterbart, J Pept Sci 2018, Manuscript accepted

[2] L. Malavolta, M. R. S. Pinto, J. H. Cuvero, C. R. Nakaie, Protein Science 2006, 15, 1476-1488.

A general method towards ADPr ribosylated peptides and proteins

Gerbrand van der Heden van Noort

Leiden University Medical Centre, Netherlands

https://doi.org/10.17952/35EPS.2018.024

Ubiquitination of target proteins is a post-translational modification involved in almost all aspects of eukaryotic biology including the regulation of immune responses, cell proliferation and cell survival. The attachment of the 76 amino acid long protein Ubiquitin (Ub) to a nucleophilic functionality in the amino acid side chain of proteins alters the fate of the modified protein. The distinct signals that are brought about by this modification are mostly invoked by poly ubiquitination. Ub has the ability to form polymers by connecting one of eight amines of the first Ub molecule and the C-terminal carboxylic acid of a second Ub molecule etc. This will lead to eight differently linked homotypic poly Ub chains. All of these linkages have been detected in cells and their abundance changes during specific cellular events, indicative of their various functions. Extensive investigations of Ub led to the discovery of proteins able to assemble, recognize and break down Ub chain types specifically.[1] In general Ub is activated and ligated to its target protein by a cascade of E1-, E2-, and E3-enzymes. This process is reversed by a large group of proteases called deubiquitinating enzymes (DUBs) that break down the ubiquitin modification, liberating the substrate protein and recycling Ub. The basis of the Ub system thus is fairly simple, but complexity comes from the large amounts of possible combinations of E2/E3 enzymes that dictate substrate specificity and formation of the appropriate poly Ub linkage type. We thought we understood the system, but do we actually? Recently it has been discovered that Ub can be regulated on a totally new level *via* modification (and covalent attachment of) adenosine di-phosphate ribose (ADPr), a modification that is found to play crucial roles in bacterial infection[2] and DNA damage response pathways.[3] Effector enzymes from *Legionella pneumophila*, the pathogenic bacterium causing Legionnaires disease, are responsible for this ADPr-Ub conjugation and are able to activate the hosts' Ub pool and catalyze the ubiquitinationof substrates proteins independentof the regular E1–E2-E3 ubiquitinationmachinery and without the use of energy source ATP.

Synthesis of ADPr-peptides is tricky business due to the incompatibility of the nucleotide and pyrophosphate regions in the ADPr part with strong acidic conditions commonly used in peptide chemistry. Reversely the alkaline conditions commonly used in nucleotide chemistry are incompatible with peptide. Synthesis of ribosylated amino acids and their incorporationin model peptides followed by installment of the adenosine di phosphate region onto the ribosyl peptide in the past have shown the potential of chemical strategies to make such complex biomolecules.[4] In such synthesis schemes however protective groups shielding amino acid side chain functionalities were chosen to be base-labile allowing a global alkaline deprotection step after installment of the ADPr-modification onto the peptide. Expanding this methodology towards large peptides or even (synthetic) proteins would be cumbersome.

To circumvent such issues a general strategy to produce ADPribosyl peptides and synthetic ubiquitin conjugates based on a copper catalyzed click reaction was developed.[5] An ADPribose analogue was prepared carrying an alkyne functionality positioned in the correct α-orientation on the anomeric position of the ribose moiety. Using conventional Fmoc-based solid phase peptide synthesis peptides derived from proteins found to be ADPribosylated *in vitro* were prepared carrying an azide group at the site of ADPribosylation and subsequently globally deprotected using strong acidic conditions and purified. These purified peptides were used in a copper catalyzed alkyne azide reaction with the alkyne equipped ADPr-synthon resulting in ADPribosyl peptides (see Figure 1A).

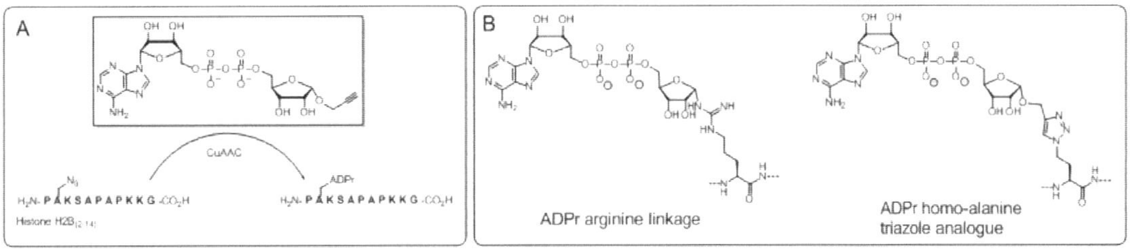

Figure 1: A) Copper mediated conjugation of azide modified peptide to alkyne modified ADPr, B.) structural comparison of native arginine-ribose linkage and triazole mimic.

After installing the ADPr-moiety no chemical reactions, such as strong acid treatment, were needed and simple dialysis and size exclusion chromatography or HPLC-purification yielded the final compounds. Of note is that the linkage between ADPr and peptide as a result of the copper mediated click reaction is a triazole linkage, that has been used before as stabilized mimic preventing proteolytic degradation[6] (see Figure 1B). This is of particular interest as ADPribosylation is a dynamic post-translational modification and by making such stabile analogues degradation of prepared constructs is prevented.

Moving forward from model peptides to more complex systems we prepared two ubiquitin mutants carrying an azide groups either at position 42 or 76, site that are found to be ADPribosylated either in *Legionella* infection or as a result of DNA-damage. Solid phase peptide synthesis of the two azide mutants followed by global deprotection and purification led to functionalized fully synthetic proteins (see Figure 2). Of note is that introduction of azide mutations in other proteins can also be performed using biochemical techniques not limiting the scope of this methodology to synthetically attainable (small) proteins. Conjugation of the ADPr-alkyne and azide modified ubiquitin mutants followed by size exclusion chromatography resulted the triazole containing ADPr-Ub analogues. Validation of biological activity was performed using recombinantly expressed *Legionella* effector protein SdeA, that shows an auto-ubiquitinating activity based on ADPr-Ub. It was shown that, although at reduced rate, the chemically prepared an stabilized ADPr-Ub mimics, were accepted by the enzyme as auto-ubiquitination was observed.

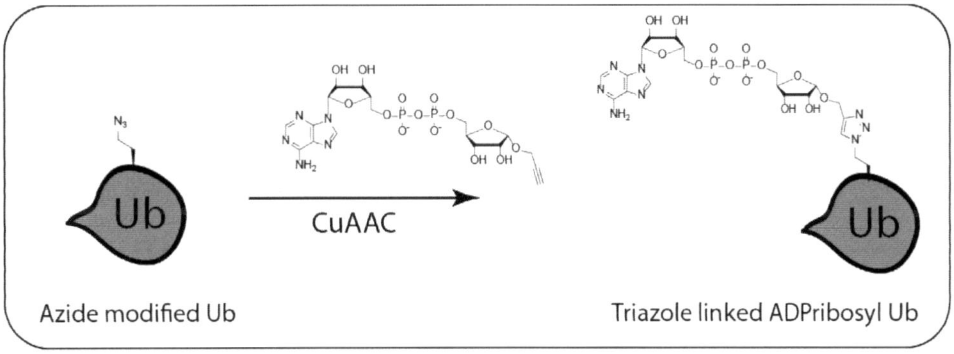

Figure 2: Synthesis of ADPr-ubiquitin analogues.

In conclusion, this method gives easy access to the first example of ADPribosylated peptides and proteins, an otherwise difficult to prepare bio-conjugate. Such post-translationally modified proteins retain their biologic activity and can be applied in studies investigating the interesting post-translational interplay of ubiquitination and ADPribosylation.

References

[1] D. Komander, M. Rape, in Annual Review of Biochemistry, Vol 81, Vol. 81 (Ed.: R. D. Kornberg), 2012, pp. 203-229.

[2] a K. M. Kotewicz, V. Ramabhadran, N. Sjoblom, J. P. Vogel, E. Haenssler, M. Zhang, J. Behringer, R. A. Scheck, R. R. Isberg, Cell Host & Microbe 2017, 21, 169-181; b J. Qiu, M. J. Sheedlo, K. Yu, Y. Tan, E. S. Nakayasu, C. Das, X. Liu, Z.-Q. Luo, Nature 2016, 533, 120-124.

[3] C.-S. Yang, K. Jividen, A. Spencer, N. Dworak, L. Ni, L. T. Oostdyk, M. Chatterjee, B. Ku´smider, B. Reon, M. Parlak, V. Gorbunova, T. Abbas, E. Jeffery, N. E. Sherman, B. M. Paschal, Molecular Cell 2017, 66, 503-516.e505.

[4] a G. J. van der Heden van Noort, M. G. van der Horst, H. S. Overkleeft, G. A. van der Marel, D. V. Filippov, Journal of the American Chemical Society 2010, 132, 5236-5240; b H. A. V. Kistemaker, G. J. van der Heden van Noort, H. S. Overkleeft, G. A. van der Marel, D. V. Filippov, Organic Letters 2013, 15, 2306-2309.

[5] Q. Liu, H. A. V. Kistemaker, S. Bhogaraju, I. Dikic, H. S. Overkleeft, G. A. van der Marel, H. Ovaa, G. J. van der Heden van Noort, D. V. Filippov, Angewandte Chemie International Edition 2018, 57, 1659-1662.

[6] D. Flierman, G. J. V. van Noort, R. Ekkebus, P. P. Geurink, T. E. T. Mevissen, M. K. Hospenthal, D. Komander, H. Ovaa, Cell Chemical Biology 2016, 23, 472-482.

A novel Evans blue organic compound-appended PSMA-617 peptide as SPECT molecule imaging agent for human prostate LNCaP animal model

Ming-Hsin Li, Sheng-Nan Lo, Shih-Min Wang, Ming-Wei Chen, Shih-Ying Lee, Chun-Fang Feng, Shih-Wei Lo, Su-Jung Chen, Chih-Hsien Chang

Institute of Nuclear Energy Research, Taiwan

https://doi.org/10.17952/35EPS.2018.027

Background

The PSMA-617 is a high binding affinity ligand for prostate-specific membrane antigen (PSMA) which is strongly expressed in prostate cancer and up-regulated in poorly differentiated, metastatic and hormone-refractory carcinoma compared with health tissue (1). To conjugate with albumin is a new method for improving the pharmacokinetic profile of peptides. We used a new R&D strategy that changed the linking site of tetraazacyclododecane-1, 4, 7, 10-tetraacetic acid in PSMA-617 and modified with truncated Evans blue which was a good human serum albumin binder (2).

Method

The modified PSMA-617 was abbreviated to DOTA-EB-PSMA-617. The DOTA-EB-PSMA-617 were prepared by linking a truncated Evans blue to the lysine, attaching the DOTA chelator to the α-amine, attaching a maleimide to the ε-amine, and finally attaching thiolated PSMA-617 peptide to the maleimide. To determine radiolabeling parameters, the DOTA-EB-PSMA-617 (peptide amount 10-25μg) was labeled with ~6mCi indium-111 chloride (from INER, Institute of Nuclear Energy Research in Taiwan) in sodium acetate (NaOAc) with pH 6.0 at 95°C for different incubation time (5 to 20mins). The radiochemical purity (RCP) of In-111-DOTA-EB-PSMA-617 was qualified by radio-TLC with citric acid/ sodium citrate buffer. The solvent front was free indium-111 and the origin spot was In-111-DOTA-EB-PSMA-617. Then we used prostate carcinoma cells LNCaP (PSMA +) and PC-3 (PSMA-) to determine the *in vitro* bio-activity of radiolabeled products by cell binding assay. And we used LNCaP or PC-3 tumor bearing mice to determine the *in vivo* bio-distribution of In-111-DOTA-EB-PSMA-617 *via* nanoSPECT/CT.

Result

The radiochemical purity of In-111-DOTA-EB-PSMA-617 was greater than 95%. In cell binding assay, the peptide concentration of In-111-DOTA-EB-PSMA-617 was 2.5nM to 250nM and incubated for 45mins at 37°C. With the amount of treated peptide increase, there had been strong and higher radio-counts in LNCaP cells than PC-3 cells analyzed by γ-counter (Fig. 1). And we found that radio-activity accumulated in the LNCaP tumor site at 48 hours post-injection of 18.5MBq In-111-DOTA-EB-PSMA-617. There has longer circulating time of In-111-DOTA-EB-PSMA-617 compared with In-111-PSMA-617 (Fig. 2).

Conclusion

We altered the pharmacokinetic/pharmacodynamic (PK/PD) profile of PSMA-617 successfully by conjugating PSMA-617 with EB. And PSMA-617 modified with EB motif has potential to be a theranostics agent to change the therapeutic cycle of PSMA-617 in shorter time.

References

1. Benešová M, Schäfer M, et al., J Nucl Med. 2015 Jun;56(6):914-20. doi: 10.2967/jnumed.114.147413.
2. Chen H, Jacobson O, et al., J Nucl Med. 2017 Apr;58(4):590-597. doi: 10.2967/jnumed.116.182097.

Sol-gel and peptides: An attractive route to unprecedented biomaterials

Gilles Subra[1], Laurine Valot[1], Manon Maurel[1], Titouan Montheil[1], Cécile Echalier[1], Muriel Amblard[1], Danièle Noël[2], Marie Maumus[2], Tao Jia[3], Jean-Luc Coll[3], Ahmad Mehdi[4], Jean Martinez[1]

[1]Institute of Biomolecules Max Mousseron, France
[2]Institute of Regenerative Medicine and Biotherapies, France
[3]Institute of Advanced Biotechnologies, France
[4]Institute Charles Gerhardt, France

https://doi.org/10.17952/35EPS.2018.028

Affording biological properties to a material is one of the challenges the chemists have to tackle, to improve the efficiency of existing devices and also to propose new biomaterials. Most of synthetic approaches to obtain functional materials rely on non-covalent coatings or post-grafting realized by multistep ligation chemistry. In this context, we envisioned a bottom-up strategy to introduce bioactive peptides in polymers or materials and in particular to prepare hydrogels. [1–3]

A bottom-up approach to mimic extracellular matrix

Extracellular matrix (ECM) is a complex and bioactive scaffold offering a physical support for cells but also providing the chemical and mechanical signals cells need to migrate, differentiate and grow properly. Hydrogels have been investigated as artificial ECM for 3D-cell culture, tissue engineering or regenerative medicine and are of paramount interest for bioink design.[4]

Natural biopolymers (e.g. collagen, gelatin, hyaluronic acid, alginate) are bioactive and biocompatible. However due to their natural origin, they present a poor batch-to-batch reproducibility and may lead to safety issues (immunogenicity, biocontamination). Their large-scale production can be difficult. Moreover, biopolymers are mostly used as physical hydrogels meaning that they have to be chemically cross-linked for long-lasting tissue engineering applications where the hydrogel is supposed to behave as support of cells for weeks. In contrast, several biocompatible synthetic polymers (e.g. PEG, PVA, Pluronic) are produced (even in large scale) with a good reproducibility and offering a large panel of molecular weight/chain sizes in order to tune their properties. Unfortunately they are bioinert (they are called 'permissive hydrogels') in a sense that they do not give any biologically relevant signaling to cells and have to be further functionalized to acquire suitable bioactivity. They also need to be cross-linked to form a 3D hydrogel network. By the way, it requires selective chemical cross-linking reactions (e.g. thiol-ene reaction, acrylate cross-linking) which may ideally proceed biorthogonally in the presence of other fragile biomolecules and eventually cells (if the hydrogel is a bioink for 3D biofabrication). At last, small organic gelators and in particular self-assembling peptides have also been described (betasheets KLD-12 [5]; RADA16 [6]; FEFQFK [7]; betahairpin VKVKVKVKvPPTKVEVKVKV [8] ; foldamers (β 3AA)3 [9]). They present many advantages such as easy synthesis, shear-thinning behavior (injectability), biocompatibility and could be used as drug delivery platform. Unfortunately, they are bioinert and their further functionalization may disrupt their self-assembly. Moreover, once implanted, dilution in biological fluids induce rapid disassembly of the structure which is a strong limitation as their mechanical properties cannot be easily tuned.

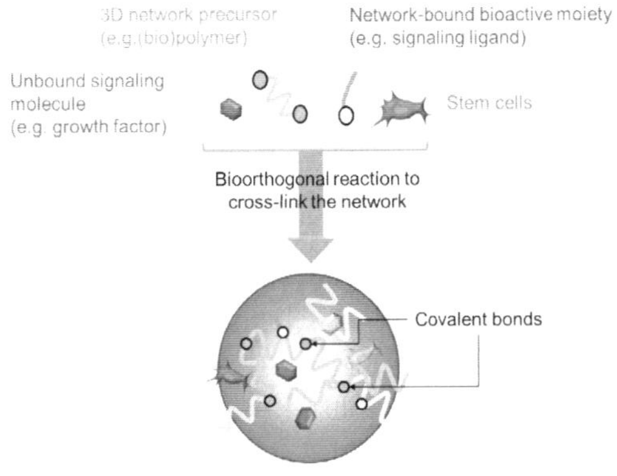

Figure 1: Hybrid hydrogel composition

In that context, we developed chemical hydrogels, obtained *via* a block-wise approach offering a degree of flexibility required to mimic different ECM compositions (Fig. 1). The bioorthogonal reaction creating the 3D network is driven by hydrolysis and condensation of alkoxysilanes moieties connected the different hydrogel components. The hydrolysis and condensation of such moieties proceeds chemoselectively, at room temperature or 37°C and is compatible with the side-chains of biomolecules. Interestingly, any type of peptides but also biopolymers and dyes can be silylated and thus can participate to the covalent formation of the material. We already used this sol-gel process to prepare peptide-functionalized fluorescent silica nanoparticles [10,11], grafting of wound-healing or antibacterial peptides on silicone catheter,[12] dressings [13] and titanium [14] and comb-like and silicone linear polymers were prepared. [15,16]

At pH 7, the sol gel process needs to be catalyzed with a nucleophilic compound to faster the gelation. We selected sodium fluoride at low and non-toxic concentration (< 0.3 mg/ml) to perform the hydrolysis of alkoxysilane and subsequent condensation reactions (Fig. 2).

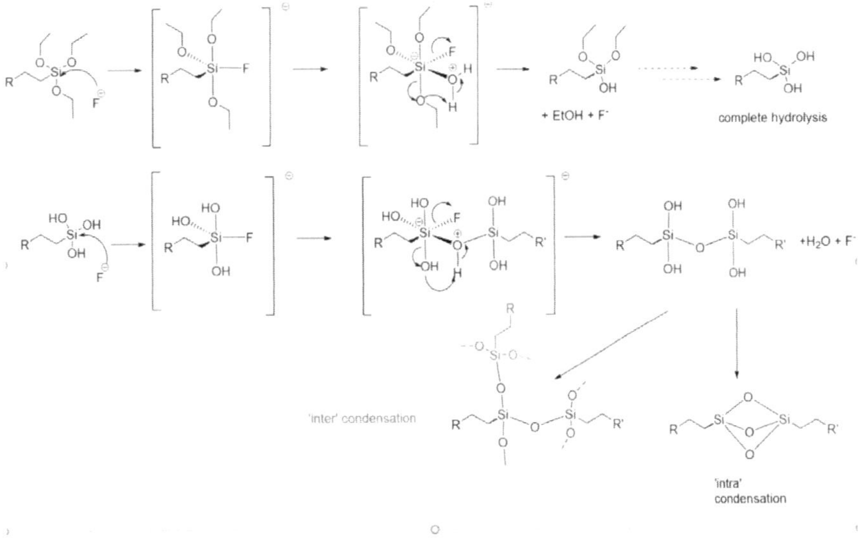

Figure 2: Proposed mechanism of fluoride catalyzed sol-gel process

Peptide-modified PEG based hydrogels

We first demonstrate the approach on PEG-based hydrogels. We synthesized bisilylated PEG, for the 'bricks' of the hydrogel network, and monosilylated RGD peptide (Fig. 3) as adhesion ligand for cells.[1]

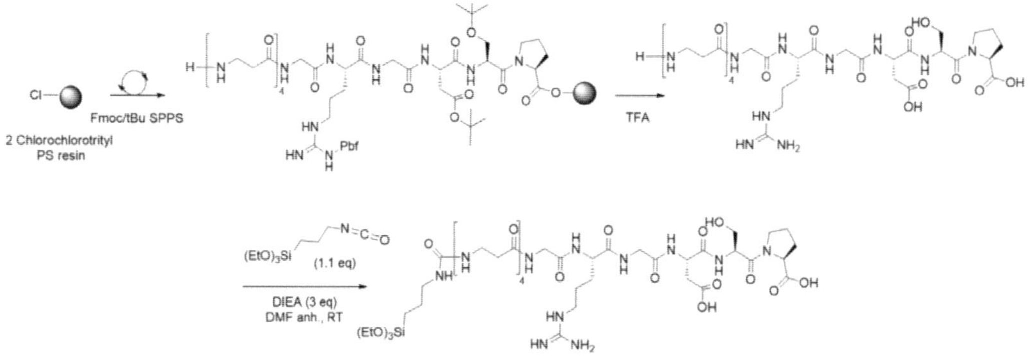

(EtO)$_3$Si-(CH$_2$)$_3$-NHCO-(βAla)$_4$GlyArgGlyAspSerPro-OH

Figure 3: Silylation of PEG and synthesis of silylated RGD peptide

Peptide-functional hydrogel was simply obtained by solubilizing the bisilylated PEG at 10%wt and the silylated RGD peptide at different concentrations (20 and 10 mol %) in phosphate buffer and in presence of NaF. After overnight gelation at 37°C, L929 fibroblasts were seeded on the surface of RGD-functionalized hydrogel and non-functionalized hydrogel as a negative control. We observed that the cell adhesion of fibroblasts after 2 hours was strongly improved (over 5 times) on peptide functionalized-hydrogel.[1] Interestingly, we demonstrated that this hydrogel could be 3D-printed by an extrusion process (Biobot 1 3D printer), after reaching the appropriate viscosity (between 2500 and 4500 mPa/s) during sol-gel condensation (Fig. 4).

Figure 4: 3D printing of hybrid PEG hydrogel

Collagen peptide-based hydrogels

We wanted to replace the PEG by a more biomimetic building block. Collagen being the most abundant protein found in ECM, we chose to synthesize several peptides presenting the proline-hydroxyproline-glycine repetition unit (from 3 to 9 units) found in natural collagen sequence. Flanked by two lysine residues, these peptides were bis-silylated (Fig. 5).[3]

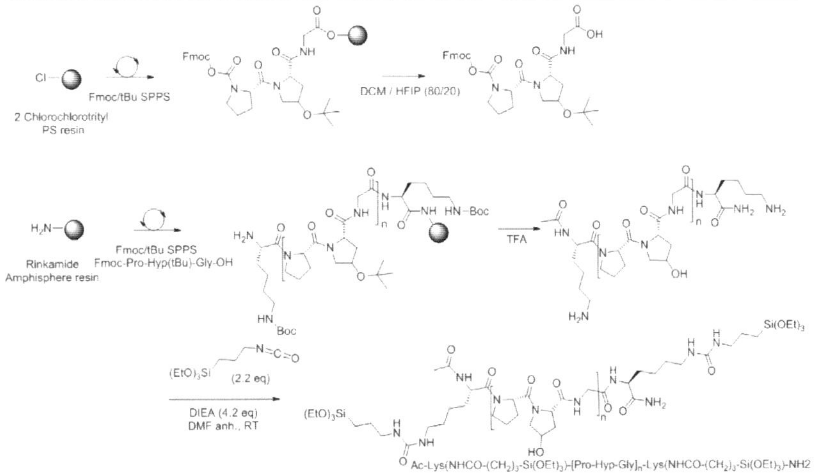

Figure 5: Synthesis of bisilylated collagen-like peptides

Hybrid peptide hydrogels were obtained by sol-gel process at 6-10 %weight concentration. Noteworthy, we demonstrated that human mesenchymal stem cells (hMSC) could be seeded on the surface of hydrogels but also could be embedded directly in the solution during gelation with a good viability. The incorporation of silylated hyaluronic acid, peptide ligands and growth factors (TGFβ -3) to mimic the ECM composition of cartilage is under investigation to induce the hMSC differentiation into chondrocytes for cartilage repair applications.

References

1. Echalier C, Pinese C, Garric X, Van Den Berghe H, Jumas Bilak E, Martinez J, Mehdi A, Subra G. Chem Mater. 2016; 28: 1261–1265.

2. Echalier C, Levato R, Mateos-Timoneda MA, Castano O, Déjean S, Garric X, Pinese C, Noël D, Engel E, Martinez J, Mehdi A, Subra G. RSC Adv. 2017; 7: 12231–12235.

3. Echalier C, Jebors S, Laconde G, Brunel L, Verdié P, Causse L, Bethry A, Legrand B, Van Den Berghe H, Garric X, Noël D, Martinez J, Mehdi A, Subra G. Mater Today.

4. Bronzino JD, Peterson DR. 2015; Fourth Edition: 114.

5. Kisiday J, Jin M, Kurz B, Hung H, Semino C, Zhang S, Grodzinsky AJ. Proc Natl Acad Sci. 2002; 99: 9996–10001.

6. Cheng T-Y, Chen M-H, Chang W-H, Huang M-Y, Wang T-W. Biomaterials. 2013; 34: 2005–2016.

7. Martin C, Oyen E, Mangelschots J, Bibian M, Ben Haddou T, Andrade J, Gardiner J, Van Mele B, Madder A, Hoogenboom R, Spetea M, Ballet S. Med Chem Commun. 2016; 7: 542–549.

8. Haines-Butterick L, Rajagopal K, Branco M, Salick D, Rughani R, Pilarz M, Lamm MS, Pochan DJ, Schneider JP. Proc Natl Acad Sci. 2007; 104: 7791–7796.

9. Luder K, Kulkarni K, Lee HW, Widdop RE, Del Borgo MP, Aguilar M-I. Chem Commun. 2016; 52: 4549–4552.

10. Jia T, Choi J, Ciccione J, Henry M, Mehdi A, Martinez J, Eymin B, Subra G, Coll J-L. Biomaterials. 2017.

11. Ciccione J, Jia T, Coll J-L, Parra K, Amblard M, Jebors S, Martinez J, Mehdi A, Subra G. Chem Mater. 2016; 28: 885–889.

12. Pinese C, Jebors S, Echalier C, Licznar-Fajardo P, Garric X, Humblot V, Calers C, Martinez J, Mehdi A, Subra G. Adv Healthc Mater. 2016; 5: 3067–3073.

13. Pinese C, Jebors S, Stoebner P-E, Humblot V, Verdié P, Causse L, Garric X, Taillades H, Martinez J, Mehdi A, Subra G. Mater Today Chem. 2017; 4: 73–83.

14. Masurier N, Tissot J-B, Boukhriss D, Jebors S, Pinese C, Verdie P, Amblard M, Mehdi A, Martinez J, Humblot V, Subra G. J Mater Chem B. 2018.

15. Jebors S, Ciccione J, Al-Halifa S, Nottelet B, Enjalbal C, M'Kadmi C, Amblard M, Mehdi A, Martinez J, Subra G. Angew Chem Int Ed. 2015; 54: 3778–3782.

16. Jebors S, Pinese C, Nottelet B, Parra K, Amblard M, Mehdi A, Martinez J, Subra G. J Pept Sci. 2015; 21: 243–247.

Ghrelin receptor ligands: From the bench to the drug on the market

Jean-Alain Fehrentz[1], Vincent Guerlavais[2], Céline M'Kadmi[1], Marjorie Damian[1], Sophie Mary[1], Didier Gagne[1], Mathieu Maingot[1], Khoubaib Ben Haj Salah[1], Séverine Denoyelle[1], Jean-Louis Banères[1], Jacky Marie[1], Jean Martinez[1]

[1]IBMM, UMR 5247, Faculte de Pharmacie, Montpellier, France
[2]Aileron Therapeutics, Cambridge, MA, United States

https://doi.org/10.17952/35EPS.2018.033

Growth hormone is essential for the development of the child, but also for the hormonal balance of the adult. To play its role, it must be produced and released into the body. This secretion is controlled by another hormone, ghrelin, which is partly produced in the brain (pituitary gland) and which, by binding to its receptor named growth hormone secretagogue receptor or GHSR, releases growth hormone.

Ghrelin receptor or GHSR is a G protein-coupled receptor (GPCR) that mediates, among others, ghrelin-induced growth hormone(GH) secretion, food intake, and reward-seeking behaviors. Because of its possible implication in several physiological disorders such as obesity and drugs/alcohol addictions, GHS-R1a represents a major target for the development of therapeutic small molecules.

As part of a European research program to identify molecules that would allow, like ghrelin, to trigger the secretion of growth hormone, we designed and synthesized a new series of compounds including a gem-diamino moiety.[1] After an *in vivo* structure activity relationship study, a compound - named JMV1843 - able to mimic the action of ghrelin, was chosen for clinical trial. Phase I clinical trial revealed that compound JMV 1843 was well tolerated and no adverse events were reported. Maximal GH release was achieved using a 0.5mg/kg dose by oral route. Stimulation of GH appears to be selective as no other hormones measured were affected by administration of JMV 1843 (no effects on ACTH; cortisol; ghrelin; prolactin; insulin and glucose).[2]

After twenty years, from the first *in vitro* tests to clinical studies on humans, Æterna-Zentaris received at the end of 2017 a marketing authorization by the Food and Drug Administration, the American drug agency, for its application in the diagnosis of growth hormone deficiency, in adults. Its arrival will improve the diagnosis of this pathology that occurs in adults due to a hypothalamic-pituitary disorder (pituitary tumor for example) or brain trauma, which affects about 70,000 people worldwide. While the test used so far (insulin tolerance test) involves restrictive methods including painful injections, the new compound will be ingested orally. By stimulating the release of growth hormone(provided that it is always manufactured by the patient), it will be possible to dose it by blood test. The test will thus make possible to detect a deficiency, but also to dissociate a deficit linked to a non-release of the hormone or to its non-production by the cells, and thus to adapt the therapeutic approaches.

Renamed Macrilen™, the compound is now marketed in North America by Strongbridge Ireland Ltd. laboratories. It could find other applications, such as the treatment of cachexia (phase 2 clinical study in progress) or stunting in children (application for evaluation filed with the European Medicines Agency). If the IBMM-derived molecule JMV1843 is now tracking its course as a pharmaceutical, researchers continue to explore the GHSR receptor. Their goal: to develop compounds that cannot stimulate but block it, and thus propose new approaches to treat obesity, diabetes or addictions.

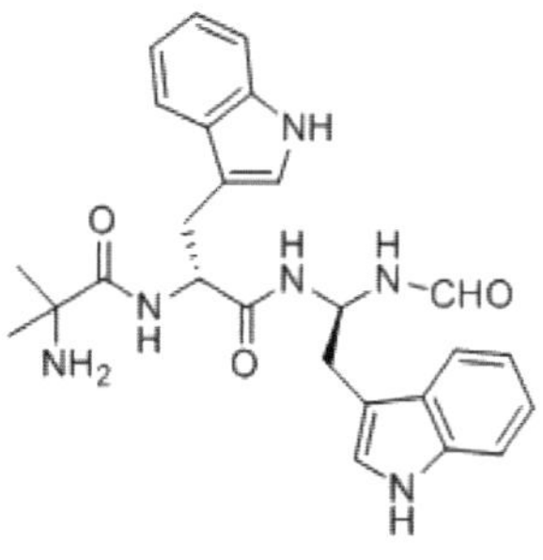

Structure of compound JMV 1843

Figure 1: Structure of compound JMV 1843

Figure 2: 3D model of compound JMV 1843

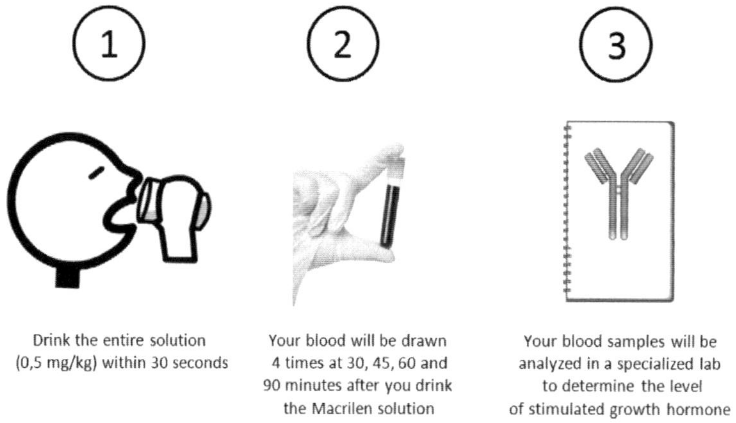

Figure 3: Macrilen test: a real improvement in Growth Hormone dosage!

References

1. V. Guerlavais et al., A new active series of Growth hormone secretagogues, J. Med. Chem. 2003, 46, 1191-1203.

2. F. Piccoli et al., Pharmacokinetics and Pharmacodynamic Effects of an Oral Ghrelin Agonist in Healthy Subjects, J. Clin. Endocrinol., & Metabol., 2007, 92, 1814–1820.

Rationally designed peptidomimetics

Andrew Jamieson

University of Glasgow, United Kingdom

https://doi.org/10.17952/35EPS.2018.035

Peptides are well known to act as potent and selective ligands for a variety of bimolecular targets. Although there are a many peptide drugs in clinical use, their potential has been limited by unfavorable physicochemical properties including limited plasma stability and a lack of cell permeability. Peptidomimetics are molecules that can be rationally designed to overcome many of the issues associated with peptides and provide tool molecules for chemical biology and lead compounds for drug discovery.[1]

Disulfide bonds play an important structural and/or catalytic role in many proteins. However the synthesis of peptides containing more then one disulfide-bond is not trivial.[2] Solid phase synthesis of the linear peptide and subsequent oxidative folding under thermodynamic conditions generally produces a mixture of structural isomers. Synthesis of specific isomers requires a solid phase approach and orthogonal protection strategy. Disulfide bond rich peptides and proteins are susceptible to disulfide bond shuffling and reduction catalysed by glutathione. As such, an effective surrogate that accurately mimics a disulfide bond would be a powerful tool in chemical biology with a range of applications from understanding protein-folding mechanisms to peptide drug discovery.

A number of synthetic disulfide bond mimetics have been developed. The most successful strategies to date include diselenide (Se–Se), selenylsulfide (Se–S), cystathionine (CH2–S, Ctt), and ditelluride (Te–Te) bonds. However, each of these methods has disadvantages such as multiple step synthesis required for their preparation such that they have not be adopted up by the peptide community. Other methods to mimic disulfide bonds include olefin-based isosteres and thioethers. However, thioethers require multistep synthesis utilizing complex orthogonal protecting group strategies. Olefin-based isosteres produced by ring-closing metathesis (RCM) reactions give complex mixtures of cis/trans isomers, and required a subsequent palladium-catalyzed hydrogenation step to access a suboptimal, conformationally flexible alkane bridge. Kolmer was first to describe the use of a 1,5-disubstitiuted 1,2,3-triazole as a disulfide bond mimic.[3] In this seminal work the disulfide bond within a macrocyclic sunflower trypsin inhibitor-I was replaced by a 1,5-triazole and the resulting peptidomimetic retained biological activity.

Urotensin-II (U-II) is the endogenous cyclic peptide ligand of the urotensin receptor (UTR).[4] The U-II peptide is the most potent vasoconstrictor known and so is a promising lead compound that targets the UTR for development as a potential therapeutic for the treatment of heart failure and atherosclerosis. U-II is an eleven-residue peptide that adopts a bioactive beta-turn conformation that is constrained by a cross-strand disulfide bond. An issue with studying the interaction of U-II with the UTR is that the peptide acts as a pseudo-irreversible agonist. Previously developed peptide ligands are based on U-II and so display the undesired effect of irreversibility. The development of reversible UTR ligands would be a useful addition to the biochemical toolbox to investigate the physiological role of the U-II/UTR system.

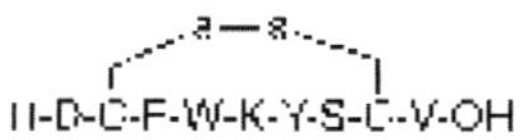

1 U-II

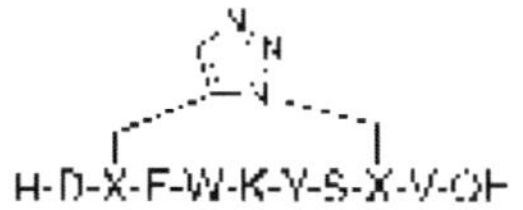

2 1,5-triazole

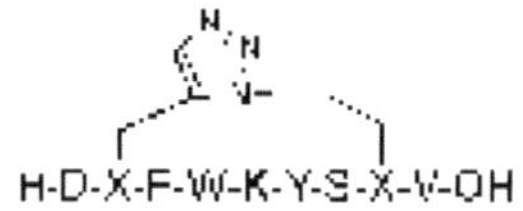

3 1,5 triazole

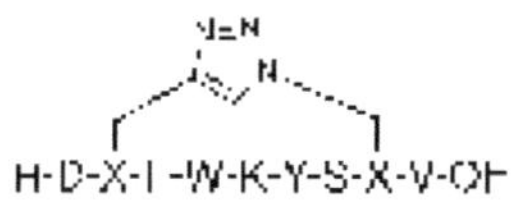

4 1,4-triazole

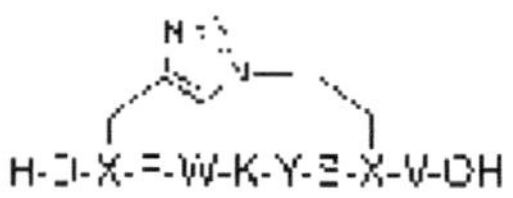

5 1,4-triazole

Figure 1: Urotensin-II peptide 1 and peptidomimetic analogues 2-5.

A series of U-II analogues 2-5 were designed to incorporate either a 1,4- or 1,5-triazole bridge as a disulfide bond surrogate (Figure 1).[5] The four analogues 2-5 were assessed by computational overlay with the U-II disulfide bond. Compound 3 incorporating a 1,5-triazole and 6-atom bridge proved to be the most accurate surrogate for the disulfide bridge, giving a room-mean-square-deviation (RMSD) of 0.182 Å. The linear peptides required to produce the peptidomimetics were synthesised by automated microwave assisted solid phase peptide synthesis using a Fmoc/tBu protection strategy. Commercially available building blocks Fmoc-L-propargylglycine (Fmoc-Pra-OH) and Fmoc-L-azidoalanine (Fmoc-Aza-OH) or Fmoc-L-azido-homoalanine (Fmoc-Aha-OH) provided the required functionality to prepare the triazole bridges. 1,2,3-Triazoles were then prepared as either the 1,4-isomer using a Copper-Catalyzed Azide–Alkyne Cycloaddition (CuAAC); or as the 1,5-isomer using a Ruthenium-Catalyzed Azide–Alkyne Cycloaddition (RuAAC) reaction. The progress of the reaction was monitored by IR spectroscopy (disappearance of the azide absorbance at 2110 cm-1). The peptides were cleaved from solid support under acidic conditions and purified by reverse-phase HPLC. Of the four peptidomimetics produced, two retained good activity relative to U-II. U-II 1 give pIC_{50} = K_D of 9.36 nM and peptidomimetics 2 and 3 also had essentially identical IC_{50} = K_D values of around 8 nM. Importantly, binding of compounds 2 and 3 but not U-II to UTR in CHOhUT cells was reversible.

With this successful application of the 1,5-triazole as a disulfide bond surrogate we decided to apply this strategy to a more challenging conotoxin peptide that incorporates two-disulfide bonds.

Conotoxin α-GI 6 is a nAChR antagonist that elicits its effect at the skeletal neuromuscular junction so has potential as a lead compound for the development of a new class of muscle relaxant.[6] Two analogues of globular native α-GI were produced in which the two disulfide bonds were replaced in turn by the 1,5-triazole bridge designed previously.[7] The peptidomimetics were incubated with human muscle-type nAChRs and their activity determined *in vitro* by assessing antagonism of the nAChR mediated increase of $[Ca^{2+}]$ in CN21 cells. One of the analogues 8, with the (Cys3-Cys13) disulfide bond replaced, retained full antagonist activity (IC_{50} 8.2 nM) compared to native α-GI 6 (IC_{50} 9.8 nM). We next assessed the blood plasma half-life of the bioactive peptidomimetic vs the native α-GI. Incorporation of the triazole bridge in place of the native disulfide significantly improves its plasma half-life by ~10 fold c.f. the native 6. To investigate if the triazole acts as an accurate structural mimic of the disulfide bond we developed bespoke force field descriptions of the triazole mimetic that allowed us to determine the solution structure by NMR. Significant similarities in conformation between the mimetic and the peptidic bioactive toxin can be observed. It is evident that the conformation of the β-turn pharmacophore of the native peptide 6 is preserved in peptidomimetic 8. Our NMR structure supports a previously reported hypothesis that the pharmacophore of the α-GI is located within the first half of the peptide as a β-turn.

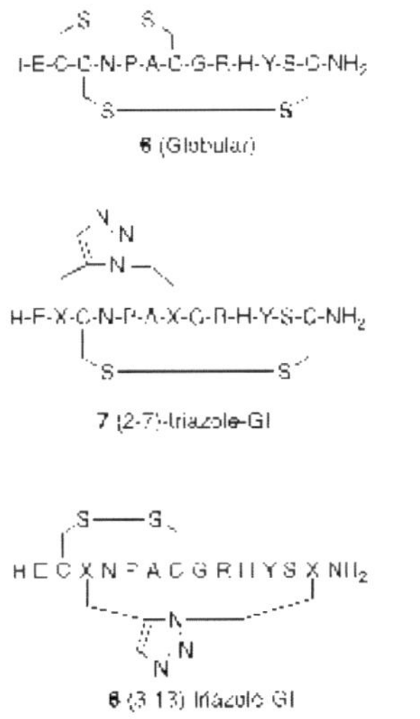

Figure 2: a-Conotoxin GI 6 and 1,5-triazole peptidomimetic analogues 7 & 8.

In summary, a highly accurate disulfide bond surrogate has been developed based on a 1,5-trizole that is synthetically tangible from commercially available staring materials. Given the effectiveness of the strategy described we believe this approach is broadly applicable to other peptides containing disulfide bonds that have potential for development as novel probes and therapeutics.

References

1. M. J. Adler, , A. G. Jamieson and Hamilton, A. D., Curr. Top. Microbiol. Immunol., 2011, 348, 1-23.

2. A. Gori, P. Gagni, S. Rinaldi, Chem. Eur. J. 2017, 23, 14987-14995.

3. M. Empting, O. Avrutina, R. Meusinger, S. Fabritz, M. Reinwarth, M. Biesalski, S. Voigt, G. Buntkowsky and H. Kolmar, Angew. Chem., Int. Ed., 2011, 50, 5207–5211.

4. N. D. Doan, T. T. M. Nguyen, M. Létourneau, K. Turcotte, A. Fournier and D. Chatenet, Br. J. Pharmacol., 2012, 166, 243–257.

5. S. Pacifico, A. Kerckhoffs, A. J. Fallow, R. E. Foreman, R. Guerrini, J. McDonald, D. G. Lambert, A. G. Jamieson, Org. Biomol. Chem., 2017, 15, 4704-4710.

6. I. G. Marshall , A. L. Harvey, Toxicon, 1990, 28, 231–234.

7. A. Knuhtsen, C. Whitmore, F. S. McWhinnie, L. McDougall, R. Whiting, B. O. Smith, C. M. Timperley, C. Green, K.I. Kinnear, A. G. Jamieson, Chem. Sci., 2018.

Amyloidogenicity of regions of Nucleophosmin 1: A direct link between protein misfolding and Acute Myeloid Leukemia

Daniela Marasco, Ettore Novellino, Giancarlo Morelli, Sara La Manna

University of Naples Federico II, Italy

https://doi.org/10.17952/35EPS.2018.038

Nucleophosmin (NPM1) is a multidomain protein involved in a variety of biological processes and identified as the most frequently mutated gene in Acute Myeloid Leukemia (AML). Its C-terminal domain (CTD) is endowed with a three helix bundle tertiary structure: H1 (243-259), H2 (264-277) and H3 (280-294) helices fold through a compact transition state and unfold keeping a residual secondary structure, they are held together by an aromatic core at the interface between H2 and H3 helices [1]. AML mutations cause the loss of tryptophans288 and 290 in the third helix of the CTD, that, in the wt form constitute a NOLS (nucleolar localization signal), and the creation of an additional leucine-rich nuclear export signal (NES), inducing an aberrant cytoplasmic dislocation (named NPM1c+). AML cells expressing NPM1c+ present differentiation arrest, apoptosis blockage, impaired DNA repair, and induced cell proliferation. Many important tumor suppressors are deregulated either directly or indirectly by NPM1c+ namely- p14ARF, p53, PTEN and also caspases-6 and -8 resulted inhibited with altered apoptotic pathways [2]. There are many different strategies in targeting NPM1 such as i) NPM1 oligomerization, ii) its interctome, iii) its post-translational modifications, iv) its selective displacement from nucleoli that causes nucleolar stress and consequent apoptotic cell death, following the so-called "nucleolar starvation hypothesis" [2].

Few years ago, through a protein dissection approach we started to analyse the structural and functional independence of each helix of the CTD. The first study revealed that the first helix holded for an uncommon helical content intrinsically endowed with an unusual thermal stability [3]. More surprisingly the neighbouring regions H2 [4] [5] and H3 in AML mutated variants [6] separately revealed able to form amyloid-like assemblies endowed with fibrillar morphology and β -sheet structure. These aggregates proved to be toxic in cell viability assays.

Furthermore, a short stretch (nine residues) spanning 264-272 region, derived from the aromatic core of the bundle and located within H2, revealed the most amyloidogenic fragment of the entire protein. Solution and fiber state investigations through Circular Dichroism, Fluorescence spectroscopies, amyloid seeding assay (ASA), isothermal titration calorimetry (ITC) and Electrospray ionization (ESI) mass analyses, demonstrated that this peptide is able to self-aggregate and that helical conformation plays a crucial role in the aggregation mechanism. Indeed related nanostructures, investigated by means of Scanning Electron Microscopy (SEM) and Wide-Angle X-ray Scattering (WAXS), revealed flat with twisted profiles and typical cross-β-structures [7].

Amyloidogenic CTD regions demonstrated able to interact with model systems of membranes with different mechanisms and the presence of cholesterol revealed of crucial importance. Indeed the interaction of CTD-NPM1 amyloidogenic regions with lipid membranes was analysed through fluorescence, SPR, CD and ESR spectroscopies and by immune-fluorescence in leukemic cells. Resulting data indicate that H2 peptide showed a direct interaction with cholesterol within the membrane causing a sensitive fluidification of the bilayer. Noticeably, NPM1mutA (the most common AML mutation) thicken at the plasma membrane, differently from wt NPM1. These findings are in accordance with diverse mechanisms of interaction of wt CTD and Cterm mutA toward membrane models analyzed *in vitro* [8].

More recently, we found that all AML mutations, from A to F, without both or one Trp of the NOLS (nucleolar localization signal) induce amyloid-like aggregation and cytotoxicity. Using a combination of biophysical techniques including ThT-fluorescence, CR absorbance, CD, SEM (figure 1) and WAXS on a series of peptides bearing mutations we evidenced that the amyloidogenicity is directly linked to AML. The type of mutation influence the cytotoxicity in neuroblastoma cells and the morphology of amyloid fibrils.

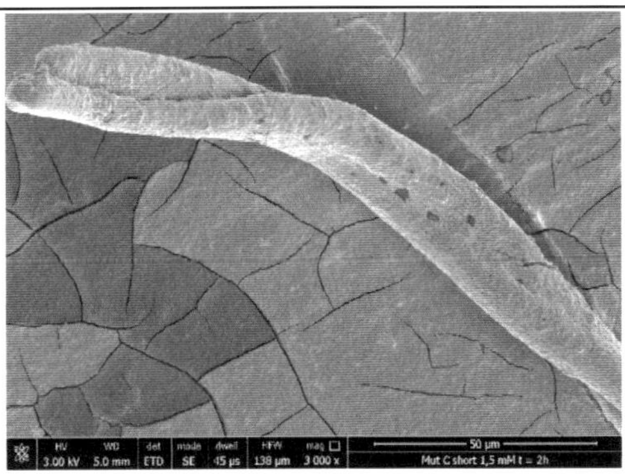

Figure 1: Micrographs of mutC short morphologies of surfaces at 3000x (50µm scale bar).

Our results demonstrate, unequivocally, that separated regions of the CTD of NPM1 are prone to aggregate to amyloid states: thus we formulated a mechanicistic hypothesis where the destabilization of the helical bundle by AML-mutations predispose it to the formation of toxic aggregates causing the exposure of the H2 and H3 regions.

Since the oligomeric state of NPM1 is of outmost importance for its biological roles in its wild type form, in the light of our results we can argue that the contribution of the CTD to the oligomeric/aggregated form could be directly linked the leukomogenic potentials of AML mutations. Departing away from the fragmentation approach for the explanation of the behaviour of the entire mutated protein, we investigated the polypeptide covering the third and second helices of the bundle of type A mutated CTD. Also this region, spanning 264-298 residues, demonstrated an amyloid aggregation resulting from a self-recognition mechanism and soluble assemblies resulted cytoxic in MTT assay corroborating a therapeutic strategy in AML consisting in the self-degradation of mutated protein [9].

Since structural studies are crucial in drug discovery process focused on NPM1c+/AML, these studies are of utmost importance in innovative strategies to identify more potent NPM1-targeted drugs.

References

[1] C.G. Grummitt, F.M. Townsley, C.M. Johnson, A.J. Warren, M. Bycroft, Structural consequences of nucleophosmin mutations in acute myeloid leukemia, The Journal of Biological Chemistry 283(34) (2008) 23326-32.

[2] A. Di Matteo, M. Franceschini, S. Chiarella, S. Rocchio, C. Travaglini-Allocatelli, L. Federici, Molecules that target nucleophosmin for cancer treatment: an update, Oncotarget (2016).

[3] P.L. Scognamiglio, C. Di Natale, M. Leone, M. Poletto, L. Vitagliano, G. Tell, D. Marasco, G-quadruplex DNA recognition by nucleophosmin: new insights from protein dissection, Biochimica et Biophysica Acta 1840(6) (2014) 2050-9.

[4] C. Di Natale, P.L. Scognamiglio, R. Cascella, C. Cecchi, A. Russo, M. Leone, A. Penco, A. Relini, L. Federici, A. Di Matteo, F. Chiti, L. Vitagliano, D. Marasco, Nucleophosmin contains amyloidogenic regions that are able to form toxic aggregates under physiological conditions, FASEB journal : official publication of the Federation of American Societies for Experimental Biology (2015).

[5] A. Russo, C. Diaferia, S. La Manna, C. Giannini, T. Sibillano, A. Accardo, G. Morelli, E. Novellino, D. Marasco, Insights into amyloid-like aggregation of H2 region of the C-terminal domain of nucleophosmin, Bba-Proteins Proteom 1865(2) (2017) 176-185.

[6] P.L. Scognamiglio, C. Di Natale, M. Leone, R. Cascella, C. Cecchi, L. Lirussi, G. Antoniali, D. Riccardi, G. Morelli, G. Tell, F. Chiti, D. Marasco, Destabilisation, aggregation, toxicity and cytosolic mislocalisation of nucleophosmin regions associated with acute myeloid leukemia, Oncotarget 7(37) (2016) 59129-59143.

[7] S. La Manna, Roviello V., Scognamiglio, P. L., Diaferia, C., Giannini, C., Sibillano, T., Morelli, G., Novellino

E., Marasco D., Amyloid fibers deriving from the aromatic core of C-terminal domain of nucleophosmin 1, International Journal of Biological Macromolecules, under review (2018).

[8] A. De Santis, S. La Manna, I.R. Krauss, A.M. Malfitano, E. Novellino, L. Federici, A. De Cola, A. Di Matteo, G. D'Errico, D. Marasco, Nucleophosmin-1 regions associated with acute myeloid leukemia interact differently with lipid membranes, Biochimica et Biophysica Acta 1862(4) (2018) 967-978.

[9] C. Di Natale, La Manna, S., Malfitano, A., Di Somma, S., Florio D., Scognamiglio, P. L., Novellino E., Netti, P. A., Marasco D., Structural and kinetic insights into amyloid aggregation of a polypeptide covering the second and the third helices of the type A mutation of Nucleophosmin 1 in Acute Myeloid Leukemia, Bioorg Chem, under review (2018).

Quaternary ammonium isobaric labeling for a relative and absolute quantification of peptides

Mateusz Waliczek, Bartosz Setner, Monika Kijewska, Remigiusz Bąchor, Dorota Gaszczyk, Piotr Stefanowicz, Zbigniew Szewczuk

Faculty of Chemistry, Wrocław University, Poland

https://doi.org/10.17952/35EPS.2018.041

Isobaric tags for relative and absolute quantitation (iTRAQ) has been a labeling method used in quantitative analysis of peptides by tandem mass spectrometry. This technique allow to determine the amount of peptides in complex mixtures in a single experiment. Isobaric tag usually consists of three functional groups: a chemically reactive group (active ester based on N-hydroxysuccinimide group), an isotopically modified balance group and a reporter group, which is the third part of the tag. The proper isotopic substitution in the balance and reporter group results in formation of isobaric tags, which have identical molecular weight.

The conventional iTRAQ workflow relies on simultaneous digestion and labeling of a few samples (e.g. 12-plex). Finally, the samples are pooled and analyzed by LC-MS. During MS/MS experiment, the reporter ions of differential masses are released from peptide to give sample-specific quantitation of a particular peptide.

Although the isobaric labeling quantification has become a widely used method in proteomics studies, the currently available isobaric tags are characterized by relatively high price and do not offer significant improvement of detection limit. Tagging of peptides with quaternary ammonium salt (QAS) has been widely used technique allowing to increase sensitivity of detection by ESI-MS[1]. Recently, we proposed a linear[2] and bicyclic[3] QAS as ionization enhancers for analysis of peptides at the attomole level[4].

Herein, we proposed two novel approaches, representing new classes of quaternary ammonium isobaric tags for relative and absolute quantitation (QA-iTRAQ 2-plex)[5]. The first procedure is based on application of N-(4-iodobenzylpiperidinylcarbonyl)glycine tag containing two deuterium atoms incorporated into the benzyl group (characterized by high stability) or glycine (balance group) residue (Fig. 1).

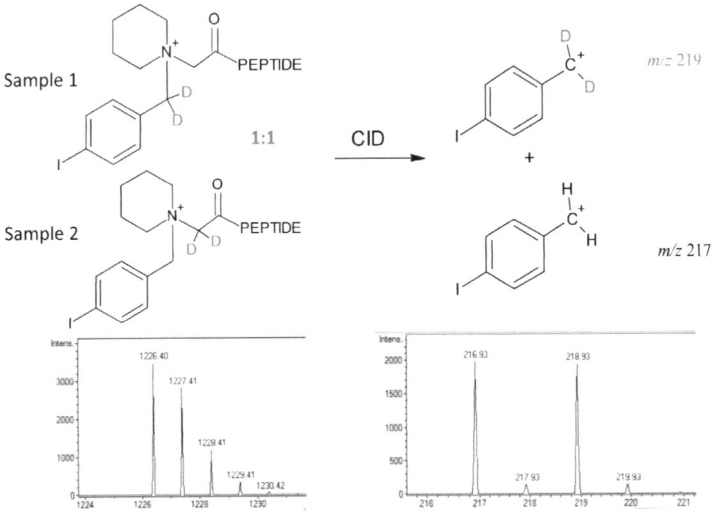

Figure 1: MS spectrum of the isobars and formation of reporter ions at m/z 219 and 217 after CID fragmentation.

The second approach is based on the combination of enzymatic ^{16}O / ^{18}O exchange followed by chemical modification of C-terminal lysine side chain with isotopically labeled pyrylium salt[6], which in turn allows for isobaric labeling of tryptic digests (Fig. 2). Each of modified peptide contain zero or four ^{13}C atoms in the reporter group, which results in a unique reporter mass during tandem MS/MS for sample identification and relative quantitation.

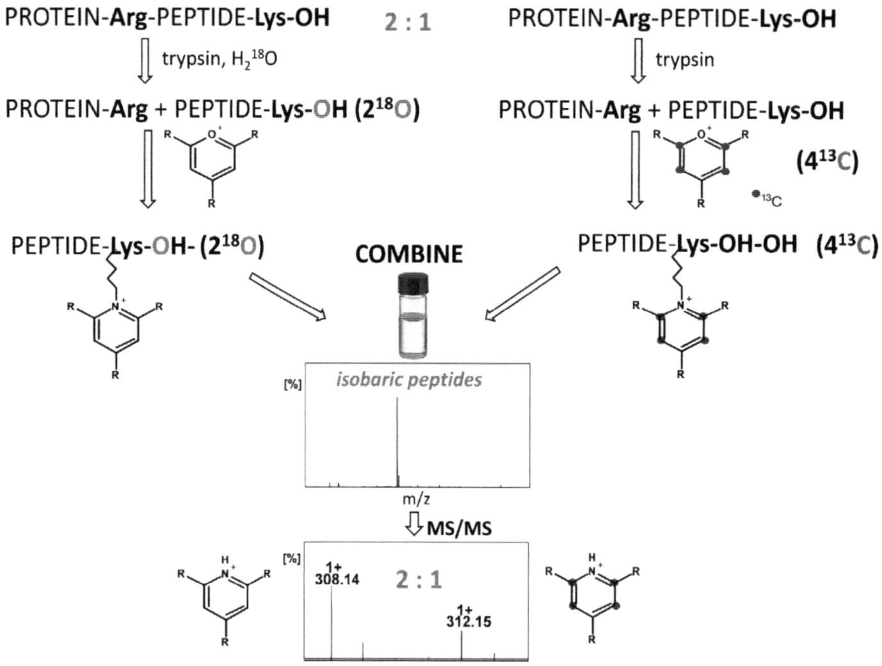

Figure 2: Scheme presenting the procedure of duplex formation.

We tested two different types of QAS allowing for formation of 2-plex on synthetic peptides, protein tryptic digests as well as on a more complex sample - podocyte cells digest. The obtained results suggest usefulness of the isobaric ionization tags for relative and absolute quantification of trace amounts of peptides.[7]

This work was supported by the National Science Centre in Poland grant [UMO-2016/23/B/ST4/01036]

References

[1] P. Stefanowicz, A. Kluczyk, Z. Szewczuk, Amino Acids, Pept. Proteins, 2016, 40, 36.

[2] M. Cydzik, M. Rudowska, P. Stefanowicz, Z. Szewczuk, J. Am. Soc. Mass Spectrom., 2011, 22, 2103.

[3] B. Setner, M. Rudowska, E. Klem, M. Cebrat, Z. Szewczuk, J. Mass Spectrom., 2014, 49, 995.

[4] R. Bąchor, P. Mielczarek, M. Rudowska, J. Silberring, Z. Szewczuk, Int. J. Mass Spectrom., 2014, 362, 32.

[5] B. Setner, P. Stefanowicz, Z. Szewczuk, J. Mass Spectrom., 2018, 53, 115.

[6] M. Waliczek, M. Kijewska, M. Rudowska, B. Setner, P. Stefanowicz, Z. Szewczuk, Sci. Rep., 2016, 6, 37720.

[7] M. Waliczek, R. Bąchor, M. Kijewska, D. Gąszczyk, K. Panek-Laszczyńska, A. Konieczny, K. Dąbrowska, W. Witkiewicz, K. Marek-Bukowiec, J. Tracz, M. Łuczak, P. Stefanowicz, Z. Szewczuk, Anal. Chim. Acta, 2018, in press: doi.org/10.1016/j.aca.2018.10.012

Design, synthesis and study of multimeric peptidic conjugates for a new approach of anti-tumoral immunotherapy

Benjamin Liet, Eugénie Laigre, Claire Tiertant, Didier Boturyn, Nathalie Berthet, Olivier Renaudet

Départementde Chimie Moléculaire, France

https://doi.org/10.17952/35EPS.2018.043

Abstract

Recruitment of natural human antibodies for targeted destruction of tumoral cells has emerged as a very promising immunotherapy strategy. Although several compounds have been proposed in this area, efforts have to be made to increase the specificity towards cancer cells and to optimize the response of immune system. Here we reported a new bimodal compound named ARM, combining four cRGD-based peptidomimetics anchored on a cyclodecapeptidic platform to target cells displaying the $\alpha_V\beta_3$ integrin overexpressed on a majority of tumor cells, with another peptidic scaffold displaying a tetravalent α-L-Rhamnose, for the recruitment of endogenous anti-Rhamnose antibody present in human serum.

Introduction

Although significant progress has been made these last years in treatment modalities to combat cancer, current therapeutic regimens are still deficient due to severe side effects for the patient [1][2]. Recently, an original antitumoral approach-based immunotherapy has emerged, consisting in the recruitment of pre-existing natural antibodies present in human bloodstream and their redirection against tumor cells thereby activating antibody-mediated immune responses such as complement-mediated cytotoxicity (CDC) or antibody-dependent responses, i.e. antibody-dependent cellular phagocytosis (ADCP) or antibody-dependent cellular cytotoxicity (ADCC)[3][4] (Fig. 1). While several antibody recruiting molecules (ARMs) have been reported today, the immune response activated by such a compound has to be improved. Here we evaluated a novel bifunctional conjugate comprising a cluster of α-L-rhamnose (Rhm) as antibody binding module (ABM) to recruit high titers of natural antibody and a cluster of cRGD as tumor binding module (TBMs) displayed on a molecular scaffold to simultaneously target multiple tumoral membrane proteins (Fig. 2). The cRGD peptidomimetic has been used to target tumor cells displaying the $\alpha_V\beta_3$ integrin while the expression is restricted in normal tissues but highly expressed in tumor tissues [5]. The rhamnose is a monosaccharide not expressed in human cells but prevalent in microbes. Studies have shown that anti-Rhm antibodies are widely present in human serum and are able to form relatively stables complexes with rhamnose indicating that this carbohydrate might be very attractive for activation of endogenous immune response for targeted anti-tumoral immunotherapy [6]. A cyclodecapeptide scaffold, largely described by our group as well as chemoselective ligations (copper-catalyzed alkyne-azide cycloaddition) were used to elaborate modules decorated with four cRGD ((cRGD)$_4$-TBM) or tetravalent rhamnose units ((Rhm)$_4$-ABM). Both modules were first tested independently of each other with human serum containing anti-Rhm antibodies for ABM and with U-87 MG glioma cells expressing $\alpha_V\beta_3$ integrins or MCF-10a deficient in integrin for TBM. (cRGD)$_4$ TBM and (Rhm)$_4$ ABM were then assembled and the cytotoxic property of the bimodal compound was evaluated.

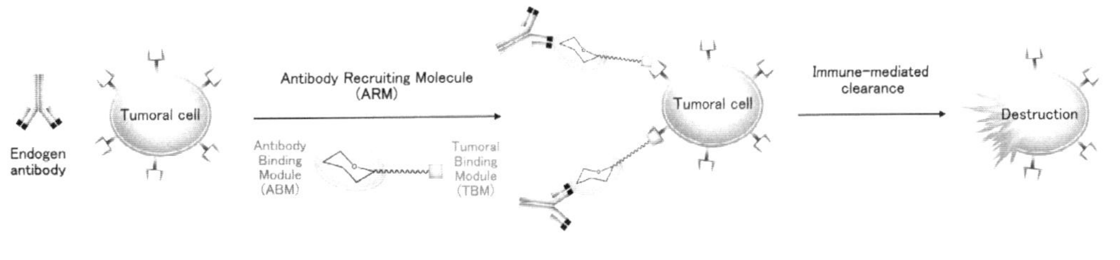

Figure 1: Antibody recruiting molecule (ARM) forming a ternary complex between a tumoral cell and endogenous antibodies leading to the immune-mediated clearance of the targeted cells.

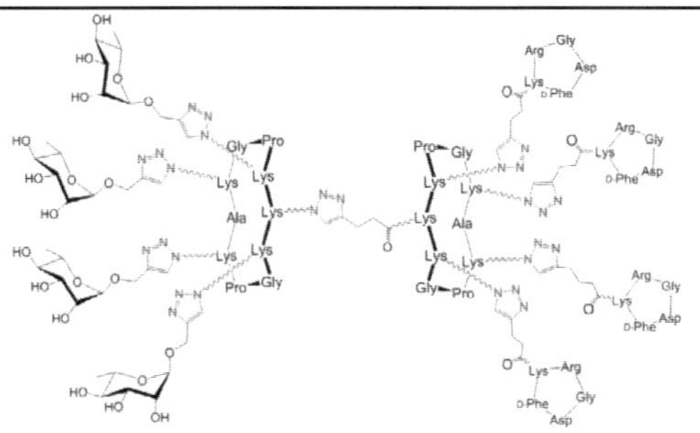

Figure 2: Chemical structure of ARM with tetravalent cRGD as TBM and tetravalent rhamnose as ABM

Integrin-bindingand antibody-bindingassays

To assess the ability of our tetravalent cRGD module to bind specifically cancer cells, we first quantified by flow cytometry the $\alpha_v\beta_3$ integrin expression at the surface of U-87 MG cells and MCF-10a using a commercial monoclonal anti- $\alpha_v\beta_3$ integrin antibody conjugated to the R-Phycoerythrin fluorochrome(23C6 mAb-R-PE). Figure 3a shows the level of $\alpha_v\beta_3$ integrin expression in the two cell lines after immunostaining with the labeled antibody. The binding property of our tetravalent (cRGD)$_4$-TBM towards $\alpha_v\beta_3$ integrins was then evaluated by measuring the fluorescence of the U-87 MG cells and MCF-10a used as control, when incubated with the cyanine 5-(cRGD)$_4$-TBM labeled module. Cells were first incubated 1 hour at 37°C in presence of 5 μM of the fluorescent compound. At this temperature, a 15-fold binding improvement for the U-87MG cells was measured, while binding for MCF-10a was only enhanced by a factor 5 (Fig. 3b). In order to minimize the internalization of the (cRGD)$_4$-TBM, incubation with cells was investigated at 4°C and the cells fluorescence analyzed. At this lower temperature, the binding of (cRGD)$_4$-TBM for the U-87MG cells was only 7-fold increase, while a 3-fold binding improvement was measured for the MCF-10a cells (Fig. 3c). In regard to these experiments, we decided next to evaluate the cytotoxicity of the (cRGD)$_4$-TBM-(Rhm)$_4$-ABM conjugate at 37°C.

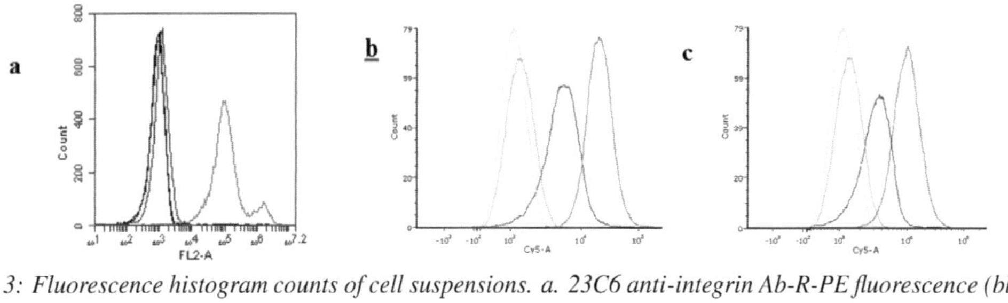

Figure 3: Fluorescence histogram counts of cell suspensions. a. 23C6 anti-integrin Ab-R-PE fluorescence (black: MCF-10a and U-87 MG autofluorescence; blue: MCF-10a with antibody; red: U-87 MG with antibody) b. Cy5 fluorescence after incubation with (cRGD)4-TBM at 37°C and c. at 4°C (cyan: MCF-10a autofluorescence; yellow: U-87 MG autofluorescence; blue: MCF-10a with compound; red: U-87 MG with compound)

The ability of (Rhm)$_4$-ABM to recruit endogenous antibodies present in serum was then evaluated in an indirect ELISA-type competition assay. In this test, a polymer functionalized with rhamnose (PAA-Rhm) was immobilized in a multi-well plate. A binding competition of the natural anti-Rhm antibodies present in human serum with the PAA-Rhm or (Rhm)$_4$-ABM present at different concentrations was performed. Binding of the anti-Rhm antibody to the PAA-Rhm was revealed using a secondary antibody conjugated to the Horse Radish Peroxydase which in presence of its substrate led to an absorbance signal at 490 nm. An IC_{50} of 122 μM, measuring the ability of the compound to inhibit 50% of antibody binding to the PAA-Rhm was determined for (Rhm)$_4$-ABM which correspond to a 30 times increase in the ability of the compound to recruit antibodies in serum compared to the rhamnose monomer, or an 8-fold binding improvement per rhamnose unit.

Cytotoxicity of the (cRGD)$_4$ TBM - (Rhm)$_4$ ABM bimodal compound

The cytotoxicity of the bimodal construct (ARM) was evaluated using the commercial kit CellTiter-Glo® (Promega). This test determine the number of viable cells in culture in multi-well plates based on the quantification of the ATP present, an indicator of metabolically active cells. The assay consists in measuring a bioluminescent signal proportional to the amount of the intracellular ATP. U-87MG cells displaying the $\alpha_v\beta_3$ integrins have been incubated with or without human serum (HS) and in presence of ARM (5 or 20 μM). The higher bioluminescence signal was measured with cells not incubated with ARM and serum, while the bioluminescence signal was almost undetectable for the dead cells obtained by treatment with hydrogene peroxide. For the cells treated with the lowest concentration of ARM (5 μM) and in presence of human serum, a decrease of the bioluminescent signal of 28% was observed indicating a cytotoxicity induced by HS and the ARM (Fig. 4a). An increase by four of the ARM concentration induced a rise in the cytotoxicity by the same factor showing a dose dependent effect (Fig. 4b). We clearly showed that our ARM conjugate is able to recruit antibodies in human serum and activate an immune leading to the tumor cell destruction.

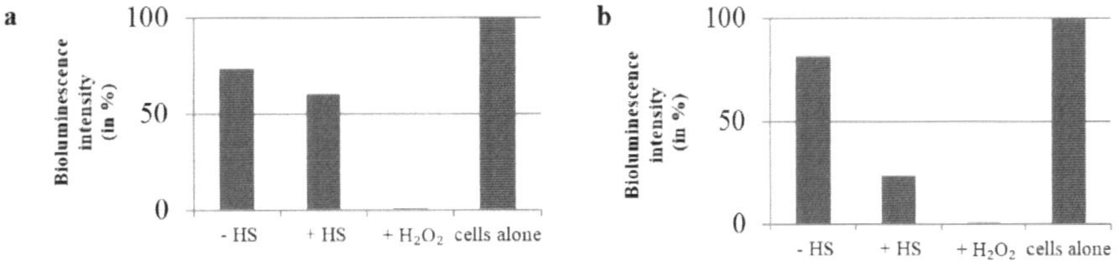

Figure 4: U-87 MG cell cytotoxicity induced by the bimodal ARM in presence (+HS) or in absence (-HS) of human serum, calculated using the bioluminescence titration of the intracellular ATP. a. 5 μM b. 20 μM

In conclusion, we elaborated a bimodal molecule displaying four cRGD motifs for binding to $\alpha_v\beta_3$ integrins expressed on tumor cells and a tetravalent rhamnose units for recruitment of anti-Rhm antibodies naturally present in human serum. The tetravalent cRGD displaying module showed strong binding for tumor cells expressing the $\alpha_v\beta_3$ integrins. Similarly, the tetravalent rhamnose scaffold showed the higher ability to recruit anti-rhamnose antibodies than the rhamnose monomer. Finally, the bifunctional molecule resulting from the assembly of the the two modules showed cytotoxicity when tested with U-87 MG tumor cells. These results shows that increasing the valency of the two binding modules is an efficient way to increase the destruction of cancer cells by bifunctional antibody recruiting molecules.

Materials and Methods

Cell lines and cultivation

Cell lines U-87 MG was purchased from the American Type Culture Collection (Manassas, VA, USA). Cell line MCF-10a was provided by Dr. Y Chebloune from the Grenoble Alpes University, France. U-87 MG was cultured in MEM supplemented with 10% heat-inactivated FBS and 1% penicillin/streptomycin. MCF-10a was cultures in MEM supplemented with 10% heat-inactivated FBS, 1% penicillin/streptomycin, 100 ng/ml chelora toxin B subunit, 10 μg/ml insulin and 20 ng/ml EGF. Cells were cultured adherently in complete medium at 37°C in an humidified atmosphere with 5% CO_2. Cells harvest and passaging were done after reaching 80% confluency by washing cells with PBS and detaching them from tissue culture flasks after incubation for 6 minutes at 37°C with trypsin/EDTA. The cells are then counted and diluted in complete medium or experiment buffer.

Flow cytometry

Evaluation of the $\alpha_v\beta_3$ integrins expression was assessed by using a monoclonal antibody 23C6 conjugated to R-Phycoerythrin. Evaluation of the TBM was assessed by determining the amount of tetravalent cRGD conjugated with Cy5 binding to the cells. Each cell line were harvested and 1.0 x 106 cells were resuspended in 1 ml of ice-cold binding buffer HBSS during 30 minutes. After a wash with HBSS, the cells were incubated at 37°C and 4°C with 5 μM of fluorescent tetravalent cRGD-ABM for 1 h in the dark, or with 23C6 mAb PE conjugated (1:10) at 37°C during 30 minutes. Subsequently, the cells were centrifuged, wash once in PBS and resuspended in 1 ml of PBS and finally kept on ice until being analyzed. Flow cytometry was performed on a BD FACSCalibur platform (Becton Dickinson, Heidelberg, Germany).

Cell Titer Glo assay

A cytotoxicity test was performed in a white bottom 96-well-plate. The cells were harvested by trypsin/EDTA treatment, counted and suspended in complete medium to a density of 5.0 x 105 cells/ml. Subsequently, the cells were seeded at 100 μl/well and incubated for 16 h at 37°C. Glycopeptide ARM was diluted from stock solution (1 mM) into MEM at 5 and 20 μM concentrations. After incubation of the peptide for 1h at 37°C and 1 wash with MEM, 100 μl of MEM with 20% v/v of Human Serum was added to each well, following by a final incubation for 2 h at 37°C. The amount of live cells in each well was determined by luminescence using the CellTiterGlo viability assay per the manufacturer's instruction (Promega). The maximal killing values were untreated Dead cells obtained by treatment with MEM with 5% v/v hydrogen peroxide were used as control.

References

1. Adams GP, Weiner LM. Monoclonal Antibody Therapy of Cancer. 2005; 23: 1147–1157; DOI:10.1038/nbt1137.

2. Hansel TT, Kropshofer H, Singer T, Mitchell JA, George AJT. The Safety and Side Effects of Monoclonal Antibodies. Nat. Rev. Drug Discov. 2010; 9: 325.

3. Carlson CB, Mowery P, Owen RM, Dykhuizen EC, Kiessling LL. Selective Tumor Cell Targeting Using Low-Affinity, Multivalent Interactions. ACS Chem. Biol. 2007; 2: 119–127; DOI:10.1021/cb6003788.

4. Murelli RP, Zhang AX, Michel J, Jorgensen WL, Spiegel DA. Chemical Control over Immune Recognition: A Class of Antibody-Recruiting Small Molecules That Target Prostate Cancer. 2009; 17090–17092.

5. Garanger E, Boturyn D, Coll J-L, Favrot M-C, Dumy P. Multivalent RGD Synthetic Peptides as Potent?V?3 Integrin Ligands. Org. Biomol. Chem. 2006; 4: 1958; DOI:10.1039/b517706e.

6. Sheridan RTC, Hudon J, Hank JA, Sondel PM, Kiessling LL. Rhamnose Glycoconjugates for the Recruit-ment of Endogenous Anti-Carbohydrate Antibodies to Tumor Cells. ChemBioChem 2014; 15: 1393–1398; DOI:10.1002/cbic.201402019.

Engineered protein-protein interaction regulators for therapeutic applications

Nir Qvit

Bar-Ilan University, Israel

https://doi.org/10.17952/35EPS.2018.047

Protein-protein interactions (PPIs) are intimately involved in almost all biological processes, including inter- and intracellular signal transduction,gene expression, cell proliferation and apoptosis. Therefore, they are important phenomena in basic research and promising targets for treating human disease. Nevertheless, targeting PPIs is challenging, as PPI interfaces are large, flat, and are usually endowed with a significant degree of conformational flexibility. Peptides are ideal candidates for targeting PPIs as they have demonstrated high conformational flexibility, increased selectivity, and are generally inexpensive. Moreover, peptide limitations, such as poor stability and inefficient crossing of cell membranes, can be overcome using peptidomimetics (modified peptides)[1-4].

Here we combine two techniques for the development of peptidomimetics. The first is an approach for the identification of the two critical PPI interfaces. We developed novel algorithms to detect specific PPI interfaces between a vital protein and only one of its many partners, using rational design, based on unique interaction sites (Fig. 1) [5-10]. A second technique, backbone cyclization, is an already established approach used for the development of peptidomimetics that retain the functional groups of the substrate side chain residues essential for their bioactivity. We thus convert active sequences into peptidomimetics while retaining their conformational flexibility (Fig. 2) [11, 12]. Together, these techniques allow the development of selective inhibitors of specific binding sites that are metabolically stable and highly active.

One example of inhibition of specific PPIs is the selective inhibition of the phosphorylation of pyruvate dehydrogenase kinase (PDK) by delta protein kinase C (δ PKC), which demonstrated its correlation with cellular processes following cardiac ischemia. Since δ PKC phosphorylates many substrates, the extent to which this specific phosphorylation is critical for cardiac damage could not be previously explored. We developed an inhibitor of the δ PKC/PDK PPI, ψPDK, which selectively inhibits PDK phosphorylation. In animal models of heart attack, ψPDK reduced infarct size by 55% and tripled ATP levels. ψPDK was found to be highly active (EC50 5 nM) and specific, and did not affect the phosphorylation of other δ PKC substrates even at 1 μM [13, 14].

A second example of specific PPI inhibition involves proteins that regulate mitochondrial fission, dynamin-related protein 1 (Drp1) and mitochondrial outer membrane protein, Fis1. Excessi ve mitochondrial fission results in oxidative stress and contributes to the pathology of neurodegenerative diseases. We developed a selective inhibitor of the Drp1/Fis1 PPI, p110. The selective inhibition reduced ROS production, improved mitochondrial membrane potential and reduced apoptosis. Our overall findings demonstrated that P110 prevents mitochondrial fragmentation and cell death under pathological conditions [15].

Inhibition of δ PKC/PDK and Drp1/Fis1 PPIs are both examples of specific and selective inhibition by rationally designed peptidomimetics. These novel, highly selective PPI inhibitors may be used as lead compounds for therapeutic applications. Furthermore, similar approaches can be applied for the development of various other specific PPI regulators.

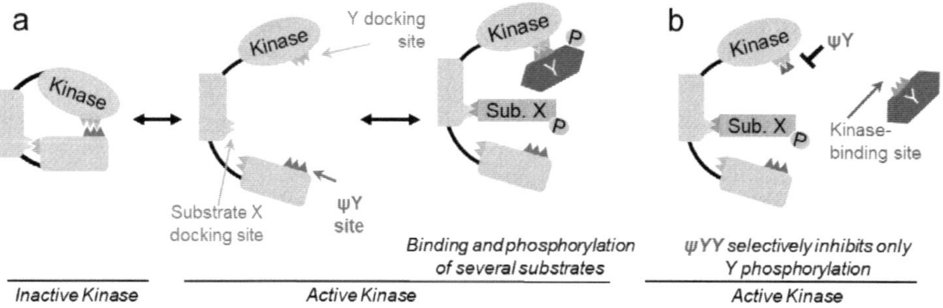

Figure 1: A scheme representing the design of a PPI modulator that is selective for the interaction between one substrate and a multi-substrate kinase. (a) In its inactive state (aka 'closed', or auto-inhibited conformation) (left), the substrate-docking site on the kinase forms an intramolecular interaction with a sequence similar to the kinase-binding site on the substrate, termed the pseudo-specific substrate site (ψY, brown). Upon activation, the kinase undergoes a conformational change, disrupting the intramolecular interaction and exposing its active site. As a result, the substrate-specific docking sites are available for binding (shown are docking sites for Y and substrate X (Sub. X) on the kinase, two of several protein substrates of this kinase). Specific PPIs between the kinase and its substrates results in substrate phosphorylation (P). (b) A peptide corresponding to the kinase-like sequence on Y, ψY (blue), is a competitive inhibitor for docking to and phosphorylation of Y by the kinase without inhibiting docking and phosphorylation of other kinase substrates (e.g. Sub. X).

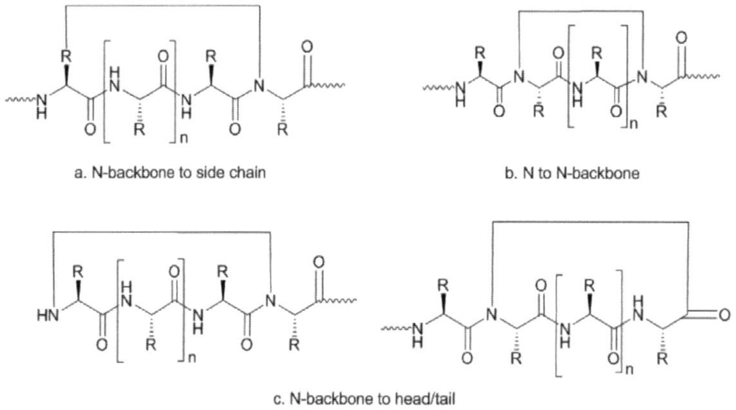

Figure 2: Backbone cyclization approach for alternative peptide cyclization strategies. Backbone cyclization is a technique that enables development of cyclic peptides without utilizing the residues that are part of the natural linear peptide, which may be essential for the biological activity of the peptide. In backbone cyclization, a building unit composed of a functional atom with an orthogonal protecting group, covalently bonded via spacer to the backbone, is used to form a cyclic peptide from a linear parent molecule. The main advantage of this method is that the cyclization linkage is formed between backbone atoms and not the atoms of the side chain functional groups, which are typically critical for binding and biological function. There are eight modes of cyclization amenable to backbone cyclization technique, including four natural modes: head-to-tail (C-terminus to N-terminus), side chain-to-tail, side chain-to-side chain, and side chain-to-head; and four modes of backbone cyclization: backbone-to-side chain (a), backbone-to-backbone (b), and backbone-to-N or -to-C terminus (c).

References

[1] Rubin, S.; Qvit, N., Cyclic peptides for protein-proteininteraction targets: Applications to human disease. Critical Reviews in Eukaryotic Gene Expression, 2016, 26, (3), 199-221.

[2] Cunningham, A.D.; Qvit, N., Specific kinase inhibition using peptides to target kinase-substrate docking. Chimica Oggi/Chemistry Today, 2016, 34, (5), 22-25.

[3] Qvit, N.; Rubin, S.J.S.; Urban, T.J.; Mochly-Rosen, D.; Gross, E.R., Peptidomimetic therapeutics: Scientific approaches and opportunities. Drug Discovery Today, 2017, 22, (2), 454-462.

[4] Cunningham, A.D.; Qvit, N.; Mochly-Rosen, D., Peptides and peptidomimetics as regulators of protein-protein interactions. Current Opinion in Structural Biology, 2017, 44, 59-66.

[5] Qvit, N.; Crapster, J.A., Peptides that target protein-protein interactions as an anti-parasite strategy. chimica Oggi / CHEMISTRY today, 2014, 32, (6), 62-66.

[6] Qvit, N.; Joshi, A.U.; Cunningham, A.D.; Ferreira, J.C.; Mochly-Rosen, D., GAPDH (glyceraldehyde-3-phosphate dehydrogenase) protein-protein interaction inhibitor reveals a non-catalytic role for GAPDH oligomerization in cell death. Journal of Biological Chemistry, 2016, 291, (26), 13608-13621.

[7] Qvit, N.; Kornfeld, O.S.; Mochly-Rosen, D., Engineered substrate-specific delta PKC antagonists to enhance cardiac therapeutics. Angewandte Chemie, International Edition in English, 2016, 55, (50), 15672-15679.

[8] Qvit, N.; Schechtman, D.; Pena, D.A.; Berti, D.A.; Soares, C.O.; Miao, Q.; Liang, L.; Baron, L.A.; Teh-Poot, C.; Martínez-Vega, P.; Ramirez-Sierra, M.J.; Churchill, E.; Cunningham, A.D.; Malkovskiy, A.V.; Federspiel, N.A.; Gozzo, F.C.; Torrecilhas, A.C.; Manso Alves, M.J.; Jardim, A.; Momar, N.; Dumonteil, E.; Mochly-Rosen, D., Scaffold proteins LACK and TRACK as potential drug targets in kinetoplastid parasites: Development of inhibitors. International Journal for Parasitology: Drugs and Drug Resistance, 2016, 6, 74-84.

[9] Qvit, N.; Kornfeld, O.S., Development of a backbone cyclic peptide library as potential antiparasitic therapeutics using microwave irradiation. Journal of Visualized Experiments, 2016, (107), e53589.

[10] Franco, A.; Kitsis, R.N.; Fleischer, J.A.; Gavathiotis, E.; Kornfeld, O.S.; Gong, G.; Biris, N.; Benz, A.; Qvit, N.; Donnelly, S.K.; Chen, Y.; Mennerick, S.; Hodgson, L.; Mochly-Rosen, D.; Dorn, G.W., Correcting mitochondrial fusion by manipulating mitofusin conformations. Nature, 2016, 540, (7631), 74-79.

[11] Rubin, S.J.; Tal-Gan, Y.; Gilon, C.; Qvit, N., Conversion of Protein Active Regions into Peptidomimetic Therapeutic Leads Using Backbone Cyclization and Cycloscan–How to Do it Yourself! Current Topics in Medicinal Chemistry, 2018, 18, (7), 556-565.

[12] Rubin, S.J.S.; Qvit, N., Backbone-Cyclized Peptides: A Critical Review. Current Topics in Medicinal Chemistry, 2018, 18, (7), 526-555.

[13] Qvit, N.; Mochly-Rosen, D., Highly specific modulators of protein kinase C localization: Applications to heart failure. Drug Discovery Today: Disease Mechanisms, 2010, 7, (2), e87-e93.

[14] Qvit, N.; Disatnik, M.H.; Sho, J.; Mochly-Rosen, D., Selective phosphorylation inhibitor of delta protein kinase C-pyruvate dehydrogenase kinase protein-protein interactions: Application for myocardial injury *in vivo* . Journal of the American Chemical Society, 2016, 138, (24), 7626-7635.

[15] Qi, X.; Qvit, N.; Su, Y.C.; Mochly-Rosen, D., A novel Drp1 inhibitor diminishes aberrant mitochondrial fission and neurotoxicity. Journal of Cell Science, 2013, 126, (Pt 3), 789-802.

Utilizing combinations of new approaches to peptide and peptidomimetic design for G-Protein Coupled Receptors

Victor J. Hruby, Minying Cai, Yanh Zhou, Jose Rios-Monterrosa

Departmentof Chemistry and Biochemistry, University of Arizona, Tucson, Arizona, United States

https://doi.org/10.17952/35EPS.2018.050

Living systems are very smart chemists and about 700 million years ago when animal life started most small molecule synthesis was discontinued. Now we get most of our small molecules from our diet. Instead, animal life evolved much more robust nucleic acid, peptide and protein, sugar and lipid chemistries. Not surprisingly, most of our current small molecule drugs have toxicities and mostly treat symptoms of our degenerative diseases. Drugs for the future need to be composed from peptides, proteins, nucleic acids, sugars and lipids to minimize toxicities. Here we will briefly illustrate this approach utilizing the melanocortin system (five receptors) and its native precursor peptide ligands derived from proopiomelanocortin (POMC), a primordial animal system involved in most of the biological activities, critical to our survival and good health.

The melanocortin system is composed of five receptors (MC1R, MC2R, MC3R, MC4R and MC5R) and are found throughout the body and brain. They are critical for most of the major biological functions critical for survival and good health including pigmentation, response to stress, feeding behavior, sexual behavior, immune response, inflammatory response, cardiovascular and kidney function and many others, some still being discovered. The endogenous peptide ligands for these receptors are derived from a single primordial gene POMC and include ACTH (specific for MC2R), α-, β- and γ-MSHs which interact with MC1R, MC3R, MC4R and MC5R, but they are nonselective and have very short half lives *in vivo* (a few minutes) and all are agonists. A major goal in this research is to develop novel, potent, receptor selective agonists and antagonists (orthosteric and allosteric) that are selective for these receptors and are more stable and bioavailable, and that can (or cannot) cross the blood brain barrier (1). Here we will briefly discuss our recent efforts toward these goals, using the full repertoire of approaches to peptide and peptidomimetic design we have developed over the past 40 years (2,3). These include computer based drug design, biophysical methods, conformational constraints, novel amino acids, novel cyclic systems, N-methylation and peptide mimetic design.

In earlier studies, we designed a number of peptide analogues of α-MSH using these methods including [Nle^4, D-Phe^7]α-MSH(MT-I, NDP-α-MSH) (Ac-Ser-Tyr-Ser-Nle-Glu-His-DPhe-Arg-Trp-Gly-Lys-Pro-Val-NH_2, MT-II-(Ac-Nle-c[Asp-DPhe-Arg-Trp-Lys]-NH_2) and Shu-9119-(Ac-Nle-c[Asp-His-D-Nal(2')-Arg-Trp-Lys]-NH_2. These are highly potent (nanomolar to subnanomolar), stable (2 hours to 2 days) and bioavailable. MT-II crosses the blood brain barrier, MT-I does not. Neither are receptor selective except that neither interacts with the MC2R. SHU-9119 is very unique being a nanomolar agonist at the MC1R and the MC5R and a potent antagonist at the MC3R and MC4R. These ligands have been widely used by us and worldwide to make many of the biological discoveries for the system including biological activities in the brain.

An example of a novel allosteric cyclopeptidomimetic we replaced the arginine residue in a cyclic disulfide analogue of MT-II with a guanindinylbutyl group at various backbone locations to give for example Ac-c[Cys-His-D-Phe-Cys-guanidinylbutyl-Cys]-Trp-NH_2 which was a potent (IC_{50}=1.8 nM), highly selective, MC1R antagonist (4).

In an extensive collaboration with Professor Horst Kessler and his group (5), we examined a very comprehensive library of backbone N-methylated cyclic lactam analogues related to MT-II. A number of highly selective ligands were obtained, the selectivity depending exclusively on the backbone N-methylation. For example, the Ac-Nle-c[Asp-N-MeHis-D-Phe-NMeArg-N-MeTrp-NMeLys]-NH_2 ligand was a potent (IC_{50}, 14nM) agonist for the MC1R. Comprehensive NMR studies were done to examine the conformational effects of N-methylation on the biological activity profile which has been used to further design novel constrained ligands for the melanocortin receptors.

In an effort to obtain nonpeptide peptidomimetic ligands, we have prepared a number of designed novel bicyclic heterocyclic ligands such as 7 in Figure 1. Surprisingly, very few such bicyclic heterocycles have been designed and published. Compound1 was found to be a very selective and potent (IC_{50}=0.4nM) at the MC5R, while its enantiomer was inactive at the MC5R (6).

A combination of site specific mutagenesis, domain swapping, computational chemistry, and computer assisted docking of receptor selective ligands at the various melanocortin receptors has been developed for this project. These experiments have provided three dimensional receptor models for the 5 melanocortin receptors which distinguish different critical residues for binding of agonist and antagonist ligands at these receptors. One approach to testing the validity of these models for use in computer assisted drug design is to test these models by designing a melanocortin 1 receptor agonist ligand using only eukaryotic α-amino acids. This led us (7) to the design of [Leu^3, Leu^7, Phe^8]-α-MSH-NH_2 which was found to be a highly MC1R selective agonist for the hMC1R, both *in vitro* and *in vivo* .

The above and many other studies have provided us with the excellent structural tools for agonist and antagonist design for 4 human melanocortin receptors. This provides us with the potential of applying these tools for a variety of medical needs, including pigmentation without sun for the prevention of melanoma cancer and other pigmentary disorders, for treatment of melanoma cancer, for treatment of stress related disorders, for feeding behavior disorders, for sexual behavior and dysfunction, for neurodegenerative diseases, immune response disorders and many others. These are all major current goals.

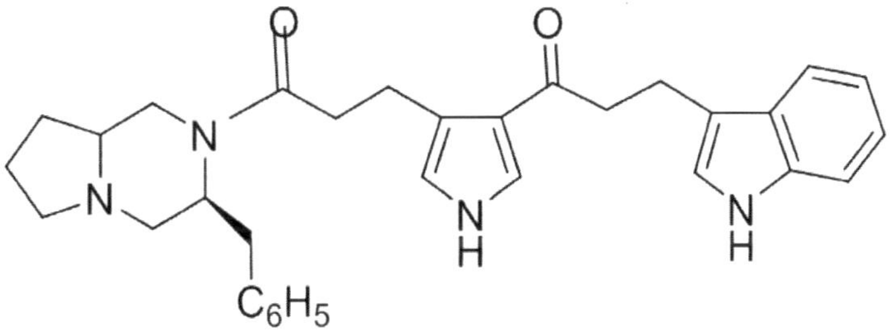

Figure 1: Structure of Heterocyclic Ligand for Melanocortin Receptors

Acknowledgments

We thank the U.S. Public Health Service, National Institutes of Health and Tech Launch Arizona for partial support of these studies, and to the many undergraduate, graduate, and postdoctoral students who have participated in this research and especially Professor Mac Hadley (deceased) who collaborated with us for over 40 years on this research.

References

1. Cai, M., and Hruby, V.J., Current Protein and Peptide Science, 17 (2016) 488.

2. Hruby, V.J., Nature Rev. Drug Discov., 1 (2002) 847.

3. Hruby, V.J. and Cai, M., Ann. Rev. Pharmacol. and Toxicol., 53 (2012) 557.

4. Ying, J., Gu, X., Cai, M., Dedek, M., Vagner, J., Trivedi, D., and Hruby, V.J., J. Med. Chem., 49 (2006) 6888.

5. Doedens, L., Opperer, F., Cai, M., Beck, J.G., Dedek, M., Palmer, E., Hruby, V.J. and Kessler, H., J. AM. Chem. Soc., 132 (2010) 8115.

6. Cain, J.P., Mayorov, A.V., Cai, M., Wang, H., Tan, B., Chandler, K., Lee, Y.S., Petrov, R.R., Trivedi, D. and Hruby, V.J., Bioorg. Med. Chem. Letts., 16 (2006) 5462.

7. Zhou, Y., Haghighi, S.M., Zoi, I., Sawyer, J.R., Hruby, V.J. and Cai, M., J. Med. Chem., 60 (2017) 9320.

Cathepsin B-activatable cell-penetrating peptides for imaging cancer-related Cathepsin B

Konstantin Kuhne[1], Lydia Behring[1], Birgit Belter[1], Robert Wodtke[1], Lara von Krüchten[2], Max Keller[2], Jörg Steinbach[1], Jens Pietzsch[1], Reik Löser[1]

[1] Helmholtz-Zentrum Dresden-Rossendorf, Institute of Radiopharmaceutical Cancer Research, Germany
[2] Institute of Pharmacy, Faculty of Chemistry and Pharmacy, University of Regensburg, Germany

https://doi.org/10.17952/35EPS.2018.052

In our understanding of the many drivers of malignant progression and cancer metastasis, proteases are increasingly drawn into the spotlight as crucial modulators in cancer angiogenesis, invasion, and metastasis.[1] Elevated activities of multiple members of the cathepsin family have been shown to correlate with increased metastasis and high therapy resistance.[2, 3] In particular, high expression levels of extracellular Cathepsin B (CatB) indicate poor prognosis in various neoplastic diseases, which makes CatB an interesting target for activity-based molecular imaging in cancer diagnostics as well as in cancer treatment monitoring for personalised therapies. It is our aim to develop such an imaging probe for CatB by combination of a polyarginine-based, activatable cell-penetrating peptide (ACPP), as first described by R. Tsien and coworkers, and an optimised endopeptidase substrate for CatB in one molecule.[4] Substrate optimisation proofed to be challenging as two opposite factors needed to be balanced: stability against serum proteases to prevent premature cleavage of the activation sequence, while retaining efficient and specific endoproteolytic cleavability by CatB.

We have generated a CatB-endoprotease substrate by C-terminally elongating the CatB carboxydipeptidase substrate AbzGIVR*AK(Dnp)OH (Abz aminobenzoyl; Dnp dinitrophenyl; * denominates CatB cleavage site), described by Cotrin *et al.* in 2004, to the octapeptide AbzGIVR*AK(Dnp)GXCONH $_2$, which can be used as activation site in the final ACPP.[5] Introduction of any amino acid other than glycine at the P4' position resulted in hysteretic kinetics for the CatB-catalysed hydrolysis of the octapeptides, which might indicate the displacement of the occluding loop from the active site upon interaction with the substrate. Valine was identified as the most efficient amino acid in the P4' position (Fig. 1).

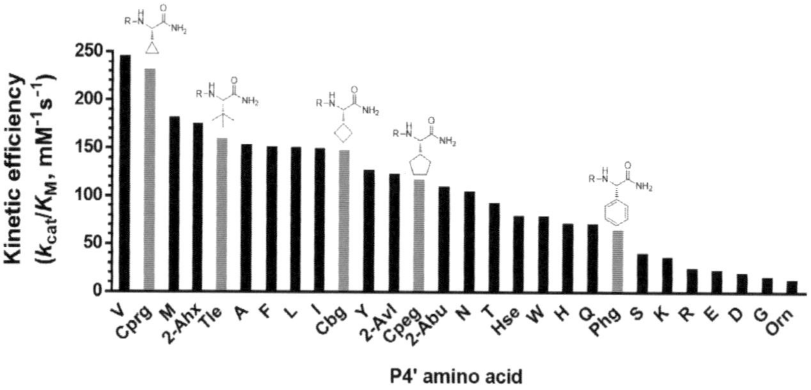

*Figure 1: Effect of the P4' amino acid on the kinetic efficiency for the proteolytic cleavage by CatB. Experiments were performed using the model substrate Abz-GIVR*AK(Dnp)GX-CONH$_2$, with Abz-Dnp forming a FRET-pair and X indicating the variability of the P4' position. Abbreviations: 2-Ahx 2aminohexanoic acid; 2-Avl 2-aminovaleric acid; Cbg cyclobutylglycine; Cpeg cyclopentylglycine; Cprg cyclopropylglycine; FRET fluorescence-resonance energy transfer; Hse homoserine; Orn ornithine; Phg phenylglycine; Tle tert-leucine.*

Using LC-ESI-MS-based analysis of substrates incubated in human serum, the positions P1 and P3' where determined to be primary determinants of serum stability. After suppression of the unspecific cleavage at the P3' by N$^{\alpha}$-methylation and optimisation within the positions P1-P3 (Fig. 2A), we were able to increase serum half-life from < 5 min to > 24 h under concomitant improvement of kinetic substrate efficiency towards CatB for substrates carrying the GFLG sequence at P4-P1 (Fig. 2B, Compounds 3-5).[6]

Based on these results, we have synthesised an ACPP mirroring Tsien's original design: A nonaDarginine CPP, carrying 6-carboxytetramethylrhodamine (6TAM) as reporter group, is linked *via* our optimised CatB-substrate (as activation sequence) to a nonaDglutamate attenuation sequence (Fig. 3A). Cellular uptake was evaluated using CatB-expressing human glioblastoma cells (U87MG).[7] After activation, the ACPP-6TAM accumulated rapidly within the nucleolus, independent of experimental conditions (Fig. 3B). This was in good accordance with literature, as Martin *et al.* described nucleolus-specific accumulation for polyarginine CPPs containing at least six arginines.[8]

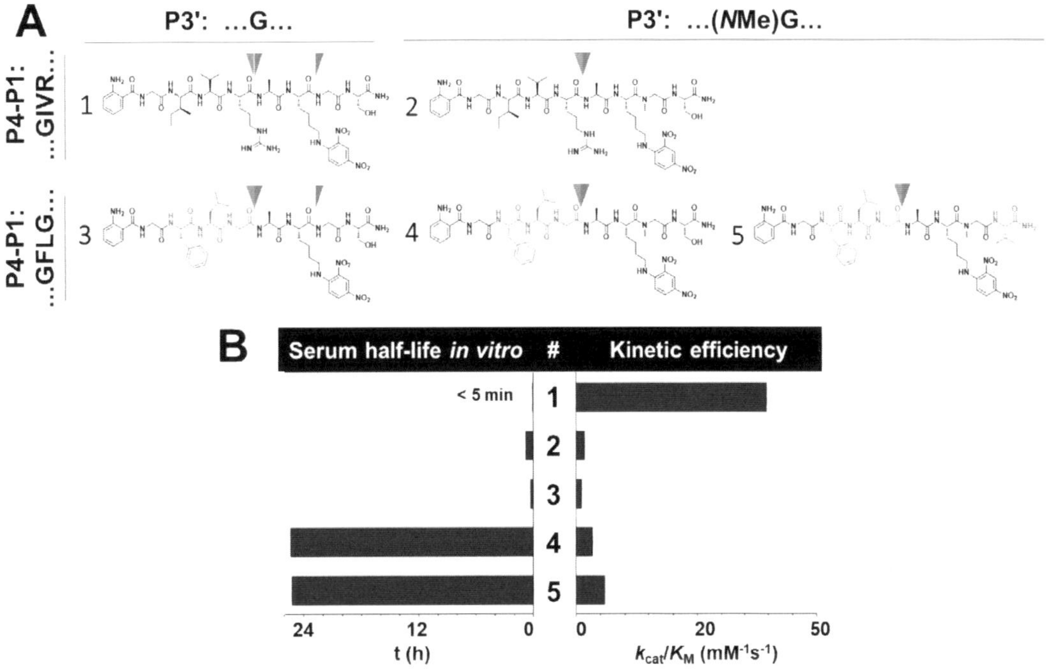

Figure 2: CatB-substrates evaluated for optimised stability by in vitro serum stability assay using human serum. Green half-arrow heads indicate CatB cleavage site, red half-arrow heads indicate unspecific cleavage site from human serum proteases identified by LC-ESI-MS fragment analysis. (B) Combination plot of serum half-life in vitro versus kinetic efficiency for substrate cleavage by CatB. Identifiers correspond to substrate structure in (A).

ACPP-6TAM activation and subsequent uptake was found to be dependent on both temperature (Fig. 3C, blue vs. red boxes) and the presence of 0.5 μM dithiothreitol (DTT, Fig. 3C, filled vs. unfilled red boxes), indicating a biocatalytic process which is additionally dependent on a reductive environment - as is necessary for continuous CatB-activity during *in vitro* assays.[5] Performing the cellular uptake experiment in the presence of either broad-range Cys-protease and Cys-cathepsin inhibitors (Leupeptin and E64[9]) as well as the CatB-specific inhibitor CA074[10] resulted in significantly reduced uptake, while the addition of different inhibitors for cathepsin S (Gue2313[11]), cathepsin K (Odanacatib[12]), Asp-proteases (Pepstatin A) or matrix metalloproteases (Ilomastat[13]) had no effect on the uptake of the ACPP-6TAM. We therefore assume that the activation and subsequent uptake of the ACPP-6TAM is a specific, CatB-dependent process.

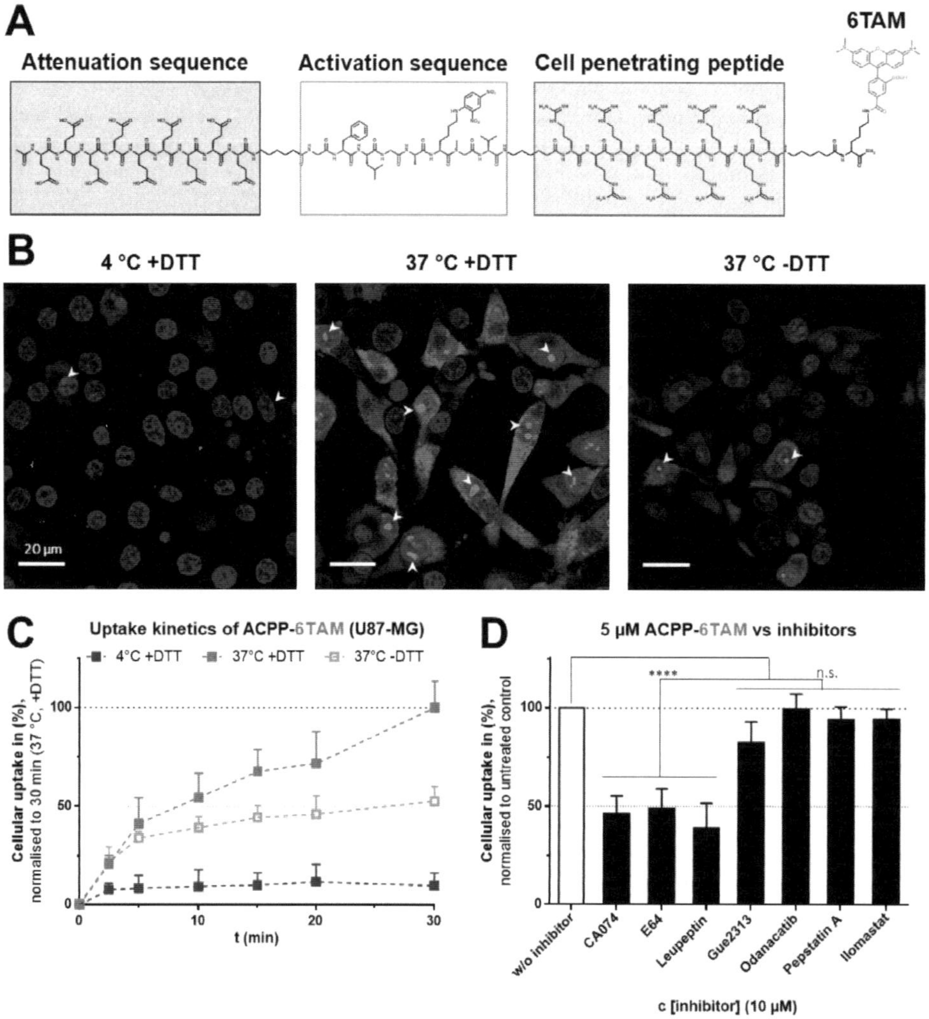

Figure 3: (A) Structure of the activatable cell-penetrating peptide. The N-terminal nona-d-glutamate attenuation sequence is highlighted by a red box, the central activation sequence (CatB substrate sequence) by a grey box and the cellpenetrating nona-d-arginine sequence by a green box, with 6TAM (red) linked to a C-terminal lysine residue. 6-Aminohexanoic acid spacers were used to confer flexibility between subunits. (B) Cellular uptake of ACPP-6TAM in U87MG cells. Intra-nucleolar accumulation of CPP(6TAM)-fragment indicated by white arrowheads. Nuclear stain: Hoechst 33342. (C) Kinetics of the cellular uptake of 5 µM ACPP-6TAM conjugate in U87MG cells. Uptake is strongly dependent on temperature (4 °C, +DTT vs. 37 °C, +DTT) and on the addition of 0.5 mM DTT (37 °C, +DTT vs. 37 °C, -DTT), both indicating a biocatalytic process responsible for the ACPP activation and uptake. Mean 6TAM-fluorescence values were normalised to nuclear stain for each image. Corresponding ratios were then normalised to untreated cells (= 0 %) and treatment 37 °C, +DTT (= 100 %) for each experiment. $n = 3, N \geq 6$. (D) CatB-selective inhibition of the cellular uptake of ACPP-6TAM in U87MG cells. Cellular uptake (15 min, 37 °C, +DTT) of 5 µM ACPP-6TAM in U87MG cells was significantly reduced by the addition of CA074 (CatB-selective inhibitor), E64 (broad-range cathepsin inhibitor) as well as Leupeptin (broad-range Cys-, Ser- and Thr-protease inhibitor). In contrast, the addition of either Gue2313 (cathepsin S-selective inhibitor), Odanacatib (cathepsin K-selective inhibitor), Pepstatin A (Asp-protease inhibitor) or Ilomastat (broad-range MMP inhibitor) did not result in a significant reduction of cellular uptake. Inhibitors were used at c = 10 µM. Data analysis performed as described in previous section. $n = 3, N \geq 6$.

References

[1] Yang *et al.* Cancer Growth Metastasis 2009, 2, 13-27;

[2] Aggarwal and Sloane Proteomics Clin. Appl. 2014, 8, 427-437;

[3] Löser and Pietzsch Front. Chem. 2015, 3, article 37;

[4] Jiang *et al.* Proc. Natl. Acad. Sci. USA 2004, 101, 17867-17872;

[5] Cotrin *et al.* Anal. Biochem. 2004, 335, 244-252;

[6] Duncan *et al.* Br. J. Cancer 1987, 55, 165-174

[7] Yan *et al.* DNA Cell Biol. 2000, 19, 79-91;

[8] Martin *et al.* Nucleus 2015, 6, 314-325;

[9] Hanada *et al.* Agric. Biol. Chem. 1978 42, 523-528;

[10] Murata *et al.* FEBS Lett. 1991, 280, 307-310;

[11] Frizler *et al.* J. Med. Chem. 2012, 55, 5982-5986;

[12] Gauthier *et al.* Bioorg. Med. Chem. Lett. 2008, 18, 923-928;

[13] Grobelny *et al.* Biochemistry 1992, 31, 7152-7154

Fluorogenic peptides for the detection of programmed cell death

Ramon Subiros-Funosas[1], Nicole Barth[1], Jesus Sot[2], Felix Goni[2], Lorena Mendive-Tapia[1], Rodolfo Lavilla[3], Rodger Duffin[1], Adriano Rossi[1], Ian Dransfield[1], Marc Vendrell[1]

[1]The University of Edinburgh, United Kingdom
[2]Instituto Biofisika, Universidad del Pas Vasco, Spain
[3]The University of Barcelona, Spain

https://doi.org/10.17952/35EPS.2018.056

Programmed cell death is an essential process in the development and progression of numerous diseases [1]. Therefore, there is a need for sensitive methods for the detection of apoptotic cells as mechanistic tools to unravel disease pathogenesis. The current gold standard for imaging apoptosis are fluorescently-labelled annexins, which display strong binding to phosphatidylserine (PS), a membrane phospholipid that is translocated to the external leaflet of the plasma membrane during the early stages of the apoptotic process [2,3]. However, annexins require high concentrations of Ca^{2+} for binding, which are not found in many physiological environments, including injured tissues [4]. Herein we developed peptide-based fluorogenic probes for Ca^{2+}-independent imaging of cells undergoing programmed death.

We used amphipathic peptides as templates to generate optical probes binding PS in apoptotic cells while showing marginal binding to viable cells. Notably, fluorogenic staining was achieved by incorporation of the environmentally sensitive Trp-BODIPY amino acid, which fluoresces only in close proximity to the binding target [5]. We have recently reported this amino acid as an optimal fluorophore to label peptides with minimal disruption of their binding properties and capabilities for wash-free imaging in real time [6]. Moreover, Trp-BODIPY displays excellent photophysical properties including emission in the green region of the visible spectra, matching GFP filters found in conventional microscopes.

We designed different fluorogenic peptides to study the hydrophilic and aromatic sequence requirements that defined optimal binding to PS. All peptides contained the Trp-BODIPY fluorogen while the nature of the amphipathic residues varied and included a broad range of polar and hydrophobic amino acids. We screened the peptides using lipid films *in vitro* , where the most promising binders displayed remarkable fluorescence emission in PS when compared to uncharged lipids or buffer. From a synthetic point of view, Trp-BODIPY is compatible with standard Fmoc/tBu SPPS protocols and can be incorporated using conventional coupling methods. However, the intrinsic lability of the BODIPY core under acid media require the use of acid labile solid supports (e.g. 2-chlorotritylchloride polystyrene resin) and side-chain protecting groups that can be removed in neutral conditions, such as hydrogenolysis.

All peptides were screened in cell mixtures containing ~50% apoptotic neutrophils and ~50% viable neutrophils, and we compared their signal to background fluorescence in viable cells. Fluorescence fold increase (as a measure of selectivity for apoptotic over viable cells) and fluorescence after washing (as a measure of binding strength) were measured for all peptides. From these studies, we concluded that peptides containing positively-charged residues bound strongly to apoptotic cells with fast kinetics, whereas negatively-charged ones did not show any binding to apoptotic or viable cells. We also observed that a balanced ratio of positive charges and hydrophobic residues was necessary to achieve rapid and selective binding to apoptotic cells. Some peptides exhibited strong binding to apoptotic neutrophils and low fluorescence emission in viable cells.

We further validated the peptides for detection of apoptotic cells from multiple origins and species. Furthermore, in contrast to Annexin V, these peptides stained apoptotic cells independently of Ca^{2+} concentration, with optimal performance even in the presence of the Ca^{2+} chelator EDTA. We further characterized cell populations based on their fluorescence staining, and confirmed that peptide-positive cells displayed features that are characteristic of apoptosis [7,8]. Finally, we used the peptides to detect apoptosis in mouse models. Peptide-stained cells were quantified by fluorescence confocal microscopy in tissue sections, revealing significant increase of apoptosis in inflamed tissues compared to healthy tissues. Notably, the peptides did not show any cytotoxic effect and did not impair clearance of dead cells by professional phagocytes, such as macrophages. In summary, we developed fluorogenic peptides for the detection of apoptosis under a broad spectrum of physiological environments and avoiding the need for high levels of extracellular Ca^{2+} that might facilitate their potential use for *in vivo* applications.

References

[1] Poon, I.K., Lucas, C.D., Rossi, A.G. & Ravichandran, K.S. Apoptotic cell clearance: basic biology and therapeutic potential. Nat Rev Immunol., 2014, 14, 166-180.

[2] Koopman, G., *et al.,* Annexin V for flow cytometric detection of phosphatidylserine expression on B cells undergoing apoptosis. Blood, 1994, 84, 1415-1420.

[3] Subiros-Funosas, R., et al., A Trp-BODIPY cyclic peptide for fluorescence labelling of apoptotic bodies. Chem. Commun., 2017. 53(5), 945-948.

[4] Baird, G.S. Ionized calcium. Clin Chim Acta, 2011, 412, 696-701.

[5] Mendive-Tapia, L., et al., Spacer-free BODIPY fluorogens in antimicrobial peptides for direct imaging of fungal infection in human tissue. Nat. Commun., 2016. 7, 10940.

[6] Mendive-Tapia, L., et al., Preparation of a Trp-BODIPY fluorogenic amino acid to label peptides for enhanced live-cell fluorescence imaging. Nat. Protocols, 2017. 12(8), 1588-1619.

[7] Knepper-Nicolai, B., Savill, J. & Brown, S.B. Constitutive apoptosis in human neutrophils requires synergy between calpains and the proteasome downstream of caspases. J Biol Chem, 1998, 273, 30530-30536.

[8] Morris, R.G., Hargreaves, A.D., Duvall, E. & Wyllie, A.H. Hormone-induced cell death. 2. Surface changes in thymocytes undergoing apoptosis, Am J Pathol, 1984, 115, 426-436.

Development of the MC1R selective ligands for the melanoma prevention

Minying Cai, Yang Zhou, Saghar Mowlazadeh Haghighi, Ioanna Zoi, Victor Hruby, Jixun Dai

University of Arizona, United States

https://doi.org/10.17952/35EPS.2018.058

Introduction

The incidence of melanoma, the most commonly fatal form of skin cancer, is increasing faster than any other potentially preventable cancer in the United States [1]. Estimates for 2016 are that 76,100 invasive melanomas will be diagnosed in the United States and that melanoma could claim 9710 lives.[2] Melanoma is the fifth most common cancer in men and seventh in women in the United States. Melanoma lesions generally evolve in a stepwise fashion, beginning as a typical nevi and progressing to melanoma in situ, *via* a radial growth, vertical growth malignant melanoma and metastasis. Although pre-cancerous nevi are easy to diagnose and curable with resection, they often go unnoticed and the majority of melanoma patients present with advanced disease. Once the disease is metastatic it is uniformly fatal with a median survival rate of only 4-6 months. Many approaches to treatment have been explored including radiotherapy, chemotherapy, immunotherapy and hormonal therapy, with minimal success. Therefore, developing melanoma prevention agent is becoming critical important. Normally, people with lighter skin color, light hair and light eyes are more at risk, because they have less melanin in their skin to protect them. Current efforts to prevent UV damage miss leading people exposed more under the sun, which in many cases leads to melanoma and other skin cancers. This proposal will provide a new way of melanoma prevention.

Results

Melanoma progression is associated with altered expression of cell surface proteins, including adhesion proteins and receptors.[3-6] It is estimated that over 80% of malignant melanomas express higher levels of melanocyte stimulating hormone (α-MSH) receptors, human melanocortin 1 receptor (hMC1R). The hMC1R is associated with skin pigmentation. Upon activation, the hMC1R in melanocyte and keratinocytes will form the pigmentation to block the UV radiation to prevent skin damage. The endogenous hMC1R ligands, α, β, γ-stimulate hormone, (α, β, γ-MSH) are derived from the POMC gene,[7] a part of a primordial system which is critical for survival. Endogenous MSH peptides exist *in vivo* and therefore, developing MSH based melanoma therapeutic agents are unlikely to run into issues of rejection and drug resistance. Several previous studies have investigated tumor targeting with non-selective radiolabeled MSH analogs, resulting in high quality images with sharp contrast. However, these studies all lack specificity. A motivation for the current study is developing highly selective hMC1R melanotropins and conjugates to reach the highest specificity to melanoma cells. To date, several compound of our discovery are already in Phase 2/3 clinical trials; the linear peptide [Nle^4, D-Phe^7]α-melanocyte (MT-1) stimulating hormone (NDP-α-MSH, ScenesseTM) has advanced to the status of a marketed drug (for congenital erythropoietic protoporphyria) in Europe and skin disorder prevention agent in Australia. However, NDP-α-MSH (MTI-1) represents a non-selective MCR agonist.[8] Nonselective MSH analogues, such as MTI/MTII, as a single target for drug delivery and imaging are problematic as hMCRs are also highly expressed in a number of normal tissues, including in the colon and the lung. Hence, coupling the target of a highly selective MC1R ligand to melanoma cells should provide improved specificity for earlier diagnosis, treatment and ultimately the prevention of melanoma.

We have been very successful in developing selective hMC1R ligands in the past twenty years. Recently, using NMR and computational aid drug design combined with chimeric receptor studies we can design more bioavailable and druggble selective hMC1R ligands rationally. As an example, we successfully developed only natural amino acid made peptide, [Leu3, Leu7, Phe8]-γ-MSH-NH_2, which is a potent selective hMC1R agonist with 24nM binding affinity and 4.5 nM functional activity[9] (Figure 1 .) and many other hMC1R selective agonists such as Ac-His-DPhe(4-CF_3)-Nle-Trp-NH_2 which are more druggable and bioavailable with 339nM binding affinity and 10nM functional activity.[10] NMR structure demonstrated that Ac-His-DPhe(4-CF_3)-Nle-Trp-NH_2 is β-turn structure. *In vivo* studies demonstrated these peptides can cause immediately pigmentation. The natural skin color can be resumed less than 20 hours. The high selectivity of peptides [Leu^3, Leu^7, Phe^8]-

-γ-MSH-NH_2 and Ac-His-DPhe(4-CF_3)-Nle-Trp-NH_2 for the hMC1R, and shorter half-life provides a safer and reduced side-effect agent for the prevention of melanoma skin cancer. Since binding affinity of [Leu^3, Leu^7, Phe^8]-γ-MSH-NH_2 is 20 times than peptide Ac-His-DPhe(4-CF_3)-Nle-Trp-NH_2 the pigmentation is stronger. This research will be more applicable and will be benefit for most people for skin cancer prevention.

Supported by GM108040, NIH

Figure 1: Computational aid drug design combined with chimeric melanocortin receptor studies led to selective hMC1R agonist: [Leu3, Leu7, Phe8]-γ-MSH-NH_2 with functional activity of pigmentation.[9]

Figure 2: Conformational studies of pharmacophore of MSHs led to selective hMC1R agonist Ac-His-DPhe(4-CF3)-Nle-Trp-NH_2 with functional activity of pigmentation.[10]

References

1. Silveira, F. L.; Pacheco, M. T.; Bodanese, B.; Pasqualucci, C. A.; Zangaro, R. A.; Silveira, L., Jr., Discrimination of non-melanomaskin lesions from non-tumorhuman skin tissues *in vivo* using Raman spectroscopy and multivariate statistics. Lasers in surgery and medicine 2015, 47 (1), 6-16.

2. Siegel, R. L.; Miller, K. D.; Jemal, A., Cancer statistics, 2016. CA: a cancer journal for clinicians 2016, 66 (1), 7-30.

3. Wang, Y.; Rao, U.; Mascari, R.; Richards, T. J.; Panson, A. J.; Edington, H. D.; Shipe-Spotloe, J. M.; Donnelly, S. S.; Kirkwood, J. M.; Becker, D., Molecular analysis of melanoma precursor lesions. Cell growth & differentiation: the molecular biology journal of the American Association for Cancer Research 1996, 7 (12), 1733-40.

4. Wang, R.; Kobayashi, R.; Bishop, J. M., Cellular adherence elicits ligand-independentactivation of the Met cell-surface receptor. Proc Natl Acad Sci U S A 1996, 93 (16), 8425-30.

5. Lehmann, J. M.; Holzmann, B.; Breitbart, E. W.; Schmiegelow, P.; Riethmuller, G.; Johnson, J. P., Discrimination between benign and malignant cells of melanocytic lineage by two novel antigens, a glycoprotein with a molecular weight of 113,000 and a protein with a molecular weight of 76,000. Cancer Res 1987, 47 (3), 841-5.

6. Siegrist, W.; Solca, F.; Stutz, S.; Giuffre, L.; Carrel, S.; Girard, J.; Eberle, A. N., Characterization of receptors for alpha-melanocyte-stimulating hormone on human melanoma cells. Cancer Res 1989, 49 (22), 6352-8.

7. Cone, R. D., The melanocortin receptors. In The Receptors [Online] Humana Press: Totowa, N.J., 2000; pp. 1 online resource (x, 551 p.). http://ezproxy.library.arizona.edu/login?url=https://doi.org/10.1007/978-1-59259-031-5.

8. Sawyer, T. K.; Sanfilippo, P. J.; Hruby, V. J.; Engel, M. H.; Heward, C. B.; Burnett, J. B.; Hadley, M. E., 4-Norleucine, 7-D-Phenylalanine-Alpha-Melanocyte-Stimulating Hormone - a Highly Potent Alpha-Melanotropin with Ultralong Biological-Activity. P Natl Acad Sci-Biol 1980, 77 (10), 5754-5758.

9. Zhou, Y.; Haghighi, S. M.; Zoi, J.; Sawyer, J. R.; Hruby, V. J.; Cai, M. Y., Design of MC1R Selective gamma-MSH Analogues with Canonical Amino Acids Leads to Potency and Pigmentation. J Med Chem 2017, 60 (22), 9320-9329.

10. Mowlazadeh Haghighi, S.; Zhou, Y.; Dai, J.; Sawyer, J. R.; Hruby, V. J.; Cai, M., Replacement of Arg with Nle and modified D-Phe in the core sequence of MSHs, Ac-His-D-Phe-Arg-Trp-NH_2, leads to hMC1R selectivity and pigmentation. Eur J Med Chem 2018, 151, 815-823.

Substitution of aromatic residues in the macrocyclic opioid peptide [D-Trp]CJ-15,208 alters the opioid activity profile *in vivo*

Jane Aldrich[1], Dmitry Yakovlev[1], Nicolette Ross[2], Shainnel Eans[1], Heather Stacy[1], Thomas Murray[3], Jay McLaughlin[1]

[1]University of Florida, United States
[2]University of Kansas, United States
[3]Creighton University, United States

https://doi.org/10.17952/35EPS.2018.061

Kappa opioid receptors (KOR) are involved in responses to stress, and KOR antagonists have potential therapeutic application for the treatment of substance abuse and mood disorders.[1] The macrocyclic peptide natural product CJ-15,208 was reported to be a KOR antagonist *in vitro*,[2] but the stereochemistry of the tryptophan residue was not reported. Therefore we synthesized both the L- and D-Trp isomers of CJ-15,208 [3] and evaluated their opioid activity profile *in vitro* and *in vivo*.[4] *In vivo* [D-Trp]CJ-15,208 antagonizes KOR after both central (intracerebroventricular, i.c.v.) [4] and oral [5] administration. It can penetrate the blood-brain barrier after oral administration to antagonize central KOR and prevent stress-induces reinstatement of extinguished cocaine-seeking behavior.[5] Therefore [D-Trp]CJ-15,208 is a promising lead peptide for structural modification to enhance its *in vivo* activity after oral administration.

Initially the alanine analogs of [D-Trp]CJ-15,208 were prepared and evaluated both *in vitro* and *in vivo*.[6] Unexpectedly all of the alanine analogs exhibited potent full agonist (antinociceptive) activity *in vivo* in the mouse 55 °C warm-water tail withdrawal assay, in contrast to the parent peptide that exhibits only modest antinociception (40% of the maximum response) at the highest dose tested (30 nmol i.c.v.). Therefore subsequent modifications focused on conservative modifications to the aromatic residues, including modifications that may decrease or prevent metabolism by liver oxidative enzymes.

Methods

The peptides were synthesized by a combination of solid phase peptide synthesis of the linear precursors as previously described, followed by cyclization in solution and purification by normal phase column chromatography.[3, 7]

The peptides were evaluated *in vitro* and *in vivo* using methods described previously.[6] Opioid receptor affinity was determined in radioligand binding assays using cloned receptors.[8] The peptides were evaluated *in vivo* for agonist (antinociceptive) and antagonist activity in C57BL/6J mice in the 55 °C warm-water tail withdrawal assay. To determine opioid receptor involvement in observed agonist activity the antinociception was measured in KOR and mu opioid receptor (MOR) knockout mice, as well as in wild-type mice pretreated with the delta opioid receptor antagonist naltrindole (0.5 mg/kg i.p.). To determine KOR antagonist activity mice were pretreated with peptide prior to the administration of the KOR selective agonist U50,488 (10 mg/kg, i.p.), and antinociception measured 40 min later.

[D-Trp]CJ-15,208, [Ala[1],D-Trp]CJ-15,208 and nor-BNI were evaluated for their ability to prevent reinstatement of extinguished morphine conditioned place preference (CPP) using procedures similar to those described previously for evaluation of compounds for their ability to prevent reinstatement of extinguished cocaine CPP.[6] Mice were subjected to 4 days of place conditioning in a counterbalanced morphine CPP paradigm. The mice were then evaluated for their place preference twice weekly until extinction was established, which required 3-6 weeks. Following extinction mice were pretreated with either vehicle or peptide and subsequently subjected to forced swim stress for two days as previously described.[4] Mice were tested for place preference on the day after stress exposure. The results are presented as the difference in the time spent in the morphine-paired vs. vehicle-paired compartments.

Results

Substitutions on the phenyl ring of Phe^3 in [D-Trp]CJ-15,208 (Figure 1) can alter the *in vivo* opioid activity profile of the resulting analogs. The effects of incorporation of a fluorine onto this ring depended on the position of this heteroatom on the ring. The m-fluoro-substituted analog retained KOR antagonism, but also exhibited antinociception following i.c.v. administration (ED_{50} (95% confidence interval) = 33 (12-102) nmol, Figure 2); evaluation in knockout mice indicated that both MOR and KOR are involved in the observed antinociception. In contrast, the para-substituted analog exhibited minimal antinociception (<35%) and weak KOR antagonist activity only at the highest dose tested (100 nmol i.c.v.). These results are consistent with the KOR affinities of the peptides (Ki = 43 ± 12 vs. 134 ± 63 nM for the meta and para substituted analogs, respectively). Other substitutions for Phe^3 also resulted in mixed agonist/KOR antagonist activity, although the receptors involved in the antinociception varied with the different analogs. However, some substitutions (e.g. His in place of Phe^3) resulted in analogs that produced antinociception but lost KOR antagonist activity.

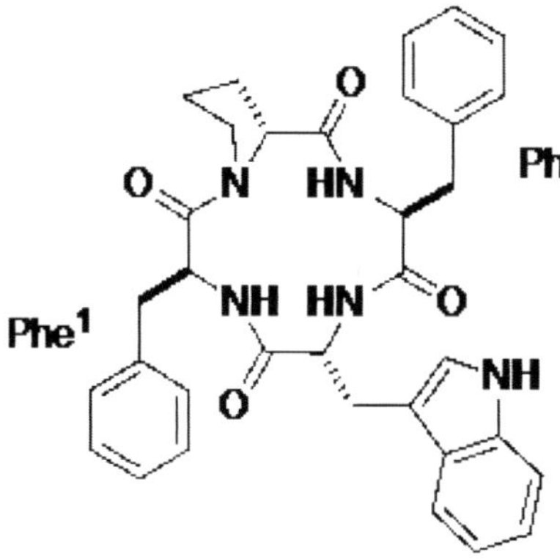

Figure 1: Structure of [D-Trp]CJ-15,208 with numbering of Phe residues.

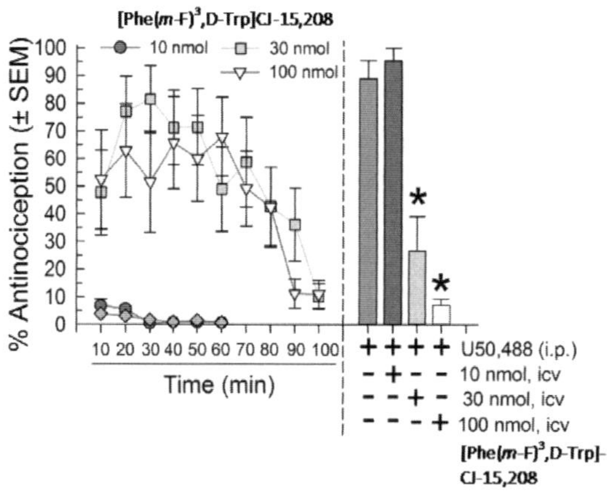

*Figure 2: Antinociceptive and KOR antagonist activity of [Phe(m-F)3,D-Trp]CJ-15,208 in the mouse 55 oC warm-water tail assay. * significantly different (p<0.05) from U50,488 alone.*

KOR antagonists can prevent reinstatement of extinguished cocaine-seeking behavior,[1] and a "functional KOR antagonist" (buprenorphine plus naltrexone) significantly improved drug abstinence in heroin-dependent patients compared to patients treated with naltrexone alone.[9] Therefore we examined the ability of [D-Trp]CJ-15,208, the alanine analog [Ala^1,D-Trp]CJ-15,208, which also exhibits KOR antagonism and prevents stress-induced reinstatement of extinguished cocaine seeking behavior,[6] and the small molecule KOR antagonist nor-BNI to prevent stress-induced reinstatement of morphine seeking behavior (Figure 3). Pretreatment with all three of the compounds significantly decreased the reinstatement of morphine seeking behavior in the conditioned place preference assay. Other analogs of [D-Trp]CJ-15,208 that exhibit KOR antagonism also prevented stress-induced reinstatement of morphine CCP, demonstrating the potential of these macrocyclic peptides in the treatment of drug abuse.

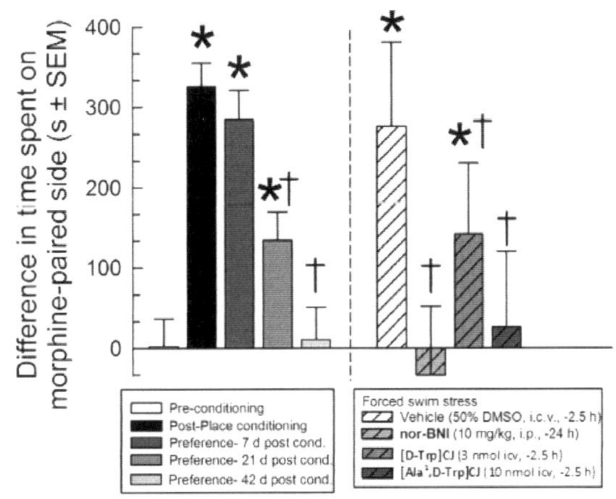

*Figure 3: [D-Trp]CJ-15,208, [Ala1,D-Trp]CJ-15,208 and nor-BNI prevented reinstatement of stress-induced morphine conditioned place preference. Following conditioning with morphine, extinction occurred over the next 6 weeks (left bars). Exposure to forced swim stress reinstated morphine CPP (yellow bar). Pretreatment with nor-BNI (10 mg/kg i.p., orange bar), [D-Trp]CJ-15,208 ([D-Trp]CJ, 3 nmol i.c.v., red bar) or [Ala1,D-Trp]CJ-15,208 ([Ala1,D-Trp]CJ, 10 nmol i.c.v., purple bar) significantly decreased reinstatement of CPP. * and †, significantly different from preconditioning and post-conditioning, respectively.*

In conclusion, substitutions for Phe^3 in [D-Trp]CJ-15,208 can alter the opioid activity profile of the resulting peptides. Most of the analogs retained KOR antagonism, with several also exhibiting antinociception that was mediated by multiple opioid receptors. Consistent with their KOR antagonism [D-Trp]CJ-15,208 and selected analogs also prevented stress-induced reinstatement of morphine-seeking behavior. Further studies are ongoing in our laboratories to further characterize and develop these promising macrocyclic peptides.

Acknowledgements

This research was supportedby grants PR141230 and PR141230P1 from the U.S. Army and grant R01 DA018332 from the National Institute on Drug Abuse.

References

1. J. V. Aldrich and J. P. McLaughlin, AAPS J., 2009, 11, 312-322.

2. T. Saito, H. Hirai, Y.-J. Kim, Y. Kojima, Y. Matsunaga, H. Nishida, T. Sakakibara, O. Suga, T. Sujaku and N. Kojima, J. Antibiot., 2002, 55, 847-854.

3. N. C. Ross, S. S. Kulkarni, J. P. McLaughlin and J. V. Aldrich, Tetrahedron Lett., 2010, 51, 5020-5023.

4. N. C. Ross, K. J. Reilley, T. F. Murray, J. V. Aldrich and J. P. McLaughlin, Br. J. Pharmacol., 2012, 165, 1097-1108.

5. S. O. Eans, M. L. Ganno, K. J. Reilley, K. A. Patkar, S. N. Senadheera, J. V. Aldrich and J. P. McLaughlin, Br. J. Pharmacol., 2013, 169, 426-436.

6. J. V. Aldrich, S. N. Senadheera, N. C. Ross, K. J. Reilley, M. L. Ganno, S. O. Eans, T. F. Murray and J. P. McLaughlin, Br. J. Pharmacol., 2014, 171, 3212-3222.

7. S. N. Senadheera, S. S. Kulkarni, J. P. McLaughlin and J. V. Aldrich, in Peptides: Building Bridges, ed. M. Lebl, American Peptide Society, San Diego, CA, 2011, pp. 346-347.

8. S. Arttamangkul, J. E. Ishmael, T. F. Murray, D. K. Grandy, G. E. DeLander, B. L. Kieffer and J. V. Aldrich, J. Med. Chem., 1997, 40, 1211-1218.

9. G. Gerra, A. Fantoma and A. Zaimovic, J. Psychopharmacol., 2006, 20, 806-814.

Dipeptide-derived alkynes as novel irreversible inhibitors of Cathepsin B

Lydia Behring[1], Christian Trapp[2], Robert Wodtke[1], Konstantin Kuhne[1], Birgit Belter[1], Jörg Steinbach[1], Jens Pietzsch[1], Reik Löser[1]

[1]Helmholtz-Zentrum Dresden-Rossendorf, Germany
[2]Hochschule Zittau/Görlitz, Germany

https://doi.org/10.17952/35EPS.2018.064

Introduction

Until recently, alkynes were considered to be bioinert. Thus, they are popular reaction partners in bioorthogonal click reactions *in vitro* and *in vivo* . Despite the virtual chemical inertness of the alkyne moiety, two research groups observed the irreversible inhibition of a cysteine protease by an alkyne-functionalised substrate derivative: both Ekkebus *et al.* and Sommer *et al.* independently described the unexpected inactivation of deubiquitinating enzymes by ubiquitin or ubiquitin-like modifiers bearing propargylamine in place of C-terminal glycine [1, 2]. We intended to harness these findings for the design of inhibitor-based probes for molecular functional imaging of tumour-associated cysteine proteases, particularly cathepsin B.

Tumour progression is a complex process involving an extensive number of enzymes. Amongst others, the cysteine protease cathepsin B was identified to be involved in tumour progression and metastasis. In healthy tissues, cathepsin B is located in the lysosomes, but it is secreted into the extracellular space in a broad variety of tumours. There it is, inter alia, able to degrade components of the extracellular matrix and activate different proteolytic cascades. Elevated extracellular levels are linked to increased malignancy and poor prognosis [3]. Therefore, this enzyme represents a promising target for the imaging and therapy of tumours.

The aim of our work was the synthesis and charakterisation of dipeptide-derived alkynes as novel inhibitors of cathepsin B. If the design of suitable alkyne derivatives would be successful, such inhibitors should exhibit very low unspecific reactivity, thus enabling clinical application.

Greenspan *et al.* reported a potent and highly selective dipeptidyl nitrile-based cathepsin B inhibitor (1, Figure 1) [4]. Based on that lead compound, a cathepsin B-targeting dipeptide alkyne was designed by isoelectronic replacement of the nitrile nitrogen atom by a methine group (2).

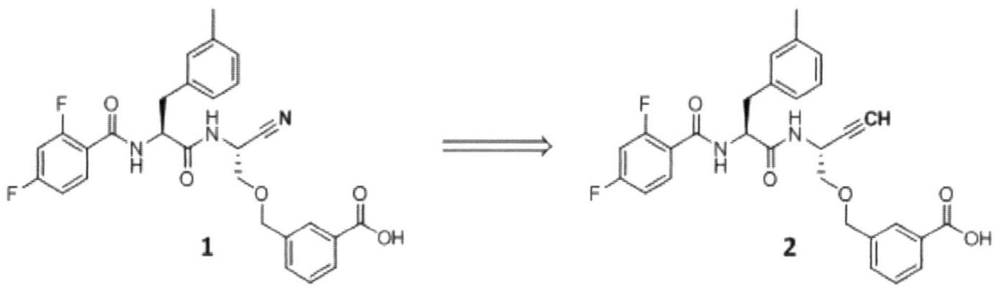

Figure 1: Inhibitor reported by Greenspan et al. (left) and irreversible inhibitor designed by isoelectronic replacement.

Results and Discussion

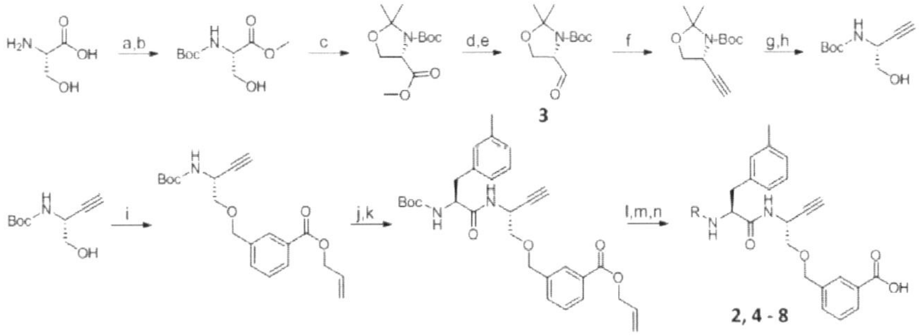

Scheme 1: Synthesis of the dipeptide-derived alkynes.

Reagents and conditions: a) acetyl chloride, MeOH, reflux; b) Boc2O, TEA, THF; c) 2,2-dimethoxypropane, BF3 · OEt_2, acetone; d) $LiAlH_4$, THF; e) oxayl chloride, DiPEA, DMSO, CH_2Cl_2, - 78 °C rt; f) Bestmann-Ohira reagent, K_2CO_3, MeOH, 0 °C rt; g) HCl (4 M), MeOH/H_2O, reflux; h) Boc_2O, TEA, THF; i) NaH, DMF, 0 °C; j) TFA, CH_2Cl_2; k) N-Boc-3-methylphenylalanine, DiPEA, PyBOP, THF; l) TFA, CH_2Cl_2; m) acyl chloride (RCl), TEA, CH_2Cl_2 or carboxylic acid (ROH), DiPEA, PyBOP, THF; n) morpholine, $Pd(PPh_3)_4$, CH_2Cl_2

In first experiments, formation of the C-C triple bond by reaction of the corresponding open-chain serine-derived aldehyde with the Bestmann-Ohira reagent was accompanied by partial enantiomerisation. Therefore, the synthesis was performed *via* Garner's aldehyde (3, Scheme 1). Due to conformational fixation in the five-membered ring, enantiomerisation *via* deprotonation/enolisationat the C_α atom was suppressed. Hence, the chosen synthesis route shown in Scheme 1 accounted for a high stereochemical purity of the final compounds.

To determine the inhibitory potential, a fluorimetric cathepsin B activity assay was performed. Over a time course of 15 min we observed the cleavage of the cathepsin B standard substrate Z-Arg-Arg-AMC (Z = Benzyloxycarbonyl; AMC = 7-amino-4-methylcoumarin) in the presence of different inhibitor concentrations (Figure 2A). The calculated kobs values were plotted against the inhibitor concentrations (Figure 2B) giving the specificity constant $\frac{k_{inact}}{K_I}$ arising from $\frac{k_{obs}}{[I]} * (1 + \frac{[S]}{K_M}) = \frac{k_{inact}}{K_I}$.

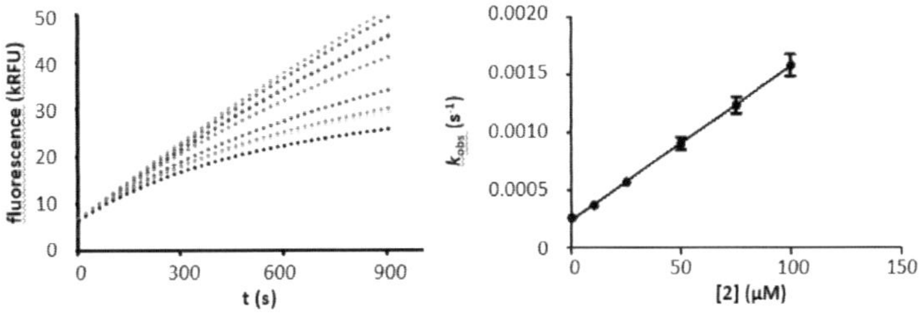

Figure 2: A) Exemplary substrate turnover curves for different concentrations of 2 (0 – 100 μM) in the presence of cathepsin B. B) kobs versus inhibitor concentration-plot for 2 including linear regression.

Whilst the decreasing substrate turnover velocities already indicated irreversible inhibition, we performed a reactivation experiment (referred to as "jump-dilution" experiment [5]) to verify the mechanism of inhibition. As expected for reversible inhibitors, the literature-known inhibitor 1 showed complete activity recovery in the assay. In contrast, for the dipeptide-based alkyne 2 no activity recovery could be observed after dilution of the enzyme-inhibitor complex. These results proof that dipeptide-derived alkynes are indeed able to irreversibly inhibit cathepsin B.

To improve the inhibitory potential, we performed consecutive variation of the 2,4-difluorobenzoyl residue. All synthesized inhibitors were tested not only against cathepsin B, but also towards cathepsins S, L and K to determine their selectivity within the cathepsin family. The results are graphically depicted in Figure

3. Compound 8 exhibited the most potent irreversible cathepsin B inhibition with an inactivation constant (kinact/KI) of 771 M-1s-1. Values for cathepsins L and S were significantly lower; no irreversible inhibition was observed for cathepsin K. In addition, inhibition of cathepsin B activity in human glioblastoma cell lysates and living cells (U87MG and U251MG cells) has been demonstrated. Surprisingly, compound 6 turned out to be a highly potent and highly selective inhibitor for cathepsin L ($\frac{k_{inact}}{K_I} = 1968 M^{-1}s^{-1}$). As the involvement of cathepsin L in tumour progression and a correlation with poor prognosis have been demonstrated in the past [3], compound 6 will be a promising lead compound for the development of cathepsin L selective alkyne-based probes in further studies.

For values marked with " Ø " no irreversible inhibition could be observed within the feasible concentration range.

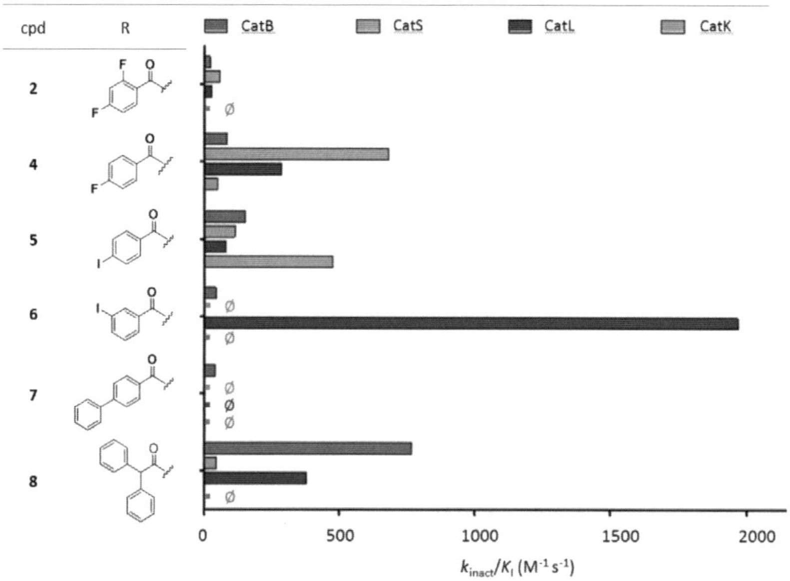

Figure 3: Selectivity profiles of assayed inhibitors with cathepsins B (CatB), S (CatS), L (CatL) and K (CatK).

Conclusion

We were able to synthesize alkyne-based irreversible inhibitors not only for cathepsin B, but also for cathepsins S, L and K including a distinct selectivity profile for each enzyme by variation of the N-terminal residue. Based on these initial results, dipeptidyl alkynes have the potential to become a valuable tool for molecular functional imaging approaches due to the expected low activity towards other cysteine proteases. In further studies, selected inhibitors for cathepsin B will be labelled with suitable radionuclides to obtain an inhibitor-based probe directed towards cathepsin B. The fact that alkynes are indeed capable to irreversibly inhibit cysteine proteases is important not only for the design of inhibitors, but also for the *in vivo* application of alkynes e.g. as reaction partners in click reactions, since the possibility of unwanted side reactions must be taken into account.

References

[1] Ekkebus et al., J. Am. Chem. Soc., 2013, 135, 2867-2870.

[2] Sommer et al., Bioorg. Med. Chem., 2013, 21, 2511-2517.

[3] Löser and Pietzsch, Front. Chem., 2015, 3:37.

[4] Greenspan et al., J. Med. Chem., 2001, 44, 4524-4534.

[5] Copeland, Evaluation of Enzyme Inhibitors in Drug Discovery (2nd Edition), John Wiley & Sons, 2013, 318-325

Light-controlled inhibition of MLL1 methyltransferase by azo-containing peptides: Towards optoepigenetic leukemia regulation

Lea Albert[1], Jing Xu[2], Vasundara Srinivasan[1], Yali Dou[2], Olalla Vázquez[1]
[1]Philipps-Universitat Marburg, Germany
[2]University of Michigan, United States

https://doi.org/10.17952/35EPS.2018.067

Epigenetics studies the variation in gene expression unrelated to changes in the DNA sequence. Epigenetic regulation includes chemical modulation at both, nucleic acid and protein level. Thus, one of the main mechanisms of control are the post-translational modifications (PTMs) on the histone proteins, which have context-dependent effects, together with the complex cross-talks between the responsible protein machineries—histone language.[1] Following the language metaphor, such epigenetic proteins can be divided into three classes: writers, erasers and readers that add, remove or reorganize these PTMs, respectively. Consequently, gene expression is ruled by a dy-namic signaling network of protein-protein interactions (PPIs). Particularly, we are interested in the PPI between the mixed-lineage leukemia 1 (MLL1) histone methyltransferase and the WD40-repeat protein 5 (WDR5) since it is key for MLL1 activity (Figure 1). In particular, this interaction is established between the arginine binding pocket of WDR5 and the WDR5-INteracting (WIN) motif of MLL1 (Ac-GSARAEVHLRKS-NH_2).[2]

MLL1 is mostly found in the promoter regions of actively transcribed genes and is essential for hematopoiesis during embryogenesis.[3] Importantly, disregulation of MLL1 as well as overexpression of MLL1 target genes are related to acute myeloid leukemia (AML), acute lymphoblastic leukemia (ALL) and some solid tumors.[4,5] Thus, MLL1 is a promising therapeutic target in cancer research.

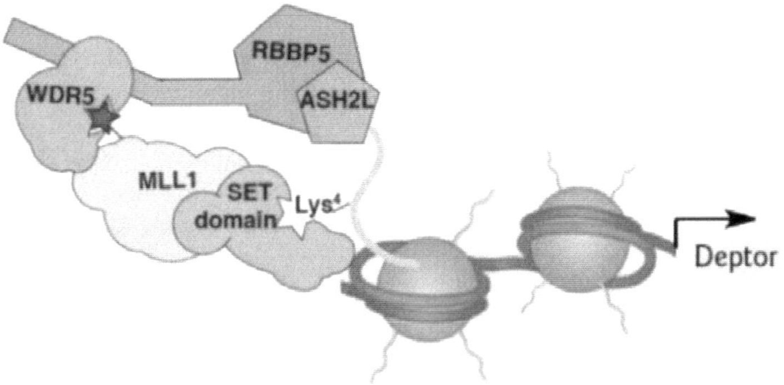

Figure 1: MLL1 core complex, whose formation is needed for H3K4 methylation activity of MLL1 enzyme.

Photopharmacology, which aims at using light as an external non-invasive trigger to modulate drug activity, has recently grown into a brilliant field. In the context of epigenetics and before our work,[6] there have only been two precedents of reversible regulation of a histone-modifying enzyme: the histone deacetylase (HDAC).[7,8] To our knowledge, there were not any photocontrollable probes for methyltransferases at that time and in both former examples photoswitchable small-molecule inhibitors bind to the HDAC catalytic domain directly.

In contrast to theses precedents, we envisioned the reversible control of MLL1 indirectly through targeting the essential MLL1-WDR5 PPI. Since we were targeting PPIs, designing photoswitchable peptidomimemtics as inhibitors was advantageous over small-molecules (Figure 2). Our strategy is based on the truncation of the WIN peptide: NH_2SARAEVHLRKSNH_2 (1), which interacts with higher affinity than the parent WIN peptide (Ki = 160 nM for WIN, Ki = 20 nM for 1).[9] The –ARA– sequence was identified as the crucial motif for WDR5 recognition.[9]

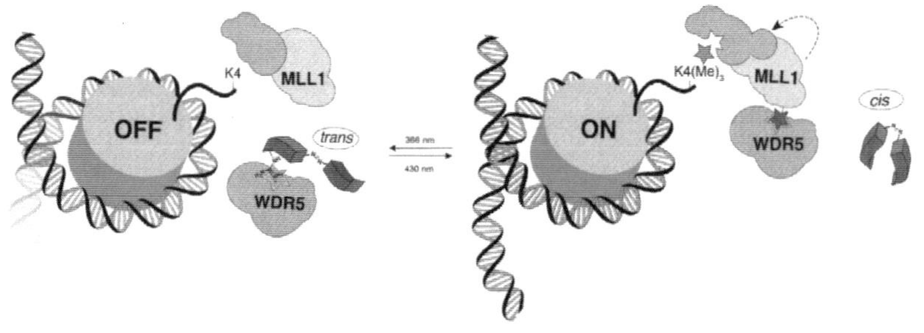

Figure 2: Strategy for MLL1 control through photoswitchable peptidomimetics via indirect targeting of MLL1-WDR5 PPI.

As an initial proof of concept, we chose the classic (4-aminomethyl)phenylazobenzoic acid (AMPB) (Figure 3) as a molecular transducer, due to its synthetic simplicity, acceptable spectroscopic properties and the possibility of direct introductioninto solid phase peptide synthesis (SPPS). Since the parental peptide does not have a defined structure, we decided to introduceAMPB following an amino acid scan strategy (Figure 3). The set of peptidomometics was prepared following the standard Fmoc-solid phase methodology and the synthesis of AMPB was carried out following literature procedures.[10] All the peptidomimetics showed fast and reversible photoisomerization and the thermal cis to trans relaxation was a rather slow process compatible with our biological experiments (data not shown).

To evaluate the binding affinities of the peptidomimetics to WDR5, a fluorescence polarization-based assay was used. The binding affinities are summarizedin Figure 3, demonstratingthat the trans isomer has always a higher affinity to WDR5 than its correspondingcis isomer. In addition, the influence of the introduced AMPB decreased when it is located far from the -ARA- motif. The peptidomimetic 5 shows the best binding affinity and one of the best differences between isomers. Gratifyingly, we observed that compound5 had even one magnitude higher affinity than the parental peptide 1.[6]

Peptides	Non-irr. (*trans*) K_i (nM)	Irr. at 366 nm (*cis*) K_i (nM)	*cis*/ *trans*
H_2N-SARAEVHLRKS-CONH2 **1**	20.0 ± 0.98	24.5 ± 0.49	**n.c.**
H_2N-SXARAEVHLRKS-$CONH_2$ **2**	n.c	n.c	**n.c.**
H_2N-SXRAEVHLRKS-$CONH_2$ **3**	66.8 ± 10	432 ± 73	**6.48**
H_2N-SARXEVHLRKS-$CONH_2$ **4**	n.c	n.c	**n.c.**
H_2N-SARAXVHLRKS-$CONH_2$ **5**	1.25 ± 0.36	6.50 ± 1.4	**5.00**
H_2N-SARAEXHLRKS-$CONH_2$ **6**	2.40 ± 0.60	9.10 ± 2.2	**3.79**
H_2N-SARAEVXLRKS-$CONH_2$ **7**	40.4 ± 3.9	39.7 ± 0.35	**0.99**
H_2N-SARAEVHXRKS-$CONH_2$ **8**	17.0 ± 1.9	33.3 ± 4.2	**1.96**
H_2N-SARAEVHLXKS-$CONH_2$ **9**	32.8 ± 4.5	70.4 ± 6.6	**2.15**
H_2N-SARAEVHLRXS-$CONH_2$ **10**	19.8 ± 1.7	40.6 ± 2.5	**2.05**

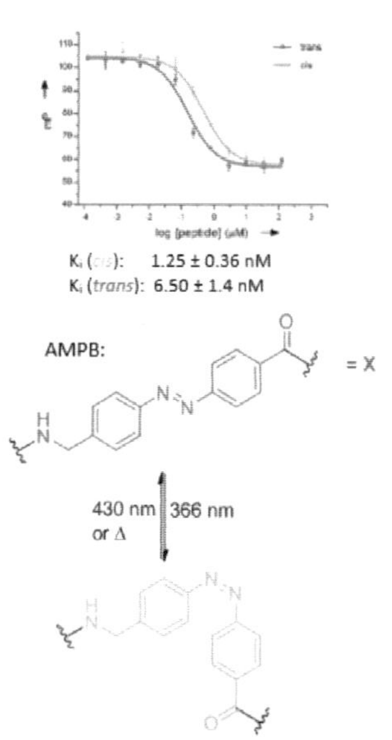

Figure 3: Left: Table with summarized Ki values for peptidomimetics 1-10. Mean values from three indepen-dent measurements. n.c. = not calculable (low binding affinity). Right, top: dose-dependent curve of best peptidomimetic 5. Bottom: photoswitching of AMPB molecule.

To have a molecular interpretationof this increase in affinity, we determinedthe co-crystal structureof WDR5 in complex with trans peptidomimetic 5. In Figure 4 the interactions of the parent WIN peptide with the arginine side chain sandwiched between the two phenylalanines, are displayed. The overlay of the crystal structureof peptidomimetic 5 shows that the key interactions with the WDR5 protein agree well with the WIN peptide/WDR5 complex. However, from the glutamate onwards the orientation is different. The replacement of the solvent exposed glutamate residue by the AMPB provided additional stabilization through van der Waals interactions between the benzene ring of the AMPB with the aromatic side chain of Tyr260 and a hydrogen bond between the first nitrogen atom of AMPB and Lys259.

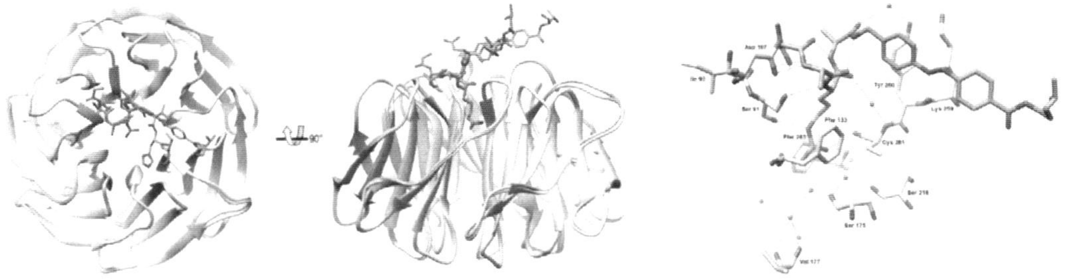

Figure 4: Co-crystal structure of WDR5 and peptidomimetic 5. Left: Overlay of WDR5/WIN peptide (purple) and WDR5/peptidomimetic 5 (pink). Right: Display of hydrogen bonds of peptidomimetic 5 to WDR5.

To determine whether our peptidomimetic 5 is not only binding to WDR5, but also able to inhibit the methylation activity of MLL1, we performed radioactive histone methyltransferase (HMT)-Assay. We could prove that the perturbationof the WDR5-MLL1 PPI by compound5 was sufficient to inhibit the activity of MLL1 and, importantly, the difference in IC_{50} has turned up to higher than 15 fold.[6]

Next, we wanted to investigate if peptidomimetic 5 could inhibit leukemic cell proliferation, wherefore we prepared the photoswitchable peptidomimetic 5 connected to a cell penetrating peptide (CPP) to enable cell uptake. We showed, that the original WIN peptide, even connected to the same CPP, is not able to inhibit cell proliferation, whereas our probe is, indeed, able to downregulate leukemic cell proliferation significantly. Moreover, we observed a modest but clear difference between isomers, even with in situ cell irradiation (Figure 6).[6]

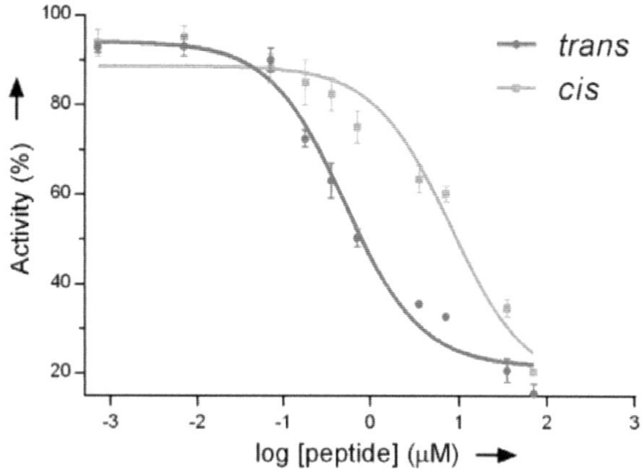

IC$_{50}$ (*cis*): 8.32 ± 0.38 μM

IC$_{50}$ (*trans*): 0.531 ± 0.023 μM

Figure 5: HMT-Assay of peptidomimetic 5 with MLL1 core complex.

Finally, we evaluated whether our probe can act as an optoepigenetic regulator and is able to control the expression of MLL1-target genes. Gratifyingly, we showed that the expression of deptor gene is significantly downregulated and observed a small but statistically significant ($p = 0.008$) difference between isomers. (Figure 6).[6]

In summary, our results are proof-of-concept of a new strategy for indirect external epigenetic control through targeting PPIs within a multi-protein complex. Our work highlights the importance of peptidomimetics as efficient alternatives to small molecule inhibitors and opens the door to PPIs as drugable targets.

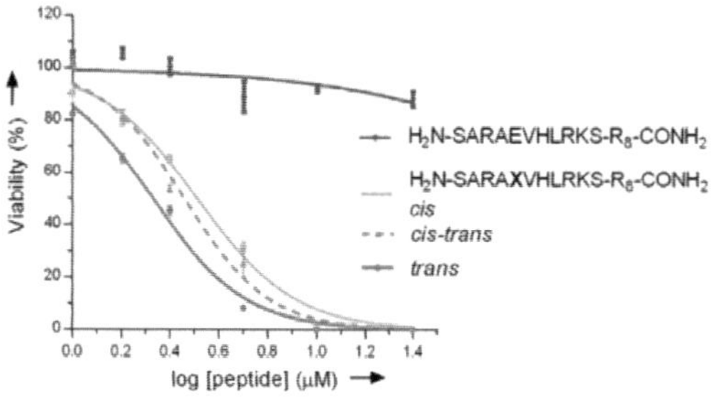

GI$_{50}$ (*cis*): 3.25 ± 0.23 μM

GI$_{50}$ (*cis-trans*): 2.88 ± 0.24 μM

GI$_{50}$ (*trans*): 2.14 ± 0.10 μM

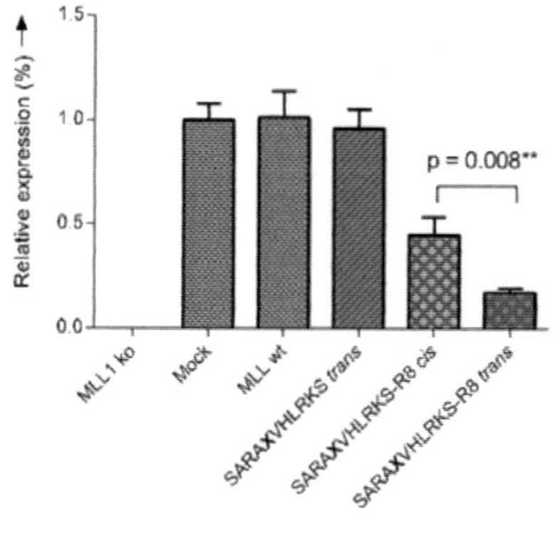

Figure 6: Left: Cell proliferation of MAF9 leukemia cells. Right: qPCR of MLL1 target gene deptor (GAPDH used as internal standard)

References

[1] T. A. Milne, S. D. Briggs, H. W. Brock, M. E. Martin, D. Gibbs, C. D. Allis, J. L. Hess, Mol. Cell 2002, 10, 1107–1117.

[2] Y. Dou, T. A. Milne, A. J. Ruthenburg, S. Lee, J. W. Lee, G. L. Verdine, C. D. Allis, R. G. Roeder, Nat. Struct. Mol. Biol. 2006, 13, 713–719.

[3] M. Vedadi, L. Blazer, M. S. Eram, D. Barsyte-Lovejoy, C. H. Arrowsmith, T. Hajian, Protein Sci. 2017, 26, 662–676.

[4] G. De Vita, P. Barba, N. Odartchenko, J. C. Givel, G. Freschi, G. Bucciarelli, M. C. Magli, E. Boncinelli, C. Cillo, Eur. J. Cancer 1993, 29, 887–893.

[5] D. Waltregny, Y. Alami, N. Clausse, J. De Leval, V. Castronovo, Prostate 2002, 50, 162–169.

[6] L. Albert, J. Xu, R. Wan, V. Srinivasan, Y. Dou, O. Vázquez, Chem. Sci. 2017, 8, 4612–4618.

[7] W. Szymanski, M. E. Ourailidou, W. a. Velema, F. J. Dekker, B. L. Feringa, Chem. - A Eur. J. 2015, 21, 16517–16524.

[8] S. a Reis, B. Ghosh, J. A. Hendricks, D. M. Szantai-Kis, L. Törk, K. N. Ross, J. Lamb, W. Read-Button, B. Zheng, H. Wang, et al., Nat. Chem. Biol. 2016, 12, 6–15.

[9] H. Karatas, E. C. Townsend, D. Bernard, Y. Dou, S. Wang, J. Med. Chem. 2010, 53, 5179–5185.

[10] B. Priewisch, K. Rück-Braun, J. Org. Chem. 2005, 70, 2350–2352.

Zn-catalyzed tert-butyl nicotinate-directed amide cleavage for applications in peptide synthesis and peptidomimetic design

Karlijn Hollanders[1], Charlène Gadais[1], Evelien Renders[2], Laurent Van Raemdonck[2], Clarence Wybon[2], Bert Maes[2], Steven Ballet[1]

[1]Research Group of Organic Chemistry, Departments of Chemistry and Bioengineering Sciences, Vrije Universiteit Brussel, Belgium
[2]Organic Synthesis, Department of Chemistry, University of Antwerp, Belgium

https://doi.org/10.17952/35EPS.2018.072

Introduction

The amide bond is the key functional group that makes up peptides and proteins. Consequently, the selective formation of this bond has been extensively studied by peptide scientists with the development of a well-established catalog of so-called coupling reagents as a result, and the general method remains the direct acylation of an amine by means of an in situ activated carboxylic acid.[1] To date, amides have rarely been considered as a carboxylic acid surrogate because of its poor electrophilicity and intrinsic stability. Recently, we presented the transformation of primary amides into esters through use of a tert-butyl nicotinate directing group.[2] Logically, the amide to ester conversion represented the ideal precedent to attempt transamidation reactions in a next step.

The activation mechanism of the amide can be considered as a 'biomimetic' protocol: the C3-ester substituent of the pyridine in the directing group populates a trans amide conformer suitable for Zn-chelation, $C=O_{amide}$-Zn-$N_{directing\ group}$, and the incoming nucleophile, the Zn-coordinated alcohol, for example, is additionally activated as a nucleophile by hydrogen bonding with the acetate ligand of the catalyst. Additionally, the acetate ligand assists in an intramolecular O-to-N proton transfer.

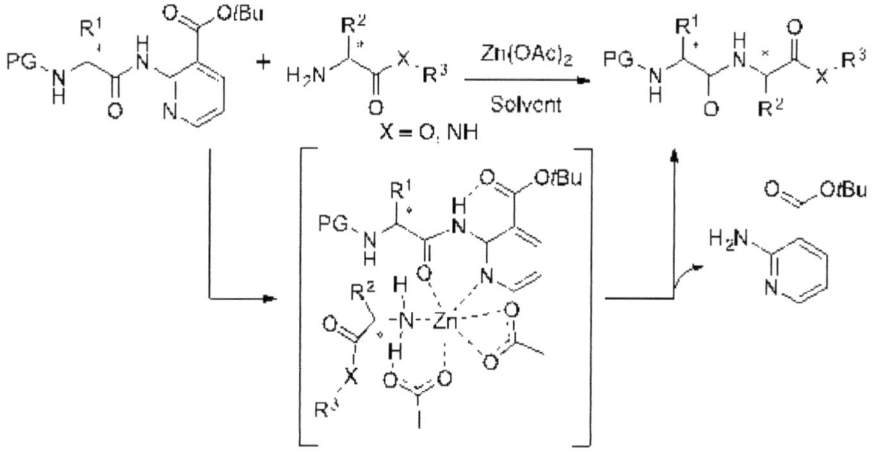

Figure 1: Peptide Coupling by Transamidation

We developed reaction conditions which allow the use of the directing group activated amino acid amides in transamidation reactions involving the α-amines of amino acid esters/amides as a method for diverse applications in peptide research. Next to amino acid – amino acid couplings, we exemplify the methodology's utility on more complex systems by segment condensations and macrocyclisations.[3]

Results and Discussion

The directing group is introduced *via* a previously developed Pd-catalyzed amidation protocol onto the N-Boc-protected amino acid amides. Here, tert-butyl 2-chloronicotinate 2 is coupled to the primary amides in the presence of 3rd generation palladacycle with a dcpf-ligand 3 and a carbonate base (Table 1).

Table 1: Synthesis of N-Boc protected tBu nic amino acid amides 4

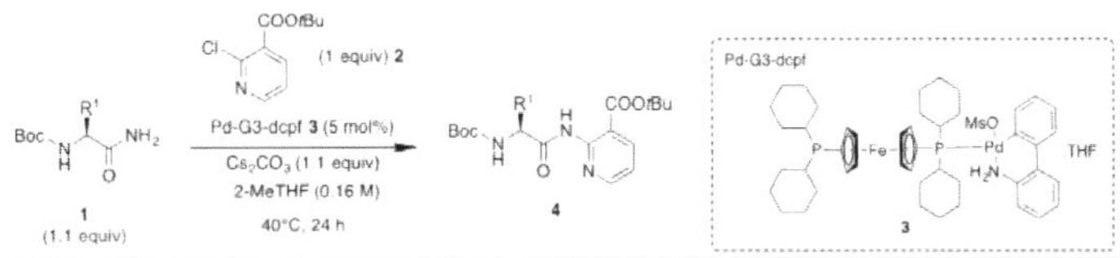

Entry	Boc-AA-NH-*t*Bu *nic* **4**	Yield [%][a]
1	Boc-L-Phe-NH-*t*Bu *nic* **L-4a**	93[b]
2	Boc-L-Ala-NH-*t*Bu *nic* **L-4b**	71[b]
3	Boc-L-Tyr(*t*Bu)-NH-*t*Bu *nic* **L-4c**	94[b]
4	Boc-L-Pro-NH-*t*Bu *nic* **L-4d**	80[c]
5	Boc-L-Met-NH-*t*Bu *nic* **L-4e**	85[b,d]
6	Boc-βAla-NH-*t*Bu *nic* **L-4f**	99
7	Boc-L-Val-NH-*t*Bu *nic* **L-4h**	98[e]
8	Boc-L-Ile-NH-*t*Bu *nic* **L-4i**	99[e]
9	Boc-L-Ser(*t*Bu)-NH-*t*Bu *nic* **L-4j**	72[b,f]

[a] Isolated yield. [b] K2CO3 (7.5 equiv), Pd-G3-dcpf (10 mol%) in 1,4-dioxane, 50°C, 30h. [c] Cs2CO3 (1.2 equiv). [d] 60°C. [e] Cs2CO3 (2 equiv). [f] K2CO3 (6 equiv).

For transamidations yielding 'peptide bonds', the *t*Bu *nic* amino acid amides 4 were used in the presence of 20 mol% $Zn(OAc)_2$ at 70°C for 24 h in THF. Since α-amino esters of type 5, used herein as a nucleophile in slight excess (1.1 equivalents), are often sold as HCl salt, NaOAc was added in a 1:1 ratio to liberate the free amine in situ. Under these conditions, the synthesis of dipeptides was realized (Table 2). The transamidation tolerated variable sterical hindrance in the lateral chains, as shown for both the tBu nic amino acid amides and the amino acid-based nucleophiles. Regarding the α-amino ester functionality, methyl, allyl, tert-butyl and benzyl esters remain unaffected. For amino acid amides as the nucleophile, featuring a primary amide, 3 equivalents were required to obtain full conversion. A similar observation was made for H-L-Pro-OMe.HCl. In the latter case, however, an increase in temperature to 100°C was additionally required to obtain full conversion within 24 h.

Table 2: Scope of the transamidation reaction for the synthesis of various dipeptides

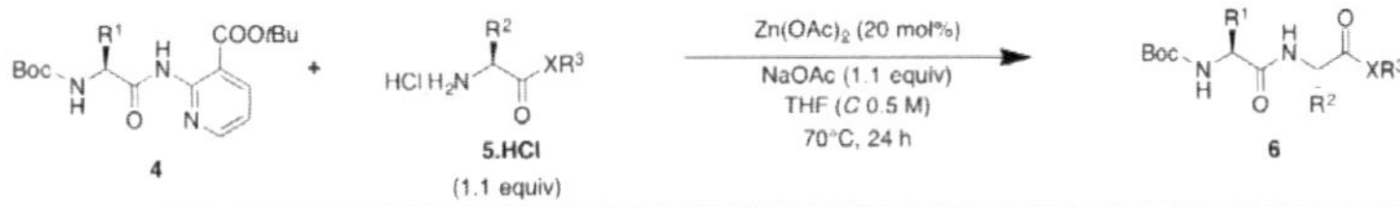

Entry	Boc-AA-NH-*t*Bu *nic* **4**	Amine **5.HCl**	Product	Yield [%][a]
1	Boc-L-Phe-NH-*t*Bu *nic* **L-4a**	H-L-Phe-OMe.HCl **L-5a.HCl**	**L,L-6a**	95
2	Boc-L-Ala-NH- *t*Bu *nic* **L-4b**	H-L-Leu-OAll.HCl **L-5b.HCl**	**L,L-6b**	99
3	Boc-L-Tyr(*t*Bu)-NH-*t*Bu nic **L-4c**	H-L-Val-O*t*Bu.HCl **L-5c.HCl**	**L,L-6c**	83
4	Boc-L-Pro-NH-*t*Bu *nic* **L-4d**	H-L-Leu-OMe.HCl **L-5d.HCl**	**L,L-6d**	95
5	Boc-L-Met-NH-*t*Bu *nic* **L-4e**	H-L-Trp-OBn.HCl **L-5e.HCl**	**L,L-6e**	80
7	Boc βAla NH *t*Bu *nic* **L-4f**	H L Phe NH$_2$.HCl **L-5f.HCl**	**L,L-6f**	74[b]
8	Boc-L-Met-NH-*t*Bu *nic* **L-4e**	H-L-Phe-NH$_2$.HCl **L-5f.HCl**	**L,L-6g**	89[b]
9	Boc-L-Ala-NH-tBu *nic* **L-4b**	H-L-Pro-OMe.HCl **L-5g.HCl**	**L,L-6h**	68[b,c]

[a] Isolated yield; [b] 3 equiv of the amine; [c] T = 100°C

Upon use of the reported transamidation in segment condensations and cyclizations, the N-Boc tBu nic amino acid amides were used as building blocks in peptide synthesis (e.g., synthesis of L,L-7 in Scheme 1). The Boc deprotection was performed using HCl in dioxane, followed by a coupling in solution with EDC-HOBt, and resulting in the dipeptide in excellent yield without loss of our directing group.

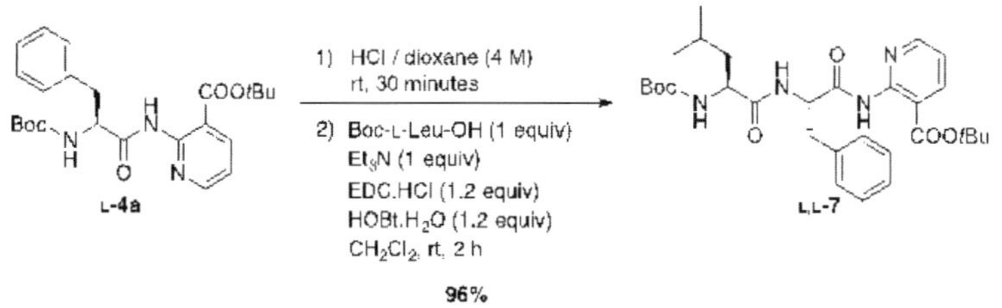

Scheme 1: Use of N-Boc tert-butyl nicotinate amino acid amide in peptide synthesis

Following the dipeptide synthesis, the transamidations were evaluated for the assembly of small segments. The *t*Bu *nic* dipeptides l,l7 and l,l-8 were engaged in a transamidation with, respectively, HlLeu-l-Val-OtBu.HCl (l,l-9.HCl) and H-l-Ala-l-Phe-OtBu.HCl (l,l-10.HCl) under standard conditions (Scheme 2a). Both tetramers L,L-11 and L,L-12 could be isolated in a very good yield.

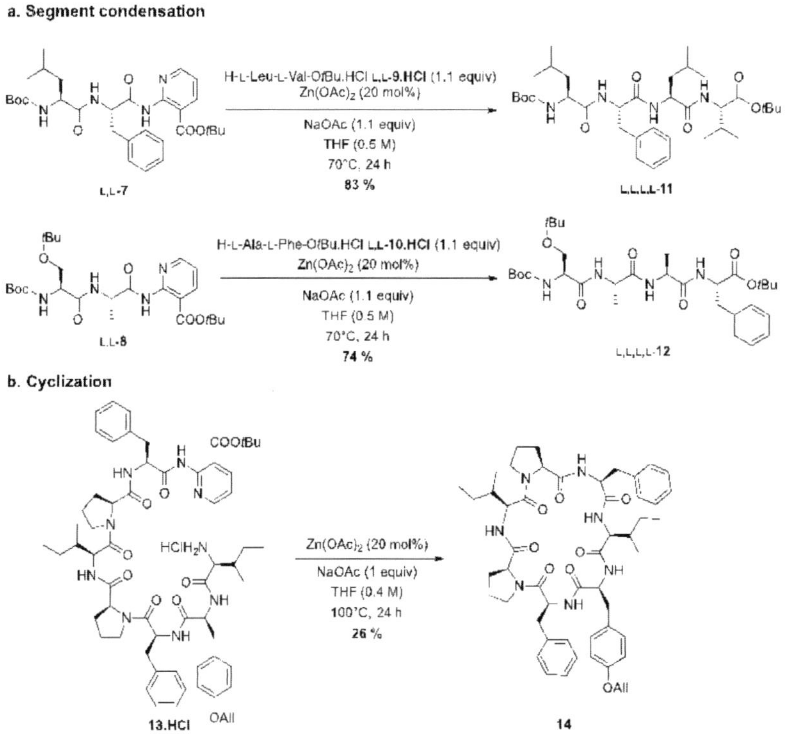

Scheme 2: Application of the methodology on a) dipeptide segment condensations and b) the macrocyclization of 13.HCl

Macrocyclization was demonstrated by the synthesis of an allylated analog of Stylissatin A. Similar to the segment syntheses, the linear heptamer 13.HCl was prepared starting from N-Boc protected tBu nic phenylalanine in solution under classical coupling/deprotection conditions. The cyclization was achieved smoothly within 24 hours at 100°C, allowing the isolation of the cyclic heptamer 14 by means of preparative HPLC purification (Scheme 2b).

Conclusion

To conclude, we have developed a transamidation protocol, proceeding under mild conditions, that is applicable on amino acids and small peptides. As the N-Boc protected *t*Bu *nic* amino acids amides are compatible with standard deprotection/coupling conditions, this methodology has added to the chemical toolbox of peptide chemists. Importantly, the mild condensation and cyclization conditions are orthogonal to standard peptide synthesis conditions

Acknowledgements

This work was financially supported by the Research Foundation Flanders (FWO; Research Project and Scientific Research Network (WOG)), the Hercules Foundation and the University of Antwerp (BOF).

References

[1] a) A. El-Faham, F. Albericio, Chem. Rev. 2011, 111, 6557-6602; DOI: 10.1021/cr100048w b) F. Albericio, A. El-Faham, Org. Process Res. Dev. 2018, 22, 760-772; DOI: 10.1021/acs.oprd.8b00159.

[2] C. C. D. Wybon, C. Mensch, K. Hollanders, C. Gadais, W. A. Herrebout, S. Ballet, B. U. W. Maes, ACS Catal. 2018, 8, 203-218; DOI: 10.1021/acscatal.7b02599

[3] K. Hollanders, C. Gadais, E. Renders, L. Van Raemdonck, C.C. D. Wybon, D. Masullo, B. U. W. Maes, S. Ballet, Sumitted.

Synthesis of peptides glycated at Lys residues

Daria Roshchupkina[1], Yulia Filatova[1], Alexey Chistov[1], Tatana Melnikova[2], Ekaterina Kolesanova[1]

[1]Institute of Biomedical Chemistry (IBMC), Moscow, Russia
[2]N.D. Zelinsky Institute of Organic Chemistry, Moscow, Russia

https://doi.org/10.17952/35EPS.2018.076

Summary

Since the most common *in vivo* glycated protein products represent the ones glycated at Lys residues, we studied several approaches for the preparation of HSA fragments and artificial glycated peptide haptens containing glycated (1-deoxyfructosyl-, or Amadori product (AM), carboxymethylated (CM) and carboxyethylated (CE)) Lys residues. The studied approaches were: 1) modification by glucose in methanol solution [1]; 2) Schiff base formation with further reduction [1-2]; 3) alkylation (for CM-and CE-modifications) [3-4]. Peptide modification versus Fmoc-Lys modification with further application of the latter in SPPS evaluated. Yields and purity of glycated peptides obtained with the use of different approaches were compared.

Introduction

Diabetes represents a serious medico-social and economic problem in all countries over the world. Despite the fact that there are systems for early diagnosis of this disease, there is still no adequate control of treatment efficiency and prognosis of complications. Elevated blood glucose levels result in increased protein non-enzymatic glycation in diabetic patients. Glycated hemoglobin relative content is recognized as a "gold standard" in diabetes diagnostics by WHO. However, in certain cases this parameter does not correlate well with the severity of hyperglycemia and, because of the long life of erythrocytes and hemoglobin in blood, cannot be used for a midterm evaluation of a treatment efficiency. Glycated human serum albumin (HSA) represents another possible marker of diabetes, suitable for the midterm treatment efficiency evaluation, as well as for diabetes complication prognoses [5-6]. The problem of the glycated HSA use as a diagnostic marker lies in its poorly studied *in vivo* glycation. Nowadays HSA glycation studies by proteomic methods are in progress, but in order to get accurate quantitative results, glycated HSA fragments are needed as standards and calibrants for LC-MS analyses and as antigens for producing specific antibodies for the selective detection of glycated proteins.

Experimental

Fragments of HSA used for the preparation of glycated peptides are shown in Table 1.

Peptides were prepared by SPPS by the FastMoc procedure on 433A synthesizer (Applied Biosystems), were purified by HPLC and analyzed by LC-MS and MS/MS.

Table 1: Peptides, sites and types of modifications.

Peptides	Sequence	Modified site	Type of modification	Peptide MH^+; Da
HSA fragment 549-558 Shown to be highly glycated in diabetes	KQTALVELVK	$\mathbf{K_{AM}}^{549}$QTALVELVK $\mathbf{K_{CM}}^{549}$QTALVELVK $\mathbf{K_{CE}}^{549}$QTALVELVK	AM-Amadori product (1-deoxyfructosylated) CM-carboxymethylation CE-carboxyethylation	1290.74 1186.70 1200.71
Artificial peptides for producing anti-(glycated Lys) antibodies	GSGSGK(amide)	GSGSG$\mathbf{K_{AM}}$(amide) GSGSG$\mathbf{K_{CM}}$(amide) GSGSG$\mathbf{K_{CE}}$(amide)	AM-Amadori product CM-carboxymethylation CE-carboxyethylation	653.40 549.29 563.30

Three approaches were used for the preparation of ε-Lys-Amadori products (3).

1. Incubation of synthesized (1) and deprotected (except N-terminal Fmoc-group) Lys-containing peptide (2) in saturated methanolic glucose solution at 60-70oC [1] (Scheme 1, Approach A).

2. Reductive amination of Lys ε-amino group in a free α-N-Fmoc-protected peptide (2) by the protected aldoketose (4, 5) in the presence of Nax(CN) xBH4 in methanol at room temperature with further peptide and

1-deoxyfructosyl moiety total deprotection [1] (Scheme 1, Approach B).

Protected 2,3:4,5-di-O-isopropylidene-aldehydo-b-D-arabino-hexosulo-2,6-pyranose (4) was obtained by the method described in [7].

3. Reductive amination of Lys ε-amino group in the resin-attached peptide with selectively deprotected Lys ε-amino-group (6, 7) [2] (Scheme 1, Approach C).

The first approach gave the worst results: the lowest yield of a mixture of peptide glycation products, which were difficult to separate by HPLC after the elimination of glucose by adsorbing the peptides on C18-silicagel (ZipTips) and extensive sorbent washing by water.

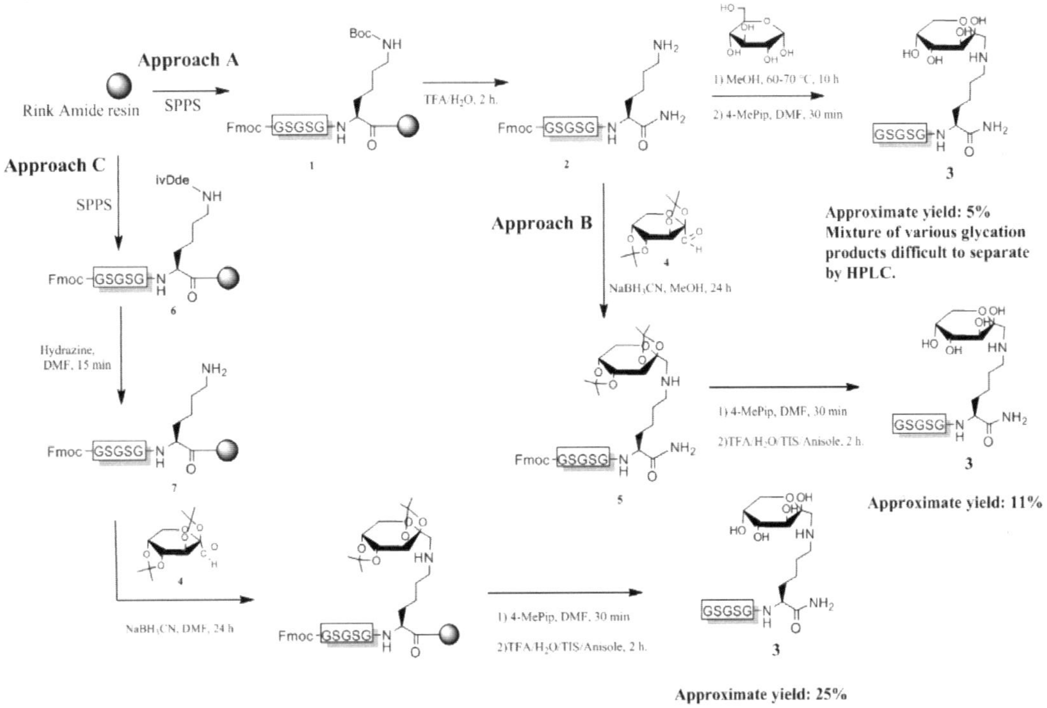

Scheme 1: Synthesis of Amadori products.

Although syntheses of Lys-CM- and CE-modified peptides are described in several papers, the majority of these approaches are time-consuming and suffer from poor yields. The most frequently used method is the alkylation of Lys ε-amino group with Boc or Nosyl (4-nitrobenzene sulfonyl, Ns) protection (to avoid dialkylation) by α-bromo-acetic or -propionic acid esters. We compared the application of this method to the preparation of Lys-CM- and Lys-CE-containing peptides. In this study, we used the 4-nitrobenzenesulfonyl (Nosyl) group for protection ε-amino group of Lysine residue [3]. This group allows specific monoalkylation of the ε-amine.

In this work, we compared and evaluated two approaches for obtaining CM- and CE-modified peptides (11).

1. Fmoc-Lysine modification (8) with its further application in SPPS [3-4] (Scheme 2, Approach D).

2. Alkylation of ε-Ns-protected Lys residue in the on-resin peptide (9) with further peptide deprotection (10) and detachment from the resin (11) [3-4] (Scheme 2, Approach E).

Approximate yields in both cases were almost the same, but the second approach was more convenient, cheaper and faster than the first one.

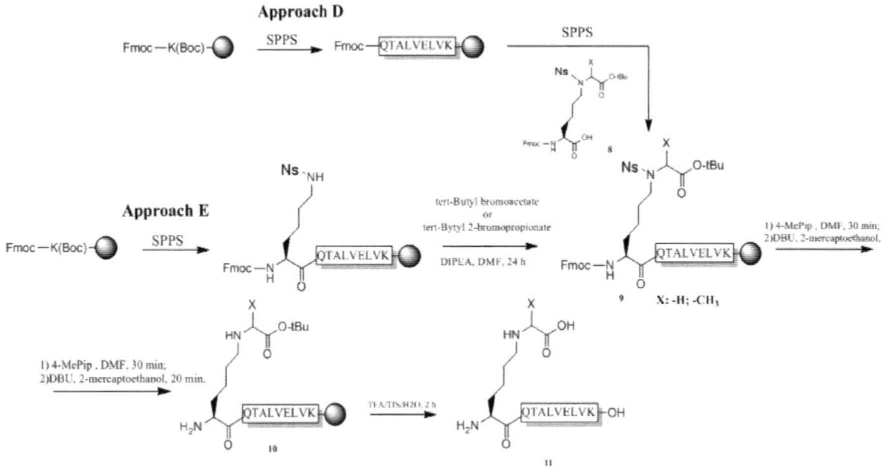

Scheme 2: Synthesis of peptides containing carboxymethylated (CM) and carboxyethylated (CE) Lys residues.

However, we could not obtain CE-modified peptides by both approaches, though the ε-CE-ε-Ns-α-Fmoc-Lys (10) was obtained and its structure was confirmed. A model experiment with the treatment of (8) with trifluoroacetic acid/water mixture 95%/5% (as for peptide detachment from the resin) showed that a possible elimination of carboxyethyl group from Lys ε-amino group during TFA treatment occurred (Scheme 3), and it didn't allow the preparation of carboxymethylated peptide by the on-resin modification.

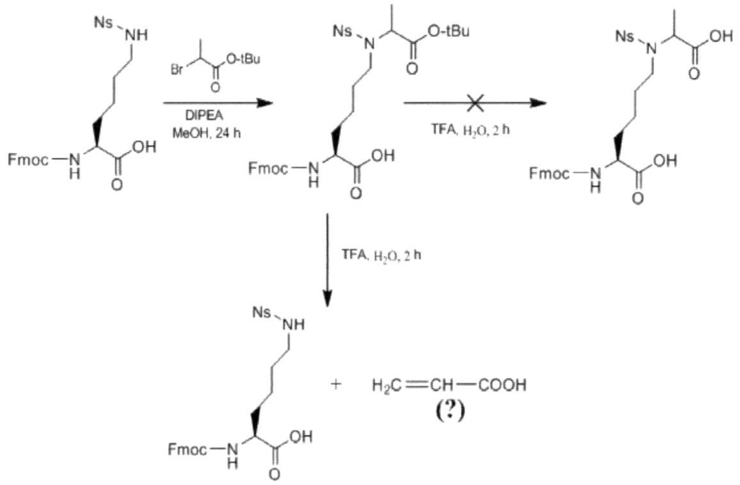

Scheme 3: Model experiment.

Results

1) Modifications of peptides by glucose in saturated solution resulted in low yields and a mixture of various glycation products difficult to separate selectively by HPLS.

2) Lys glycation in on-resin peptides was shown to be the most efficient procedure compared to a separate preparation of glycated Fmoc-Lys with its further application in SPPS and free peptide glycation.

3) Carboxyethyl group elimination from Lys ε-NH_2 group was observed during TFA treatment that did not allow the preparation of carboxyethylated peptides by on-resin modification.

4) The suitability of each studied approach depends on the type of glycation and the position of the modified Lys residue in peptide chains.

Acknowledgments

The study was performed in the framework of the Program for Basic Research of the Russian Academies of Sciences for 2013-2020. Peptide synthesis was done with the use of the equipment of the "Human Proteome" Core Facility of the Institute of Biomedical Chemistry (Moscow, Russia) supported by the Ministry of Education and Science of the Russian Federation (unique project ID RMEFI62117X0017).

References

1. A. Jakas *et al.* / Carbohydrate res. 343 (2008) 2475-2480 DOI:10.1016/j.carres.2008.07.003

2. P. Stefanowicz *et al.* / Tetrahedron Letters 48 (2007) 967-969 DOI:10.1016/j.tetlet.2006.12.022

3. M. Kamalov *et al.* / Synlett 25 (2014) 1835-1838 DOI:10.1055/s-0033-1378305

4. H. Kaur *et al.* / Acc. Chem. Res. 49 (2016) 2199–2208 DOI: 10.1021/acs.accounts.6b00366

4. T. M. Woods *et al.* /Org. Lett. Vol. 14, No. 22 (2012) 5740-5743 DOI:10.1021/ol302745f

5. A. M. Korwar *et al.* / Mol. & Cell. Proteomics 14.8, 201 DOI:10.1074/mcp.M115.050518

6. J. Anguizola *et al.* /Clin Chim Acta. (2013) October 21; 0: 64-76 DOI:10.1016/j.cca.2013.07.013

7. R.W.Lowe *et al.* /Carbohydrate res., 28 (1973) 281-293 DOI:10.1016/S0008-6215(00)82783-4

Proceedings of the 35th European Peptide Symposium
Patrick B. Timmons, Chandralal M. Hewage, Michal Lebl (Editors)
80 *European Peptide Society & PSP, 2018*

Synthesis and biochemical evaluation of GnRH-III-drug conjugates

Sabine Schuster[1], Beáta Biri-Kovács[1], Adina Borbély[2], Norbert Sewald[2], Ines Neundorf[3], Cesare Gennari[4], Gábor Mező[1]

[1]Eötvös Lorand University, Faculty of Science, Institute of Chemistry, Research Group of Peptide Chemistry, Budapest, Hungary
[2]University of Bielefeld, Department of Chemistry, Organic and Bioorganic Chemistry, Bielefeld, Germany
[3]University of Cologne, Institute of Biochemistry, Cologne, Germany
[4]Università degli Studi di Milano, Dipartimento di Chimica, Milano, Italy

https://doi.org/10.17952/35EPS.2018.080

1. Introduction

Gonadotropin releasing hormone-III(GnRH-III, <EHWSHDWKPG-NH_2; <E – pyroglutamic acid, isolated from sea lamprey), a native isoform of the human peptide hormoneGnRH-I, represents a promising starting point for the development of efficient peptide-based drug delivery systems (DDS) for targeted cancer therapy. Beneficially, GnRH-III specifically binds to GnRH receptors on cancer cells without revealing a significant endocrine effect [1]. Due to that, a variety of GnRH-III-drugconjugates have been designed and characterized in our laboratories, in which the anthracycline daunorubicin (Dau) was linked to GnRH-III *via* oxime bond by insertion of an aminooxyacetyl moiety [1]. To achieve an improved antitumoractivity, we synthesized a set of oxime-linked GnRH-III-Dau conjugates containing different unnaturalamino acids within the sequence and studied their *in vitro* anticancer activity [2,3]. The best compoundswere chosen for further biochemical evaluation and as targeting moiety of novel drug conjugates containing a self-immolative p-aminobenzyloxycarbonyl (PABC) spacer between a cathepsin B cleavable dipeptide (Val-Ala or Val-Cit) and the drug Dau or paclitaxel (PTX). For comparative purpose, non-cleavable GnRH-III-drug conjugates were also synthesized. All bioconjugates have been analyzed for their *in vitro* cytostatic effect and the release of the drug was followed by lysosomal degradation studies in presence of rat liver lysosomal homogenate. Furthermore, the highly efficient drug candidate cryptophycin was conjugated to a GnRH-III homing device by using different cleavable and non-cleavable linker systems and the corresponding *in vitro* antitumor activity of the compounds was determined.

2. Results and Discussion

2.1. Oxime bond containing GnRH-III-daunorubicinconjugates

All GnRH-III-[^{8}Lys(Dau=Aoa)] derivatives were prepared by Fmoc-SPPS, whereby Dau was conjugated in solution to an aminooxyacetic acid linker at the side chain of ^{8}Lys by formation of an oxime bond. The cytostatic effect was determined by Alamar blue® assay and compared to the control compoundsK1 and K2 (Scheme 1).

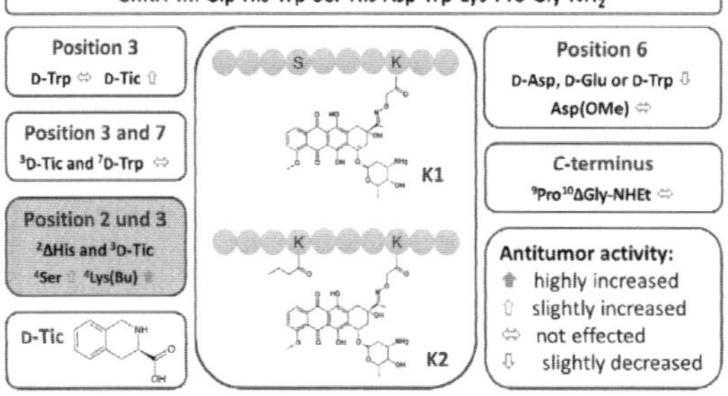

Scheme 1: Influence of sequence modification on the anticancer activity (on MCF-7 and HT-29 cancer cells) in comparison to the controls K1 and K2. D-Tic - (R)-1,2,3,4-tetrahydroisoquinoline-3-carboxylic acid.

All GnRH-III-Dau derivatives displayed an *in vitro* cytostatic effect on MCF-7 breast and HT-29 colon can-cer cells in a low micromolar range, whereby only compound P19-H (GnRH-III-[$^2\Delta$His-^{3}D-Tic-^{4}Lys(Bu)] - [8Lys(Dau=Aoa)]) displayed a highly improved antitumor activity on both human cancer cell lines (Figure 1A) [2,3]. To interpret these results, we performed additional studies of P19-H in direct comparison with our lead compound K2. Flow cytometry studies revealed an increased cellular uptake of compound P19-H on MCF7 (Figure 1B) and HT-29 cancer cells, whereby the uptake rate particularly at low compound concentrations was improved in comparison to K2. Additionally, receptor binding studies revealed similar binding affinities for both conjugates indicating that the receptor affinity is not retarded by amino acid substitution. Furthermore, we analyzed the release of the active drug metabolite in presence of lysosomal rat liver homogenate. The corresponding LC-MS data of P19-H and K2 showed that the N-terminus of P19-H was more resistant to lysosomal enzymes. However, the smallest Dau-containing metabolite H-Lys(Dau=Aoa)-OH was released within one hour in case of both conjugates. These outcomes confirmed that the release of the active metabolite is not decelerated by the N-terminal modification of P19-H, which is of great importance for the biological activity of the conjugate. Considering all findings, we assume that the improved anticancer activity of P19-H is mainly related to an enhanced cellular uptake [3].

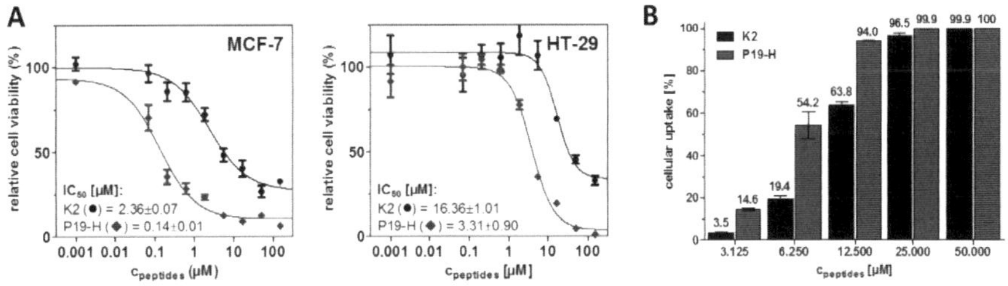

Figure 1: A) Anticancer activity of P19-H and K2 on MCF-7 and HT-29 human cancer cells after 24 h treatment and additional 48 h incubation. Curves obtained by non-linear regression (sigmoidal dose responds, n=2). B) Cellular uptake of the GnRH-III conjugates K2 and P19-H on MCF-7 cancer cells after 6 h treatment determined by flow cytometry. (Error bars represent standard deviation).

2.2 Self-immolative linker containing GnRH-III-daunorubicin and paclitaxel conjugates

Based on the promising results of K2 and P19-H, novel GnRH-III-PTX and Dau conjugates were developed by using the corresponding peptide sequence as targeting moiety. To ensure the release of the free drug, self-immolative linker systems with cathepsin B cleavable sites (Val-Ala or Val-Cit) were applied to link the drugs. The peptide carriers were synthesized by Fmoc-SPPS, while the dipeptidyl-PABC-drug linkers, as well as the non-cleavable linkers were synthesized in solution as recently described [4,5]. The cytotoxic effect of the compounds was investigated on GnRH-receptor positive A2780 ovarian carcinoma cells. The corresponding IC_{50} values of the Dau-conjugates emphasize that the novel targeting moiety (GnRH-III-[^{2}ΔHis-^{3}D-Tic-^{4}Lys(Bu)] has a beneficial impact on the antitumor activity with IC_{50} values 2.5-times lower than that of the GnRH-III-[^{2}His-^{3}Trp-^{4}Lys(Bu)] conjugates. Moreover, the outcomes indicate that the Val-Ala linker containing conjugates provide a slightly improved cytotoxic effect in comparison to the Val-Cit containing conjugates. This might be mainly related to the accelerated release of Dau. The corresponding lysosomal degradation studies pointed out that the Val-Ala spacer is cleaved already within the first 5 minutes, while the cleavage of the Val-Cit linker and release of free Dau was detected after one hour of incubation with rat liver lysosomal homogenate. On the contrary, the release of the free drug could not be detected in case of the non-cleavable linker containing conjugates, even after 24 hours incubation with lysosomal enzymes. Thus, it can be assumed that the decreased biological activity of these conjugates is related to the fact that the free Dau is not released, as well as to the fact that a modification of the amino group of the sugar moiety can lead to the loss of bioavailability [6]. Recent studies demonstrated a markedly decreased *in vitro* antitumor activity of GnRH-III conjugates by using the amino function of the daunosamine sugar for amide bond formation to a glutaryl-spacer [7].

Apart from that, the IC_{50} values of the PTX-conjugates display a similar anticancer activity for all cleavable conjugates which might be related to the releasing mechanism of the PTX. It has been reported that by using the PABC spacer in combination with the diamine cyclization linker, at first a diamine linker containing prodrug is released. The following cyclisation of the diamine-linker to 1,3-dimethyl-2-imidazolidinone and the corresponding release of PTX has been reported to be the rate-limiting step of the self-immolative process [4,8]. Taking this into account, it can be assumed that the stability of the prodrug and the release of the free PTX has a higher impact on the antitumor activity of the GnRH-III conjugates than the targeting sequence or the cathepsin cleavable dipeptide spacer. In contrast, the non-cleavable GnRH-III-PTX conjugates displayed a highly reduced cytotoxic effect on GnRH-receptor positive A2780 ovarian cancer cells which clearly demonstrates that the release of the cytotoxic payload is of high importance for the biological activity of the bioconjugates and mainly related to the dipeptidyl spacer in combination with the self-immolative linker system.

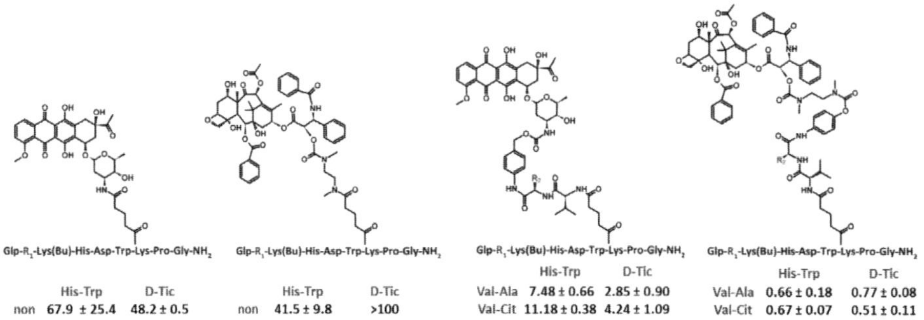

Figure 2: Structure of GnRH-III-Dau and PTX-conjugates and corresponding IC50 values [μM] on A2780 ovarian cancer cells. IC50 values were determined after 6 h (PTX) or 24 h (Dau) treatment and additional 66 h or 48 h incubation.

2.3 Cathepsin cleavable and non-cleavable GnRH-III-cryptophycin conjugates

The highly efficient cryptophycin analog cryptophycin-55-glycinate (Cry-55-Gly) was used for the development of a novel GnRH-III based drug delivery system. The antiproliferative activity of cryptophycins, is based on their ability to destabilize microtubules, whereby these highly cytotoxic cyclo-depsipeptides provide a much higher potency than other tubulin effectors like paclitaxel or vinblastine [9]. The functionalized Cry-55-glycinate, which is suitable for conjugation to peptide-based homing devices, was prepared by total chemical synthesis as recently reported [10]. Moreover, the Cry-55-glycinate was combined with three different alkyne functionalized linker systems to ligate the payload to an azide-containing GnRH-III targeting moiety by copper catalyzed azide–alkyne cycloaddition. The designed linker systems contain either a non-cleavable spacer or a cathepsin-cleavable Arg-Asp-Arg-Val-Cit peptide spacer either in combination with the PABC self-immolative moiety or without. The resulting GnRH-III-Cry-55-Gly conjugates were analyzed for their cytotoxic activity on GnRH-receptor positive A2780 ovarian carcinoma cells (Figure 3). The resulting IC_{50} values of all three conjugates are in a low nanomolar range indicating that the linker structure did not substantially affect the biological activity of the compounds. Due to these results, we assume that the activity of the GnRH-III-Cry-55-Gly conjugates is not related to a cathepsin mediated drug release. A possible explanation might be that the conjugation of Cry-55-Gly to the peptide carrier did not affect the microtubule destabilization or the related biological activity of the drug.

In conclusion, all GnRH-III-Cry-55-Gly conjugates display highly efficient antiproliferative activity on GnRH-receptor positive A2780 ovarian cancer cells, whereby the biological activity was approximately 20-times higher in comparison to the related GnRH-III PTX conjugates. This emphasizes the potential of GnRH-III-cryptophycin conjugates for targeted tumor therapy and encourage us to further investigate this approach.

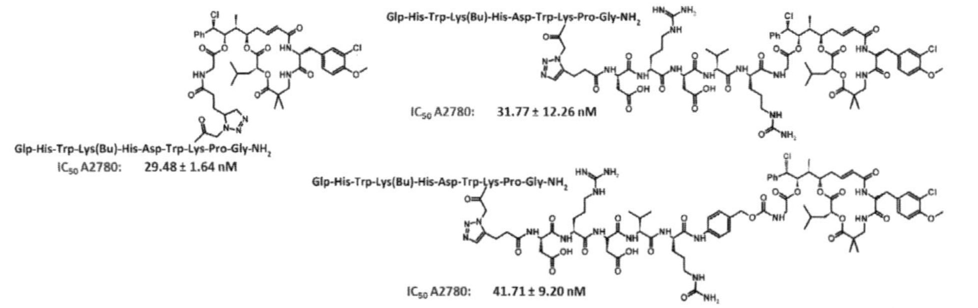

Figure 3: Structure of GnRH-III-Cryptophycin-conjugates and corresponding IC50 values. The cytotoxic effect was determined by cell-viability assay on A2780 ovarian cancer cells after 6 h treatment and additional 66 h incubation.

3. Conclusion

To study the influence of sequence modification in GnRH-III on the efficiency of tumor targeting, we synthesized and characterized 20 novel GnRH-Dau conjugates. The results reveal one highly promising candidate with a substantially improved antitumor activity in comparison to our lead compound. Thus, this compound was chosen for further investigation in vivo on tumor bearing mice. Moreover, the two most potent targeting sequences were successfully used for the development of novel GnRH-III-Dau and PTX conjugates containing two different cathepsin B labile dipeptides and a self-immolative moiety. The results of the cell viability assay on A2780 ovarian cancer cells confirm the beneficial impact of the novel targeting sequence on the anticancer activity of GnRH-III conjugates. Furthermore, we developed and evaluated three different GnRH-III-based cryptophycin conjugates. It was shown that all conjugates display IC_{50} values in a low nanomolar range, which further emphasizes the high potential of GnRH-III based drug conjugates for targeted tumor therapy.

This project has received funding from the European Union's Horizon 2020 research and innovation program under the Marie Sklodowska-Curie grant agreement No 642004, and from the National Research, Development and Innovation Office (NKFIH K119552 and NVKP_16-1-2016-0036), Hungary.

References

[1] G. Mező, et al., 2007, Peptides, 28, 806–820

[2] S. Schuster et al., 2018, Beilstein J. Org. Chem., 14, 756–771

[3] S. Schuster et al., 2018, Pharmaceutics (under revision)

[4] A. Dal Corso et al., 2015, Chem. Eur. J., 21, 6921-6929

[5] G.M. Dubrowchik et al., 2002, Bioconjugate Chem., 13, 855-869

[6] Y-G. Gao et al., 1991, Proc. Natl. Acad, Sci. USA, 88, 4845-4849

[7] G. Mező, et al., 2008, Current Medicinal Chemistry, 15, 2366-2379

[8] F.M.H. de Groot et al., 2001, J. Org. Chem., 66, 8815-8830

[9] C. Weiss et al., 2017, J. Pept Sci., 23, 514-531

[10] Cazzamalli et al., 2018, ACS Omega (accepted)

Proceedings of the 35th European Peptide Symposium
Patrick B. Timmons, Chandralal M. Hewage, Michal Lebl (Editors)
84 European Peptide Society & PSP, 2018

Convenient method of peptide hydrazide synthesis using a new hydrazone resin

Thomas Bruckdorfer[1], Pavel S. Chelushkin[2], Ksenia V. Polyanichko[2], Maria Leko[2], Marina Dorosh[2], Sergey V. Burov[2]

[1]Iris Biotech GmbH, Germany
[2]Institute of Macromolecular Compounds, St. Petersburg State University, Russia

https://doi.org/10.17952/35EPS.2018.084

Introduction

Peptide hydrazides are useful building blocks for the synthesis of conjugates with different carrier molecules such as proteins, dendrimers and polymers[1-3]. Attachment of peptide vectors by means of hydrazone ligation is widely applied for the design of targeted drug delivery systems. Recently it was shown that peptide hydrazides can be easily converted to corresponding isocyanates or thioesters[4,5]. These reactive species are especially useful for the conjugation with different nucleophiles, peptide cyclisation and synthesis of proteins using chemical ligation approach.

Hydrazinolysis of peptide esters can be accompanied by the formation of side products while their solid phase synthesis commonly demands application of rather expensive starting materials and/or multistage synthetic procedure. Previously we described simple synthetic protocol based on the acylation of aminomethyl polystyrene resin by the Fmoc-protected hydrazone of pyruvic acid[6]. In this study we present the optimized loading protocol and application of novel resin for the synthesis of linear and branched peptides.

Results and Discussion

The ability of pyruvic acid to form relatively stable hydrazones in reasonable yields determines their utility as anchoring groups for peptide hydrazides synthesis using Fmoc/Bu^t strategy. Aminomethyl polystyrene (AMPS) resin, widely applied for solid phase synthesis, seems to be a cheap and convenient starting material for subsequent modification.

Investigation of different protocols for the synthesis of hydrazone resin has shown the utility of AMPS polymer acylation by the Fmoc-hydrazone of pyruvic acid. Attachment of a preformed hydrazone linker ensures high loading level along with the possibility of its control by UV spectroscopy at the stage of Fmoc group removal.

It should be mentioned that attachment of the first amino acid using different coupling agents (DIC, HCTU or SA) is accompanied by partial loss of polymer capacity. These results cannot be explained by incomplete acylation or partial decomposition of the hydrazone resin and the nature of this process is not completely clear.

In the course of loading protocol optimization it was shown that application of high capacity AMPS resin (1.2 mmol/g) permits to achieve the desirable loading level (up to 0.9 mmol/g) using only 1 eqv of acylating agent. However, the standard procedure of residual amino groups capping (Ac_2O/DIEA) seems to be inefficient. Moreover, acetylation in DMF resulted in remarkable decrease of loading level. An optimized capping protocol presumes application of Ac_2O/DIEA/DCM mixture and acetylation in swelling volume during 1-2 h.

Attachment of the first amino acid demands thorough control of acylation completeness and proper selection of reaction conditions. In general, HCTU coupling increases the acylation yield, especially in the case of bulky amino acid derivatives as compared to DIC/Cl-HOBt protocol.

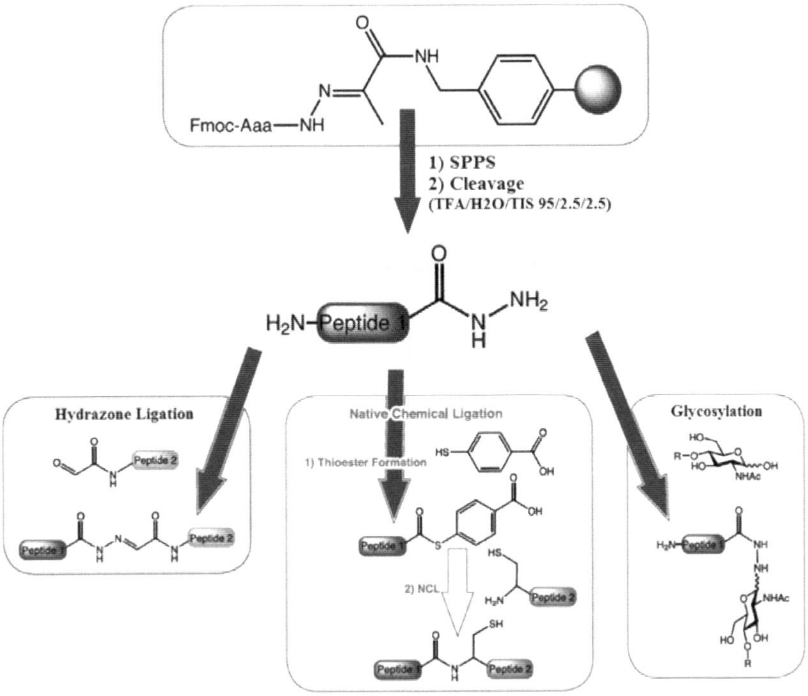

Figure 1: Application of hydrazone resin.

It was shown that the hydrazone bond is completely stable in the course of Fmoc group removal and DIC mediated coupling in the presence of acidic additives such as HOBt. Moreover, it is tolerant to treatment with 5% TFA/DCM for 3 h thus permitting selective removal of Mtt and related acid-labile groups.

To verify the possibility of selective Mtt deprotection we synthesized model peptide sequence: Fmoc-Arg(Pbf)-Gly-Asp(OBut)-D-Phe-Lys(Mtt) attached to the hydrazone resin. Then Mtt group was removed by repeating treatment with 5% TFA and 5% TIS in DCM (5 x 2 min). Subsequent acylation of the ε-amino group with 5(6)-carboxyfluorescein (Flu) followed by cleavage/deprotection procedure resulted in the desired peptide: H-Arg-Gly-Asp-D-Phe-Lys(Flu)-N_2H_3.

The practical utility of the suggested resin was proved by the synthesis of different peptides including the nuclear localization signal (NLS) of large T-antigen of SV-40 virus, tuftsin, Szeto-Schiller peptide, and building blocks for the design of self-assembling biomaterials containing RADA units.

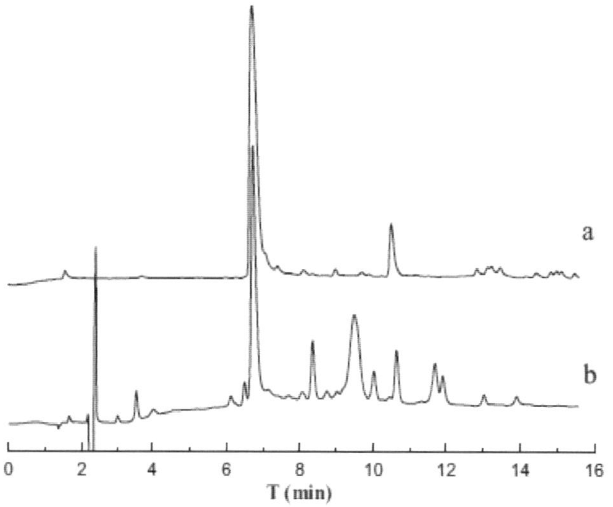

Figure 2: RP-HPLC analysis of crude NLS hydrazide synthesized using hydrazone resin (a) and protected peptide precursor (b).

The advantages of new resin can be illustrated by the synthesis of NLS hydrazide as compared to HCTU-mediated acylation of BOC-hydrazide by protected peptide precursor synthesized on Cl-Trt polymer (Fig. 2). It should be mentioned that application of hydrazone resin permits to improve both purity of crude peptide and the yield of final product (55% as compared to 19% in the case of BOC-hydrazide acylation).

In conclusion, the suggested approach permits to prepare peptide building blocks containing a hydrazide group useful for the conjugation with different carrier molecules using hydrazone ligation technique.

References

1. Brunel F.M., Lewis J.D., Destito G., Steinmetz N.F., Manchester M., Stuhlmann H., Dawson P.E.

Nano Lett. 2010, 10, 10931097. DOI: 10.1021/nl1002526.

2. Shao, J., Tam, J. P. J. Am. Chem. Soc. 1995, 117, 38933899. DOI: 10.1021/ja00119a001.

3. Gauthier M.A., Klok H.-A. Chem. Commun. 2008, 21, 2591–2611. DOI: 10.1039/b719689j.

4. Vinogradov, A.A., Simon, M.D., Pentelute B.L. Org. Lett. 2016, 18, 1222–1225.

DOI: 10.1021/acs.orglett.5b03625.

5. Huang, Y.-C., Fang, G.-M., Liu L. Natl. Sci. Rev. 2016, 3, 107-116. DOI: 10.1093/nsr/nwv072.

6. Chelushkin P.S., Polyanichko K.V., Leko M.V., Dorosh M.Yu., Bruckdorfer T., Burov S.V.

Tetrahedron Lett. 2015, 56, 619-622. DOI: 10.1016/j.tetlet.2014.12.056.

Peptide couplings by reactive extrusion: Efficient, continuous and green

Thomas-Xavier Métro[1], Yves Yeboue[1], Benjamin Gallard[2], Nicolas Le Moigne[2], Jean Martinez[1], Frédéric Lamaty[1]

[1]Institut des Biomolecules Max Mousseron, France
[2]IMT Mines Ales, France

https://doi.org/10.17952/35EPS.2018.087

Therapeutic peptides exhibit a wide array of advantageous characteristics that place them among the most promising active pharmaceutical ingredients.[1] Yet, industrial production of peptides is hampered by the large amounts of toxic organic solvents that are required during the synthesis and purification steps.[2-3] Although presenting highly concerning environmental issues, polar aprotic solvents such as DMF and NMP are regularly used.[4] These problematic solvents could be avoided since it is known that solvent-less/solvent-free synthesis of small peptides is possible by using ball-milling.[5-8] Enabling to produce 4g of a dipeptide for the best case,[9] this approach was not demonstrated to be further scalable. In addition, the ball-milling process was limited to discontinuous batch production, thereby hampering a wide utilisation by the peptide industry.

In this work, we envisioned overcoming these hurdles by using reactive extrusion. Extruders are composed of a barrel containing one or two rotating screws enabling the efficient transport and mixing along the barrel. Of note, the presence of a recirculation pipe can be utilised to increase the residence time and hence the mixing time of the material to be extruded (Figure 1).

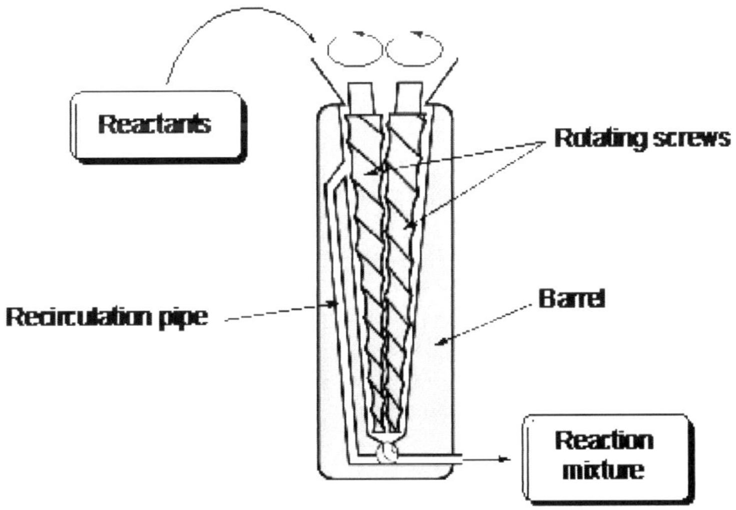

Figure 1: Schematic representation of a twin-screw extruder.

This equipment enables to work under continuous flow conditions while drastically reducing the amount of solvent required for contacting the reactants (if not completely discarding). On the contrary to classical solution-based flow chemistry, extruders allows for the efficient mixing of highly concentrated mixtures that contain a high proportion of solids. Although widely used in the food and plastic industry and identified as a key research area by the pharmaceutical industry,[10] production of high added-value chemicals by reactive extrusion has been scarce. For our part, we first studied the capacity of a co-rotating twin-screw extruder to produce the dipeptide Boc-Trp-Gly-OMe. After a thorough screening of reaction conditions, it appeared that the peptidic bond of Boc-Trp-Gly-OMe could be formed by recirculating inside a twin-screw extruder a mixture composed of Boc-Trp-OSu (1.0 eq.), HCl.H-Gly-OMe (1.1 eq.) and $NaHCO_3$ (1.2 eq.) along with a small amount of acetone (1.5 mL for a total mass of 10g of reactants) while operating the extruder at 40 °C with a screw speed set 150 rpm.

Table 1: Synthesis of dipeptides and tripeptides by using a twin-screw extruder.

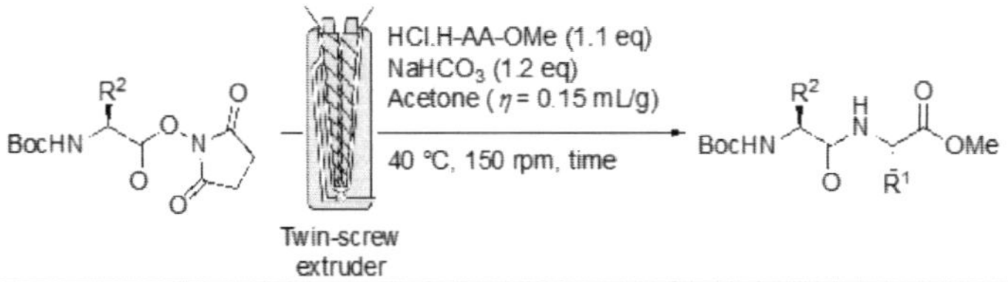

Entry	Boc-AA-OSu	HCl H-AA-OMe	Mixing time (min)	Yield (%)	Purity (%)	Product
1	Trp	Gly	10	85	>99[a]	Boc-Trp-Gly-OMe
2	Trp	Phe	10	61	>99[a]	Boc-Trp-Phe-OMe
3	Asp(OBzl)	Phe	1.5[b]	92	>99[a]	Boc-Asp(OBzl)-Phe-OMe
4	Asp(OBzl)	Trp-Gly	10	86	96[c]	Boc-Asp(OBzl)-Trp-Gly-OMe
5	Asp(OBzl)	Trp-Phe	10	89	94[c]	Boc-Asp(OBzl)-Trp-Phe-OMe

[a] >99% ee or de determined by chiral HPLC. [b] Reaction mixture was extruded without recirculation. Residence time of 1.5 min. [c] Purity determined by HPLC.

As such an amount of acetone is not sufficient to completely solubilize the reactants, acetone would be more appropriately described here as a liquid additive than a solvent. After EtOAc solubilisation and aqueous washings of the extrudate, Boc-Trp-Gly-OMe was isolated in 85% yield and >99% enantiomeric excess (Table 1, entry 1). When HCl.H-Gly-OMe was replaced with HCl.H-Phe-OMe, Boc-Trp-Phe-OMe could be produced in 61% yield and >99% diastereomeric excess (Table 2, entry 2). To our delight, Boc-Asp(OBzl)-Phe-OMe could be formed without recirculating the reaction mixture in the extruder: Boc-Asp(OBzl)-OSu was fully converted after the 1.5 min residence time inside the barrel, leading to the production of Boc-Asp(OBzl)-Phe-OMe in 92% yield and >99% de (Table 2, entry 3). Tripeptides could also be formed by using this approach. Dipeptides hydrochlorides HCl.H-Trp-Gly-OMe and HCl.H-Trp-Phe-OMe were first synthesized by solvent-free gaseous HCl treatment. After being reacted with both NaHCO3 and Boc-Asp(OBzl)-OSu in the extruder, Boc-Asp(OBzl)-Trp-Gly-OMe and Boc-Asp(OBzl)-Trp-Phe-OMe were isolated in 86% and 89% yield and excellent purity (96% and 94% respectively; Table 1, entries 4 & 5). As an illustration of the potential application of this strategy to the production of industrially-relevant peptides, Boc-Asp(OBzl)-Phe-OMe was transformed into the renowned sweetener Aspartame by hydrogenation, Boc removal under solvent-free acidic conditions before final precipitation at the isoelectric point. By doing so, Aspartame was produced in 81% yield in three steps starting from Boc-Asp(OBzl)-OSu.

In conclusion, dipeptides and tri-peptides can be produced in high yields, high stereoisomeric excesses and very short reaction times by using reactive extrusion.[11] The capacity to implement synthesis under continuous conditions is clearly setting the path to intensified industrial peptide production. On the contrary to synthesis in solvent-based continuous flow and in SPPS, the presence of solids could be very easily handled by the mechanical forces occurring in the extruder. Thus, this strategy enabled to work under highly concentrated reaction conditions, while avoiding the use of highly problematic solvents and bases (such as DMF and Et_3N).

Acknowledgements

Université de Montpellier, Centre National de la Recherche Scientifique (CNRS), and LabEx CheMISyst (through ANR programme ANR-10-LABX-05-01) are acknowledged for financial support.

References

(1) Lau, J. L.; Dunn, M. K. Therapeutic peptides: Historical perspectives, current development trends, and future directions. Biorg. Med. Chem. 2017, 26, 2700-2707, DOI: 10.1016/j.bmc.2017.06.052.

(2) Thayer, A. M. Making Peptides At Large Scale. Chem. Eng. News 2011, 89, 21-25.

(3) Patel, P. Executive reveals how Amgen shifted to greener chemistry. Chem. Eng. News 2017, 95, 27-28.

(4) According to Regulation (EC) No 1272/2008, DMF is classified as presenting reproductive toxicity and DCM , THF and 1,4-dioxane as toxic and/or carcinogenic (Category 2).

(5) Declerck, V.; Nun, P.; Martinez, J.; Lamaty, F. Solvent-Free Synthesis of Peptides. Angew. Chem. Int. Ed. 2009, 48, 9318-9321, DOI 10.1002/anie.200903510.

(6) Maurin, O.; Verdié, P.; Subra, G.; Lamaty, F.; Martinez, J.; Métro, T.-X. Peptide synthesis: ball-milling, in solution, or on solid support, what is the best strategy? Beilstein J. Org. Chem. 2017, 13, 2087-2093, DOI 10.3762/bjoc.13.206.

(7) Porte, V.; Thioloy, M.; Pigoux, T.; Métro, T.-X.; Martinez, J.; Lamaty, F. Peptide Mechanosynthesis by Direct Coupling of N-Protected α-Amino Acids with Amino Esters. Eur. J. Org. Chem. 2016, 3505-3508, DOI 10.1002/ejoc.201600617.

(8) Bonnamour, J.; Métro, T.-X.; Martinez, J.; Lamaty, F. Environmentally benign peptide synthesis using liquid-assisted ball-milling: application to the synthesis of Leu-enkephalin. Green Chem. 2013, 15, 1116-1120, DOI 10.1039/c3gc40302e.

(9) Métro, T.-X.; Colacino, E.; Martinez, J.; Lamaty, F. In Ball Milling Towards Green Synthesis: Applications, Projects, Challenges, The Royal Society of Chemistry: 2015, 114-150, DOI 10.1039/9781782621980-00114.

(10) Jiménez-González, C.; Poechlauer, P.; Broxterman, Q. B.; Yang, B.-S.; am Ende, D.; Baird, J.; Bertsch, C.; Hannah, R. E.; Dell'Orco, P.; Noorman, H.; Yee, S.; Reintjens, R.; Wells, A.; Massonneau, V.; Manley, J. Key Green Engineering Research Areas for Sustainable Manufacturing: A Perspective from Pharmaceutical and Fine Chemicals Manufacturers. Org. Process Res. Dev. 2011, 15, 900-911, DOI 10.1021/op100327d.

(11) Yeboue, Y.; Gallard, B.; Le Moigne, N.; Jean, M.; Lamaty, F.; Martinez, J.; Métro, T.-X. Peptide couplings by reactive extrusion: solid-tolerant and free from carcinogenic, mutagenic and reprotoxic chemicals. ACS Sustainable Chem. Eng. 2018, in press, DOI: 10.1021/acssuschemeng.8b04509.

Therapeutic peptides: Synthesis optimization using parallel automated SPPS

Andrew Kennedy[1], James Cain[2], Cyf Ramos-Colon[2], Daniel Martinez[2]

[1]Gyros Protein Technologies, Sweden
[2]Gyros Protein Technologies, United States

https://doi.org/10.17952/35EPS.2018.090

Process development for the manufacture of therapeutic peptides remains a vital part of the commercial peptide production process. An optimized solid-phase synthesis protocol can be challenging to develop. Many of the recent advances in peptide therapeutics focus on peptides with greater structural complexity compared to naturally occurring peptides, allowing for the design of more physiologically stable products with increased target specificity and membrane permeability [1]. The advent of the neoantigen and personalized medicine sectors adds a further requirement for faster peptide synthesis protocols and reduced processing times, including adherence to regulatory guidelines and GxP requirements during manufacture.

Solid-phase peptide synthesis (SPPS) of linear peptides using heat (>50°C) and highly reactive coupling reagents have resulted in products with increased crude purities and yields when using significantly shorter coupling cycles compared to room-temperature methods[2]. Development of automated solid-phase (on-resin) protocols for cyclization, potentially using heat, may provide numerous advantages for the development of complex peptide products [3]. Parallel automated synthesis is a useful tool for quickly and efficiently synthesizing multiple peptide analogs simultaneously for structure-activity relationship (SAR) studies of synthetically challenging sequences.

Solid-support screening, reagent screening and temperature screening are demonstrated using an automated peptide synthesizer as part of the optimization process for difficult peptides such as Aib-Enkephalin (Aib-Enk) and Jung-Redemann (JR 10-mer). A number of automated solid-phase protocols are also shown to be more efficient replacements for traditionally manual or solution-phase manipulations, resulting in reduced processing times and high crude product purities and yields. We report the use of heat to increase the crude purity of Melanotan II peptide from 43% to 60% using solely on-resin synthetic methods.

Table 1: Synthesized peptide sequences.

Peptide	Sequence
Melanotan II (MT-II) Sequence	Ac-Nle-cyclo[Asp-His-D-Phe-Arg-Trp-Lys]-NH_2
Jung-Redemann Sequence	H-Trp-Phe-Thr-Thr-Leu-Ile-Ser-Thr-Ile-Met-NH_2
Aib-Enkephalin Sequence	H-Tyr-Aib-Aib-Phe-Leu-NH_2

Methods

Protocol A: The MT-II linear peptides were synthesized at 100 μmol scale using Rink Amide resin (loading 0.22 mmol/g). Deprotection was performed with 20% piperidine in DMF 2 x 2 min at 25°C. Couplings were performed at a final concentration of 100 mM AA (5 eq.), 100 mM COMU (5 eq.) and 200 mM DIPEA (10 eq.) for 30 min at 25°C. Alloc and Allyl side chain protection was used for Lys and Asp. Cyclization: Following Pd-mediated removal of the side chain protecting groups and washing, a solution of PyClock (5 eq) and DIEA (10 eq) in DMF was added to the resin. After cyclization, the resin was washed with DMF and DCM.

Protocol B: JR10 and Aib Enkephalin were synthesized using Rink Amide ChemMatrix Resin (0.47mmol/g substitution) at a 50 umol scale. Deprotection time was 2 minutes at 25°C, 60°C, or 90°C using 20% Piperidine in DMF. Amino acids were coupled using a six-fold excess and final concentration of 100mM for Amino Acids and Activators/Additive and 200mM for DIEA. Coupling time was 3 min at 25°C, 60°C, or 90°C. DMF washes followed both deprotect and coupling steps with 3 repetitions at 30 seconds each.

Cleavage: Final cleave used TFA:TIS:EDT:Water (95:1:2.5:2.5) for 2 hours at 25°C.

Analysis: MT-II was analyzed on a U3000 Thermo HPLC using a C18, 180 Å, 5 um, 250 x 4.6 mm column (Agilent Polaris), over 30 minutes with a flow rate of 1 mL/min, and using a gradient of 5-95% B, where Buffer A is 0.1% TFA in water, and Buffer B is 0.1% TFA in acetonitrile. Detection was at 214 nm. Mass analysis was performed on a Shimadzu LCMS-2020 Single-Quad mass spectrometer, equipped with a C18, 100 Å, 2.6 um,

50 x 2.1 mm column (Phenomenex Kinetex), over 7 min with a flow rate of 1 mL/min and using a gradient of 5-50% B where Buffer A is 0.1% formic acid in water and Buffer B is 0.1% formic acid in acetonitrile.

JR10 and Aib enkephalin prepared using Protocol A were analyzed using a Varian Microsorb MW 300-5 C18 50x4.6 mm column with a gradient of 5-95%B in 15 min using Water (0.1%TFA):ACN(0.1%TFA) at 1ml/min. A 1:10 dilution of a standard sample of 3 mg/ml was run on a Phenomenex Kinetex 2.6 um C18 100A 50x2.1 mm column for LCMS with a gradient of 5-95%B in 15 min for JR10 and 10-20%B in 9 min for Aib-Enkephalin using Water(0.1%FA):ACN(0.1%FA) at 1 ml/min.

Results

MTII

On resin cyclization is typically done using methods of 2-18 h, here we show that by increasing the temperature, the cyclization of MTII is possible in 30 min with high crude purity. Increasing the temperature to 50°C resulted in a crude purity of 60.1% compared to 42.7% done at 25°C with a longer reaction time of 1h.

Table 2: Crude purity of cyclic MTII.

Condition	% Crude Purity
1 h @ 25°C	42.7
30 min @ 50°C	60.1

JR 10-mer and Aib-enkephalin

JR 10-mer and Aib-enk were synthesized using different coupling reagent combinations at three different temperatures (Table 3). Regardless of coupling reagent combination an increase in crude purity can be seen as the temperature is increased. COMU produced the best result on the synthesis of both JR-10mer (66.9%) and Aib-Enk (89.3%).

Table 3: Effect of coupling reagents and temperature on the crude purities of JR 10-mer and Aib-enk.

Reagent	JR 10-mer % Crude Purity			Aib-Enk % Crude Purity		
	25°C	60°C	90°C	25°C	60°C	90°C
HDMC/OxymaPure	20.8	52.9	59.8	29.2	54.2	71.3
HDMC	21.6	44.9	51.3	21.3	38.1	56.9
COMU	23.5	55.7	66.9	49.9	78.5	89.3

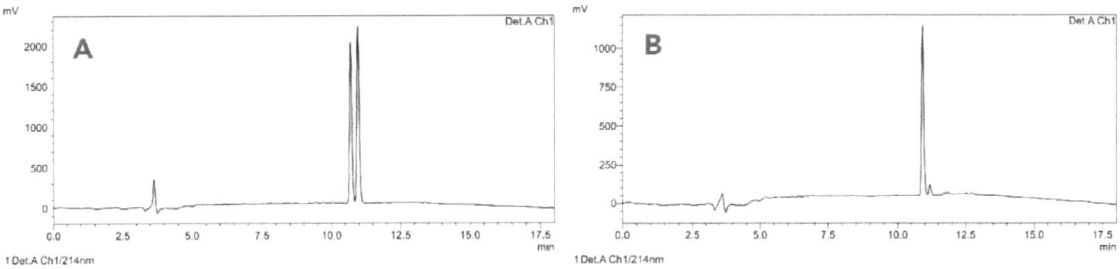

Figure 1: Crude purity profiles of Aib-Enk synthesized with COMU at A) 25°C and B) 90°C.

Conclusions

Difficult peptide sequences (JR 10-mer and Aib-Enk) and a cyclic peptide (MT-II) were successfully prepared in high purity with short synthesis times using automated peptide synthesizers. Independent induction heating on the Prelude X allowed multiple temperatures to be screened simultaneously on the synthesis of JR 10-mer and Aib-Enk. Purities of both difficult sequences improved with an increase in temperature up to 90°C. With MT-II, induction heating achieved a significant decrease of cyclization time without compromising peptide crude purity. Efficient cyclization can be obtained from 30 min cyclization reaction times. Finally, multi-variable conditions were effectively tested in parallel for the process optimization of JR 10-mer, Aib enkephalin, and MT-II syntheses.

References

1) S. H. Joo, "Cyclic peptides as therapeutic agents and biochemical tools," Biomol. Ther., 20, 1,19-26, 2012.

2) T. Redemann and G. Jung, Proc. 24th Eur. Pep. Symp., 1996, 749.

3) N. Tsomaia, "Peptide therapeutics: Targeting the undruggable space", 2015.

Chemical synthesis of venom peptides using directed-disulfide bond formation

Claude Zoukimian[1], Didier Boturyn[2]

[1]Département de Chimie Moléculaire, UMR CNRS 5250 - Smartox Biotechnology, France
[2]Département de Chimie Moléculaire, UMR CNRS 5250, France

https://doi.org/10.17952/35EPS.2018.093

Abstract

We herein describe a method to direct disulfide bond formation of peptides containing up to three disulfide bridges. This method is based on the use of several protecting groups of cysteine. The selective and sequential formation of two disulfide bridges was first developed and optimized on the conotoxin analogue NMB-1, allowing formation of disulfide bonds in a one-pot fashion. We finally successfully applied this strategy to the synthesis of the spider toxin SGTx1 that contains 3 disulfide bridges.

Introduction

Venoms are the result of thousands of years of evolution that allowed venomous species to develop complex mixtures of peptides and proteins known as toxins. Such molecules have received a growing interest due to their potential as drugs in the medical field. These toxins are particularly well folded due to the high abundance of cysteines. Usually, toxins are first isolated from venoms, but the low quantity of material that can be extracted remains a major hurdle to overcome. To address this issue, chemical peptide synthesis is needed, and the key step is then to correctly reproduce the fold of the native biomolecule. Random folding using oxidative buffers is the easiest way to fold a cysteine-containing peptide, but generally lead to a complex mixture containing undesired products.[1]

We herein describe a chemical scheme to direct disulfide bond formation that involves the introduction of cysteines containing orthogonal protecting groups during solid phase peptide synthesis (SPPS). Among the wide range of commercially available protected cysteines, we selected three protecting groups: trityl (Trt), acetamidomethyl (Acm) and methoxybenzyl (Mob). Trityl is generally used as protecting group of cysteine for Fmoc based SPPS. This acid labile group is removed using TFA and gives a pair of unprotected cysteines that can be oxidized. This group determines the first disulfide bond to be formed. The Acm group is removed using oxidative agent like iodine,[2] which forms the second disulfide bridge by a concerted mechanism. The Mob group is deprotected using trifluoromethanesulfonic acid (TFMSA) before oxidation.[3] The strategy was first developed for the synthesis of NMB-1,[4] an analogue of conotoxin containing two disulfide bonds, using Trt and Acm groups, and adapted by adding the Mob group to the synthesis of SGTx1, toxin isolated from the spider *Scodra Griseipes* , that contains three disulfide bonds.[5]

Materials and Methods

Peptide syntheses. SPPS was carried out on an automated Symphony synthesizer from Gyros Protein Technologies using standard Fmoc/t-Bu chemistry with HCTU as coupling reagent (0.1 mmol scale).

General protocol for directed-disulfide bond formation of NMB-1 in a one-pot fashion. After SPPS, crude NMB-1 was dissolved in H_2O/MeCN (1:1) at 10 mg/mL and added dropwise to a solution containing 0.1M citric acid, 2M Gn.HCl and 20% DMSO, at pH 7, to a final concentration of 0.1mg/mL in peptide. After one night under gentle stirring, pH was adjusted to 1-2, and 1 eq. of iodine 50mM in MeCN was added every five minutes, for a total of four additions. Five minutes after the last addition, the excess of iodine was quenched with sodium ascorbate and the solution was filtered and purified by preparative HPLC. Isolated yield: 57% over two steps. MS calculated 2465.17; found 2465.18

General protocol for directed-disulfide bond formation of SGTx1. After SPPS, crude SGTx1 was dissolved in H_2O/MeCN (1:1) at 10 mg/mL and added dropwise to a solution containing 0.1M citric acid and 10% DMSO, at pH 7, to a final concentration of 0.1mg/mL in peptide. After one night under gentle stirring, pH was adjusted to 1-2, and 1 eq. of iodine 50mM in MeCN was added every five minutes, for a total of four additions. Five minutes after the last addition, the excess of iodine was quenched with sodium ascorbate and the solution was filtered and purified by preparative HPLC. The freeze dried peptide was dissolved in TFA/phenol at 0°C and TFMSA was added to reach a concentration of 5mg/mL of peptide in TFMSA/phenol/TFA (1:1:8). The mixture was stirred for 10 min at 0°C and then the peptide was precipitated with ice-cold diethyl ether, recovered by centrifugation and washed twice with diethyl ether. The peptide was dissolved in H2O/MeCN (1:1) at 10 mg/mL and added dropwise to a solution containing 0.1M citric acid and 10% DMSO, at pH 2, to a final concentration of 0.1mg/mL in peptide. After 48h, the solution was filtered and purified by preparative HPLC. Isolated yield: 3% based on resin loading. MS calculated 3773.58; found 3773.59

Results and Discussion

To develop our strategy, we first decided to work on NMB-1. We synthesized this toxin with a pair of cysteine (C5 and C19) protected by Trt groups and the other one (C_6 and C_{10}) by Acm groups, according to the native pattern.[4] The position of Acm groups was judiciously chosen as several side reactions[6] can occur during SPPS such as the migration of Acm on an adjacent tyrosine, or the formation of aspartimide promoted by the sequence Asp(OtBu)-Cys(Acm). The formation of each disulfide bond was achieved sequentially with an intermediate purification. The first one was formed in a Tris buffer (pH 8) containing DMSO as oxidative agent and guanidinium as denaturantand in diluted conditions avoiding intermolecular disulfide bonds. The second one was then formed in an acidic buffer to avoid scrambling of disulfide bonds which can lead to peptide misfolding.[7] After several attempts, the combination of citrate buffer and iodine in presence of guanidinium proved to be better than other acidic solutions (acetic acid, TFA, formic acid…). To increase the overall yield, we decided to use citric acid as buffer for both steps. First disulfide bond was formed in citrate buffer (pH 7) containing DMSO as oxidative agent and guanidinium as denaturant. The complete formation of disulfide bridge was monitored by HRMS. Interestingly, the presence of DMSO did not interfere with Acm deprotection. NMB-1 was obtained after a single RP-HPLC purification improving the isolated overall yield from 31% to 57% (Scheme 1).

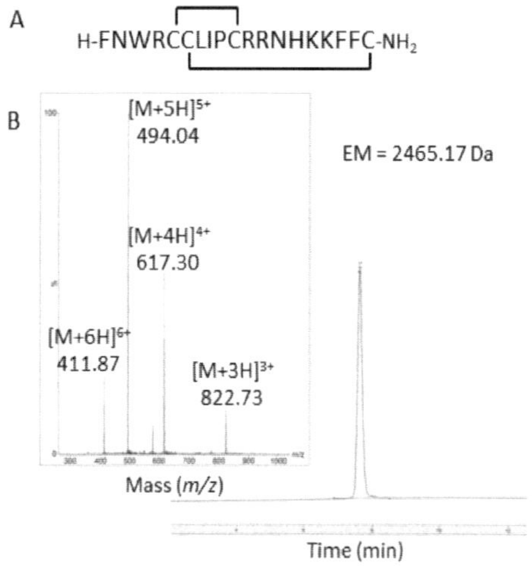

Scheme 1: (A) Sequence and disulfide pattern of NMB-1 ; (B) Analytical RP-HPLC (λ = 214 nm) and ES-MS of purified NMB-1. Observed mass 2465.18 Da vs calculated mass 2465.17 Da

We then used this strategy to prepare SGTx-1 toxin that contains three disulfide bonds. As problematic sequences (Tyr-Cys, Asp-Cys) for Acm group are present, we chose to protect cysteines, according to the native pattern describe in the literature,[5] as followed: C_2 and C_{16} with Trt, C_9 and C_{21} with Acm, C_{15} and C_{28} with Mob. The first two disulfide bridges were formed as described previously, then, Mob groups were cleaved with TFMSA to give the last cysteine pair that was finally oxidized in cystine in citrate buffer at pH 2 containing DMSO. The final product was isolated by RP-HPLC (Scheme 2).

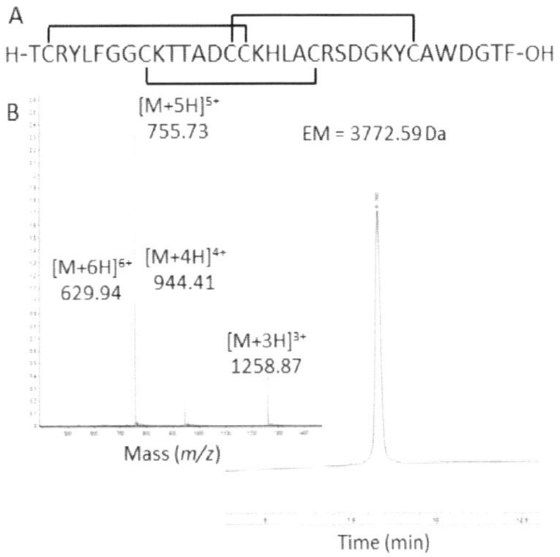

Scheme 2: (A) Sequence and disulfide pattern of SGTx1 ; (B) Analytical RP-HPLC (λ = 214 nm) and ES-MS of purified SGTx1. Observed mass 3772.59 Da vs calculated mass 3772.58 Da

Conclusion

Herein, we report a convenient strategy to direct disulfide formation of three cysteine pairs by using Trt, Acm and Mob groups. This approach has been applied with success to relevant toxins. Interestingly, regioselective formation of two disulfide bonds was carried out in one-pot, avoiding fastidious and time-consuming purification of intermediates.

References

1. Arolas JL, Aviles FX, Chang JY, Ventura S. Folding of small disulfide-rich proteins : clarifying the puzzle Trends Biochem. Sci. 2006; 31: 292-301

2. Kamber B, Hartmann A, Eisler K, Riniker B, Rink H, Sieber P, Rittel W. The synthesis of cystine peptides by iodine oxidation of S-trityl-cysteine and S-acetamidomethyl-cysteine peptides Helv. Chim. Acta. 1980; 63: 899-915

3. Akabori, S.; Sakakibara, S.; Shimonishi, Y.; Nabuhara, Y. A New Method for the Protection of the Sulfhydryl Group during Peptide Synthesis. Bull. Chem. Soc. Jpn. 1964; 37: 433-434.

4. Drew LJ, Rugiero F, Cesare P, Gale JE, Abrahamsen B, Bowden S, *et al.* High-Threshold Mechanosensitive Ion Channels Blocked by a Novel Conopeptide Mediate Pressure-Evoked Pain. PLoS ONE 2 2007; 2: e515

5. Marvin L, De E, Cosette P, Gagnon J, Molle G, Lange C. Isolation, amino acid sequence and functional assays of SGTx1 Eur. J. Biochem. 1991; 265: 572-579

6. Isidro-Llobet A, Alvarez M, Albericio F. Amino Acid-Protecting Groups Chem. Rev. 2009; 109: 2455-2504

7. Szabó I, Schlosser G, Hudecz F, Mező G. Disulfide bond rearrangement during regioselective oxidation in PhS(O)Ph/CH3SiCl3 mixture for the synthesis of α-conotoxin GI Biopolymers 2007; 88: 20-28

Synthesis of RGD peptides inducing multicellular spheroids formation

Maria Leko[1], Anna Pokhvoshcheva[1], Marina Dorosh[1], Roman Akasov[2], Elena Markvicheva[2], Markus Weishaupt[3], Thomas Bruckdorfer[3], Sergey Burov[1]

[1]Medical-Biological Research-Industrial Complex CYTOMED, Russia
[2]Shemyakin-Ovchinnikov Institute of Bioorganic Chemistry, RAS, Russia
[3]Iris Biotech GmbH, Germany

https://doi.org/10.17952/35EPS.2018.096

Introduction

Availability of physiologically relevant *in vitro* models is one of the most important requirements determining the efficiency of drug screening process. Nowadays 2D monolayer cell culture represents one of the most employed *in vitro* methodologies for drug development. However, this simplified model with lack of cell-cell interactions is not able to adequately mimic the *in vivo* response.

Sutherland et al.[1] were first to propose multicellular tumor spheroids as a 3D model of small solid tumors. In contrast to 2D cell culture, 3D spheroids are able to mimic some features of solid tumors, including their spatial organization, physiological responses and drug resistance mechanisms. In the case of normal cells, multicellular spheroids (MS) are especially useful for the screening of neuroprotective agents, 3D bioprinting and tissue regeneration.

There are several classical protocols useful for MS formation, such as agarose, rotary, hanging drop and scaffold methods; however, each of them is limited by various factors. The most critical disadvantage is the absence of cellular and ECM interactions observed *in vivo* [2].

Recently we demonstrated the practical utility of a cyclic RGD peptide containing a triphenylphosphonium (TPP) moiety, namely *cyclo*-RGDfK(TPP), in a novel and highly reproducible one-step approach for MS formation[3]. The practical application of this technique requires development of simple and reliable protocol for the production of cyclic RGD peptides, containing TPP moiety.

Here we describe an optimized procedure for the synthesis of *cyclo*-RGDfK(TPP) and related peptides using our novel hydrazone resin[4].

Results and Discussion

There are numerous publications related to the synthesis of *cyclo*-RGDfK peptide, including different variants of solution-phase and on-resin cyclization[5-7]. The suggested protocols imply cyclization of protected peptide precursor followed by subsequent removal of side-chain protecting groups.

Recently we described synthesis of inexpensive hydrazone resin and its practical utility for the preparation of peptide hydrazides[4]. These results prompted us to test possible advantages of azide method for the synthesis of *cyclo*-RGDfK(TPP) and related peptides. In order to compare the efficiency of various reaction protocols, the synthesis was performed in parallel on different polymer supports, namely Wang resin, MBH-Br resin and Fmoc-hydrazono-pyruvoyl-aminomethylpolystyrene resin (Fig. 1). In preliminary experiments it was shown that attachment of the TPP moiety to the cyclic peptide in solution complicated the purification of the final product. Therefore, it was achieved by selective deprotection of Lys(Mtt)-containing peptidyl resin, followed by acylation with 4-carboxybutyltriphenyl phosphonium bromide.

Surprisingly, using MBH resin the Mtt-deprotection with 1% TFA in DCM resulted in peptide cleavage from the polymer support, while application of a mixture of AcOH/TFE/DCM (1:2:7) was inefficient (Fig. 2).

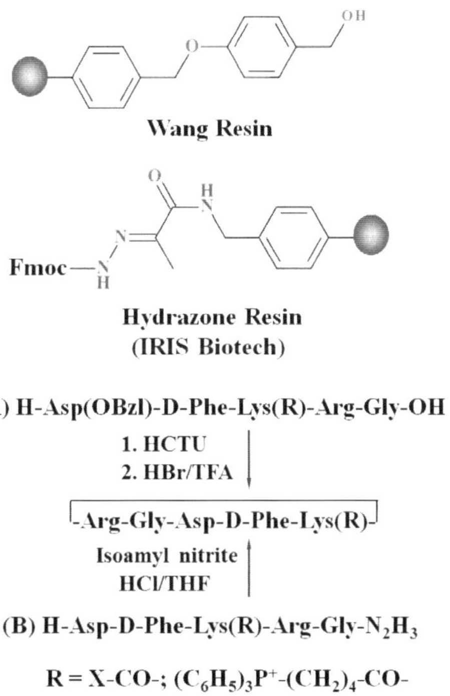

Figure 1: Synthesis of cyclic RGD peptides and their linear precursors. (A) Wang resin; (B) Hydrazone resin

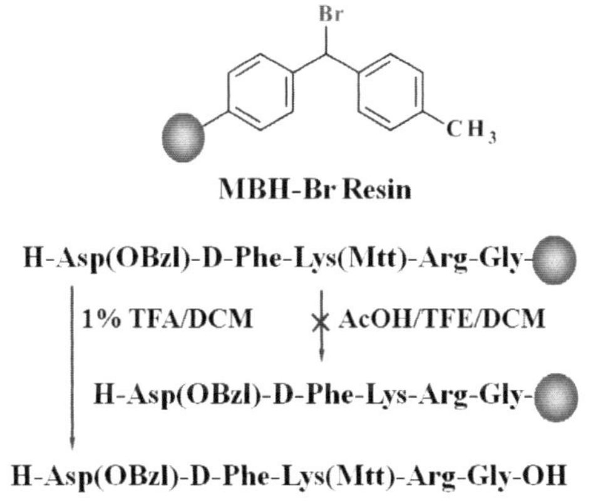

Figure 2: Deprotection of Lys(Mtt) resulted in peptide cleavage from MBH resin (1% TFA/DCM) or was inefficient (AcOH/TFE/DCM)

Synthesis on Wang resin provided linear peptide in reasonable purity. Its cyclization with HCTU proceeded smoothly without significant formation of dimeric side product. However, subsequent removal of the OBzl group from the aspartic acid residue by catalytic hydrogenolysis showed very low efficiency. Unexpectedly, the final peptide deprotection demanded drastic reaction conditions, such as two successive treatments with HBr/TFA for 30 min and 1 h.

The best results were obtained using hydrazone resin. Selective removal of the Mtt group had no influence on the stability of the hydrazone bond. Peptide cleavage from the solid support and removal of all protecting groups proceeded in one step. Cyclization of the resulting peptide hydrazide was achieved in good yield and purity using the azide method. These data evidence the practical utility of hydrazone resin for the synthesis of cyclic RGD peptides or their analogs containing cargo molecules.

The biological experiments have shown that both linear and cyclic RGD peptides containing TPP moiety efficiently suppress platelet aggregation. Withal, *cyclo*-RGDfK(TPP) is able to generate MS formation from tumor, normal or stem cell lines of various origin (Fig. 3). It was shown that suggested approach can be useful for the design of the original 3D *in vitro* co-culture models useful for the study of cell–cell interactions and invasion of different cell types. This novel one-step reproducible method of spheroid formation can be applied in the design of efficient *in vitro* models for the screening of novel drugs and drug delivery systems.

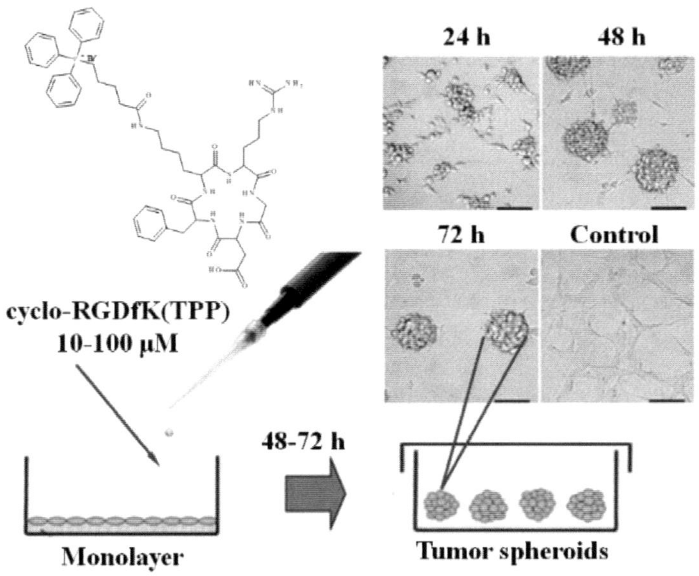

Figure 3: Experimental procedure for MTS formation using cyclo-RGDfK(TPP) method

Further experiments to clarify the mechanism of peptide action and its possible utility for tissue regeneration and wound healing are in progress.

References

[1] Sutherland R.M., McCredie J.A., Inch W.R. J. Natl. Cancer. Inst., 1971, 46, 113–120. DOI 10.1093/jnci/46.1.113.

[2] Sambi M., Qorri B., Frank S.M.S., Mouhamed Y., Kalaydina R.-V., Mendonza N., Szewczuk M.R. Curr. Top. Pept. Protein. Res. 2017, 18, 25–34.

[3] Akasov R., Zaytseva-Zotova D., Burov S., Leko M., Dontenwill M., Chiper M., Vandamme T., Markvicheva E. Int. J. Pharm. 2016, 506, 148–157. DOI: 10.1016/j.ijpharm.2016.04.005.

[4] Chelushkin P.S., Polyanichko K.V., Leko M.V., Dorosh M.Yu., Bruckdorfer T., Burov S.V. Tetrahedron Lett. 2015, 56, 619–622. DOI: 10.1016/j.tetlet.2014.12.056.

[5] Haubner R., Gratias R., Diefenbach B., Goodman S.L., Jonczyk A., Kessler H. J. Am. Chem. Soc. 1996, 118, 7461–7472. DOI: 10.1021/ja9603721.

[6] Dai X., Su Z., Liu J.O. Tetrahedron Lett. 2000, 41, 6295–6298. DOI: 10.1016/S0040-4039(00)01060-1.

[7] McCusker C.F., Kocienski P.J., Boyle F.T., Schätzlein A.G. Bioorg. Med. Chem. Lett. 2002, 12, 547–549. DOI: 10.1016/S0960-894X(01)00799-5

Solid-phase peptide synthesis: The greener, the better

Beatriz G. de la Torre[1], Ashish Kumar[1], Yahya Jad[1], Jonathan M. Collins[2], Simona Serban[3], Othman Almusaimi[1], Fernando Albericio[1]

[1]UKZN, South Africa
[2]CEM Corporation, United States
[3]Purolite, Ireland

https://doi.org/10.17952/35EPS.2018.099

Introduction

In the last years, the role of peptides in the drug discovery arena has been enormously increased as demonstrated by the number of peptide based drugs authorized by medicinal agencies.[1] The concept of solid-phase peptide synthesis (SPPS) pioneered by Merrifield in the 60's [2] and the later development of the Fmoc/tBu strategy are the cornerstones that have made peptides the subject of study in many research areas. Thus, the demand for synthetic peptides is increasing, and considerable efforts have been channeled into optimizing the synthetic process itself, i.e. the coupling reagents, protecting groups, automation of the process, etc. However, little attention has been paid for making the process greener. The "solid-phase" itself could be considered green in some aspects. For instance, the synthesis of a peptide comprising 20 amino acids involves 41 chemical steps in only one reactor, with a clear minimization of mechanical losses and mass transfer. However, SPPS involves the use of large excesses of reagents, which are removed by simple filtration and exhaustive washing steps between each reaction, and therefore the consumption of solvents is the main impediment to be considered an full eco-process. DMF, the main solvent used in SPPS, is a highly reprotoxic solvent and is classified as a Substance of Very High Concern (SVHC).[3] However, it is considered to be difficult to replace because of its properties.

The work presented here is focused to the replacement of DMF by γ-Valerolactone (GVL), a biomass derived solvent, in all steps of SPPS: coupling, Fmoc removal, and washes; and its compatibility with micro-waves conditions. Furthermore, the substitution of diethyl ether (DEE) by cyclopentyl methyl ether (CPME) in the precipitation step, after global deprotection is also described.

Results and Discussion

Initially, comprehensive review and evaluation of physical properties of selected green solvents were conducted in order to decide which was the best choice. Based on some properties, such as capacity to solubilize Fmoc-amino acids, coupling reagents and additives, aside from Fmoc removal efficiency [4], GVL has proved to be the best green solvent of choice for peptide chain elongation whereas, CPME has proved its ability as a green precipitating ether. It is worth mentioning that CPME has limited ability to solubilize Fmoc amino acids, excluding it from the elongation protocol.

For the study, four different model peptides were synthesised based in their synthetic difficulty and different length: (i) the decamer acyl carrier protein ($^{65\text{-}74}$ACP, H-VQAAIDYING-OH), (ii) the decamer Jung-Redemann (JR, H-WFTTLISTIM-NH_2), (iii) the ABC 20-mer (H-VYWTSPFMKLIHEQCNRADG-NH_2), and (iv) the 28-mer thymosin (H-SDAAVDTSSEITTKDLKEKKEVVEEAEN-NH_2). The syntheses were carried out under microwave conditions using a CEM Liberty Blue: 0.1 mmol scale, 5-fold excess of reagents, DIC/Oxyma as coupling reagent and 20% Piperidine for Fmoc removal. The same cycle was used for all syntheses: coupling time 165 s, deprotection time 95 s, both at 90°C using GVL to solubilize all the reagents and for washing steps.

For comparison of the synthetic efficiency, two syntheses of each peptide were made: one with GVL as a green solvent and another one with DMF as a standard solvent. The HPLC analysis of the obtained crude peptides shown similar purities for both solvents[5], Fig 1 shows the example of JR peptide, were the GVL synthesis even lead higher purity than DMF.

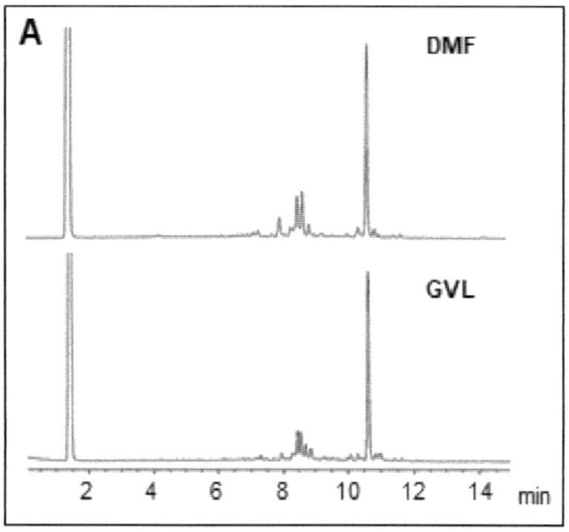

Figure 1: HPLC chromatograms for JR decamer peptide synthesized using GVL vs. DMF

Furthermore, the quality of the crude peptides obtained after precipitation by addition of three different chilled ethers namely DEE, TBME (t-butylmethylether) and CPME were compared. Thus, three identical aliquots of each peptidyl-resin were treated with the standard cleavage solution (TFA/H_2O/TIS 95:2.5:2.5), and then the corresponding ether was added for provoking precipitation. No significant differences were found in the purity of the peptides as shown by HPLC as shown in Figure 2. Additionally the percentage of peptide recovery was not affected by the ether used.[6]

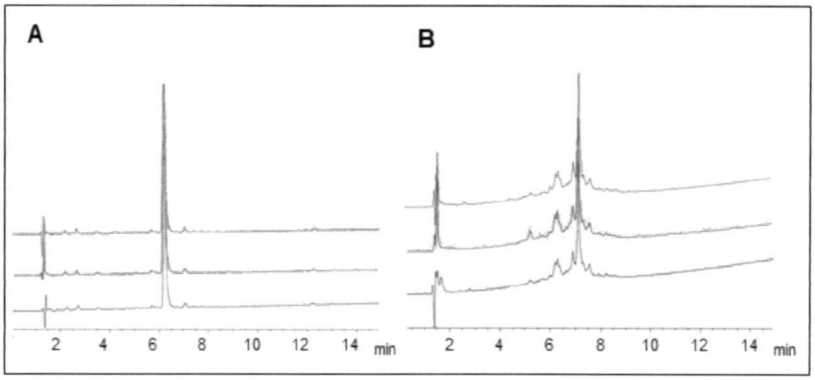

Figure 2: Comparison of 65-74ACP decamer peptide (A) and ABC 20mer (B) precipitated with CPME vs. DEE and TBME; D. peptide precipitated with CPME (green), DEE (blue) and TBME (red)

Conclusion

We have developed a very efficient green methodology for SPPS. It involves the total substitution of the hazardous DMF by GVL during the synthetic process and the substitution of DEE by CPME in the precipitation step. GVL showed full compatibility with automation and use of MW during the process. The protocols are applicable to all both PS and PEG-based solid supports.

Acknowledgment

The work was funded in part by the National Research Foundation (NRF) and the University of KwaZulu-Natal (South Africa). We thanks Luxembourg Biotech (Israel) to supply coupling reagent as a gift.

References

1. Al Musaimi, O.; Al Shaer, D.; de la Torre, B. G.; Albericio, F., 2017 FDA Peptide Harvest. Pharmaceuticals (Basel) 2018, 11 (2).

2. Merrifield, R. B., Solid Phase Peptide Synthesis. I. The Synthesis of a Tetrapeptide. J. Am. Chem. Soc. 1963, 85, 2149–2154.

3. https://echa.europa.eu/regulations/reach/understanding-reach; https://echa.europa.eu/candidate-list-table

4. Jad, Y. E.; Govender, T.; Kruger, H. G.; El-Faham, A.; de la Torre, B. G.; Albericio, F., Green Solid-Phase Peptide Synthesis (GSPPS) 3. Green Solvents for Fmoc Removal in Peptide Chemistry. Org. Process Res. Dev. 2017, 21, 365-369.

5. Kumar, A.; Jad, Y. E.; Collins, J. M.; Albericio, F.; de la Torre, B. G., Microwave-Assisted Green Solid-Phase Peptide Synthesis Using γ-Valerolactone (GVL) as Solvent. ACS Sustainable Chemistry & Engineering 2018, 6 (6), 8034-8039.

6. Al Musaimi, O.; Jad, Y. E.; Kumar, A.; Collins, J. M.; Basso, A.; de la Torre, B. G.; Albericio, F., Investigating green ethers for the precipitation of peptides after global deprotection in solid-phase peptide synthesis. Curr. Opin. Green Sustain. Chem. 2018, 11, 99-103.

Proceedings of the 35th European Peptide Symposium
Patrick B. Timmons, Chandralal M. Hewage, Michal Lebl (Editors)
102 *European Peptide Society & PSP, 2018*

Peptides going sweet: Synthesis of glycopeptide hybrid structures using Passerini and Ugi reactions

Andreja Jakas, Aleksandar Višnjevac, Ivanka Jerić

Rudjer Bošković Institute, Croatia

https://doi.org/10.17952/35EPS.2018.102

α-Acyloxy and α-acylamino carboxamides are the building blocks for the synthesis of various natural products, such as peptide and peptoid derivatives. Multicomponent reactions (MCRs) offer an attractive one-pot strategy for generating a library of these highly functionalized and complex organic compounds. The Passerini and Ugi reactions are isonitrile-based MCR that yields α-acyloxy and α-acylamino carboxamides, respectively (Scheme1). Passerini reaction involves an aldehyde, an isonitrile, and a carboxylic acid, while Ugi reaction involves an aldehyde, an isonitrile, a carboxylic acid and an amine component.[1]The utilization of carbohydrates and their derivatives in MCRs is highly desirable, owing to their polyfunctional character and stereochemical diversity.[2]

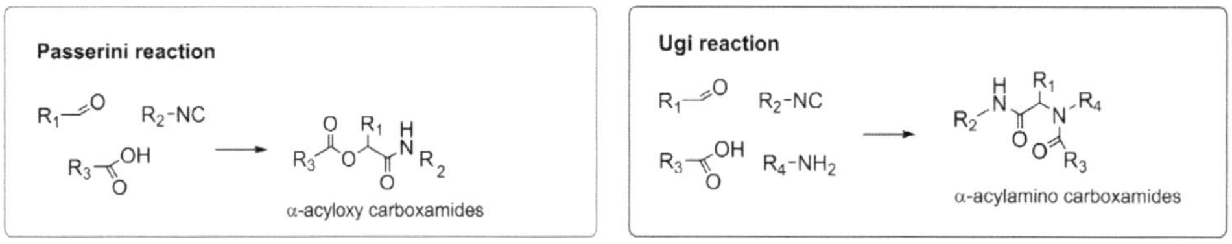

Scheme 1: Reactants and products of Passerini and Ugi reaction.

Our efforts were directed toward exploitation of carbohydrate derivatives as components in the Passerini and Ugi reactions. We used for the first time highly strained carbohydrate-derived aldehydes, isocyanides and amines in the Passerini and Ugi reaction with different commercially available carboxylic acids. As acidic non-sugar components, acetic acid, benzoic acid and Boc-protected phenylalanine were used, and as a sugar component isopropylidene protected gulonic acid (GulA) was used (Scheme 2). Three different sugar moieties were used as isopropylidene protected carbohydrate-derived components (aldehydes, isocyanide, and amine) in the Passerini and Ugi reaction: fructose (Fru), galactose (Gal) and sorbose (Sor). Combination of these components gave eleven (1-11) Passerini products (Scheme 3) and nine (12-20) Ugi products (Scheme 4).The prepared Passerini products bear up to three sugar components, while Ugi products comprise up to four sugar components.

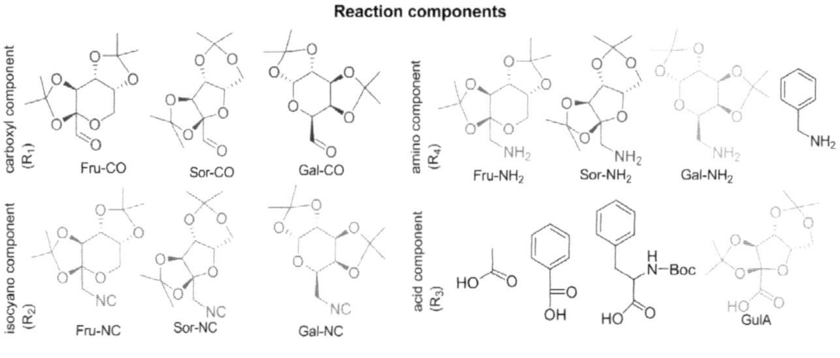

Scheme 2: Used isocyano, carboxyl, amino and acid components in Passerini and Ugi reactions.

The Passerini reactions conducted with prepared sugar aldehydes were highly diastereoselective, with d.r. 9:1, and the crystal structure analysis revealed that the stereochemistry of newly formed chiral center is S. The diastereoselectivity of Ugi reactions depends on component's structure and varies from d.r. 1:1 to d.r. 9:1 with S isomer being predominant one.

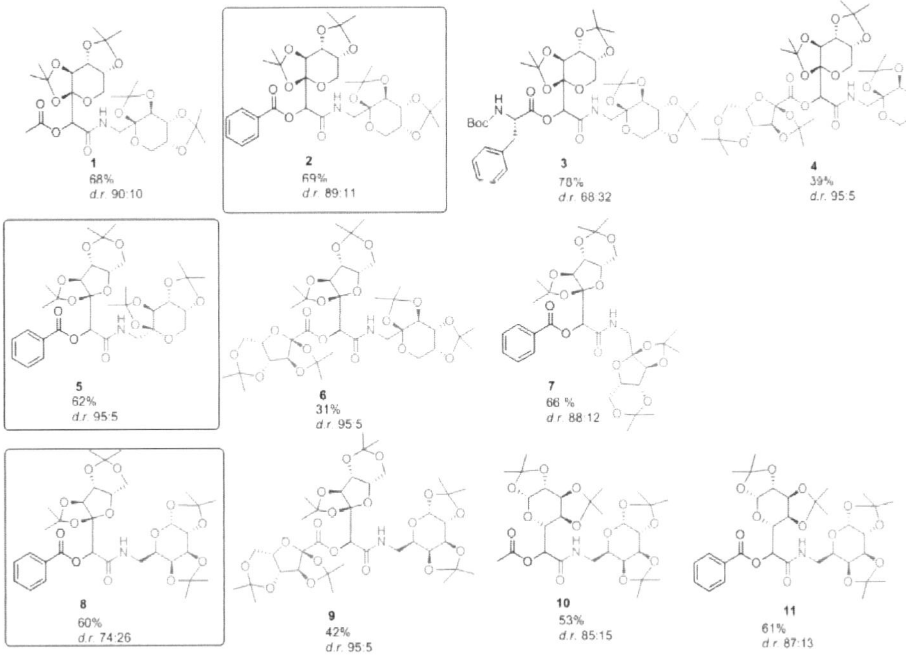

Scheme 3: Prepared Passerini reaction products. Crystal structures were determined for compounds in squares.

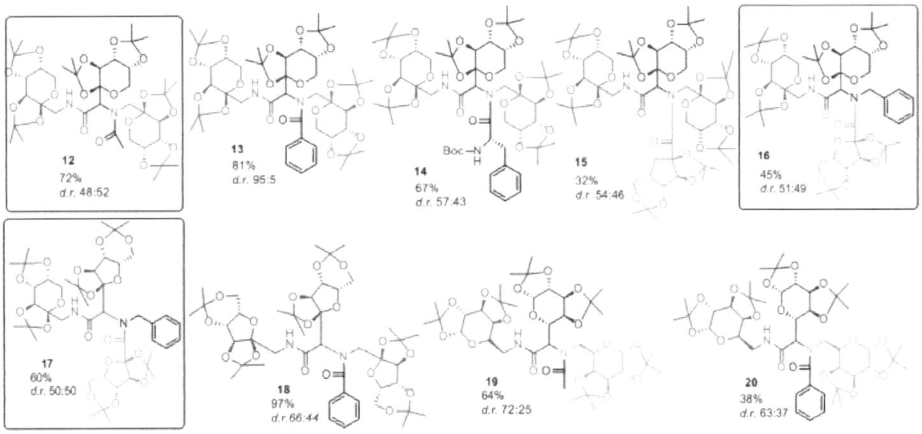

Scheme 4: Prepared Ugi reaction products. Crystal structures were determined for compounds in squares.

Acknowledgements

We are grateful for financial support from the Croatian Science Foundation, Grant number 3102.

References

1. D. Fürniss *et al.* Beilstein J. Org. Chem. 2013, 9, 56–63.
2. Md. M. Khan *et al.* RSC Adv. 2015, 5, 57883–57905.

Proceedings of the 35th European Peptide Symposium
Patrick B. Timmons, Chandralal M. Hewage, Michal Lebl (Editors)
104 *European Peptide Society & PSP, 2018*

Quality peptide versus speed: Conventional synthesis versus microwave

Hossain Saneii, Farshad Karimi, William Bennett, Mandar Maduskar, Mostafa Hatam

AAPPTec, United States

https://doi.org/10.17952/35EPS.2018.104

Abstract

We have conducted studies comparing microwave heating versus conventional heating in solid phase peptide synthesis. We have found that very rapid heating, as in microwaves, speeds the rate of synthesis, but it accelerates epimerization and other undesirable reactions as well. We have compared conventional synthesis with microwave synthesis for a number of peptides. In general, conventional methodology takes 32% longer, but no epimerization occurs. Also it provides higher yield by delay gradient heating and reduces the purification difficulties.

Discussion

The following results show that not only higher quality but almost zero epimerization is obtained with room temperature or delay gradient heating even though they take more time. ACP (65-74), ABC-20 mer, bivalirudin, dynorphin A, exenatide, GPR, JR 10-mer, liraglutide, magainin I and thymosin were prepared by standard solid phase peptide synthesis at room temperature and delayed gradient heating. The delay gradient profile began with 5 minutes at room temperature, then gradient to 65 °C for 10 to maximum 15 minutes. Fmoc deprotection was performed with delay gradient heating in 20% piperidine/DMF. The delay gradient for deprotection started with 2 minutes at room temperature, then gradient heating to 55 °C in 3 minutes.

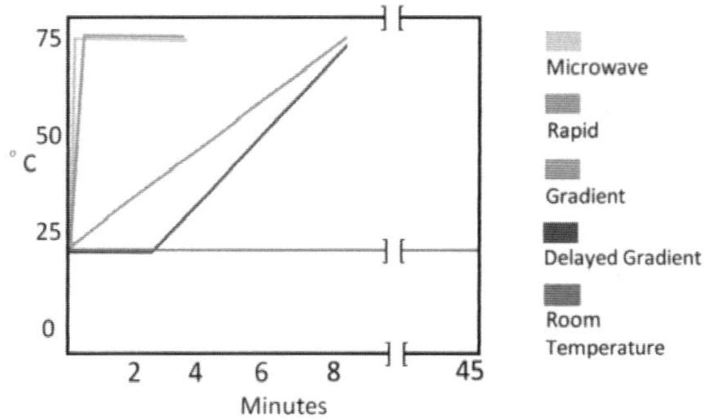

Figure 1: Comparison of Heating Methods

As an example, ACP(65-74) provided a quality of 95% by delay gradient heating which is close to the results obtained with microwave. HPLC of the crude ACP showed single major peak for both syntheses. However, through further investigation, crude ACP from microwave synthesis was revealed to contain an epimerization product that co-eluted with the desired peptide (Figure 2). With further purification pure ACP was obtained as a perfect single peak (Figure 4). However, the peptide produced at room temperature showed only one peak under all conditions.

The results in Table 1 show purities from 78% to 95.19%. Delay gradient heating shows results better than microwave heating or room temperature synthesis. Compared to microwave synthesis, the delayed gradient protocol only increased cycle time by 32% (22 min per cycle).

Table 1

Peptide	Purity (Delay Gradient)	Mass Spectrum
ABC-20 mer[a]	95.19%	$[M]^+$: 1062.22
ACP(65-74)[b]	95%	
Bivaliruden[c]	90.75%	
Dynorphin A[d]	91.75	
JR 10-mer[e]	91.22%	$[M+Na]^+$: 1233.6
Magainin I[i]	78%	
Thymosin[j]		

Crude Peptides

a) Sequence: VYWTSPFMKLIHEQCNRADG-NH_2 ; b) Sequence: VQAAIDYING-NH_2; c) Sequence: fPRPGGGGNGDFEEIPEEYL; d) Sequence: YGGFLRRIRPKLKDNQ; e) Sequence: WFTTLISTIM -NH_2; f) Sequence: HAEGTFTSDVSSYLEGQAAK(γ-Glu-palmitoyl)EFIAWLVRGRG; i) Sequence: GIGKFLHSA GK-FGKAFV GEIMKS; j) Sequence: SDAAVDTSSEITTKDLKEKKEVVEEAEN

General Procedure

All peptides were synthesized on a Focus XC 2RV or a Focus XC 6RV from AAPPTec, LLC. Both instruments were equipped with heating capabilities. Each synthesis was performed on 100 mg of Rink amide resin with a substitution of 0.5 mmol/g. Fmoc-protecting groups were removed utilizing a 20% piperdine in DMF solution containing 0.2 M Oxyma Pure. The resin was washed with DMF to less than 0.02% piperidine using two or a single wash.

Couplings reactions were conducted utilizing 3 equivalents of amino acid, 3 equivalents of diisopropylcarbodi-imide (DIC), and 3 equivalents of Oxyma Pure. The difficulty of each coupling was predicted using an algorithm we developed. Extended coupling times and double couplings were utilized based on this assessment (Figure 3). The crude peptides were analyzed by HPLC and the identity of the crude peptides was confirmed by mass spectroscopy/HPLC. (Table 1)

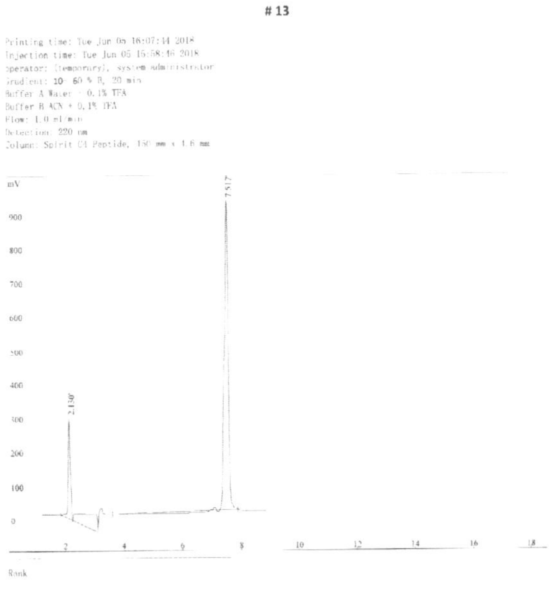

Figure 2: ACP Crude from Delayed Gradient

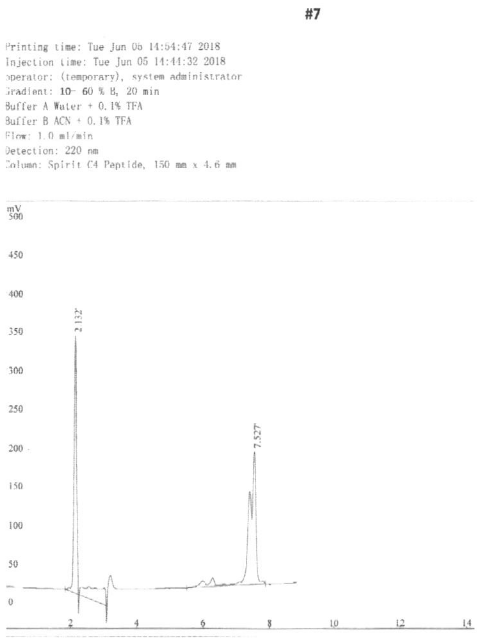

Figure 3: ACP Modified Conditions

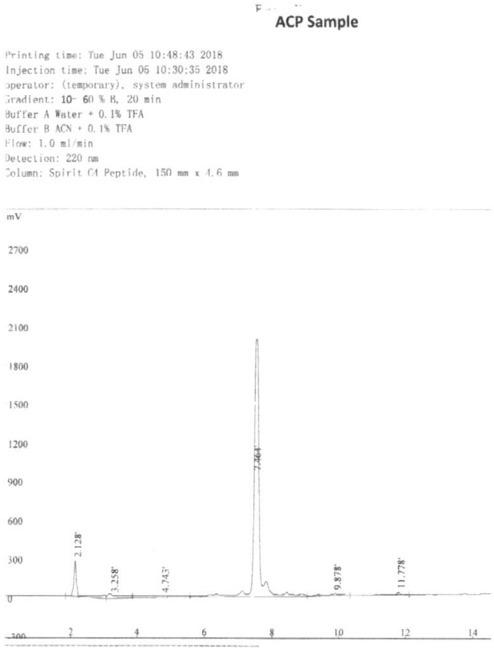

Figure 4: Purified ACP

Conclusions

Peptide synthesis with delay gradient heating produces better quality peptide than microwave heating. Delay gradient heating takes more time (22 minutes per cycle), but the improved quality of the crude peptide with little or no epimerization results in less time spent in purification.

Fluorogen-based amino acids with Aggregation-Induced Emission features for bioprobes design

Mathieu Arribat[1], Emmanuelle Rémond[1], Sebastien Richeter[2], Philippe Gerbier[2], Sébastien Clément[2], Florine Cavelier[1]

[1]IBMM, France
[2]ICGM, France

https://doi.org/10.17952/35EPS.2018.107

Introduction

Fluorescently labelled peptides are widely used as tools for molecular imaging to provide biological information on living systems. However, most of the conventional organic fluorophores used for such purpose aggregate when dispersed in aqueous media.[1] This aggregation generally implies a self-quenching and thus, a drastic decrease of the fluorescence ("turn-off"), which is a thorny problem for developing efficient bioprobes. Luminophores with Aggregation-Induced Emission (AIE) features are found to be weakly or not fluorescent when molecularly dissolved but highly fluorescent when aggregated.[2] Although fluorophores displaying AIE-effects are increasingly used in sensing and bioimaging studies,[3] they generally suffer from a lack of selectivity. To develop highly specific AIE fluorogen-based bioprobes, we decided to exploit the previous work of our laboratory on silylated amino acids[4] and on hydrosilylation[5, 6] in view of generating silole amino acids and peptides to investigate their fluorescent properties.

Results and Discussion

The synthesis of silole amino acids was achieved by hydrosilylation of protected amino acids 2 and 4 bearing alkene or alkyne on side chains. Among the different catalysts tested, Karstedt's revealed to be the most efficient and silole amino acids 3 and 5 were isolated with very good yields of 95 and 88%. Completion of the hydrosilylation was inferred from IR spectroscopy since the Si-H band of the silole 1 at v = 2118 cm^{-1} disappeared in the IR spectrum of 3. The optical purities of 3 and 5 were verified on HPLC using a chiral column by comparison with racemic samples. Formation of silole amino acid 3 was also confirmed by 1H, 13C{1H}, 29Si{1H} NMR spectroscopy in $CDCl_3$. Effectively, a clear difference is observed between the chemical shift of Si-H precursor 3 (δ = -11 ppm) and the products of hydrosilylation 3 (δ = -17 ppm) and 5 (δ = -2 ppm).

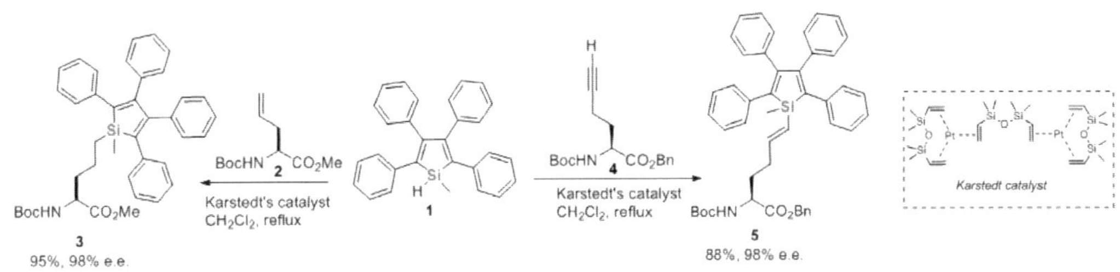

Scheme 1: Hydrosilylation of alkene 2 and alkyne 5 on amino acids side chains

The fully protected silole amino acid 3 was selectively N- and C- deprotected in order to prove the coupling feasibility (Scheme 2). The amine function was deprotected under acidic conditions (TFA) to afford the corresponding ammonium 6 in 80% yield. The methyl ester was saponified with lithium hydroxide to afford the free carboxylic acid 9 in 85% yield. The peptide coupling of 6 and 9 was achieved under classical conditions (HATU, DIPEA, DMF) with alanine moiety 7 and 10 as model. A tripeptide with two alanine residues was also synthesized by SPPS on Wang resin starting using the corresponding N-Fmoc protected silole amino acid.

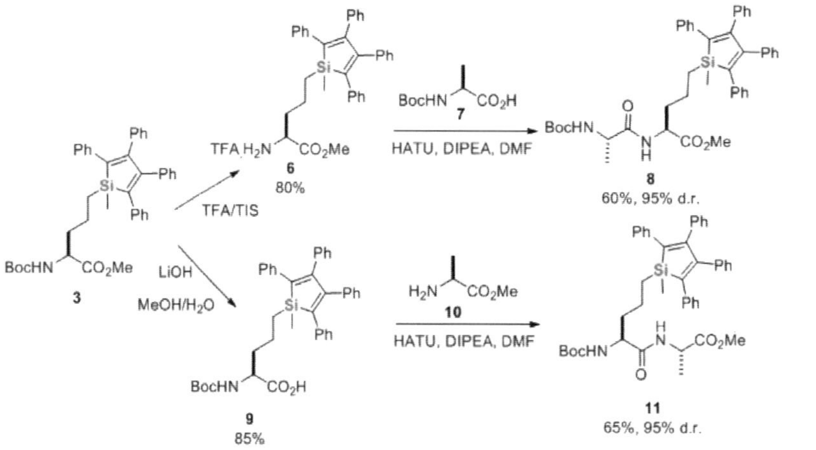

Scheme 2: Synthesis of silole dipeptides

As shown in Figure 1 for compound3, when fw varies from 0 to 60 vol %, the fluorescence remains very low. A dramatic increase in the fluorescence was noted after the water fraction reached 70 and 80 vol %, typical from AIE fluorogens leading to I_{agg}/I_{THF} ~6 (except for 5, I_{agg}/I_{THF} ~44).

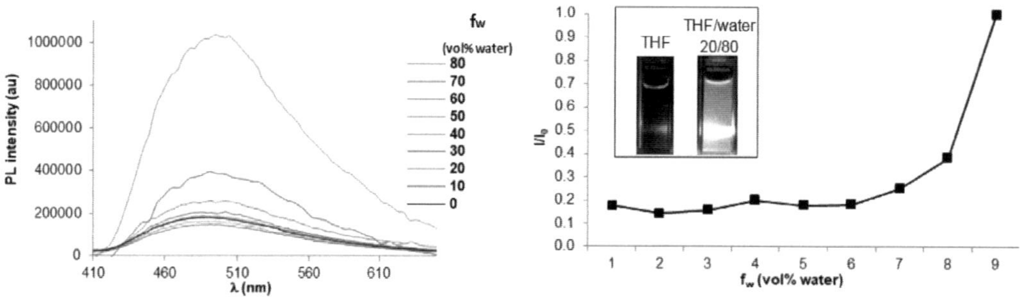

Figure 1: Emission spectra of 3 in THF/water mixtures with different water fractions (fw) (Concentration: 5.10-4M, Excitation wavelength: 360 nm) (left). Plot of I/I0 values vs. water fractions. I0 is the PL intensity for the highest fw. Inset: photos of silole 4 under the illumination of a UV lamp at 405 nm in THF/water mixture (fw = 0 and 80%) (right).

New silole amino acids were synthesized by hydrosilylation of alkyne or alkene amino acids, catalyzed by Karstedt platinum complex. After selective deprotection of carboxylic function or amine, C- or N- peptide coupling in solution and on SPPS proved the possible incorporationinto peptides. Such silole amino acids and peptides exhibit AIE properties with δem around 500 nm and $\Delta\delta$ ~100 nm. The quantumyield for the aggregated state is 5 to 10 times higher than in solution. They constitute a new class of promising fluorophores for the development of labelled peptides.

Acknowledgements

The authors are grateful to the University of Montpellier, the CNRS and the French Ministry of Research for financial support. M.A. thank University of Montpellier for his PhD grant.

References

[1] Birks J., Photophysics of Aromatic Molecules, Wiley, London 1970.

[2] Mei J., Legung N., Kwok R., La, J., Tang B., Chem. Rev., 2015, 115, 11718.

[3] Ding D., Li K., Liu B., Tang B., Acc. Chem. Res., 2013, 46, 2441.

[4] Rémond, E., Martin, C., Martinez, J., Cavelier, F. Chem. Rev., 2016, 116, 11654.

[5] a) Marchand D., Martinez J., Cavelier F., EurJOC, 2008, 3107. b) Cavelier, F.; Marchand, D.; Martinez, J. Chem. Biodiversity, 2008, 5, 1279.

[6] Fanelli R., Berthomieu D., Didierjean C., Doudouh A., Lebrun, A., Martinez J., Cavelier F., Org. Lett., 2017, 19, 2937.

High-throughput parallel synthesis optimization of Glucagon-like Peptide 1 receptor agonists

Cyf Ramos-Colon[1], Andrew Kennedy[2], Daniel Martinez[1], James Cain[1]

[1]Gyros Protein Technologies, United States
[2]Gyros Protein Technologies, United Kingdom

https://doi.org/10.17952/35EPS.2018.110

Introduction

GLP-1 related peptides are an important class of peptides due to their effects on metabolic diseases. GLP-1 receptor agonists are used for treatment and management of type 2 diabetes [1]. The GLP-1 agonist class improves glycemic control through multiple mechanisms with a low risk of hypoglycemia [1]. In 80-90% of type 2 diabetes cases, obesity is also present thus targeting multiple receptors to tackle both is of great interest [2]. Evers *et al.* have recently described peptides with dual agonist activity at the GLP-1 receptor and the glucagon receptor for this purpose [1].

Many SAR studies have been performed on GLP-1 related peptides in the search for stability and target affinity improvements, including introduction of cyclizations and attaching PEG based monomers. Published patents describe the SPPS of these type of peptides as being optimized by fragment synthesis and solution phase condensation in order to reduce impurities and maximize yields for pharmaceutical productions. Here we show the rapid automated synthesis optimization screen of two GLP-1 receptor agonists (Lixisenatide and Pramlintide; Figure 1 and 2) and one dual GLP-1 and glucagon receptor agonist (Figure 3).

H-HGEGTFTSDLSKQMEEEA VRLFIEWLKNGGPSSGAPPSKKKKKK-NH_2

Figure 1. Lixisenatide

H-KCNTATCATQRLANFL VHSSNNFGPILPPTNV GSNTY-NH_2

Figure 2. Pramlintide

H-H-Aib-QGTFTSDLSKQK-(yE-yE-C16)-DEQRAKLFIEWL-Aib- AGGPPS-Aib- KPPPK-NH_2

Figure 3. Dual GLP/GlucagonR agonist [2]

Methods

GLP-1 agonists were synthesized by Fmoc-chemistry SPPS on a Symphony® X peptide synthesizer at a 25 μmol scale. Using the parallel synthesis capability on the Symphony X, three resins and two coupling reagents were tested in parallel. The resins tested were:

R Ram Tentagel® Resin (0.19 mmol/g)

Rink Amide ChemMatrix® resin (0.55 mmol/g)

Low-loaded Rink Amide MBHA resin (0.22 mmol/g)

Deprotections were done using 20% piperidine in DMF at 25°C for 2 x 30 s. The coupling reactions were done at 25°C for 2 x 1 min with a 12- fold excess and final concentrations of 100 mM AA and 100 mM HCTU or COMU® with 200 mM DIPEA. For the synthesis of GLP/glucagonR agonist, the Fmoc-Lys(ivDde)-OH was coupled for 2 x 5 min.

The cleavage was done using TFA/thioanisole/H2O/phenol/EDT (84.5:5:5:5:2.5) for 2 h at 25°C on the Symphony X followed by precipitation in diethyl ether. The resulting peptides were dissolved in water and analyzed on a Thermo Scientific Ultimate 3000 HPLC using a C18, 100 , 2.6 um, 50 X 2.1 mm Kinetex Evo column (Phenomenex), over 4.5 or 5.5 min with a flow rate of 0.8 mL/min and a gradient of 5-95% B or 0-70% B, respectively; where A is 0.1% TFA in water and B is 0.1% TFA in acetonitrile. Detection was done at 214 nm. Mass analysis was done on a Shimadzu LCMS-2020 Single-Quad mass spectrometer, equipped with a C18, 100 , 2.6 um, 50 x 2.1 mm Kinetex column (Phenomenex), over 30 min with a flow rate of 1 mL/min and a gradient of 5-95% B where A is 0.1% formic acid in water and B is 0.1% formic acid in acetonitrile.

Results

PEG based Rink Amide ChemMatrix resin produced the best crude purities in the synthesis of all three peptides using either HCTU or COMU as the coupling reagent at room temperature. This confirms the advantage of using PEG based resins for the synthesis of long peptides.

In the synthesis of Lixisenatide the highest crude purity was achieved with COMU (Table 1) and for Pramlintide and Dual GLP-1/GluR agonist was achieved with HCTU (Table 2 and Table 3). Both HCTU and COMU are highly reactive coupling reagents able to provide high crude purity with coupling reactions of 1 min.

Table 1: Crude purity of Lixisenatide.

Resins	% Purity	
	HCTU	COMU
R Ram Tentagel	54.7	49.5
Rink ChemMatrix	56.4	68.2
Rink MBHA Low Loaded	45.1	41.0

Table 2: Crude purity of Pramlintide.

Resins	% Purity	
	HCTU	COMU
R Ram Tentagel	35.1	43.0
Rink ChemMatrix	62.7	47.0
Rink MBHA Low Loaded	62.3	42.2

Table 3: Crude purity of Dual GLP/GluR agonist.

Resins	% Purity	
	HCTU	COMU
R Ram Tentagel	59.6	59.1
Rink ChemMatrix	63.0	49.6
Rink MBHA Low LOaded	62.1	57.6

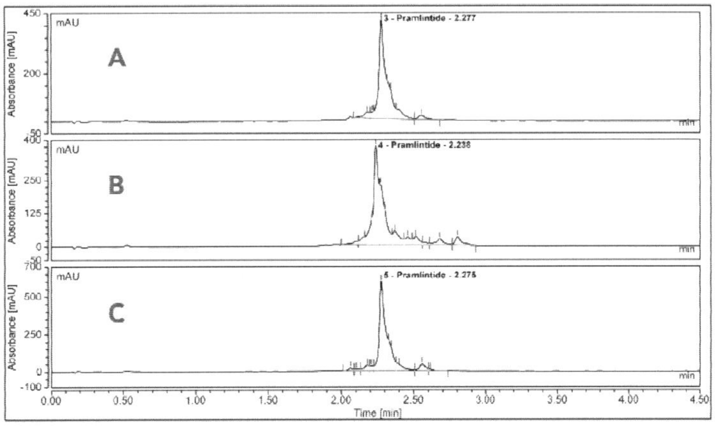

Figure 4: Crude purity profiles of Pramlintide using HCTU/DIPEA at 25°C and synthesized on different resins: A) Rink Amide MBHA, B) R Ram TentaGel, C) Rink Amide ChemMatrix resins

Conclusion

Multi-variable conditions were successfully tested in parallel for the high-throughput optimization of GLP-1 receptor agonists getting valuable synthesis information in ~2 days. ChemMatrix resin provided the best crude purity in the synthesis of GLP-1 agonists. HCTU and COMU are highly reactive coupling reagents effective in achieving high purity peptides with 2 x 1 min coupling reaction times. Synthesis of Lixisenatide resulted in optimal crude purity using Rink Amide ChemMatrix resin and COMU as the coupling reagent with 2 x 1 min couplings. Synthesis of Pramlintide and the linear dual GLP/GLuR agonist resulted in high crude purities with multiple conditions with the Rink ChemMatrix and HCTU combination providing crude purities above 62%. Next steps include the optimization of the Lys^{14} elongation after ivDde removal of dual GLP-1/GluR agonist.

References

1. Kalra, S. *et al.* Indian J. Endocrinol. Metab. 20, 254–67 (2016).

2. Evers, A. *et al.* J. Med. Chem. 1, (2018).

Proceedings of the 35th European Peptide Symposium
Patrick B. Timmons, Chandralal M. Hewage, Michal Lebl (Editors)
European Peptide Society & PSP, 2018

Synthesis of an HIV-targeted stapled peptide: Fully automated on-resin peptide cyclization *via* olefin metathesis

James Cain[1], Daniel Martinez[1], Andrew Kennedy[2], Cyf Ramos-Colon[1]

[1] Gyros Protein Technologies, United States
[2] Gyros Protein Technologies, United Kingdom

https://doi.org/10.17952/35EPS.2018.113

Introduction

Cyclic peptides are an important tool for development of peptide therapeutics. Their relative rigidity allows for increased membrane permeability and stability to proteases, addressing two major drawbacks of peptide drug development. Peptide cyclization can be achieved several ways, including head-to-tail cyclizations or side chain-to-side chain cyclizations, which can be done by lactam bridge formation, disulfide bonds or hydrocarbon stapling *via* olefin metathesis, to name a few. Hydrocarbon stapling stabilizes a-helical structures and thus increases cell permeability and stability. Other advantages of stapled peptides include the targeting of protein-protein interactions, important for targeting many disease states such as cancer.

Zhang *et al.* have described the synthesis and properties of an i, i+4 stapled peptide, NYAD-1 (Figure 1), which targets the capsid and inhibits HIV-1 in cell culture [1].

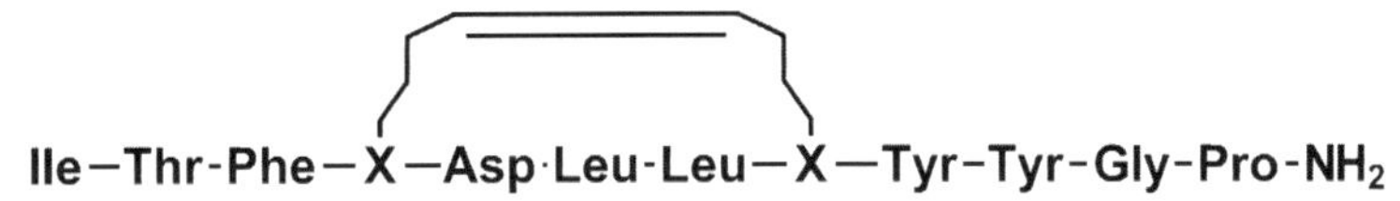

Figure 1: NYAD-1 stapled peptide structure. X = (S)-2-(4'-pentenyl)alanine

A small GTPase RAB25, has been found to have both pro-oncogenic and anti-oncogenic phenotypes in specific cellular contexts. Mitra *et al.* have shown several cell permeable peptides, including RFP14 peptide (Figure 2), that disrupt RAB25:FIP complex formation *in vitro* and in situ and compete with the context-specific phenotypes associated with RAB25 function in ovarian and breast cancer cell lines [2].

Figure 2: RFP14 peptide structure.

Here we show the fully automated synthesis, from linear, on-resin cyclization, and resin cleavage, of NYAD-1 and RFP14.

Methods

Protocol A: NYAD-1 was synthesized on Rink MBHA resin (0.33 mmol/g) at 10 μmol scale. Deprotection was done with 20% piperidine in DMF for 2 x 5 min. The coupling reaction was done with 100 mM amino acid and HATU, 400 mM NMM for 2 x 20 min with ten-fold excess for natural amino acids, 1 x 30 min with five-fold excess for unnatural amino acid.

Protocol B: RFP14 was synthesized using Rink Amide ChemMatrix Resin (0.54 mmol/g substitution) at a 25 umol scale. Deprotection time was 2 x 1:30 min at 25°C, 60°C, or 90°C using 20% Piperidine in DMF. Amino acids were coupled using a six-fold excess and final concentration of 100 mM for amino acid and HCTU and 200 mM for DIEA. Coupling time was 2:30 min at 25°C, 60°C, or 90°C. A capping step was done after every coupling reaction for 5 min. DMF washes followed deprotect, coupling, and capping steps with 3 repetitions at 30 seconds each.

Metathesis Reactions: The resin bound Fmoc-protected peptide was treated with a 10 mM solution of Grubbs catalyst (Bis(tricyclohexylphosphine) benzylidene ruthenium(IV)dichloride) in 1,2-Dichloroethane (2 mL) for 2 X 2 hours.

Cleavage: Final cleave used TFA:TIS:EDT :Water (95:1:2.5:2.5) for 2 hours at 25°C on the instrument followed by precipitation in ether.

Analysis: Peptides were analyzed using a C18, 300 Å, 5 mm, 250 x 4.6 mm column (Varian Microsorb-MV) or C18, 180 Å, 4.6 um, 250 X 4.6 mm Polaris column (Agilent), with a gradient of 5-95%B in 60 or 30 min, respectively, using Water (0.1%TFA):ACN(0.1%TFA) at 1ml/min. A 1:10 dilution of a standard sample of 3 mg/ml was run on a Phenomenex Kinetex 2.6 um C18 100A 50x2.1 mm column for LCMS or a Polaris column with a gradient of 5-50%B in 7 or 30 min, respectively, using Water(0.1%FA):ACN(0.1%FA) at 1 ml/min.

Results

NYAD-1

NYAD-1 was successfully synthesized on an automated peptide synthesizer with a 31.4% crude purity using HATU and a conservative protocol (Protocol A).

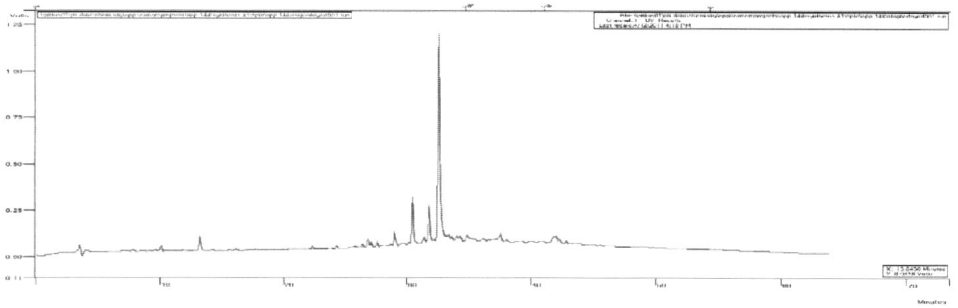

Figure 3: Crude purity profile of stapled NYAD-1 using HATU and NMM.

RFP14

RFP14 was synthesized in parallel at three different temperatures, 25°C, 60°C, and 90°C using a fast protocol of 2:30 min coupling time using HCTU. The best linear crude purity was seen at the higher 90°C temperature (Table 2, Figure 4A).

Table 1: Effect of temperature on crude purity of linear RFP14.

Peptide	25°C	60°C	90°C
RFP14	28.7%	27.7%	59.6%

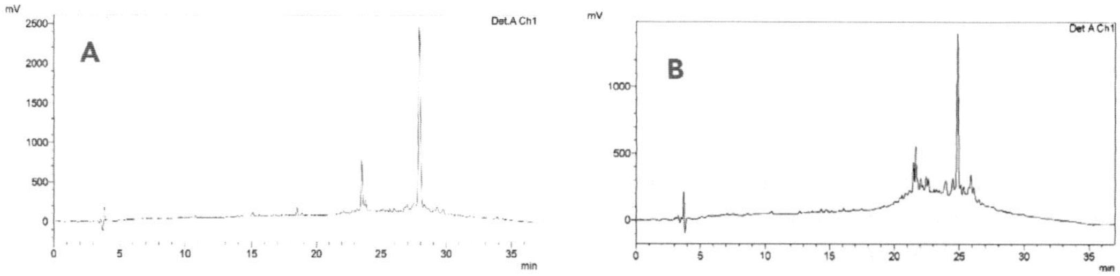

Figure 4: Crude purity profiles of: A) linear RFP14 (at 90°C) and B) stapled RFP14.

Conclusions

Therapeutic stapled peptides were successfully synthesized under variable conditions on automated peptide synthesizers in over 30% crude purity. Independent induction heating on the Prelude X allowed multiple temperatures to be screened simultaneously on the synthesis of the linear RFP14 peptide with improved purity with increase in temperature up to 90°C.

References

1. Zhang, H. *et al.* Retrovirology. 2013. 10, 1–20. 2013.

2. Mitra, S. *et al.* Nat. Commun. 2017. 8 (1).

Evaluation of the "ResPep continuous flow synthesizer" with real-time UV-monitoring, automated feedback & heating in solid phase peptide synthesis

Felix Niethammer, Daniel Maisch, Marcus Rothe

INTAVIS Bioanalytical Instruments AG, Germany

https://doi.org/10.17952/35EPS.2018.116

Evaluation of the "ResPep continuous flow synthesizer" with real-time UV-monitoring, automated feedback and heating in solid phase peptide synthesis

Automation in assembly of peptides *via* solid phase synthesis following the Fmoc-strategy [1] is a well-established method and commonly used for peptide synthesis today. Mostly, due to length of the peptide, the presence of hydrophobic stretches or sterically hindered amino acids, difficulties during peptide synthesis are sequence inherent.

Therefore, the choice of the proper conditions like coupling reagent and temperature have tremendous influence on the stepwise amide bond formation yielding the crude product in the highest purity possible. Here, we present examples of automated peptide synthesis of so called "difficult sequences" [2-5] under demanding conditions on the new ResPep continuous flow (CF) synthesizer with real-time UV monitoring and feedback.

Test sequences:

1. ACP-10mer: H-VQAAIDYING-NH_2
2. JR-10mer: H-WFTTLISTIM-NH_2
3. Exenatide: H-HGEGTFTSDLSKQMEEEA VRLFIEWLKNGGPSSGAPPPS-NH_2
4. Bivalirudin: H-(D)-Phe-PRPGGGGNGDFEEIPEEYL-NH_2
5. Asn15-FBP28: H-YYNNRTLESTWEKPQELK-OH

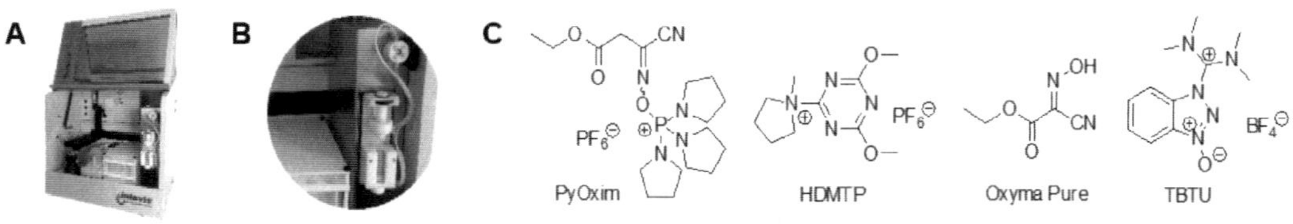

Figure 1: (A) ResPep CF synthesizer. (B) 1-Column module with piston pump and heating block. (C) Structures of applied coupling reagents used in this study.

Table 1: Synthesis conditions: coupling reagents, RP-HPLC-purity of crude, lyophilized peptides. 1) deprotection times were automatically elongated through feedback based on online UV monitoring; 2) coupling got automatically repeated if deprotection rates are slow (based on online UV monitoring)

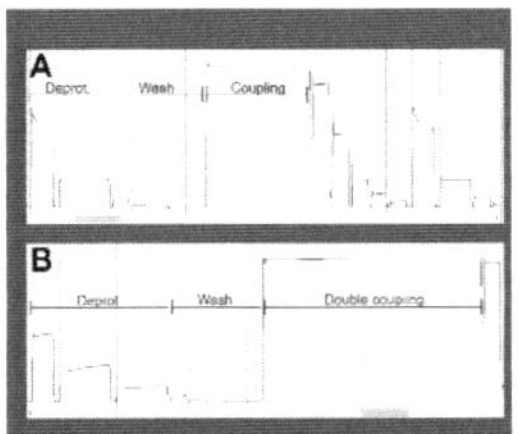

Figure 2: UV traces at 301 nm from synthesis of JR-10 using PyOxim. (A) Standard cycle with double deprotection and single coupling. (B) Difficult deprotection and coupling, automated adaption to three deprotection steps and automated repetition of coupling.

High quality of crude JR-10 product.

Peptide	Coupling reagent	Deprotection [min][1]	Coupling time [min][2]	T [°C]	HPLC purity [%]
ACP-10mer	PyOxim	1,5	0	RT	71
	PyOxim	1,5	8	90	89
	PyOxim	3	8	90	90
	PyOxim	1,5	15	90	90
	HDMTP	1,5	8	90	82
	Oxyma Pure	1,5	8	90	79
	TBTU	1,5	8	90	71
JR-10mer	PyOxim	1,5	8	90	78
	HDMTP	1,5	8	90	72
	Oxyma Pure	1,5	8	90	66
Exenatide	PyOxim	1,5	8	90	52
	HDMTP	1,5	8	90	47
	Oxyma Pure	1,5	8	90	42
Bivalirudin	PyOxim	1,5	8	90	84
	HDMTP	1,5	8	90	82
	Oxyma Pure	1,5	8	90	78
Asn15-FBP28	HDMTP	1,5	8	90	70
	PyOxim	1,5	8	90	83
	Oxyma Pure	1,5	8	90	86

RT vs. 90 °C
Heating has tremendous effect on purity.

Deprotection time 1,5 min vs. 3 min
No significant improvement.

Coupling time 8 min vs. 15 min
No improvement.

Influence of choice of coupling reagent on quality

PyOxim > HDMTP > Oxyma Pure > TBTU

BUT.
For synthesis of Asn15-FEP8:
Performance of Oxyma Pure is significant better compared to the PyOxim

Choice of most effective reagent is dependent on sequence.

References

[1] Atherton, E., Sheppard, R.C., Solid Phase peptide synthesis: a practical approach. 1989, Oxford, England: IRL Press.

[2] Redemann, T., Jung, G., Peptides 1996. Proceedings of the 24th European Peptide Symposium, Ramage, R., Epton, R., Eds., Mayflower Scientific Ltd: Kingswinford, UK, 1998, 749.

[3] Carpino, L. A., Krause, E., Sferdean, C. D., Schümann, M., Fabian, H., Bienert, M., Beyermann, M., Tetr. Lett., 2004, 45, 7519-7523. DOI: 10.1016/j.tetlet.2004.07.162

[4] GEN, July 1, 2012, 32. No. 13. DOI: 10.1089/gen.32.13.15

[5] Coin, I.,Dölling, R., Krause, E., Bienert, M., Beyermann, M.; Sferdean, C. D., Carpino, L. A., J. Org. Chem, 2006, 71, 6171-6177. DOI: 10.1021/jo060914p

Proceedings of the 35th European Peptide Symposium
Patrick B. Timmons, Chandralal M. Hewage, Michal Lebl (Editors)
118 *European Peptide Society & PSP, 2018*

3-Substituted isocyanopyridines as mildly convertible isocyanides

Mathias Elsocht[1], Karlijn Hollanders[1], Olivier Van der Poorten[1], Mouhamad Jida[2], Bert Maes[3], Steven Ballet[1]

[1]Vrije Universiteit Brussel, Belgium
[2]University of Strasbourg, France
[3]Universiteit Antwerpen, Belgium

https://doi.org/10.17952/35EPS.2018.118

Introduction

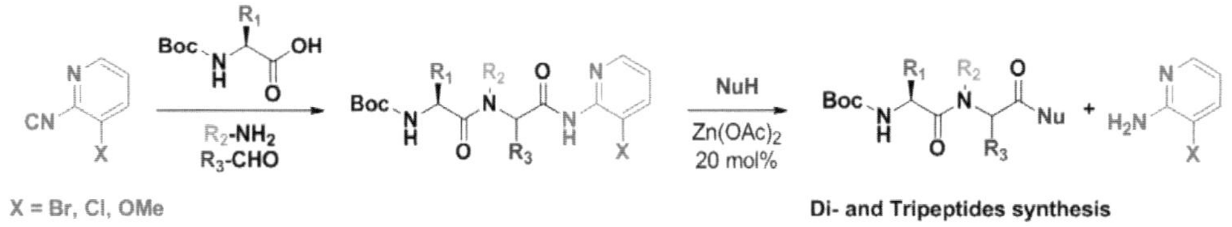

Scheme 1: Synthesis of Ugi products containing a directing group followed by a Zinc catalysed transamidation reaction.

Multicomponentreactions (MCRs) have become a powerful tool to generate peptidomimetics since they are easy to use and give great structural diversity. More specifically, the Ugi reaction forms an α-acylaminocarboxamide utilizing an aldehyde, an amine, a carboxylic acid, and an isocyanide. Such dipeptide-like structures have found great application, often in combination with post-condensation modifications. The downside of this reaction is the formation of an amide bond, derived from the isocyanide, at the C-terminus of the product, which cannot be cleaved under mild reaction conditions. This limitation can be circumvented by the use of convertible isocyanides that allow the chemoselective transformation of this amide bond after the MCR.[1] Throughout the years, different convertible isocyanides have been developed all with their advantages and restrictions. For example, the isocyanide is not highly accessible or even unstable, or the amide cleavage requires acidic/basic conditions which in multistep procedures might result in the undesired cleavage of protecting groups. With this in mind, we set out to develop a convertible isocyanide that can be cleaved, under neutral conditions, using a Zn-catalysed nicotinate-directed transamidation (Figure 1), a previously developed strategy by our groups.[2] Hence, the synthesis of different 3-substituted-2-isocyanopyridines was pursued, which were subsequently subjected to Ugi-4C conditions. These Ugi products were in turn used in the zinc catalysed transamidation reaction. It was demonstrated that, for all three directing groups, the amide bond at the C-terminus could easily and selectively be cleaved. As a proof of concept, we demonstrated the methodology in the synthesis of a previously published constrained dipeptide, which was obtained through Ugi-post-condensation modifications.[3]

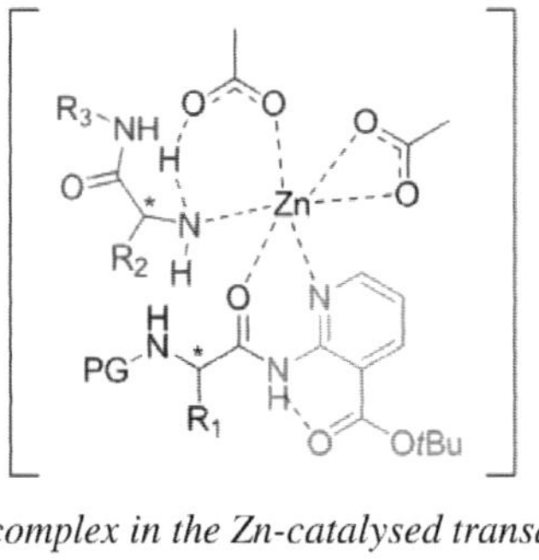

Figure 1: Starting complex in the Zn-catalysed transamidation reaction.

Results and Discussion

Synthesis of isocyanides 3 was achieved through the formylation and dehydration of the corresponding amines 1. The dehydration was performed using PhPO2Cl2, which is a green variant of POCl3 and more suitable for the synthesis of isocyanides of pyridine-based formamides 2.[5] This reagent gave access to all three desired isocyanides in good yield.

Table 1: Synthesis of 3-substituted-2-isocyanopyridines 3 via the formylation/dehydration sequence.

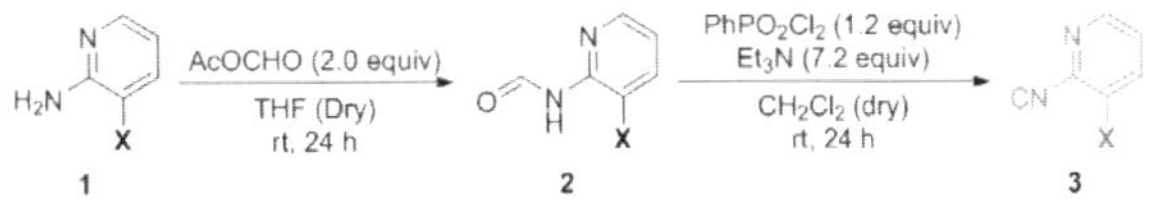

Entry	X	2	Yield of 2 (%)	3	Yield of 3 (%)
1	Br	**2a**	>99	**3a**	53
2	Cl	**2b**	>99	**3b**	63
3	OMe	**2c**	>99	**3c**	61

Subsequently, the different isocyanides 3 were used in an Ugi four component reaction towards 4. Here, isovaleraldehyde and propylamine were stirred in TFE for 2 hours prior to the addition of the N-Boc protected phenylalanine and the isocyanides 3, yielding the desired Ugi products 4 in good yield. In a next step, the Ugi products 4, equipped with the directing groups, were used in the Zn-catalysed transamidation reaction. Comparable ^{1}H-NMR yields were obtained for all three directing groups which let us belief that all three directing groups are equally efficient in the transamidation reaction. Nevertheless, as the chlorinoated directing group gave the highest yield in the Ugi reaction, we selected 3b for the synthesis of aminotriazoloazepinone 6.

Table 2: Synthesis of directing group containing Ugi product 4 and evaluation of directing groups in the Zn catalysed transamidation reaction

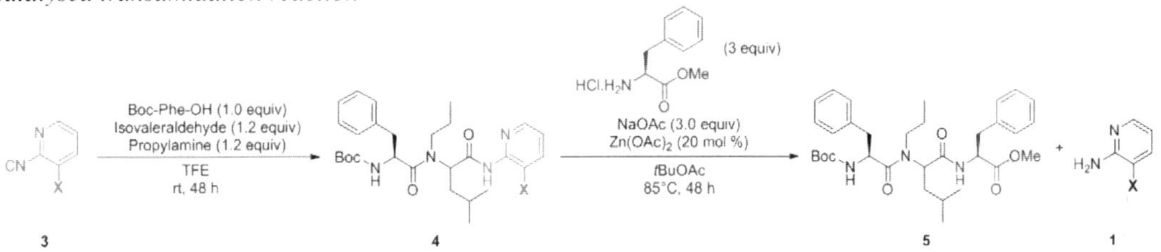

Entry	X	4	Yield of 4 (%)	5	Yield of 5 (%)[a]
1	Br	**4a**	68	**5a**	47
2	Cl	**4b**	77	**5b**	47
3	OMe	**4c**	66	**5c**	44

[a] ^{1}H NMR yield using 1,3,5-trimethoxybenzene as internal standard, corrected to a mass balance of 100%.

The 3-chloro-2-isocyanopyridine 3b was used in the synthesis of aminotriazoloazepinone-containing dipeptides 6, following the one-pot Ugi–Huisgen cyclization.[3] Afterwards, a $Zn(OAc)_2$-mediated hydrolysis was carried out towards compound 7, which was coupled, without intermediate purification, on solid-support by a standard coupling method. Final cleavage yielded the desired tripeptides 9a-b. Hence, the convertible isocyanide 3b made it possible to generate a labile amide at the C-terminus of the Ugi product, which allowed to generate, after hydrolysis, a SPPS-compatible dipeptide mimetic.

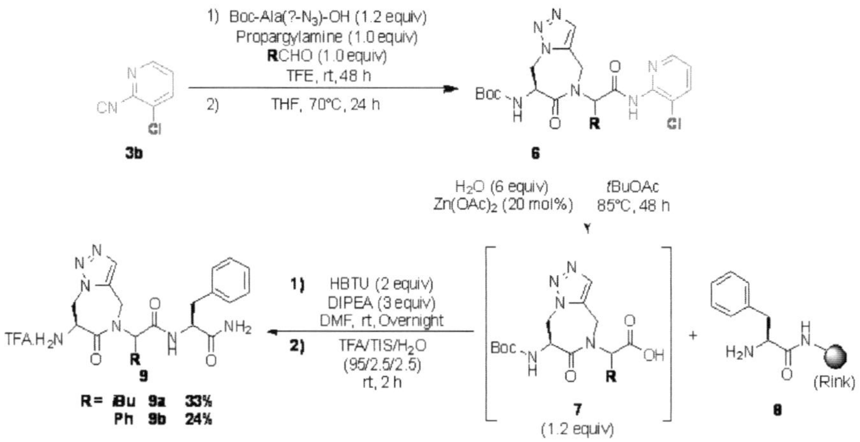

Scheme 2: Synthesis of aminotriazoloazepinone-containing dipeptides 6 followed by the anchoring on solid-support in a two-step procedure: hydrolysis – coupling.

Conclusion

Three new convertible isocyanides were developed that allow a C-terminal amide cleavage of Ugi products, under mild conditions, yielding di- and tripeptides. The practical utility of these cleavable isocyanides was demonstrated through the synthesis of a constrained dipeptide scaffold which was anchored on solid support after cleavage of the amide group utilizing a Zn-catalyzed reaction.

Acknowledgments

ME and SB are grateful to the Doctoral school (VUB) for the financial support and acknowledge the organizing committee of 35th EPS for the registration bursary.

References

[1] M. Oikawa et al, Synlett. 24 (2013) 2014–2018. doi:10.1055/s-0033-1338967.

[2] C.C.D. Wybon et al, ACS Catalysis. 8 (2018) 203–218. doi:10.1021/acscatal.7b02599.

[3] T.M.A. Barlow et al, Org. Biomol. Chem. 12 (2014) 6986–6989. doi:10.1039/C4OB01381F .

[4] G. Kobayashi et al, Synthesis. 2011 (2011) 3225–3234. doi:10.1055/s-0030-1260211.

Synthesis and characterization of new Hemorphin-5 analogue containing azobenzene moiety with potential optical switching properties

Petar Todorov, Petia Peneva, Stela Georgieva-Kiskinova, Anton Georgiev

University of Chemical Technology and Metallurgy, Sofia, 1756, blvd.Kliment Ohridski 8, Bulgaria

https://doi.org/10.17952/35EPS.2018.121

Introduction

The discovery of bio molecular control of pharmacophores and their macromolecular motion by light offers dynamic changing of the properties of molecules in biological systems with minimal disturbance to the rest of the system. Azobenzene containing peptides could be acting as bio photoswitches that interconvert between E- and Z- isomers around –N=N– group offer highly predictable light state structural properties. These molecules are relatively small and adopt predictable conformations they are well suited as tools to interrogate cellular function in a spatially and temporally controlled fashion and for applications in photopharmacology and photodynamic therapy [1,2]. Photodynamic control of peptides, proteins or drugs provides a non-invasive way to disorder these networks to investigate their effect or cause a defined outcome.

In order to evaluate the influence of azobenzene-peptides we introduced the azobenzene-4,4'-dicarboxylic acid to the N-side of Hemorphin-5 peptide analogue. This report refers to the solid-phase synthesis and characterization of novel azobenzene containing Hemorphin-5 analogue. Hemorphin-5, also known as Valorphin, is a naturally occurring, endogenous opioid peptide of the Hemorphin family with affinity for opioid receptors and morphinomimetic properties. These neuropeptides are increasingly being used in the treatment of various diseases such as hypertension, epilepsy, chronic pain, cancer, and etc. [3,4].

Results and Discussion

By considering, the scarce information about the synthesis of azo-peptides, and especially their E→Z isomerization proves the challenging of preparation and characterization such compounds. The investigation of the electrochemical oxidation and reduction mechanisms of these compounds is important since it could result in a better understanding of their physiological mechanisms in order to receive new information on biomolecular ox-red level. To assess the redox properties of azobenzene moiety in Valorphin, and to investigate the E→Z conversion, the cyclic and differential pulse voltammetry at Pt- working electrode was used.

The synthesis was achieved by a modified solid-phase peptide synthesis by Fmoc-(9-fluorenylmethoxy-carbonyl) dimerization strategy, based on the reaction of azobenzene-4,4'-dicarboxylic acid with two N-terminal amino groups of distinct Val-Val-Tyr-Pro-Trp-Thr-Gln heptapeptides directly on the resin (Fig. 1) to (Val-Val-Tyr-Pro-Trp-Thr-Gln)$_2$Azo peptide.

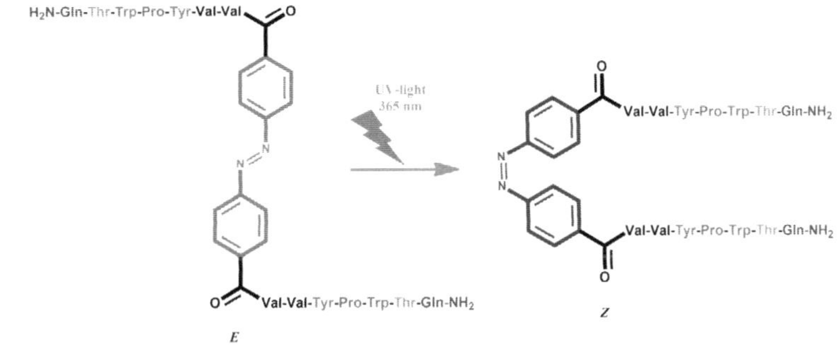

Figure 1: Schematic representation of photoswitchable (Val-Val-Tyr-Pro-Trp-Thr-Gln)$_2$Azo peptide upon long wavelength UV light as trans (E) and cis (Z) isomers.

For proving that obtained results are for azo-peptide only and to be investigate the electrochemical behaviour of such interesting azo-compound on the electrode surface the cyclic and differential pulse voltamperometry on Pt-electrode were examined. Several processes (such as oxidations and reductions) can be identified by direct analysis of the cyclic voltammograms on Pt-electrode (Fig.2). However, as previous studies have proved, it is essential to consider a working-window analysis to differentiate azo-peptide related redox processes from those of secondary reactions on azobenzene-4,4'-dicarboxylic acid and VV-hemorphin-5and among other misleading interferences. The electrochemical study of Valorphine and Valorphine derivatives were previously reported by Todorov *et al.* 2018 [5]. Anodic cyclic voltammogram for the oxidation of VV-hemorphin-5in phosphate buffer solution at Pt- electrode is shown in Fig. 2. The anode peaks of the compounds are probably due to the presence of tryptophanin the peptide molecule, which is oxidized to a platinum electrode at potential closed to +0.4 V [6]. The results on the voltammogram exhibits one anodic and cathodic peaks at 193 and -113 mV, respectively, which shows a quasi- reversible behaviour, with peak-to-peak separation ($\Delta Ep = |Epc\text{-}Epa|$) of 80 mV and it is proves the theoretical obtained data. At the same condition the voltammogramsof rest compounds were recorded (Fig. 2). The reduction potentials and nature of electrodic processes depend on the UV-irradiationof azo-peptide with $\lambda = 365$ nm and while the reduction potential have a value near to zero and quasi-reversible reduction for AzP (N) it was observed that AzP(IRr) have a more negative potential value and fully irreversible reduction (Fig. 2). The data were confirmed using DPV mode. Two well defined oxidation peaks for AzP before irradiated were observed in DPV mode; one at more negative potential and second at $Ep_a = 0.300$ V corresponding to –N=N–group and VV-hemorphin-5, respectively (Fig. 3).

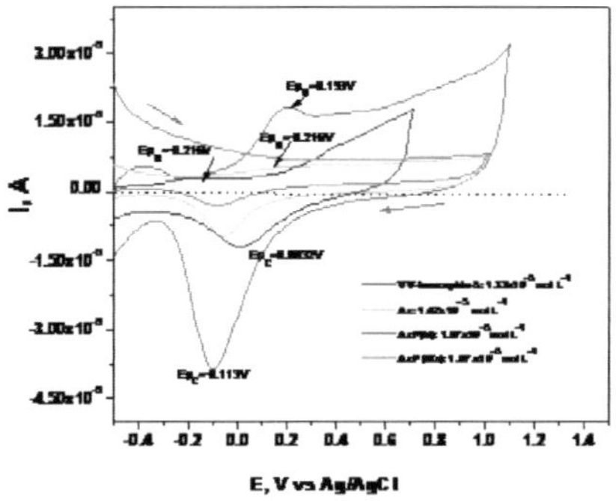

Figure 2: Cyclic voltammograms of azobenzene-4,4'-dicarboxylic acid (Az), VV-hemorphin-5 and VV-hemorphin-5 azopeptide (AzP) before (AzP(N)) and after (AzP(IRr)) illumination at $\lambda = 365$ nm whit same concentrations in pH 6.86 (phosphate buffer solution, 0.1 mol L^{-1}), scan rate 1.00 V s^{-1} , with working Pt- and Ag/AgCl,KCl as reference electrodes.

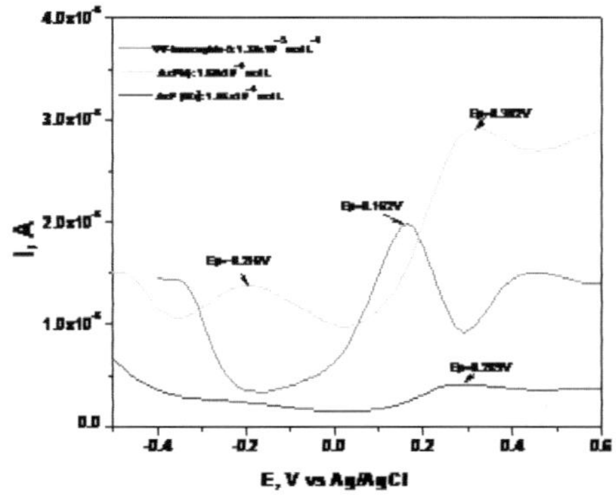

Figure 3: DPVs of VV-hemorphine-5 and azo-peptide (AzP) before and after UV irradiated at $\lambda = 365$ nm in pH 6.86 (phosphate buffer solution, 0.1 mol L^{-1}) at Pt- and Ag/AgCl,KCl as reference electrodes.

Conclusions

We describe the synthesis and electrochemical properties of new analogue of Hemorphin-5 containing azobenzene moiety. The compound was obtained by solid-phase peptide synthesis - Fmoc-dimerization strategy. The electrochemical properties are discussed and confirm in order to the E$\rightarrow$Z transformation changes on electrochemical behaviour of the compound by altering the reactivity of the redox function group.

Acknowledgments

This work was financial supported by the Bulgarian National Scientific Fund project 18/5 of the Ministry of Education and Science, Bulgaria.

References

1. Mart RJ, Allemann RK. Chem. Commun. 52:12262 (2016).

2. Renner C, Moroder L. ChemBioChem 7:869 (2006).

3. Blishchenko E, Sazonova O, Kalinina O, Yatskin O, Philippova M, Surovoy A, Karelin A, Ivanov V. Peptides 23:903–910 (2002).

4. Song CZ, Wang WQ, Song CC. Int. J. Cancer 131:1011–1012 (2012).

5. Todorov P, Peneva P, Pechlivanova D, Georgieva S, Dzhambazova E. Bioorg. Med. Chem. Lett. 28:3073-3079 (2018).

6. MacDonald SM, Roscoe SG. Hecfrochimico Acta 42:1189-1200 (1997).

Proceedings of the 35th European Peptide Symposium
Patrick B. Timmons, Chandralal M. Hewage, Michal Lebl (Editors)
124 *European Peptide Society & PSP, 2018*

Calibration of analytical HPLC to generate preparative LC gradients for peptide purification

Jack Silver

Teledyne ISCO, United States

https://doi.org/10.17952/35EPS.2018.124

Abstract

Preparative LC (liquid chromatography) is widely used to purify synthesized peptides. One bottleneck in the purification process is method development. Significant time can be required to produce an efficient preparative purification method that resolves the desired peptide from impurities and minimizes both time and solvent usage. This work describes a simple method of calibrating analytical HPLC systems to match the preparative LC system using the existing scouting gradients typically employed by a research group. After the calibration is complete, the determined delay volume is applied to the scouting gradient. This delay volume encompasses any dwell volumes, column volumes, mixing volumes, solvent mis-proportioning, and other corrections that are needed to match the analytical system to the preparative system. After the calibration is complete, the user only needs to enter the retention time of the desired compound from the analytical HPLC scouting run to calculate a preparative method. Although the calculated gradient is designed to run over 12 minutes with targeted peptides eluting at ~6 minutes, other gradient lengths may be run.

Prep System Calibration

This requires three steps:

1. Determine the dwell volume (V_{Dp}). This only needs to be done once. Replace one solvent with a solvent that absorbs UV light. Acetone is useful since it is miscible with water and easily washed from the system. Replace the column with a union. Run isocratically with no absorbing solvent, then program a step gradient. Note the delay in absorbance from the step and multiply by the flow rate to obtain the dwell volume.

2. Determine the column volume (V_{Cp}). Do this once per column size and chemistry used; replacement columns of the same type will have the same column volume. For example, all RediSep Prep 20 x 150 mm C18 columns will have the same column volume. Run the column with 10% organic, and inject a small amount of sodium iodide or sodium nitrate while monitoring215 nm. Note the peak elution time and multiply by the flow rate to obtain the column volume. Steps 1 and 2 allow the calculated gradient to be adjusted for the delays of the preparative system.

3. Set the elution time for the model compound using an isocratic run. Use the same solvent and modifiers typically run on the analytical system. Adjust the mobile phase composition to elute the compoundat the desired time for the preparative runs. Model compounds used include ethyl paraben, phenacetin, and N-benzylbenzamide. These were chosen because they elute at ~50% organic solvent. This step sets the retention time for the column used.

Analytical System Calibration

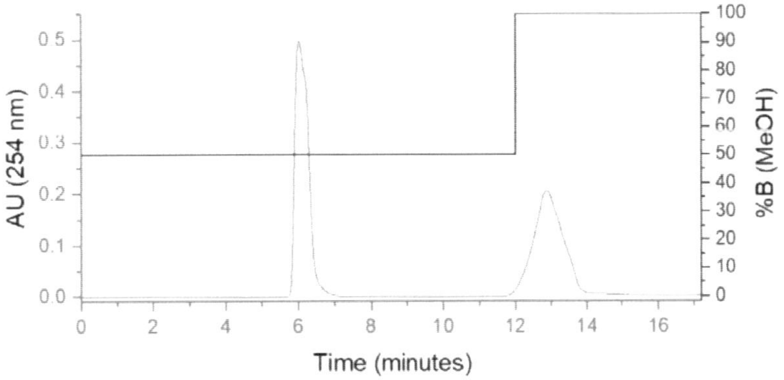

Figure 1: Calibration of a 20 x 150 mm RediSep Prep C18 column (PN 692203826) with phenacetin and N-benzylbenzamide in methanol on an ACCQPrep HP125 (PN 685230035). Model compound (phenacetin) eluted in ~6 minutes at 50% B solvent.

This only requires running the model compound with the scouting gradient used to evaluate synthesized compounds. This gradient is typically 5 or 10% to 100% organic with no isocratic hold at the start. Use the same solvent system, including modifiers, as used to calibrate the preparative system. Use columns with the same packing for preparative and analytical runs. After calibration, the programmed gradient time for the calibration compound is calculated using the equations below.

$Ma = (\%B_{Ea} - \%B_{Sa})/L_a$

$P = (\%B_I - \%B_{Sa})/M_a$

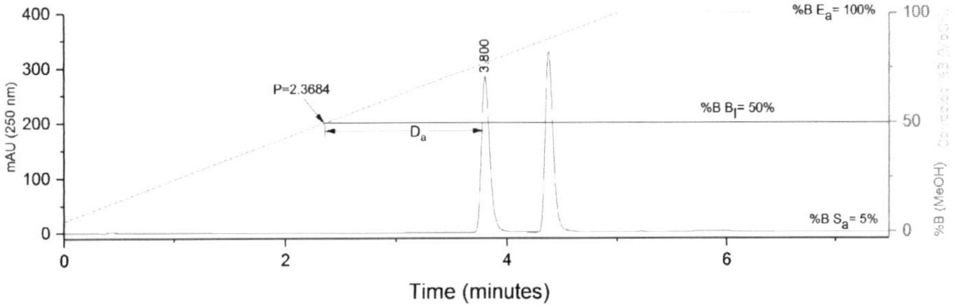

Figure 2: Sample from prep calibration run (Figure 1) on an Agilent UHPLC using an experimental 2x50 mm RediSep Prep C18 column in water/methanol. The gradient was 5-100% B over 5 minutes. The peak eluting at 3.800 minutes (phenacetin) is deemed to elute at 50% methanol.

Focused Gradient Calculation

This requires four steps:

1. Run the compound to be purified on the analytical system using the same gradient as the initial calibration.
2. Using the calculated value Da, determine the actual %B which elutes the compound.
3. Set a focused gradient encompassing the calculated %B.
4. Correct the gradient for the dwell and column volumes of the preparative system

$\%B_{Corr} = (T_{Ea} - D_a) * M_a + \%B_{Sa}$

The corrected solvent composition for the desired compound is $\%B_{Corr}$; T_{Ea} = the elution time for the desired compound in the analytical run. The other terms are the same as the earlier equations.

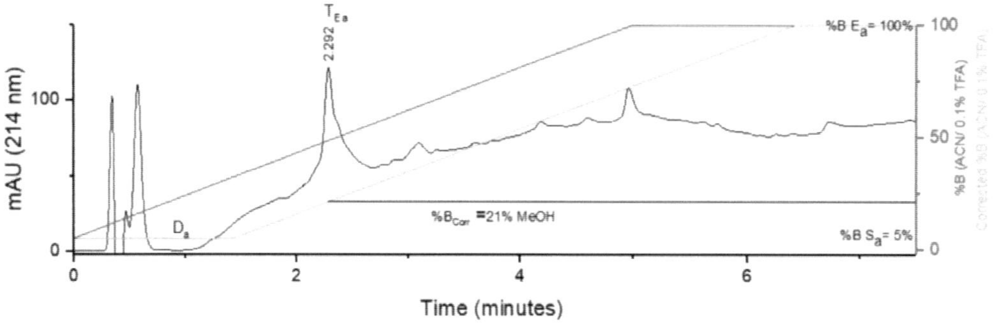

Figure 3: Analytical run of Thymosin (H2N-SDAAVDTSSEITTKDLKEKKEVVEEAEN-OH, MW 3066) using the same gradient and column as Figure 2. The solvent system was water/acetonitrile, both containing 0.1% TFA. Thymosin eluted at 2.292 minutes.

The retention of thymosin is very different from the calibration compound and thus serves as a useful example. For this example:

$\%B_{Corr} = (2.292-1.43216) * 19 + 5 = 21.3\ \%B$

Set a focused gradient with the desired range (R), centered around this number—usually a range of 10% to 20%.

$D_p = (V_{Dp} + V_{Cp})/F_p$

$\Delta\%B = R/L_p * D_p$

The final steps to calculating the gradient involve correcting the focused gradient for the preparative system. For the equations above, D_p, the preparative system delay, is determined by adding the prep system dwell volume (V_{Dp}) to the prep column volume (V_{Cp}) and dividing by the prep flow rate. The amount to increase the strong solvent concentration (Δ%B) is calculated by dividing the range (R) by the prep gradient length (L_p) then multiplying by D_p. For an ACCQPrep 125, V_{Dp} = 7.65 mL (with 5 mL loop); V_{Cp} = 24.81 mL (C18 20x150 mm RediSep Prep column), run at 18.9 mL/min. The chosen range (R) is 10%, with a length (L_p) of 12 minutes. $D_p = (7.65 + 24.81)/18.9 = 1.7174$ min; $\Delta\%B = 10/12 * 1.7174 = 1.4312\%$ The final gradient is 22.8 % ±5, or 18 to 28% methanol.

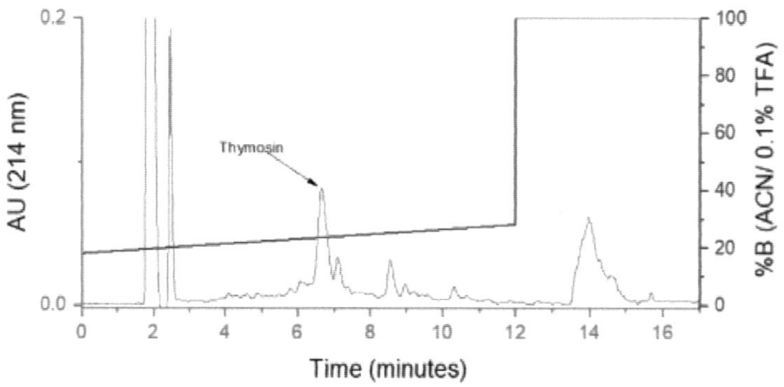

Figure 4: Thymosin (12 mg dissolved in 0.2 mL DMSO) run on a RediSep Prep C18 column in water/acetonitrile (both containing 0.1% TFA) using the calibration described in this poster eluted at the expected time. Although the calibration was performed in methanol with no modifiers, the gradient was correctly calculated when the compound was eluted from both the analytical and preparative system using acetonitrile modified with TFA.

Single step recombinant human follicle stimulating hormone purification by peptide affinity chromatography

Juan M. Gurevich Messina[1], Silvana L. Giudicessi[1], Mara C. Martinez-Ceron[1], Nicols Urtasun[1], Guillermina Forno[2], Laura Mauro[2], Osvaldo Cascone[1], Silvia A. Camperi[1]

[1]Universidad de Buenos Aires. FFyB. Cát. Biotecnologia. CONICET-UBA, Inst. NANOBIOTEC. Junin 956, 1113, Bs As., Argentina
[2]R&D Zelltek S.A., UNL, FBCB, Ciudad Universitaria, Paraje el Pozo, CC 242, Santa Fe, Argentina

https://doi.org/10.17952/35EPS.2018.127

Introduction

Human Follicle Stimulating Hormone (hFSH) is clinically used for ovulation in women and spermatogenesis induction in men, in assisted reproduction technologies (1). As FSH-based biopharmaceuticals are parenterally administered, their purity must be high. Current methods for hFSH purification include several chromatographic steps to reach the required purity. However, these involve a decrease in the hFSH total yield and rises the cost of the process. Affinity chromatography (AC) consists in the specific adsorption of target biomolecules onto ligands immobilized on chromatographic supports. Short peptides have been described as useful ligands for AC because of their low cost, simple chemical synthesis and high stability compared to protein-based ligands (2). The aim of this work was to design an affinity chromatography matrix with an immobilized synthetic peptide for rhFSH purification.

Methods

In a previous work, Sohn *et al.* (3) examined the hFSH receptor interaction with the hormone, testing each amino acid of the exoloop 3 by Ala substitution. Taking into account those works, the mutant with greater affinity: (580)KVPLITVSKAK(590) was selected to design a synthetic ligand for affinity chromatography (AC): Ac-KVPLTVSKAKV AC-NH_2. The peptide was synthesized as amide and was acetylated. The peptide was acetylated to increase its stability against possible attack by proteases present in the crude FSH sample. A Cys was incorporated at the C-termini to facilitate its subsequent immobilization to the chromatographic activated SulfoLink agarose resin. A sample of crude rhFSH was loaded to the peptide affinity column using as adsorption and elution buffers: 20 mM sodium phosphate, 0.5 mM Met, pH 5.6 and 7.2 respectively.

Results and Discussion

The affinity chromatography support with the peptide ligand immobilized was evaluated by loading crude samples composed of host cell proteins from the Chinese hamster ovary (CHO) cell and rhFSH (Fig. 1). A non-reducing SDS-PAGE was done to check rhFSH purification (Fig. 2). The purity obtained after AC purification was 94 % and the yield was 41% (Table 1). The highly glycosylated isoforms, which have the highest *in vivo* potency, were recovered. The identity of the protein band obtained in the SDS-PAGE was checked by peptide mapping analysis using LC-MS/MS. The peptide affinity column was overloaded, and the dynamic capacity obtained was 54.6 mg rhFSH/mL chromatographic resin.

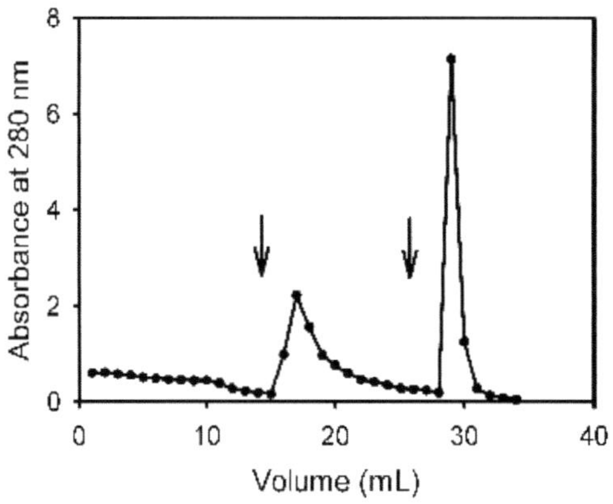

Figure 1: Chromatography of crude samples with host cell proteins and rhFSH using the affinity matrix with the peptide Ac-KVPLTVSKAKVAC-NH_2 immobilized. The mobile phases were: a) adsorption buffer (20 mM sodium phosphate, 0.5 mM Met, pH 5.6); b) elution buffer (20 mM sodium phosphate, 0.5 mM Met, pH 7.2) and c) regeneration buffer (20 mM sodium phosphate, pH 7.2, 2 M NaCl). The arrows indicate the buffer change. The rhFSH eluted when the elution buffer was added.

Table 1: Purification chart of a crude sample of rhFSH with the host cell proteins after chromatography with the affinity matrix with Ac-KVPLTVSKAKVAC-NH_2 immobilized.

Sample	Volume (mL)	Conc. rhFSH (mg/mL)	Total protein (mg/mL)	Purity % (mg/mg)	Fold purification	Yield (%)
Crude sample	17	0.727	1.203	60.4	-	-
Eluate	6	0.845	0.899	94.0	1.56	41

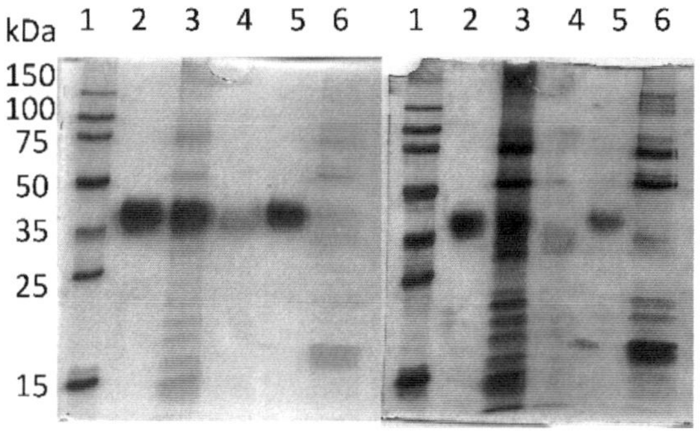

Figure 2: Non-reducing SDS-PAGE of the chromatographic step to check rhFSH purification. Left: Coomassie Blue staining. Right: Silver staining. Lane 1) protein molecular weight marker; Lane 2) standard of pure rhFSH; Lane 3) crude sample of rhFSH; Lane 4) washing fractions; Lane 5) elution fraction; Lane 6) regeneration fraction.

After its purification, rhFSH quality was analyzed. The percentage of oxidized rhFSH was 3.4 % and the percentage of free subunits was 1.2 %, both within the range established by the European Pharmacopeia as also were the sialic acid content and the isoforms profile.

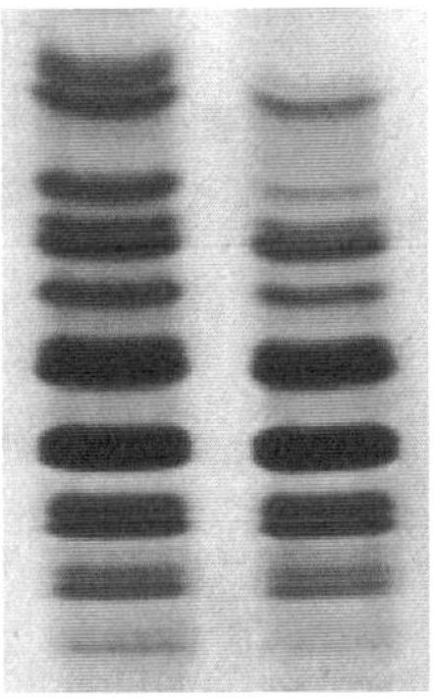

Figure 3: IEF gel for isoforms profile. 1) rhFSH standard; 2) sample of rhFSH purified by AC

Conclusions

A new affinity chromatography matrix with an immobilized peptide was synthesized for rhFSH purification. The method here designed allows obtaining a high quality rhFSH using a low-cost affinity matrix based on a short peptide ligand. The isoforms with high content of sialic acid were recovered by this new chromatographic step. These isoforms have higher plasma half-life and hence higher in-vivo potency. Furthermore, the physicochemical analysis complied with the limits established in European Pharmacopoeia for follitropin containing samples. These results evidence that affinity chromatography with the ligand peptide A c-KVPLTVSKAKV AC-NH_2 is a promising strategy for rhFSH purification from CHO crude extracts. The invention has been presented to the National Institute of Industrial Property, Argentine (Act INPI: No. 2017010281, 2017).

Acknowledgements

S.L.G, M.C.M.C, N.U., O.C. and S.A.C. are researchers of the Consejo Nacional de Investigaciones Científicasy Técnicas (CONICET). This work was partially supported by the Universidad de Buenos Aires (20020170100030B A), the Ministerio de Ciencia, Tecnologa e Innovación Productiva de la República Argentina (PICT-2014-1508) and the Consejo Nacional de Investigaciones Científicas y Técnicas de la República Argentina (PIP 11220130100119CO).

References

1. Smitz J, Wolfenson C, Chappel S, Ruman J. Reprod. Sci. 2016; 23: 706-716. DOI: 10.1177/1933719115607992

2. Camperi SA, Martínez-Ceron MC, Giudicessi SL, Marani MM, Albericio F, Cascone O. En: Methods Mol Biol. 2014;1129:277-302. DOI: 10.1007/978-1-62703-977-2_22.

3. Sohn J, Ryu K, Sievert G, Jeoung M, Ji I, Ji TH. Follicle-stimulating hormone interacts with exoloop 3 of the receptor. J Biol Chem. 2002 Dec 20;277(51):50165-75. DOI 10.1074/jbc.M207646200.

Bevacizumab purification by peptide affinity chromatography

Gabriela R. Barredo[1], Silvana L. Giudicessi[1], Mara C. Martinez-Ceron[1], Soledad L. Saavedra[1], Gustavo Mahler[2], Fernando Albericio[3], Osvaldo Cascone[1], Silvia A. Camperi[1]

[1]Universidad de Buenos Aires. FFyB. Cát. Biotecnología. CONICET-UBA, Inst. NANOBIOTEC. Junín 956, 1113, Bs As., Argentina
[2]AGC Biologics. 22021 20th Avenue SE. Bothell, WA 98021. USA, United States
[3]School of Chemistry, University of KwaZulu-Natal, Durban 4001, South Africa

https://doi.org/10.17952/35EPS.2018.130

Introduction

Therapeutic Monoclonal Antibodies (mAbs) are widely used in the treatment of many diseases. Bevacizumab (trade name: Avastin) inhibits the vascular endothelial growth factor and hence the angiogenesis process. It is used in brain, breast, colorectal, lung and renal cancer treatments (1). Due to its parenteral administration, its degree of purity must be extremely high. Nowadays, that is achieved by affinity chromatography (AC) with immobilized protein A, a highly expensive ligand that increases the cost of the purification process. On the other hand, short peptides are ideal ligands for AC due to their higher stability, easier synthesis and lower cost in comparison to protein A (2). In this work, a short peptide ligand with affinity for Bevacizumab was selected from a peptide library.

Methods

A one-bead-one-peptide combinatorial library was developed by the divide-couple-recombine method, using the HMBA-ChemMatrix resin as solid support and the Fmoc strategy (3). Bevacizumab (AGC Biologist, USA) was labeled with Texas-Red. After mixing it with the library, fluorescent beads were selected, and the peptides identified by MALDI-TOF MS/MS. Those that appeared most frequently were synthesized in larger quantities on Rink Amide resin and separated from the solid support with TFA cleavage cocktail. Chromatographies were carried out after immobilizing the peptides on PierceTM NHS-activated dry agarose resin. One of the peptides was capable to adsorb IgG from CHO cell supernatants better than the others, while all the contaminants passed through without interacting with the chromatographic matrix. Adsorption isotherms were performed at room temperature to characterize the generated affinity resin.

Results and Discussion

Chromatographies with the peptide ligand were performed using pure IgG, CHO supernatant, and CHO supernatant spiked with IgG (Fig. 1).

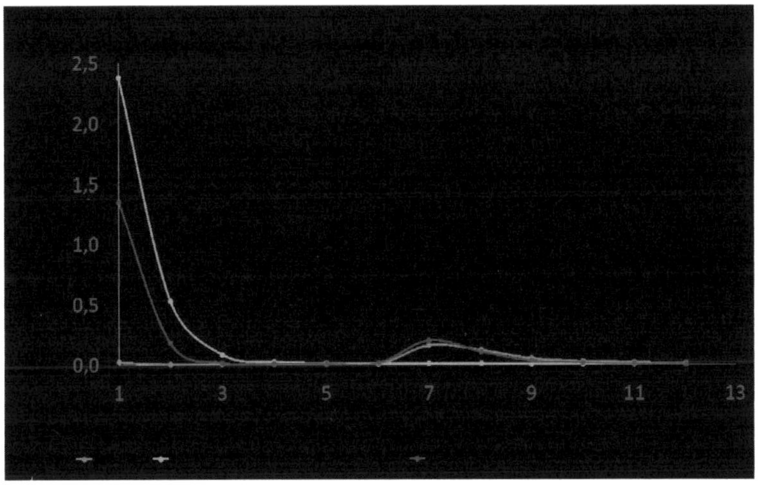

Figure 1: Chromatograms with the peptide ligand. Protein was measured by its absorbance at 280 nm. Pure IgG (red). CHO supernatant (blue). CHO supernatant spiked with IgG (purple).

Chromatographicsamples were loaded on SDS-PAGE (Fig. 2). The peptide AC showed a good selectivity for IgG and no interaction with CHO supernatantproteins was verified.

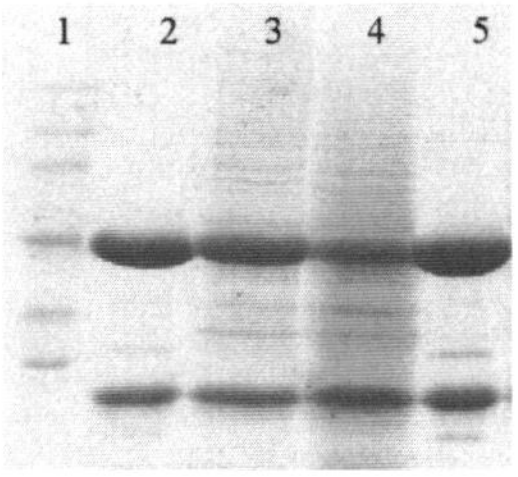

Figure 2: SDS-PAGE of the chromatographic Bevacizumab's purification. Lane 1) Protein molecular weight marker (MW); Lane 2) Standard of pure Bevacizumab; Lane 3) CHO cells supernatant spiked with IgG; Lane 4) and 5) washing and elution fractions of the chromatography respectively.

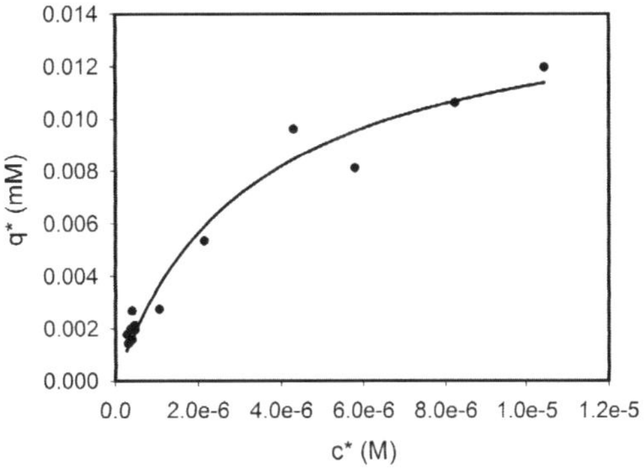

Figure 3: Isotherm carried out at 25°C for 2 h.

Fig. 3 shows the adsorption isotherm of Bevacizumab on the peptide support. The isotherm shows a good fit of experimental data to a Langmuir-type isotherm allowing the calculation of a maximum capacity (qm) of 0.015 μmol Bevacizumab/mL of matrix and a Kd of 10^{-6} M.

Conclusions

One-bead-one-peptide combinatorial libraries seemed to be a useful method to find possible ligand candidates for purifying proteins by affinity chromatography. A peptide was selected according to its frequency of appearance and its performance as an affinity ligand was studied.

The SDS-PAGE of the elution fraction of the AC with the selected peptide showed that only the IgG elutes from the complex mixture. No supernatant protein was detected in the elution fractions while some bands at the same level of IgG bands could be seen in the wash fractions, suggesting that not all IgG could be adsorbed.

The adsorption isotherm showed a Kd in the range of 10^{-6} which is higher than protein A Kd. However, as no CHO supernatant protein was found in the elution, this peptide appears to be selective.

Acknowledgements

S.L.G, M.C.M.C, O.C. and S.A.C. are researchers of the Consejo Nacional de Investigaciones Científicas y Técnicas (CONICET). This work was partially supported by the Universidad de Buenos Aires (20020170100030B A), the Ministerio de Ciencia, Tecnología e Innovación Productiva de la República Argentina (PICT-2014-1508) and the Consejo Nacional de Investigaciones Científicas y Técnicas de la República Argentina (PIP 11220130100119CO).

References

1. Poku E, Rathbone J, Wong R, Everson-Hock E, Essat M, Pandor A, Wailoo A. BMJ Open. (2014). 17;4(7):e005244. DOI 10.1136/bmjopen-2014-005244

2. Camperi SA, Martínez-Ceron MC, Giudicessi SL, Marani MM, Albericio F, Cascone O. En: Methods Mol Biol. (2014). 1129:277-302. DOI: 10.1007/978-1-62703-977-2_22.

3. Martínez-Ceron MC, Giudicessi SL, Marani MM, Albericio F, Cascone O, Erra-Balsells R, Camperi SA. Anal Biochem. (2010). 400:295-7. doi: 10.1016/j.ab.2010.01.029.

Peptide-based recovery of gallium

Nora Schönberger[1], Robert Braun[2], Sabine Matys[2], Franziska Lederer[2], Katrin Pollmann[2]

[1]Technische Universität Bergakademie Freiberg, Germany
[2]Helmholtz Institute Freiberg for Resource Technology, Germany

https://doi.org/10.17952/35EPS.2018.133

High-tech metals such as gallium, are almost ubiquitous in our everyday lives. Due to their great importance for the electronics industry, the demand is continuously growing.

Gallium is an important component of many semiconductor products and part of light-emitting diodes (LED) and photovoltaic elements as well. The supply is currently mainly covered from primary raw material sources. Efficient strategies for the recovery of gallium from secondary raw materials are missing [1].

A possible starting point to overcome this gap is the processing of industrial wastes to recover gallium. This approach offers the additional appeal of being able to establish a circular economy that enables companies to act in a more resource-efficient manner. This qualifies to more economical and sustainable company operations.

Biotechnological approaches can make a valuable contribution to the development of sustainable recycling systems. These include various methods for mobilising, complexing, concentrating and selectively separating certain metals [2]. Biosorption, i.e. the passive interaction of biomass with certain ligands in aqueous solution, is particularly promising with regard to metal recovery from wastewater containing heavy metals [3]. In particular, the targeted use of peptides as biosorbents has many advantages, since these biomolecules can exhibit high stability, metal specificity and affinity for individual ions.

In so-called biopanning experiments, different phage that express individual peptide sequences are selected against a specific target material. The phage display technology is used for this purpose. It enables the representation of certain or random amino acid sequences on the particle surface of bacteriophage. This is achieved by fusing the peptide sequence with a viral capsid protein using recombinant DNA technology. All resulting variants of a bacteriophage population form a phage display library, which can be used for selection against a specific target [4].

Here, the phage display technology was used to identify specific gallium binding peptides.

A total of 5 putative gallium binding bacteriophage clones from a commercial phage display library, that presents random peptides at the PIII protein, could be selected (Ph.D.™-C7C Phage Display Peptide Library, NEB, US). These were further characterized in single clone experiments. The binding affinity of the best gallium binding phage clones for a gallium-loaded sepharose material (Cube Biotech, Germany) was determined and compared with the binding properties of a clone without an additional expressed peptide sequence (wild type, Wt). All 5 tested clones showed a higher affinity for the target material than the wild type (see Figure 1). The clones that showed the highest binding affinity for gallium were C10.20 (HGGQTVA) with 32-fold better binding and C12.16 (SIKHAST) with 47-fold better binding.

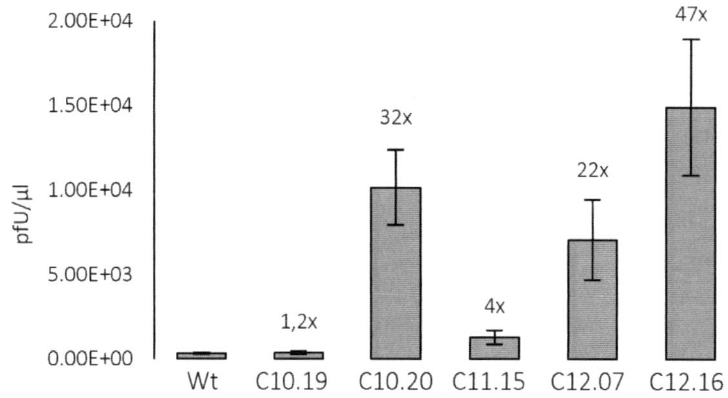

Figure 1: Single clone binding studies. In titer assay detected bacteriophage concentration bound to Ga-loaded NTA sepharose.

In further experiments, two peptides were characterized independently of the bacteriophage. Both peptides were purchased by Fmoc-based solid phase synthesis (GL Biochem, China).

In a first experiment the binding of gallium ions in aqueous solution was investigated using Raman spectroscopy (Malvern, UK). In Figure 2 Raman spectra of the putative gallium binding peptides C10.20 and C12.16 in the presence and in the absence of equimolar concentrations of gallium nitrate are shown. With the aid of Raman spectroscopy, among other things, conclusions can be obtained about non-elastic light scattering of abiotic and biotic samples in aqueous solution. The peak at 1047 cm^{-1} signals the presence of the gallium nitrate compound [5]. The amide I region between 1600 cm^{-1} and 1700 cm^{-1} states about the secondary structure of peptide samples. The Raman shift that could be observed for the investigated peptides in the presence of gallium nitrate indicates conformational changes due to the formation of a peptide complex with gallium ions.

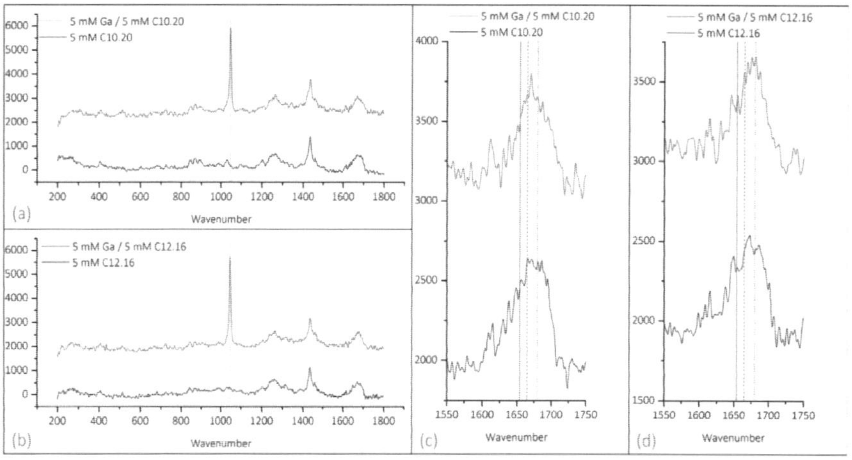

Marker bands: ( ) 1047 cm^{-1}, (—) 1654 cm^{-1}, (---) 1665 cm-1, (····) 1680 cm^{-1}

Figure 2: Raman spectra of putative gallium binding peptides in the presence and absence of equimolar concentrations of gallium nitrate. Raman spectroscopy was performed in ultrapure water for (a) 5 mM peptide C10.20 (ACHGGQTVACGGGS) and (b) 5 mM peptide C12.16 (ACSIKHASTCGGGS). A close-up those spectra is shown for the amide I region (O=C-N-H stretch) of (c) peptide C10.20 and (d) peptide C12.16. The Raman shift in this region indicates conformational changes due to the formation of a peptide complex with gallium ions.

In future experiments, an additional cysteine at the C-terminus will allow the immobilization of peptides *via* a thiol group on different surfaces such as silanized glass, gold chips or gold nanoparticles [6]. By this, peptide-based materials can be generated for the selective extraction of gallium from industrial wastewaters.

References

[1] E. Comission, Report on Critical Raw Materials and the Circular Economy, in, European Union, 2018.

[2] K. Pollmann, S. Kutschke, S. Matys, J. Raff, G. Hlawacek, F. Lederer, Bio-recycling of metals: Recycling of technical products using biological applications, 2018.

[3] B. Volesky, Z.R. Holan, Biosorption of Heavy Metals, Biotechnol. Prog, 11 (1995) 235-250.

[4] G.P. Smith, V.A. Petrenko, Phage Display, Chem Rev, 97 (1997) 391-410.

[5] W.W. Rudolph, C.C. Pye, G. Irmer, Study of gallium(III) nitrate hydrate and aqueous solutions: Raman spectroscopy andab initio molecular orbital calculations of gallium(III) water clusters, Journal of Raman Spectroscopy, 33 (2002) 177-190.

[6] J.L. Zimmermann, T. Nicolaus, G. Neuert, K. Blank, Thiol-based, site-specific and covalent immobilization of biomolecules for single-molecule experiments, Nature Protocols, 5 (2010) 975.

Proceedings of the 35th European Peptide Symposium
Patrick B. Timmons, Chandralal M. Hewage, Michal Lebl (Editors)
136 European Peptide Society & PSP, 2018

Designing a minimal synthetase

Radoslaw Piast, Rafał Wieczorek, Aleksandra Misicka

Warsaw University, Poland

https://doi.org/10.17952/35EPS.2018.136

Abstract

Minimal enzymes are enzymatic units reduced to the bare essentials with a preservation of their basic functional catalytic activity. The most famous example is Ser-His dipeptide - a minimal hydrolase. Such minimal enzymes are thought to be a missing link between prebiotic chemistry and the first organisms in the early Earth environment and hold potential for industrial use in place of enzymes in the future. As far as Ser-His dipeptide can be used to condensate substrates, this reaction is unfavorable in a water environment. In aqueous conditions direction of the reaction will be shifted from polymerization to hydrolysis. There is, however, another type of enzymes that deal with this problem by coupling unfavorable reactions with promoted ones. Synthetases are driven by hydrolysis of the pyrophosphate bond which they couple with the reaction of synthesis.

Introduction

The most widespread origin of life hypothesis is "RNA world". Although RNA molecules possess many properties crucial for life like the ability to store information, ability to evolve and catalysis their creation in the prebiotic environment remains uncertain [1]. Peptides and amino acids, on the other hand, have been proven to be produced in a variety of prebiotic reactions [2,3,4,5,6] which makes them probably first life-related organic polymers on Earth [7].

"Irreducible complexity" is partially obsoleted term that nowadays regards codependency between proteins and nucleic acids. In great simplification proteins (polymerases) produce nucleic acids and nucleic acids (ribosomes) produce proteins. This dependency is being found in all life forms across the Tree of Life thus it must have evolved very early in the history of life [8].

We believe that peptides have been present in the process of replication of nucleic acids from the very beginning. Moreover, their (peptides) enzymatic activity could have allowed nucleic acids to emerge as information polymers[9].

In order to do so, these peptides would have to possess properties related to polymerases. Since nearly all the reactions in the "irreducible complexity" can be broken down into chains of phosphorylation a and phosphate residue substitution [10] (which reactions define enzymes from the class of synthetases) what we have started to seek is a minimal synthetase.

Results and Discussion

We have synthesized a number of designed peptides. Two of them expressed similar to expected properties. We have checked for desired activities in a variety of conditions. Observation worth of notice is that the reaction course is pH dependent. In alkaline pH peptide C1 catalyzes the hydrolysis of pyrophosphate (Fig. 1A).

Peptide C1 and peptide C3 in acidic conditions are able to activate themselves by phosphorylation and subsequently substitute phosphate residue with an amino acid from the environment (Fig. 1B). Presence of Leu-NH_2 as a substituted amino acid suggests that activation takes place *via* carboxyl end.

Peptide C1 also possess the ability to dissolve and decompose pyrophosphate from calcium pyrophosphate crystals, a potential source of phosphates in the early Earth conditions thus possess chemolithotrophic abilities.

We achieved similar to desired catalytic properties. Following experiments will be conducted to better catalytic properties toward catalysis instead of autocatalysis and higher turnover speed. Minimal enzymes like the ones presented here could be a missing link between chemistry and biology and important catalytic power for the first nucleic acid replicons. Moreover, small peptidic catalysts could be used in the future as a cheaper and more stable substitute for enzymes in bioreactors and new forms of drugs.

Methods

Design. In order to design minimal synthetases, we have investigated various enzymes involved in phosphate transfer chemistry (kinases, synthetases, polymerases). In all of these enzymes, the noticeable feature was a layer-like arrangement of crucial residues and substrates. Subsequently, we have transferred these properties onto smaller, peptidic units and performed a series of energy minimalizations *in silico* using ICM-Pro and Gromacs software. Peptides with best properties were synthesized using SPPS Fmoc strategy.

Activity assays. We were incubating peptides with Mg ions, pyrophosphate, N-Me-Phe-OH and H-Leu-NH_2. As a control, we have used mixtures with Ser-His dipeptide, without peptide and without peptide and Mg ions.

To measure the catalytic properties of designed peptides we have used two types of assays: indirect (measurement of pyrophosphate utilization) and direct (peptide bond formation *via* HPLC -TOF-MS)

We have monitored the reaction by measuring the concentration of phosphates in the reaction tubes *via* micro determination of phosphates [10] and in the end, we were checking for the presence of peptide derivatives. Three such derivatives could have been produced: N-Me-Phe-peptide, peptide-Leu-NH_2, and N-Me-Phe-Leu-NH_2.

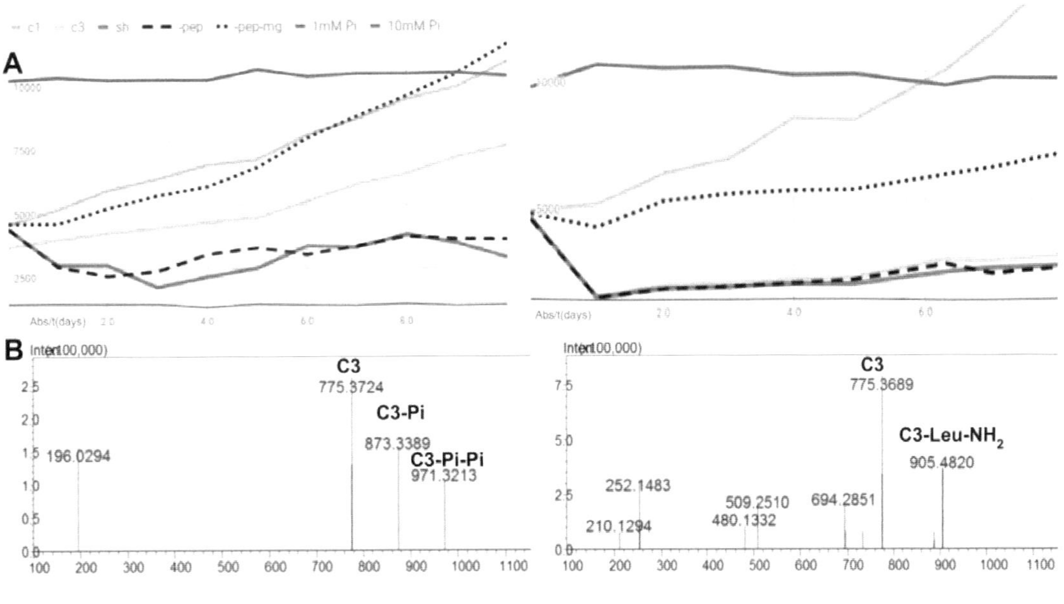

Figure 1: A: pH dependency between two types of reactions: synthetase on the left at low pH and pyrophosphate hydrolysis on the right at high pH. Only peptides C1, C3 and Ser-His shown. Dotted lines are controls without peptide and without peptide and Mg ions, solid grey lines represent known concentrations of phosphoric acid. Absorbance shown as Ax10000 and time in days. B: Mass spectra of C3 peptide reaction with Mg ions, pyrophosphate, N-Me-Phe-OH and H-Leu-NH_2. Picks corresponding to phosphorylated peptide and leucinated peptide are present.

Acknowledgments

The project was carried out with the use of CePT infrastructure financed by the European Union – the European Regional Development Fund within the Operational Programme "Innovative economy" for 2007-13.

This work was also supported by an NCN grant 501-D112-66-0005680

References

1. Muluye RA, Bian Y, Alemu PN. 2014. J Tradit Complement Med 4: 93-8

2. Xie C, Kokubun T, Houghton PJ, Simmonds MSJ. 2004. Phytotherapy Research 18: 497-500

3. Craik DJ, Swedberg JE, Mylne JS, Cemazar M. 2012. Expert Opin Drug Discov 7: 179-94

4. Weidmann J, Craik DJ. 2016. J Exp Bot 67: 4801-12

5. Pranting M, Loov C, Burman R, Goransson U, Andersson DI. 2010. J Antimicrob Chemother 65: 1964-71

6. Cao P, Yang Y, Uche FI, Hart SR, Li WW, Yuan C. 2018. Int J Mol Sci 19: 793

7. Rode, B.M. (1999). Peptides 20:773–786.

8. Orgel, L.E. (2000). Proc Natl Acad Sci USA 97:12503–12507.

9. Piast, R. W., & Wieczorek, R. M. (2017). Astrobiology, 17(3), 277-285.

10. Chen, P. S., Toribara, T. T., & Warner, H. (1956). Analytical Chemistry, 28(11), 1756-1758.

Modification of metal surfaces using TCM-derived cyclotides: An antibacterial and antibiofilm study

Pan Cao[1], Ying Yang[2], Fidelia Ijeoma Uche[2], Sarah Hart[2], Chengqing Yuan[1], Wen-Wu Li[2]

[1]School of Energy and Power Engineering, Wuhan University of Technology, China
[2]Institute for Science and Technology in Medicine, Keele University, United Kingdom

https://doi.org/10.17952/35EPS.2018.139

Introduction

Traditional Chinese Medicine (TCM) has been used in China for more than a thousand years for the prevention and treatment of various diseases including infections (1). Herbal medicine is the main component of TCM, which are grouped into many categories in the treatment of diseases. For example, the heat-clearing Chinese herbs (HCCHs) are mostly cold in nature, which "can clear away heat, purge fire, dry dampness, cool blood, and relieve toxic material."(1) Many HCCHs such as *Viola yedoensis*(2), *Coptis chinensis*, *Flos Lonicerae*, *Radix Isatidis*, and *Andrographis paniculata*, have been demonstrated to be effective in the treatment of inflammatory disease and microbial infection (1). In particular, one of the active components of *V. yedoensis* – cyclic peptides (cyclotides) are very interesting and antibacterial molecules. Cyclotides are a family of plant disulfide-rich peptides arranged in a knotted pattern which is connected *via* cysteine residues with a combinatorial cyclic backbone (3, 4). These structural features bestow on the cyclotides not only a remarkable stability, but also various biological activities including antibacterial (5, 6), antifouling effects(7) and anti-HIV(8) activities. *Viola* species (Violaceae family) contain an abundance of cyclotides (5). Recently, a series of cyclotides have also been isolated and characterized by us from several traditional Chinese herbs (6, 9, 10). *V. ordorata* and *V. arvensis* (5) are herbal medicines approved by European Medicines Agency for the treatment of skin and respiratory tract disorders and other diseases.

Bacterial biofilms are associated with more than two thirds of all infections, posing a major threat to public health. Bacteria in biofilms are highly resilient and render conventional antibiotics inefficient. Preventative strategies are appealing in comparison to allowing biofilms to form; in particular, surface modification using antimicrobial peptides may reduce biofouling by creating bactericidal, bacteria-resistant, or bacteria-repelling surfaces. However, most of peptides possess a number of disadvantages, such as susceptibility to proteolytic degradation. Hence, stable peptidomimetics need to be designed and synthesized to overcome these shortcomings (11). But these will create another obstacle for the future and wide application and commercialization due to high-cost of production, technological challenges and environmental pollution. Here, naturally occurring cyclotides, possessing remarkable stability and antimicrobial activities from a traditional Chinese Medicine (TCM) - *Viola philippica* Cav., were used for the modification of metal surfaces in order to prevent biofilm formation.

Methodology

Cyclotides were extracted and purified from *V. philippica* Cav. by HPLC, and identified using mass spectrometry. Cyclotides were subsequently utilized to modify stainless steel surfaces *via* polydopamine-mediated coupling (12). The resulting cyclotide-modified surfaces were characterized by Fourier transform infrared (FTIR) spectroscopy and contact angle analysis. The antibacterial capacity of these cyclotides against *Staphylococcus aureus* was assessed by Alamar blue assay. The antibiofilm capacity of the modified surfaces was assessed by crystal violet assay, and scanning electron microscopy (SEM) (6).

Results

A composite of Varv A, Kalata b1, Viba 15 and Viba 17 (P1), Varv E (P2), and Viphi G (P3) was isolated and identified (Table 1) . FTIR analysis of the modified surfaces demonstrated that cyclotides bound to the surfaces and induced reduction of contact angles. Antimicrobial effects showed an order P3 > P1 and P2, with P3-treated surfaces demonstrating the strongest antibiofilm capacity (Figure 1). SEM confirmed reduced biofilm formation for P3-treated surfaces (6).

Table 1: Cyclotides isolated and identified from V. philippica.

Code	Cyclotide	Sequence of amino acid residues	Theoretical monoisotopic mass	Experimental monoisotopic mass	GRAVY for their linear forms	Net charges
P1	Varv A	Cyclo-(CGETCVGGTCNTPG---CSCSWPVCTRNGLPV)	2876.17	2876.06	0.148	0
	Kalata b1	Cyclo-(CGETCVGGTCNTPG---CTCSWPVCTRNGLPV)	2890.14	2890.11	0.152	0
	Viba 15	Cyclo-(CGETCVGGTCNTPG---CACSWPVCTRNGLPV)	2860.18	2860.12	0.238	0
	Viba 17	Cyclo-(CGETCVGGTCNTPG---CGCSWPVCTRNGLPV)	2846.02	2846.08	0.162	0
P2	Varv E	Cyclo-(CGETCVGGTCNTPG---CSCSWPVCTRNGLPI)	2890.14	2890.00	0.159	0
P3	Viphi G	Cyclo-(CGESCVF I P C I SAIIGCSCSNKVCYKNGSIP)	3170.43	3170.43	0.726	+1

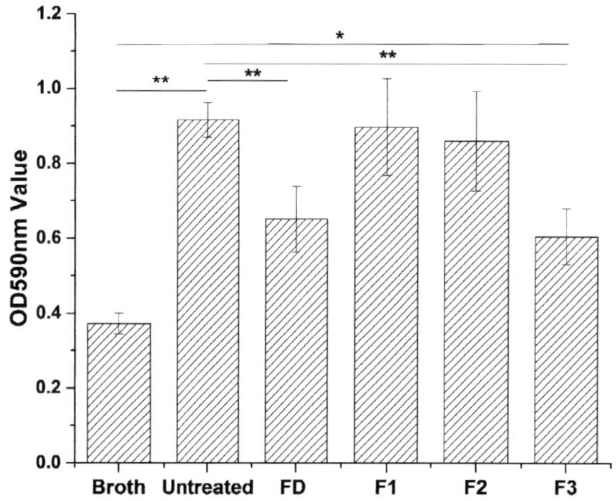

Figure 1: Comparision of antibiofilm ability of untreated, dopamine (DA)- and cyclotide-treated metal surfaces. Metal disk samples treated with DA, P1, P2 and P3 were denoted as FD, F1, F2 and F3, respectively. Optical density of the eluted acetic acid at 595 nm from the crystal violet stained metal samples after incubation with S. aureus is shown here. ANOVA was used to compare groups (, $p < 0.05$; **, $p<0.01$, $n = 4$). Error bars indicate standard deviation.*

Conclusions

In summary, cyclotides were isolated and identified from *V. philippica*. These peptides were successfully utilized to modify the surfaces of stainless steel *via* a simple, versatile and facile coupling agent, dopamine, for the first time. We therefore propose that TCM-derived cyclotides, with their unique three-dimensional structure and remarkable stability, can serve as a novel source of biological materials to be used for the modification of metal surfaces in medical devices, aquaculture, food manufacture and shipbuilding devices. This study provides novel evidence for cyclotides as a new class for development of antibacterial and antibiofilm agents (6).

Acknowledgements

This work was financially supported by the National Natural Science Foundation of China (No.51375355 and No.51422507), the Program of Introducing Talents of Discipline to Universities (B08031) and the Fundamental Research Funds for the Central Universities (2016-YB-023). We thank Mao-Hua Li for collecting plant materials. The advice on bacterial experiments from Professor Paul Horrocks and Dr Andrew Morris are highly appreciated.

References

1. Muluye RA, Bian Y, Alemu PN. 2014. J Tradit Complement Med 4: 93-8
2. Xie C, Kokubun T, Houghton PJ, Simmonds MSJ. 2004. Phytotherapy Research 18: 497-500
3. Craik DJ, Swedberg JE, Mylne JS, Cemazar M. 2012. Expert Opin Drug Discov 7: 179-94
4. Weidmann J, Craik DJ. 2016. J Exp Bot 67: 4801-12
5. Pranting M, Loov C, Burman R, Goransson U, Andersson DI. 2010. J Antimicrob Chemother 65: 1964-71
6. Cao P, Yang Y, Uche FI, Hart SR, Li WW, Yuan C. 2018. Int J Mol Sci 19: 793
7. Goransson U, Sjogren M, Svangard E, Claeson P, Bohlin L. 2004. J Nat Prod 67: 1287-90
8. Gustafson KR, McKee TC, Bokesch HR. 2004. Curr Protein Pept Sci 5: 331-40
9. Tang J, Wang CK, Pan X, Yan H, Zeng G, *et al.* 2010. Peptides 31: 1434-40
10. He W, Chan LY, Zeng G, Daly NL, Craik DJ, Tan N. 2011. Peptides 32: 1719-23
11. Andrea A, Molchanova N, Jenssen H. 2018. Biomolecules 8: 27
12. Cao P, Li WW, Morris AR, Horrocks PD, Yuan CQ, Yang Y. 2018. R Soc Open Sci 5: 172165

Proceedings of the 35th European Peptide Symposium
Patrick B. Timmons, Chandralal M. Hewage, Michal Lebl (Editors)
European Peptide Society & PSP, 2018

Modulating peptide hydrogel strength through halogenation

Wouter Vandeplassche[1], Steven Ballet [1], Charlotte Martin[1], Richard Hoogenboom[2], Annemieke Madder[2]

[1]Vrije Universiteit Brussel, Belgium
[2]Ghent University, Belgium

https://doi.org/10.17952/35EPS.2018.142

Introduction

Peptide-based hydrogels are promising delivery matrices for controlled-release. In such systems, specific peptides can self-assemble into fibers driven by non-covalent interactions (e.g. hydrogen bonding, ionic interactions, pi-pi stacking and Van der Waals interactions), a process followed by fiber entanglement, which eventually leads to the formation of the final hydrogel network. Thanks to the non-covalent nature of the fibers and their entanglements, the resulting gels are injectable (i.e. they present thixotropic or shear-thinning behavior). Fmoc-Phe and related molecules have been widely exploited as self-assembling hydrogelators. It has been found that the assembly properties of these molecules can be profoundly enhanced by the incorporation of various substituents, including halogen atoms, on the side chains. Halogenated Fmoc-Phe derivatives have a much higher propensity for spontaneous self-assembly into hydrogel fibril networks.[1,2] Similarly, halogenation of other peptide-based hydrogels has shown that these modifications can improve peptide hydrogel strength.[3,4]

Results & Discussion

Recently, we designed a new family of short amphipathic peptide-based hydrogels, which form thixotropic injectable hydrogels upon dissolution in aqueous solutions.[5,6] Based on theoretical calculations (i.e. Non-Covalent Interaction calculations[7]), the use of halogenated phenylalanines and tryptophan analogues were suggested to improve the strength of our peptidic hydrogels. The 'parent' structure in this effort is the following sequence: H-FQFQFK-NH_2. To verify whether halogenation would also impact the material properties of this sequence, a library of 14 halogenated phenylalanine-containing peptides was synthesized *via* SPPS using standard Fmoc-chemistry. The initial halogenated hydrogelator sequences were chosen in accordance with the calculated interaction energies of the distinctly substituted phenylalanines. Since meta-halogenated Phe residues (with X = Cl, Br, I), as well as the ortho-para "couple", were calculated to present the highest noncovalent interactions, sequences incorporating such amino acids were prepared. All peptides were made in good yields, whereafter gelation in saline solution and phosphate buffer solution (PBS) could be checked at weight percentages of 1% and 2%.

Beside the use of halogenated phenylalanines, the influence of halogenated tryptophans was also of interest. In contrast with the halogenated phenylalanines, the halo-tryptophans are not commercially available and therefore need to be synthesized. Based on the NCI calculations, we started with the synthesis of 4- and 7-halo-L-tryptophans. The 4-halo indole is synthesized according to the Leimgruber-Batcho method and the 7-halo indole, according to the work of Bartoli and coworkers.[8,9]

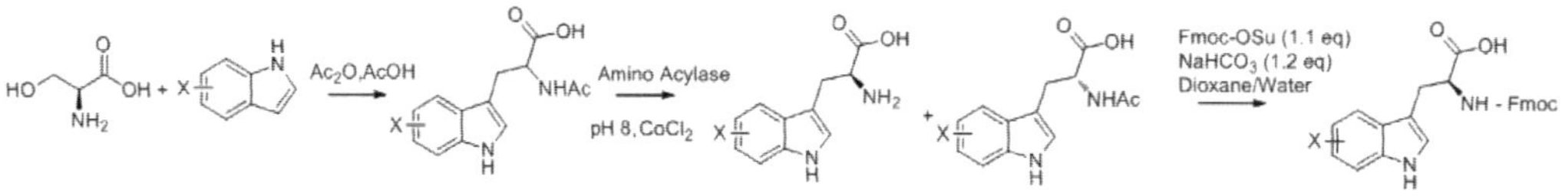

Figure 1: Synthesis of Fmoc-halo-L-tryptophans

Subsequently, the halo-tryptophans are chemically synthesized *via* a described condensation reaction, followed by enzymatic resolution.[10] Finally, the amine is Fmoc-protected in order to obtain a SPPS compatible building block. The 4- and 7-haloindole synthesis proved successful for bromine, chlorine and iodine. The Fmoc-protected brominated and chlorinated tryptophans were synthesized in moderate overall yields (3 steps, 15-29%), whereby enzymatic resolution can only present a maximum yield of 50%. The Fmoc-7-bromo-L-tryptophan and Fmoc-4-chloro-L-tryptophan were introduced into the parent hexapeptide in good yield, and subsequently, the synthesized tryptophan-containing peptide formed a hydrogel both in PBS and saline solution at weight percentages of 1 and 2%.

From a qualitative analysis (i.e. vial inversion test) it was already clear that the halogenated phenylalanines enhanced the rigidity of the hydrogels. Subsequently, quantitative results were needed to validate the material properties of the peptide hydrogels. Therefore, dynamic rheometry was applied. The following procedure was used to determine the G', G" and tan δ parameters. First a pre-shear was performed to disrupt the hydrogel. Followed by a time sweep where the gel can recover over a period of one hour. Subsequently, a frequency sweep is performed to determine if tan δ is independent of the frequency. Finally, a strain sweep determines at which strain the peptide hydrogels breaks.

Table 1: Peptide codes, sequences and average G' of the halo-peptide library.

Name	Sequence	G' (Pa)
CM63	FQFQFK-NH_2	605
WV1	FQFQ**F(3Br)**K-NH_2	12 580
WV4	FQ**F(3Br)**Q**F(3Br)**K-NH_2	3 201
WV5	**F(3Br)**Q**F(3Br)**Q**F(3Br)**K-NH_2	3 775
WV6	FQ**F(3Br)**QFK-NH_2	710
WV7	**F(4Br)**Q**F(4Br)**Q**F(4Br)**K-NH_2	2 829
WV13	**F(2Br)**QFQ**F(4Br)**K-NH_2	48 208
WV14	**F(4Br)**QFQ**F(2Br)**K-NH_2	67 952
WV8	FQFQ**F(3I)**K-NH_2	2 651
WV9	FQ**F(3I)**Q**F(3I)**K-NH_2	5 216
WV16	**F(3I)**Q**F(3I)**Q**F(3I)**K-NH_2	3 014
WV10	**F(4I)**QFQ**F(2I)**K-NH_2	1 936
WV15	**F(2I)**QFQ**F(4I)**K-NH_2	14 419
WV11	**F(4Cl)**Q**F(4Cl)**Q**F(4Cl)**K-NH_2	3 650
WV12	FQFQ**F(4Cl)**K-NH_2	2709
WV18	WQWQ**W(7Br)**K-NH2	N.D
WV19	**W(7Br)**QWQ**W(7Br)**K-NH2	N.D
WV20	WQWQ**W(4Cl)**K-NH2	N.D

These rheometry measurements, confirmed that the halopeptides have a higher rigidity, in comparison with the non-halogenated: the reference peptide (FQFQFK-NH_2) has G' of 650 Pa and the library of halogenated phenylalanine-based peptides give values varying from 917 Pa to 67 952 Pa, which is up to a hundred-fold increase in rigidity. The time sweep suggests that after one hour the hydrogel network is still gaining rigidity. Although the halopeptide hydrogels have a higher rigidity, it is observed that they break more easily because of the lower stress point looking at the strain sweep. The two peptide sequences which stand out are the partially brominated ortho-para "couples" [i.e. F(o-Br)QFQF(p-Br)K-NH_2 and F(p-Br)QFQF(o-Br)K-NH_2]. According to the present study, the theoretically calculated NCI energies give a good indication of which peptides can indeed lead to stronger gels.

Conclusion

Herein, we have conducted material studies on the effect of halogen substitution on the aromatic side chain of Phe and Trp amino acids. It has been discovered that halogen identity and ring position both exert a profound effect on self-assembly rates and on the mechanical properties of the resulting hydrogels. These results clearly demonstrate that subtle, single-atom perturbations amino acids can be used to tune the self-assembly and hydrogelation propensity of these peptide hydrogels.

References

1. Ryan, D.M., *et al.* Soft Matter 6.14 (2010): 3220-3231. DOI 10.1039/C0SM00018C

2. Liyanage, W., *et al.* Langmuir 32.3 (2016): 787-799. DOI 10.1021/acs.langmuir.5b03227

3. Pizzi, A., et al.CrystEngComm 19.14 (2017): 1870-1874. DOI 10.1039/C7CE00031F

4. Bertolani, A., *et al.* Nature Communications6 (2015): 7574 DOI 10.1038/ncomms8574

5. Martin, C., *et al.* Materials Today Chemistry 19.9 (2016) 491-502. DOI 10.1016/j.mattod.2016.01.016

6. Martin, C., et al MedChemComm7 (2016) 542-549. DOI 10.1039/C5MD00440C

7. Contreras-García, J., *et al.* (2016) 135.10 Theoretical Chemistry Accounts,242. DOI 10.1007/s00214-016-1977-7

8. Bartoli, G., *et al.* Chemical Society Reviews 43.13, (2014), 4728-4750 DOI 10.1039/C4CS00045E

9. Gribble, G. W., Indole Ring Synthesis: From Natural Products to Drug Discovery (2016): 338-348

10. Blaser, G. et al., TetrahedronLetters 49.17, (2008),2795-2798 DOI 10.1016/j.tetlet.2008.02.120

A self-assembling peptide hydrogel for ultrarapid 3D immunoassays

Alessandro Gori, Paola Gagni, Greta Bergamaschi, Alessandro Romanato, Marina Cretich

National Research Council of Italy, Italy

https://doi.org/10.17952/35EPS.2018.145

Microarray technology advent has set a new paradigm in the development of multiplexed assays for biomedical and many efforts were oriented to develop innovative surface chemistries able to balance a stable binding with retained functionality of immobilized bioprobes[1, 2, 3] . Hydrogels represent ideal systems bridging dry and wet conditions as biomolecules can be locally entrapped on analytical surfaces in their active form under solution mimetic conditions. However, some hydrogels characteristics still prevent their broad application for bioassays development. Particularly, a major issue is to match target diffusion rates through the gel matrix with stable probe entrapment since mass transport can be severely limited in dense cross-linked matrices.

In the present work, using micromolar concentrations of a self-assembling peptide we obtained a soft hydrogel matrix with controlled permeability properties for the straightforward confinement of biomolecular probes in arrayed 3D microenvironments. The self-assembling Q11 peptide (QQKFQFQFEQQ) is a fibril-forming amino acid stretch previously used as versatile nanostructured scaffold for several applications [4]. In our design we rationalized that working at low peptide concentrations was a necessary condition to obtain a low-viscosity hydrogel that could be handled for straightforward (micro)arrays production(Figure 1) [5]. Large biomolecules are entrapped within the fibrillary network due to their size whereas peptides can be immobilized by means of a co-assembly strategy with YF-Q11.

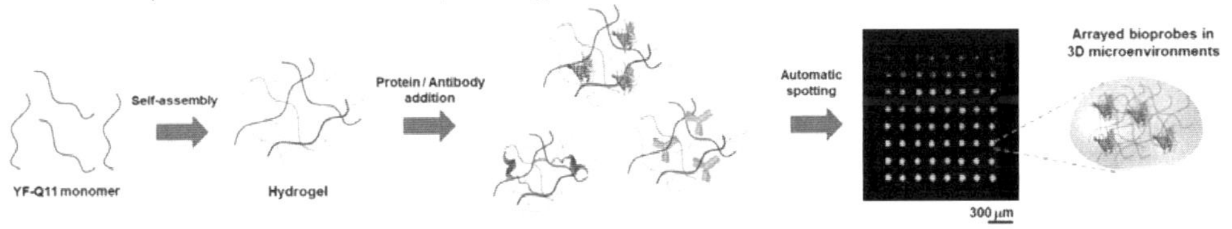

Figure 1: Overview of hydrogel bioassay strategy.

We used a slightly modified Q11 sequence (Ac-YFQQKFQFQFEQQ-coNH$_2$, YF-Q11) that formed PMMA-adhesive soft hydrogels in a wide micromolar concentration range (25-500 μM). Structural characterization of the new peptide by circular dichroism (CD) and ATR-FTIR analysis revealed that YF-Q11 preferentially shows a beta-sheet character, while spectrofluorimetric studies demonstrated that upon incubation at 40°C the spontaneous assembly proceeds rapidly and reaches a stable level within a convenient operative time frame (1-4 hours). Atomic force microscopy (AFM) confirm that peptide chains hierarchically aggregate in fibrillary structures that, in turn, self-organize into larger micrometer sized bundle of fibers (Figure 2). Rheology characterization of YF-Q11 hydrogel, showing that it behaves as a typical viscoelastic fluid under increasing strain excitation.

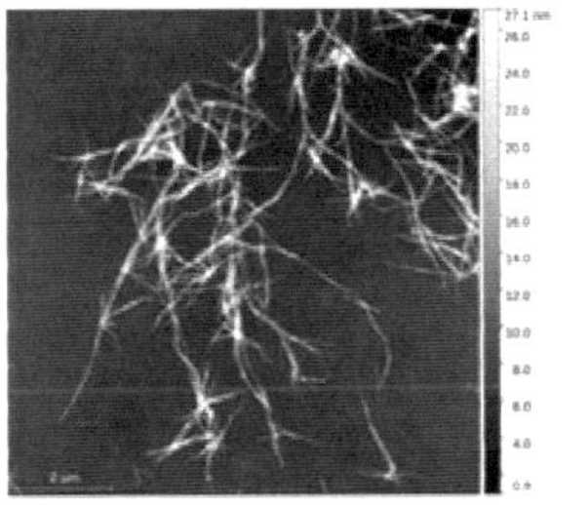

Figure 2: AFM representative image of 25μM YF-Q11 showing the formation of entangled nanofibers.

To assess the feasibility of YF-Q11 hydrogel towards bioassay applications, we designed an experiment providing combined information on spotted gel permeation to biomolecules while monitoring its stability. To this aim we exploited a dual fluorescence detection: a Cy3-labelled YF-Q11 peptide (0.1μM) was co-assembled into different YF-Q11 solutions (25-500 μM range) to monitorthe variation of fluorescence intensity of gel spots upon incubation of the slide in washing buffer at different times. Simultaneously, YF-Q11 spots containing Cy5-labelled biomolecules of representative molecular sizes (anti-α-lactalbumin IgG, streptavidin and V5 peptide) were scanned for Cy5 residual fluorescence to monitor diffusion through the fibrillar network. The whole range of YF-Q11 hydrogels tested concentrations(25-500 μM) showed a good stability over time. V5 peptide was fully permeable in all tested concentrations, while streptavidin freely diffuses below 100 μM; the same concentration allowed only for partial α-lactalbumin antibody diffusion, which however was almost quantitative at 25μM (Figure 3).

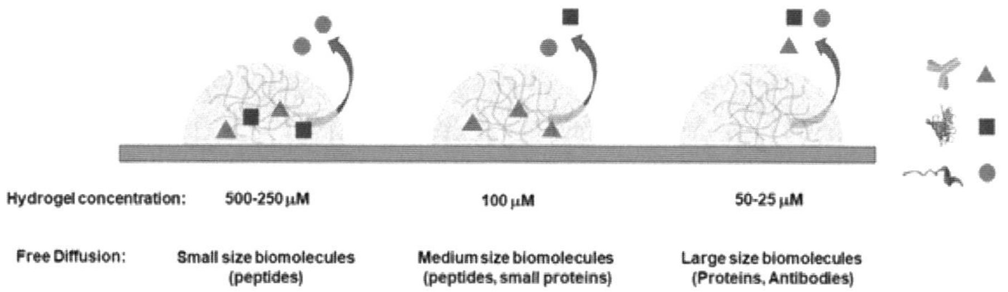

Figure 3: Graphical representation summary of hydrogel permeation to biomolecules.

Additionally, to verify that biomolecules are actually distributed in a 3D shell, confocal Raman imaging experiments were performed on an antibody-containing hydrogel with a concentration of 250 μM, results indicated that the labeled antibody was stably embedded in the hydrogel spots and uniformly distributed through the gel matrix (Figure 4). We then verified if these premises could translate into specific molecular recognition assays inside the hydrogel matrix.

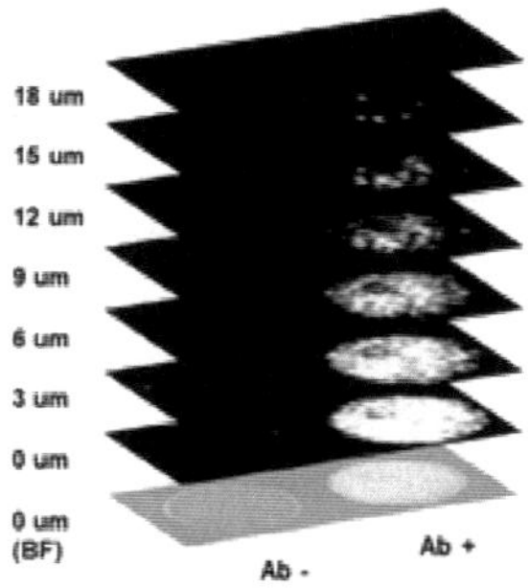

Figure 4: Confocal Raman imaging using a malachite-labelled antibody indicated it is stably embedded and uniformely distributed within the hydrogel spots.

To prove that, we designed symmetrical antibody-peptide recognition tests: in the first experiment we immobi-lized two different peptides (V5 and LAC) within the matrix and hydrogels were then probed with their specific Cy3/Cy5-antibodies. We also executed a specular assay in which we immobilized antibodies within the hydrogel. Both experiments showed that probe-specific recognition occurred only between respective antibody-peptidepairs (Figure 5).

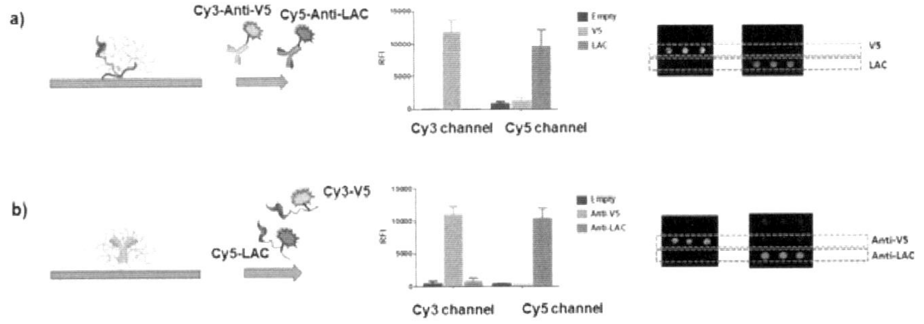

Figure 5: Fluorescence intensities detected in the Cy3 and Cy5 channels incubation of hydrogel spots with the respective probes.

As the final benchmark for our platform, we performed a real immunodiagnostic assay to detect Zika Virus (ZIKV) infection in human serum samples, confirming the actual feasibility of our system in the context of a real serological assay for infectious diseases. The total time required for this assay was <10 minutes, since the first recognition takes less than 5 minutes while incubation time for secondary antibody was kept at 2 minutes.

These results demonstrated that we obtained a stable and yet highly permeable nano-scaffold that enabled us to run fluorescence immunoassays under solution-like conditions in an ultrashort timespan. Remarkably, the supramolecular hydrogel forms rapidly and spontaneously, it is printable by a piezoelectric microarray spotter and it self-adheres on poly(methyl metacrylate) (PMMA) slides, so that neither post-spotting cross-linking nor polymerization is required for gel formation and surface attachment, greatly simplifying the fabrication process. Overall, our platform is user-friendly, robust and cost-effective and possesses a combination of favourable features which is unique in the area of 3D multiplexed bioassays in hydrogels.

References

1 A. Gori, M. Cretich, R. Vanna, L. Sola, P. Gagni, G. Bruni, M. Liprino, F. Gramatica, S. Burastero and M. Chiari, Anal. Chim. Acta, 2017, 983, 189–197.

2 A. Gori, L. Sola, P. Gagni, G. Bruni, M. Liprino, C. Peri, G. Colombo, M. Cretich and M. Chiari, Bioconjug. Chem., DOI:10.1021/acs.bioconjchem.6b00426.

3 L. Sola, P. Gagni, I. D'Annessa, R. Capelli, C. Bertino, A. Romanato, F. Damin, G. Bergamaschi, E. Marchisio, A. Cuzzocrea, M. Bombaci, R. Grifantini, M. Chiari, G. Colombo, A. Gori and M. Cretich, ACS Infect. Dis., 2018, 4, 998–1006.

4 J. S. Rudra, Y. F. Tian, J. P. Jung and J. H. Collier, Proc. Natl. Acad. Sci., 2010, 107, 622–627.

5 P. Gagni, A. Romanato, G. Bergamaschi, P. Bettotti, R. Vanna, C. Piotto, C. F. Morasso, M. Chiari, M. Cretich, A, Gori A Self-assembling Peptide Hydrogel for Ultrarapid 3D Bioassays, Nanoscale Adv., DOI:10.1039/C8NA00158H.

Proceedings of the 35th European Peptide Symposium
Patrick B. Timmons, Chandralal M. Hewage, Michal Lebl (Editors)

European Peptide Society & PSP, 2018

Development of new supramolecular nanostructured materials based on peptide hydrogelator Ac-L-Phe-L-Phe-L-Ala-NH2 with embedded liposomes for potential biomedical application

Ruža Frkanec[1], Karmen Radošević[2], Adela Štimac[1], Lucija Horvat[3], Leo Frkanec[4]

[1]University of Zagreb,Centre for Research and Knowledge Transfer in Biotechnology, Croatia
[2]Departmentof Biotechnology, University of Rijeka, Croatia
[3]Division of Molecular Biology, Ruder Boškovic Institute, Croatia
[4]Departmentof Organic Chemistry and Biochemistry, Ruder Bošković Institute, Croatia

https://doi.org/10.17952/35EPS.2018.148

In recent years design and synthesis of self-assembled nanomaterials with diversified structures and functionalities *via* fine tuning of supramolecular building blocks increased rapidly.[1] Self-assembling peptides have been widely recognized as nanomaterials with high potential for an extensive range of biomedical applications from drug delivery to tissue enginering, owing to their hydrophilic character and biocompatibility.[2,3]

Short peptide hydrogels made of two or three amino acids are of particular interest because they are cost effective and simple to prepare. It was previously shown that the newly developed self-assembling tripeptide hydrogel Ac-L-Phe-L-Phe-L-Ala-NH_2 (Ac-F-F-A-NH_2) is effective in enhancing cell proliferation and has potential for use in tissue engineering approaches.[4]

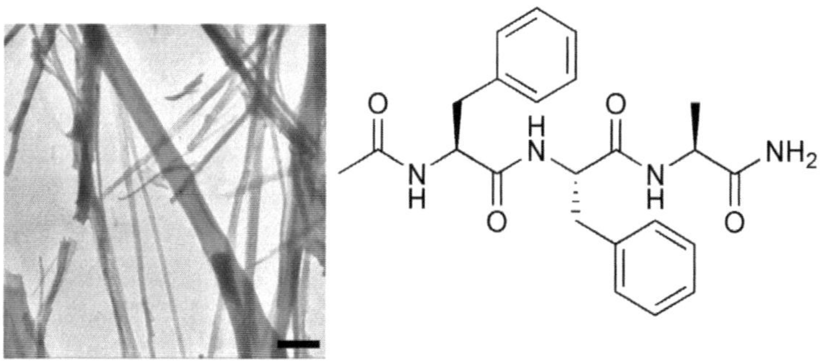

Figure 1: Chemical structure of hydrogelator Ac-L-Phe-L-Phe-L-Ala-NH_2. TEM image of hydrogel in water (2.64 mg/ml) PWK-stained. Magnification 8000x, scale bar: 1 μm.

Ac-L-Phe-L-Phe-L-Ala-NH_2 self-assembles at physiological pH without the need for any organic solvent at a minimum gelation concentration of 0.7%, w/v in saline. The tripeptide derivative contains an aromatic diphenylalanine moiety (Phe–Phe) as a minimalistic building block to drive the self-assembly of short peptides into nanostructures and hydrogels of high stability. Transmission Electron Microscopy (TEM) analysis of the nanofibrous hydrogel network has shown the presence of a mixture of fibers and straight ribbons with diameters in the range of 50–500 nm and lengths in the range of micrometers, which are stable over weeks at a very low concentration of 2.64 mg/ml (Figure 1).

The aim of the present study was preparation and characterization of supramolecular systems based on peptide hydrogelator Ac-L-Phe-L-Phe-L-Ala-NH_2 and liposomes with incorporated model proteins. The rationale behind preparation of liposomes-in-hydrogel as a delivery system is to assure sustained drug release during their prolonged presence at the administration site.

The gelling properties of the hydrogelator Ac-L-Phe-L-Phe-L-Ala-NH_2 with addition of liposomes and model proteins BSA/FITC-BSA were studied. Liposome formulations of BSA and FITC-BSA as well as empty liposomes, incorporated in the hydrogel were characterized by electron and confocal microscopy. It is known that intramolecular hydrogen bonding and noncovalent interactions between molecules of hydrogelators are responsible for the formation of the 3D gel network.[5] Therefore, we have tested the structure of the gel network after the addition of different amounts of protein and liposome.

TEM confirmed incorporation of liposomes and protein into the hydrogel (Figure 2) without disruption of the gel network. In gel samples with built-in BSA, the gel network preserved integrity, but it was noticed that the gel fibers were thinner. Confocal microscopy proved the incorporation of liposome formulations of FITC-BSA in the hydrogel (Figure 3). It was also shown that liposomes do not impair the gel network of hydrogelator at lipid concentrations lower than the gelator concentration (Table 1).

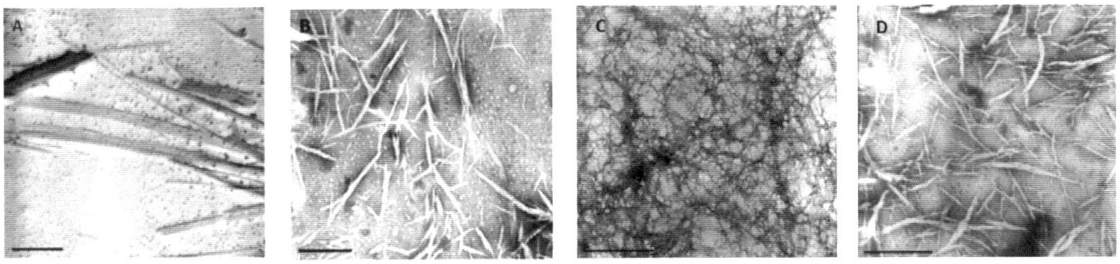

Figure 2: TEM images of (A) liposomes in hydrogel, Pd-shadowed. Lipids:gelator = 1:2.5, (B) liposomes in hydrogel, PWK-stained. Lipids:gelator = 1:5. Magnification 10 000x, scale bar= 1 μm, (C) BSA incorporated in the gel network (BSA:hydrogel = 1:1051), (D) liposome formulation of BSA embedded in gel network (lipids: hydrogel = 1:10), (C) and (D) PWK-stained, Magnification 16 000x. Scale bar: 1 μm.

Table 1: Gelation of Ac-L-Phe-L-Phe-L-Ala-NH_2 after addition of liposomes.

Mass of hydrogelator (mg)	Molar ratio (lipid:gelator)	Lipid (mg)/mg gelator	Comments
3.6	1:10	0.27	Matted, white gel, quickly formed
3.0	1:5	0.55	Matted, white gel, quickly formed
2.9	1:2.5	1.09	Matted, white gel, quickly formed
3.0	1:1	2.73	The gel is not formed

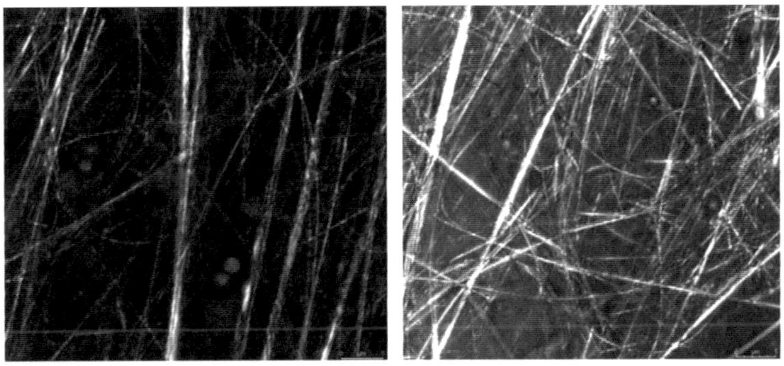

Figure 3: Confocal images of liposome formulation of FITC-BSA embedded in gel network (lipids: hydrogel = 1:10), Pd-shadowed.

We showed that use of higher amounts of protein albumin (BSA)/FITC-BSA resulted in slower gel formation or inability to transition into gel. It was also demonstrated that addition of sufficiently small amount of liposome formulation of BSA and FITC-BSA in hydrogels does not impair the gel network. The described supramolecular nanostructured material based on peptide hydrogelator Ac-L-Phe-L-Phe-L-Ala-NH_2 with embedded liposomes can be useful in various areas of application in biomedicine, including injections in situ of gelling agents for controlled release of the drug, drug delivery platforms, and tissue engineering.

Acknowledgements

This work was supported by the grant HRZZ-IP-11-2013-7387 from the Croatian Science Foundation.

References

1. J. Boekhoven , W. E. Hendriksen, G. J. M. Koper , R. Eelkema , J. H. van Esch , Science 2015, 34 , 1075.

2. X. Du, J. Zhou, J. Shi, B. Xu, Chem. Rev. 2015, 115, 13165.

3. R. Dong, Y. Pang, Y. Su, X. Zhu, Biomater. Sci., 2015, 3, 937.

4. T. Pospišil, L. Ferhatović Hamzić, L. Brkić Ahmed, M. Lovrić, S. Gajović, L. Frkanec, Biomater Sci. 2016, 4, 1412.

5. X. Du, J. Zhou, J. Shi and B. Xu, Chem. Rev., 2015, 115,13165.

Proceedings of the 35th European Peptide Symposium
Patrick B. Timmons, Chandralal M. Hewage, Michal Lebl (Editors)
European Peptide Society & PSP, 2018

Chemical labeling associated to mass spectrometry as a powerful tool for peptide detection and quantification in biology

Sonia Cantel, Mathieu Maingot, Maxime Rossato, Guillaume Miralles, David Paramelle, Didier Gagne, Jacky Marie, Guillaume Cazals, Jean Martinez, Muriel Amblard, Gilles Subra, Christine Enjalbal

IBMM, France

https://doi.org/10.17952/35EPS.2018.151

Mass spectrometry represents a method of choice to identify and characterize peptides and proteins present in complex biological mixtures at low concentrations. In the attempt to develop potent MS methodologies, we work on the design and synthesis of chemical tags able to increase detection sensitivity and specificity through direct MS detection and/or directed fragmentation.

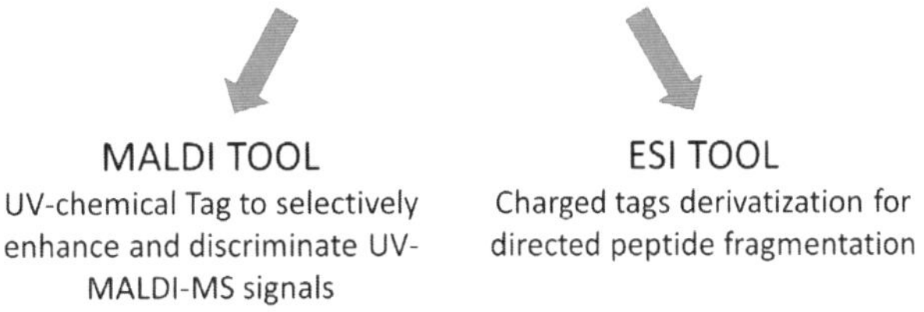

Figure 1

MALDI is well-known to provide very sensitive detection but not commonly used for quantitative measurements. On one hand, we aim at developing a novel approach that relies on the joint use of MALDI mass spectrometry and original labeling chemistries designed to specifically enhance the ionization of the tagged molecules. MALDI is well-known to provide very sensitive detection but not commonly used for quantitative measurements. Using MALDI for such purposes is challenging and has been successfully applied to track peptides at low concentrations in various media[1-3].

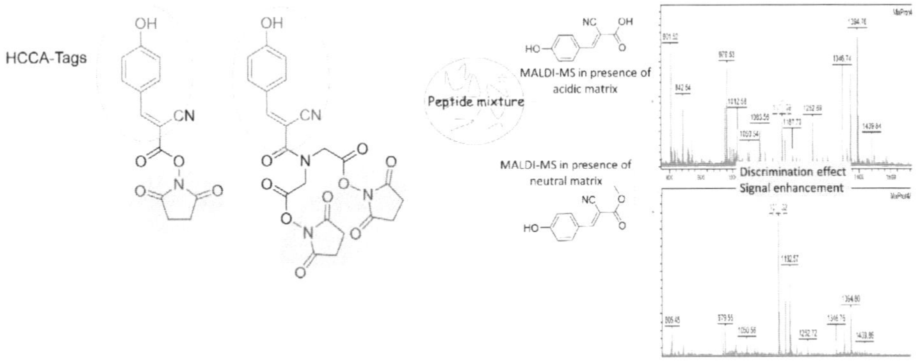

Figure 2

We developed a methodology associating HCCA (α-cyano-4-hydroxycinnamic acyl)-targeted peptide to be analyzed by MALDI-MS in a matrix such as HCCE (α-cyano-4-hydroxycinnamic methyl ester). This original approach allowed to selectively enhance and discriminate the MALDI-MS signals of targeted peptides. This concept was successfully applied to protein structure issues illustrated by the cross-linking of a model protein and peptide quantification for pharmacological studies of receptor/ligand systems[4,5].

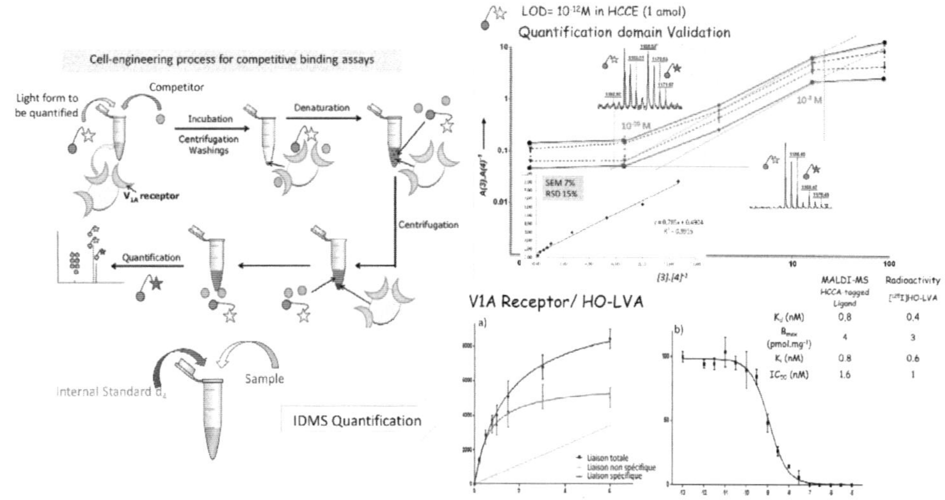

Figure 3

On the other hand, we focused on N-terminal positively charged peptide derivatization as efficient agents for directed ESI-MS fragmentation. We explored labeling by pyridinium-based molecules, well known in chemistry literature, generally for enantio-separation, pharmaceutical or biochemical analysis, showing a great tendency to meet some of the desired requirements to investigate the field of ESI-MS qualitative and quantitative analysis of biomolecules like proteins or peptides.

Two competing dissociation mechanisms can be suggested:

- Formation of α-ketene.
- Formation of an oxazolone. («a1 fragment analog »)

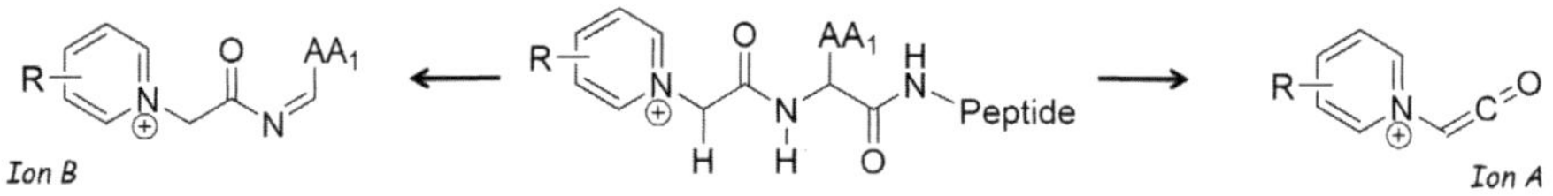

Figure 4

This technology should have a great impact in biosciences, in particular in research laboratories dealing with pharmacology. We will give you an account on the development of these methodologies.

References

[1] Lascoux, D et al., Angew Chem. Int. Ed. Engl. 2007, 46, 5594.

[2] D. Paramelle et al, Proteomics 2009, 9, 5384.

[3] D. Paramelle et al, Proteomics 2011, 11,1277.

[4] A.-L. Gassner et al, Chromatogr. A 2014, 1353, 121.

[5] M. Rossato et al, ChemMedChem 2016, 11, 2582.

[6] X. Qiao et al, Analyst, 2015, 140, 407.

Proceedings of the 35th European Peptide Symposium
Patrick B. Timmons, Chandralal M. Hewage, Michal Lebl (Editors)
European Peptide Society & PSP, 2018

Double-head lipopeptide surfactants as potential antimicrobial agents

Izabela Małuch, Oktawian Stachurski, Paulina Kosikowska-Adamus, Dariusz Wyrzykowski, Adam Prahl, Emilia Sikorska

Faculty of Chemistry, University of Gdańsk, Poland

https://doi.org/10.17952/35EPS.2018.153

Introduction

The amphipathicity and total positive charge of short poly-L-lysine lipopeptides make them an interesting group of antimicrobial compounds. Due to their surfactant-like structure they can also exhibit surface-active properties.[1-5] A group of compounds of our particular interest are double-head lipopeptides. Here in, we present the study on the novel non-symmetric lipopeptide surfactants designed by the lipidation of the side chain of L-lysine or its shorter homologs located in the 2nd position in the peptide sequence.

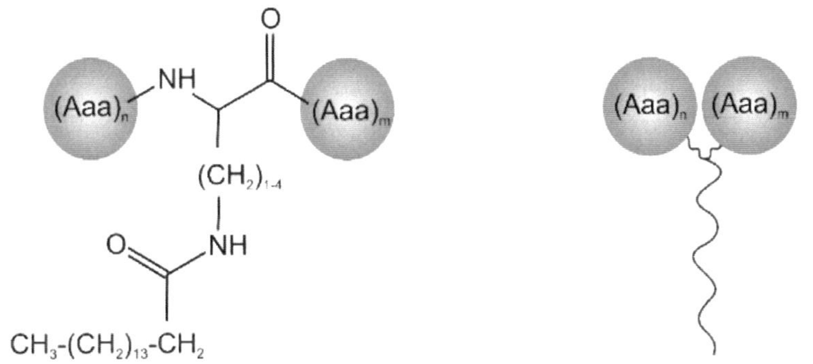

Figure 1: Double-head lipopeptide and schematic representation of the double-head surfactant.

Materials and Methods

Lipopeptides were prepared by the SPPS method (TentaGel S RAM resin) according to the Fmoc-strategy using automatic peptide synthesizer Symphony (Protein Technologies). After cleavage from the resin, crude peptides were purified by means of RP-HPLC. Final purity of all synthesized compounds was verified using analytical RP-HPLC system and MALDI-TOF MS was used to confirm their identity. All the lipopeptides were tested for antimicrobial activity. The minimal inhibitory concentration (MIC) values were determined according to standard microdilution technique using 96-well plates. Synthesized lipopeptides were studied for their antimicrobial activity against two bacterial strains (*Escherichia coli* PCM 2057, *Staphylococcus aureus* PCM 2054) and one fungal strain (*Candida albicans* PCM 2566). Mueller-Hintonbroth was used as the working medium for the experiments with bacteria, while RPMI-1640 buffered with MOPS was utilized for *C. albicans*. The effect of the tested compounds upon selected microorganisms was evaluated after 24 hours of co-incubation.

The isothermal titration calorimetry (ITC) was used to determine thermodynamic parameters of lipopeptides binding to anionic liposomes selected as a simplified model of bacterial membrane. Finally, the molecular dynamic simulations using a coarse-grained force fields (CG MD) were applied to visualize the self-assembly properties of the compounds.

Results and Discussion

We have designed and synthesized four non-symmetric double-head poly-L-lysine lipopeptides (Tab. 1). Based on the initial results of MIC values we have proved that the compounds effectively inhibit visible growth of the selected microorganisms in micromolar range of concentrations(7.8-15.6 M). There is no simple correlation between length of the lipidated amino acid side chain in position 2 and antimicrobial activity. Nevertheless, it seems that peptides modified with longer homologs, Lys and Orn, show slightly better antifungal properties.

Table 1: Sequences of synthesized lipopeptides, their analytical data and antimicrobial activity.

Peptide	t_R [min][a]	M [Da][b]	$[M+H]^+$ [Da][c]	MIC [µg/mL][d] *E.coli*	*S.aureus*	*C.albicans*
Lys-Lys(Palm)-Lys-Lys-Lys-NH_2	22.126	896.8	897.2	7.8	15.6	7.8
Lys-Orn(Palm)-Lys-Lys-Lys-NH_2	21.560	882.8	883.7	15.6	15.6	7.8
Lys-Dab(Palm)-Lys-Lys-Lys-NH_2	20.790	868.7	869.5	7.8	7.8	15.6
Lys-Dap(Palm)-Lys-Lys-Lys-NH_2	19.769	854.7	855.4	15.6	15.6	15.6

a) Jupiter 4μ Proteo 90Å , 250 x 4.6 mm; linear gradient: 15%-90% B in A for 30 minutes, flow rate: 1ml/min; A - 0.1% TFA/H2O, B - 80%ACN/H2O containing 0.1% TFA; b) calculated molecular mass; c) mass observed in MALDI-TOF MS analysis; d) Mueller-Hinton broth was used as the working medium for bacterial strains, while in case of fungi, RPMI-1640 growth medium buffered with MOPS (morpholinepropanesulfonic acid) and additionally supplemented with 2% glucose was used.

The microbiological results are in good agreement with the outcomes of ITC measurements, which confirm the general ability of the synthesized lipopeptides to interact with negatively charged artificial membrane (POPG LUVs). The titration of Lys-Orn(Palm)-Lys-Lys-Lys-NH_2 with POPG exhibits only one steep transition in the ITC curves (Fig. 2) and binding constant of $1.39 \pm 0.6 \times 10^6$ M^{-1}. The binding process is exothermic (ΔH=-0.7$\pm$0.23 kcal/mol) and stabilized by entropic factor ($|\Delta H| < |T\Delta S|$). In the case of the remaining peptides, it was found, that the binding process is accompanied by additional factors, such as lipopeptide aggregation, pore formation, micellization of phospholipid bilayer or change in the lipid phase properties during the titration.[6-8]

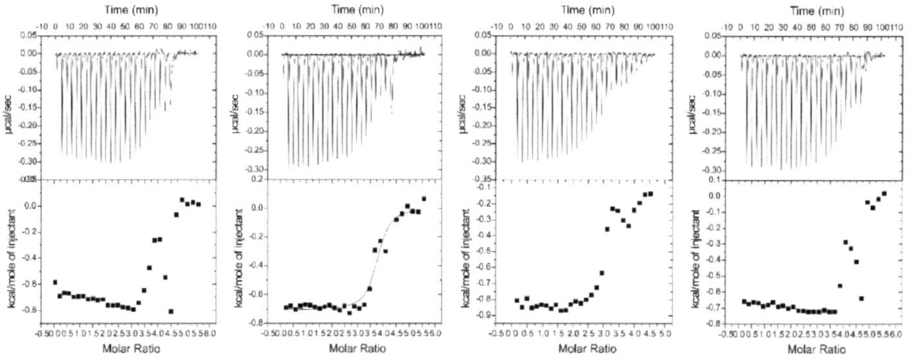

Figure 2: Isothermal titration of 0.05 mM peptide solutions with 1.31 mM POPG LUVs at 298 K. The lower curves represent the heat of reaction (measured by peak integration) as a function of the lipid/peptide molar ratio.

The preliminary results of molecular dynamics simulations indicate the tendency of tested compounds to self-assembly. Figure 3 displays the self-organization of the Lys-Lys(Palm)-Lys-Lys-Lys-NH $_2$ studied by CG MD simulations.

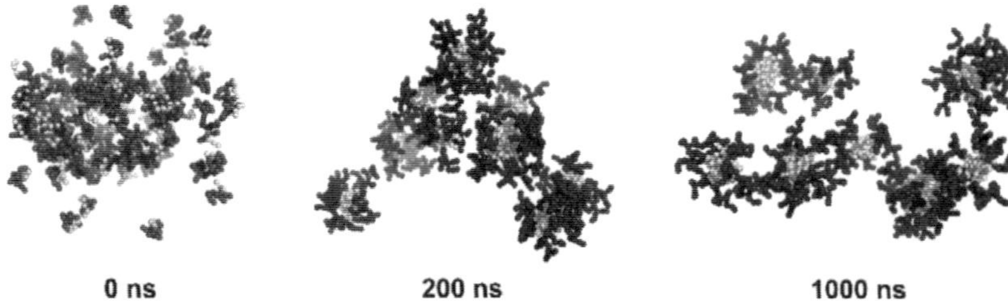

Figure 3: Snapshots from 1 μs CG MD simulations of self-assembly of Lys-Lys(Palm)-Lys-Lys-Lys-NH_2 (water and ions were removed for clarity). Fatty acyl chain is colored yellow.

In summary, the presented type of lipopeptides could serve as a novel model of membrane-targeted antimicrobial compounds. However, in order to establish their potential application additional experiments should be undertaken, like evaluation of the toxicity towards eukaryotic cells or determination of critical micelle concentration (CMC) values and size of the formed aggregates. The correlation of biological activity with self-organization would shed light on the plausible mode of action of lipopeptides and open up perspectives for designing analogues with broader antimicrobial and antifungal activities. The results of our studies can contribute to understanding of the self-assembly itself. Compared to the self-assembly of lipids, that of lipopeptides is relatively unexplored.

Acknowledgments

This work is supported by the National Science Center in Poland, contract No. UMO-2016/21/B/ST5/01375

References

1. Sikorska, E.; Dawgul, M.; Greber, K.; Iłowska, E.; Pogorzelska, A.; Kamysz, W., BBA Biomembranes 2014, 1838 (10), 2625-2634.

2. Sikorska, E.; Stachurski, O.; Neubauer, D.; Małuch, I.; Wyrzykowski, D.; Bauer, M.; Brzozowski, K.; Kamysz, W., BBA Biomembranes 2018, 1860 (11), 2242-2251.

3. Barchiesi, F.; Giacometti, A.; Cirioni, O.; Arzeni, D.; Silvestri, C.; Kamysz, W.; Abbruzzetti, A.; Riva, A.; Kamysz, E.; Scalise, G., Peptides 2007, 28 (8), 1509-13.

4. Greber, K. E.; Dawgul, M.; Kamysz, W.; Sawicki, W.; Łukasiak, J., Amino acids 2014, 46 (8), 1893-1898.

5. Makovitzki, A.; Baram, J.; Shai, Y., Biochemistry 2008, 47 (40), 10630-10636.

6. Henriksen, J. R.; Andresen, T. L., Biophys J 2011, 101 (1), 100-9.

7. Wenk, M. R.; Seelig, J., Biochemistry 1998, 37 (11), 3909-16.

8. Wenk, M. R.; Seelig, J., J Phys Chem 1997, 101 (26), 5224-5231.

Verification of mannan (polymannose)-peptide conjugation by SDS-PAGE

Areti Gkika[1], Maria-Eleni Androutsou[2], Iro Triantafyllakou[1], Alexios Vlamis-Gardikas[1], Theodore Tselios[1]

[1]University of Patras, Greece
[2]Vianex S.A., Greece

https://doi.org/10.17952/35EPS.2018.156

Abstract

Multiple Sclerosis (MS) is an autoimmune disease whereby the myelin of the central nervous system (CNS) is destroyed, leading to paralysis and serious health problems.[1] Myelin consists of proteins and lipids while in CNS, myelin basic protein (MBP), proteolipid protein (PLP), myelin oligodendrocyte glycoprotein (MOG) and myelin-associated glycoprotein (MAG) are the main proteins of the membrane sheath of myelin. [2] MOG is an autoantigen associated with the pathogenesis of MS and experimental autoimmune encephalomyelitis (EAE, animal model of MS). The 35-55 epitope of MOG has been implicated in the induction of EAE in mice.[3] Conjugation of this epitope with mannan polysaccharides *via* Schiff base formation has been extensively studied in our laboratory with the conjugates showing potential activity against EAE.[4–6] In the presented study, the conjugation reaction was investigated *via* SDS-PAGE and High-Performance Liquid Chromatography (HPLC) and the unconjugated peptide was quantified during reaction. The optimum time needed for complete conjugation was about six hours (yield more than 95 %). The developed analytical methodology could prove useful assay for the determination and study of conjugation reaction between peptides and polysaccharides.

Aim

The aim of this study was to demonstrate the decrease of $[KG]_5MOG_{35\text{-}55}$ peptide analogue during conjugation reaction and confirm the totally conjugation with mannan polysaccharide. Samples were taken during conjugation reaction to review the decrease of $[KG]_5MOG_{35\text{-}55}$ (free peptide) by the time in the conjugation process. Each sample was taken on a time-based period (t = 0, t = 30 min, t = 1h, t = 2h, t = 3h, t = 4h and t = 24h) and was tested *via* SDS PAGE and RP-HPLC. Staining of gels was performed with Coomassie Brilliant Blue.

Methods

Synthesis and coupling of the immuno dominant epitope $MOG_{35\text{-}55}$ to mannan. The $[KG]_5MOG_{35\text{-}55}$ peptide was synthesized by Fmoc/tBu methodology using 2-chlorotrityl chloride resin (CLTR-Cl). [6] The synthesized peptide was purified and identified by semi preparative HPLC and mass spectrometry (ESI-MS) respectively with a purity yield more than 95%. Mannan (poly-mannose from *Saccharomyces Cerevisiae*) was oxidized to poly-aldehyde using sodium periodate ($NaIO_4$) and purified by size exclusion chromatography (Sephadex G-25 Medium column). The purified oxidised mannan (Ox.mannan) was mixed with $[KG]_5MOG_{35\text{-}55}$ peptide and incubated at room temperature, in dark, for at least 24 hours. The conjugation reaction was achieved between the aldehydes of oxidised mannan and the free amine groups of amino acids in the peptide sequence.

Experimental

Samples which received from the Ox.mannan-$[KG]_5MOG_{35\text{-}55}$ conjugate reaction, were diluted by adding tricine sample buffer (Figure 1 - Table) and loaded on a tricine gel containing 3.5% stacking gel and 16.5% separating gel. Immediately after ending electrophoresis, gels were stained in 0.1% Coomassie Brilliant Blue G 250 (Figure 1). The free peptide was detected as a band with an apparent molecular weight of 3.0–3.5 kDa therefore the mannan–peptide conjugate appeared as a diffuse smear with an apparent molecular weight of 20–200 kDa. Quantification of unconjugated peptide was achieved by using a standard curve, constructed with known peptide concentrations in gels and the image j program. The detection limit of the method for the peptide quantification with Coomassie Blue stain was 37.5 ng. Furthermore, all samples were analyzed using RP-HPLC to demonstrate the presence of free peptide at time zero during reaction (Figure 2) and the absence of the peptide at the termination of conjugation reaction.

Results

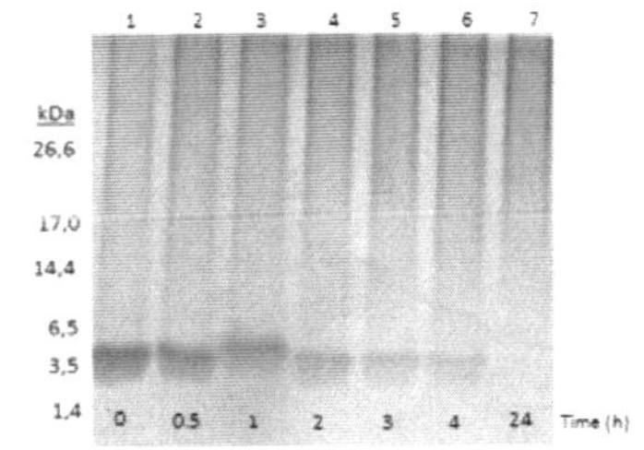

Wells	Time (h)	Conc. of unconjugated peptide (ug/ml)	Percentage of unconjugated peptide (%)
1	0	106.55	42.62
2	0.5	79.83	31.93
3	1	60.20	24.08
4	2	37.67	15.07
5	3	24.34	9.74
6	4	17.17	6.87
7	24	Not detected	Not detected

Figure 1: Time point samples of conjugation reaction via Tricine SDS-PAGE. Concentrations and % percentage of unconjugated [KG]5MOG35-55 during reaction.

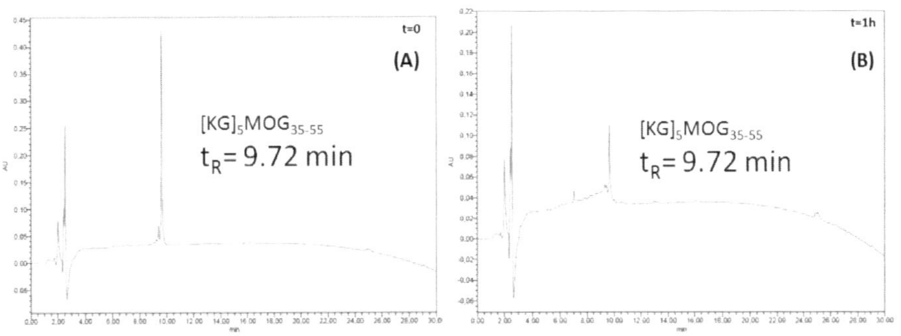

Figure 2: RP-HPLC chromatogram of Ox.mannan-[KG]5MOG35-55 solution at 214 nm: t = 0 h (A) and t = 1 h (B) after the initiation of conjugation reaction. Chromatographic separation was performed on a XBridge C18 (4.6 x 150 mm) column with 3.5 um particle size at 25 oC in a gradient separation from 5% to 100% AcN over 30min. Solvents: H2O (0.08 % TFA), AcN (0.08 % TFA).

Conclusions

The developed analytical methodology could be of great use in the chemistry and synthesis of glycopeptides. The optimum time needed for the termination of conjugation reaction was about six hours with yield more than 95%. In summary the conjugation reaction of a peptide (MOG_{35-55}) containing five extra lysines [$(KG)_5MOG_{35-55}$] with oxidised mannan polysaccharide was monitored *via* tricine gel electrophoresis. The termination of reaction was also confirmed *via* RP-HPLC. The presented methodology is a simple, fast and sensitive procedure that could be used in analytical chemistry for the detection of conjugates, peptides or proteins loaded into polysaccharides.

References

1. Mantzourani, E.D.; Mavromoustakos, T.M.; Platts, J.A.; Matsoukas, J.M.; Tselios, T.V. Current Medicinal Chemistry 2005, 12, 1521.

2. Day, S.; Tselios, T. et al., Frontiers in Immunology 2015, 6, 136.

3. Androutsou, M.E.; Tapeinou, A.; Vlamis-Gardikas, A.; Tselios, T. Medicinal Chemistry 2018, 14, 120.

4. Tseveleki, V.; Tselios, et al., L. Experimental Neurology 2015, 267, 254.

5. Kontos, C.; Androutsou, M.E.; Vlamis-Gardikas, A.; Tselios, T. Analytical Biochemistry 2017, 538, 71.

6. Ieronymaki, M.; Androutsou, M.E. et al., Biopolymers 2015, 104(5), 506.

7. Tapeinou, A.; Androutsou, M.E.; Kyrtata, K.; Vlamis-Gardikas, A.; Apostolopoulos, V.; Matsoukas, J.; Tselios, T. Analytical Biochemistry 2015, 485, 43.

Nucleus penetrating peptidomimetic synthesis and application

Magdalena Wysocka, Anita Romanowska, Irena Audzeyenka, Adam Lesner

University of Gdansk, Poland

https://doi.org/10.17952/35EPS.2018.158

Arginine rich peptides belongs to class of cell penetrating peptides (CPPs) that can transport the wide variety of attached chemical entities (cargoes) through the cell membrane. Diversity of transported molecules is impressive and includes DNAs, siRNAs, peptides, proteins and small drugs. Polyarginines comprise over five residues of l-arginine efficiently penetrate the cell membrane, and in some instances are part of more complex molecule which can deliver cargos to nucleus through nuclear localization signal (NLS).The general aim of this work is to reveal the properties of novel cell penetrating peptidomimetics which general structure is presented on figure 1. Such molecules contain several instances of l-2,3-diaminopropionic acid (DAP) connected through the peptide bond and modified on its side chain amino groups by substitution of functionalized PEG moiety. For group of compounds we proposed the abbreviation DAPEG that is merged DAP and PEG. Obtained peptidomimetic which is labelled by 5/6- carboxyfluorescein moiety efficiently penetrates cell membrane and is transported directly to nucleus of primary and transformed cell without significant toxicity.

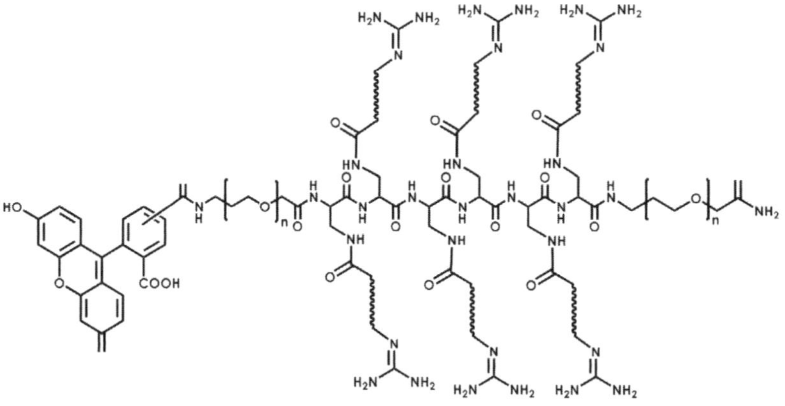

Figure 1: General formula of DAPEG mimetics.

Very recently we have synthesized a set of peptides and peptidomimetics that mimics hexaarginine one of the common cell penetrating peptide standard. All compounds were labelled by 5/6-carboxyfluorescein or rhodamine as cargo. Incubation of individual compounds with breast cell lines: HB2 and MDA-MB-231 for 2 hours at concertation of 10 μM point out that some of the compounds efficiently crossed the cell membrane (see figure 2).

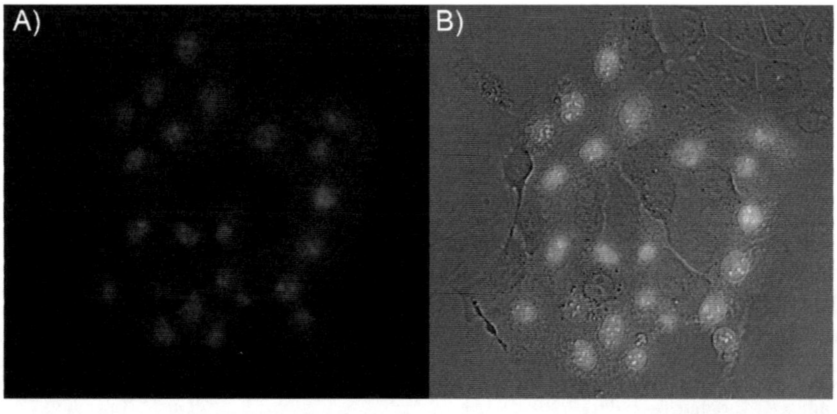

Figure 2: HB2 cell line incubated for 2 hours with 10 μM concentration of DAPEG compound. A) FITC fluorescence B) overlay of bright view and FITC fluorescence.

For evaluation DNA – DAPEG interaction we employ EMSA gels that results are presented on figure 3. The selected compound was incubated with DNA short linear fragment (76 bp) at different concentration for 1 hour. After that time, the 20 μl of DNA-DAPEG mixture was subjected to 2% gel supplemented with Sybr Safe View reagent. The gels were run for 45 minutes and further inspected under UV transilluminator

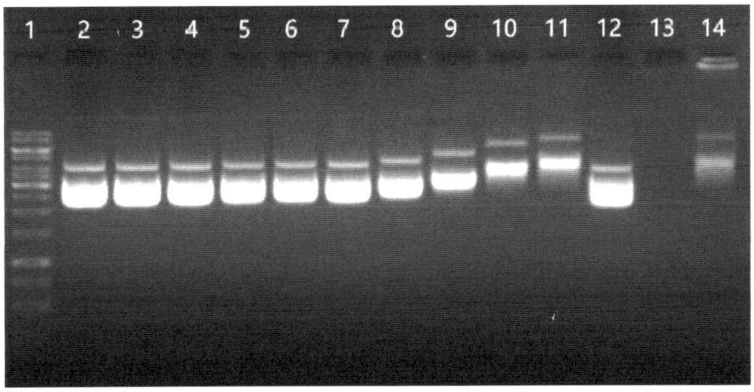

Figure 3: For evaluation DNA – DAPEG interaction we employ EMSA gels that results are presented on figure 3. The selected compound was incubated with DNA short linear fragment (76 bp) at different concentration for 1 hour. After that time, the 20 μl of DNA-DAPEG mixture was subjected to 2% gel supplemented with Sybr Safe View reagent. The gels were run for 45 minutes and further inspected under UV transilluminator.

Conclusions

We have synthesized the peptidomimetic that are rapidly cross the cell membrane and targets nucleus. This peptidomimetic display no significant cytotoxicity up to 50 μg/m and is able to bind to dsDNA in nanomolar concentration.

Acknowledgements

This work was supported by National Science Center Poland under grant UMO-2017/27/B/ST5/02061 (MW).

Design and synthesis of novel silicon-containing small molecule peptidomimetics with nanomolar anticancer activities

Lajos Gera[1], Peter Hegyes[2], Daqing Wu[3], Robert Hodges[1]

[1]University of Colorado Denver, Anschutz Medical Campus, Aurora, CO, 80045, United States
[2]Avidin Kft, Szeged, H-6726, Hungary
[3]Augusta University, Augusta, GA, 30912, United States

https://doi.org/10.17952/35EPS.2018.160

Introduction

Cancer is a major health problem around the globe. Each year, tens of millions of people are diagnosed with cancer and more than half of the patients eventually die from it. In 2018, 1,735,350 new cancer cases and 609,640 cancer deaths are projected to occur in the United States. Though there has been a steady increase in survival for most cancers, the death rate remains unacceptable for certain cancers, e.g. lung (26%), prostate (9%), colon (8%), pancreas (7%) and breast (14%). In this study, we support the concept that the ideal drug should have a multi-targeted mechanism that affects several proteins or events that contribute to the etiology, pathogenesis and progression of disease. In addition, multi-pathway targeting is one of the strategies to overcome chemo-resistance.

Results and Discussion

To design our novel anticancer drugs with unique structural properties, we have taken an innovative and nontraditional approach where we combine pharmacophoric components to create new and highly potent small peptidomimetic molecules with a simple three component A-B-C structure where each pharmacophore is known to have anticancer properties on its own or when incorporated into a multicomponent small molecule drug. Recently, we developed a new generation of this simple 3-componentA-B-C structure, a highly potent anticancer compoundGH501 [1] (Figure 1). In our new compounds,GH1501 - GH1504, the A-component was further modified to contain a silicon atom [2] (Figure 1). Using silicon over carbon has many advantages: a) compounds are more lipophilic than their carbon equivalent b) silicon can improve compound permeability c) silicon can lower compound toxicity d) silicon can change receptor selectivity e) silicon can enhance anticancer activity of compounds f) silicon compounds can treat drug-resistant tumors. The “A”-component 4-[3-(trimethylsilyl)propoxy]benzoic acid (m.p.: 172-173 °C, crystallized from ACN) and 4-[butyl-dimethylsilyl)methoxy]benzoic acid (m.p.: 57-59 °C) were synthesized *via* a modified Hegyes method using 4-hydroxybenzoic acid methylester sodium salt. [3]. Interestingly, Zaltariov *et al.* [4] claimed that they synthesized 4-[3-(trimethylsilyl)propoxy]benzoic acid for the first time. Also, their synthetic method provided only the 3-(trimethylsilyl)propyl 4-hydroxybenzoate (m.p.: 114-117 °C) in our hands. The constituent at position A of our A-B-C structures is dependent on the composition of the B-component. When the B-component is 4-(2,6-dichloro-benzyloxy)-phenyl the best silyl-alkoxy-benzoyl derivative is GH1501 (compare GH1501 and GH1503, Table 1). When the B-component is biphenyl the best silyl-alkoxy-benzoyl derivative is GH1504 (compare GH1504 and GH1502, Table 1).

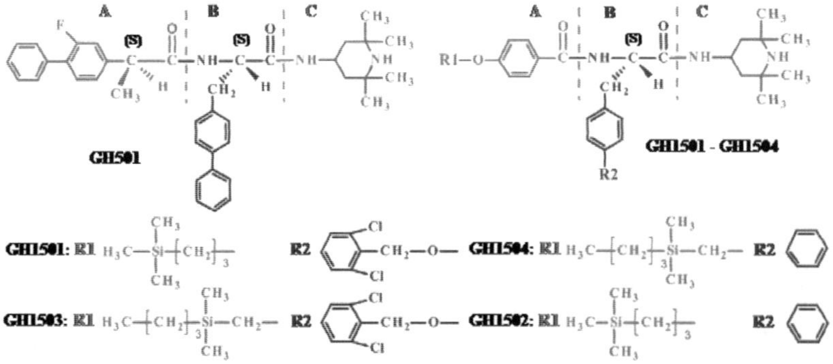

Figure 1: Structures of GH501 and GH1501 - GH1504

Table 1: Results of the NCI human colon cancer cell line. GI50: concentration in nanomolar that inhibits cancer growth by 50% Average: average of GI50 value for each compound Pink highlight is a GI50 value less than 500 nM Yellow highlight is the average GI50 value less than 500 nM

Cell Line	*GI_{50} (nanomolar) of selected anticancer compounds*				
Colon Cancer	GH501	GH1501	GH1503	GH1504	GH1502
COLO 205	184	216	342	215	392
HCC-2998	418	1050	1070	1110	1560
HCT-15	1440	357	323	309	437
HCT-116	145	206	319	218	297
HT29	178	263	342	315	430
KM12	461	363	382	376	413
SW-620	294	313	318	316	392
Average	446	395	442	408	560

GH501, GH1501 GH1504 were made using Boc-chemistry

GI50: concentration in nanomolar that inhibits cancer growth by 50% Average: average of GI50 value for each compound Pink highlight is a GI50 value less than 500 nM Yellow highlight is the average GI50 value less than 500 nM.

The diversity of the A-component can be Fluoro-biphenyl (GH501), silyl-alkoxy-benzoyl derivative (GH1501 or GH1504) depending on the "B"-component. The organo-silicon compounds showed cancer-type specific anticancer activity for colon cancer.

Acknowledgements

The authors are grateful to NCI for the NCI-60 human tumor cell line screening and to Eli Lilly for preliminary studies with our compounds on mechanism of action. We thank the NIH/NCI (Grant No. 1R41CA206725-01A1) for financial support.

References

1. Gera, L., at al. Proceedings 24th APS, p. 241-244 (2015).

2. Gera, L., Hodges, R.S., Hegyes, P. US Patent, US 9828393 B2, Nov. 28, 2017.

3. Hegyes, P., Toeroecsik, M. Hun. Pat. Appl. (2000), HU 9801407 A, March 28, 2000.

4. Zaltariov. M-F., *et al.* J. Mol. Struct. 1120, 302-316 (2016).

Proceedings of the 35th European Peptide Symposium
Patrick B. Timmons, Chandralal M. Hewage, Michal Lebl (Editors)
162 *European Peptide Society & PSP, 2018*

Synthesis and analysis of β-peptides for the specific and non-specific aggregation on model membranes

Markus Wiegand, Geralin Höger, Ulf Diederichsen

University Göttingen, Germany

https://doi.org/10.17952/35EPS.2018.162

Introduction

In biological systems protein-membrane interactions are mediated by complex structures, which fulfill numerous functions such as signaling, regulation, domain formation, molecule transport or information storage. All these tasks are based on peptides and proteins forming stable secondary structures like β-sheets or helices. They often provide further interaction or even higher aggregation.[1] Higher aggregates are responsible for the interaction between peptides/proteins and other biomolecules and are therefore also of great importance to the functional and structural integrity of biomembranes.[2,3] For example, the cytoskeletal complexes of the BAR domain containing proteins interact with spectrin and actin, which are in contact with membranes, aggregate specifically and form cell-spanning scaffolds.[4,5] Other examples for peptide-membrane interaction affecting the functional integrity of a biomembrane are the peptides Maculatin and Citropin, that have an antibiotic effect when interacting with membranes.[6] Yet many aspects of mechanism and dynamics of protein-membrane interactions remain unknown. Therefore, artificial model systems might give an insight into the complex nature of protein-membrane interactions.[7]

In this study, a synthetic approach using β^3-peptide scaffolds as model systems for aggregational peptide-peptide and peptide-membrane interactions is presented. These scaffolds have the advantage to form stable and well-defined secondary structures, such as 14-helices.[8,9] Key properties of the 14-helix are the three-residue repeating arrangement of the side chains, a pitch of approximately 5.0 and every third side chain being arranged alike. These pattern allows to define three sites of the helix with different properties.

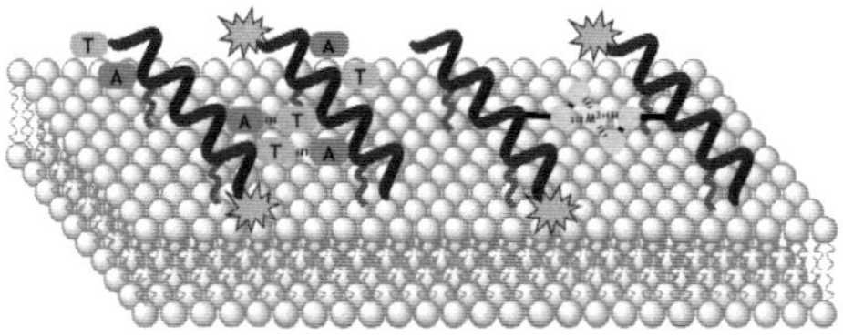

Figure 1: Model system of b-peptide helices (blue) functionalized with specific (yellow and red) and non-specific recognition units (yellow), lipidic side chains (gray), and fluorophores (green).

The peptides were obtained *via* manual solid phase peptide synthesis (SPPS). The peptides are modified with recognition units to enable specific and non-specific aggregation depending on the kind of recognition. The specific aggregation is realized by utilizing nucleobase modified amino acids and in case of non-specific aggregation a 2,2'-bipyridyl derivative was used for metal coordination (Fig. 1). The peptide-membrane interaction is induced by lipid-like amino acid side chains like C14 or C16 alkyl chains or cholesterol, thereby spontaneously anchoring within the membrane.[10]

Results

Non-specific coordination induced by metal coordination

The aggregation and organization of -peptide helices by bipyridyl ligand metal coordination was investigated by titration of peptide 1 with Cu^{2+} and Zn^{2+} metal solutions and analysis by UV/Vis titration (Fig. 2). Formation of stable bipyridyl metal complexes was indicated and taken as proof for the aggregation of peptide helices. So far, this -peptide helix aggregation was only investigated in solution. Therefore, it was required to prove the anchoring of the lipid chain modified -peptide helices within the membrane. Binding of the peptides to the surface of vesicles was indicated by a FRET assay between the NBD-peptide and labeled vesicles (Fig. 3). The

FRET pair of NBD labeled peptides and Lissamine™ Rhodamine B marked vesicles was used. Increase of fluorescence by addition of the labeled peptide is evidence for a peptide-membrane interaction, and therefore, for binding of peptide helices to the surface of the vesicle.

Specific aggregation induced by nucleobase-nucleobase recognition

For the investigation of nucleo amino acid modified peptides, oligomers 2-10 were synthesized varying in the nucleobase sequence. Anchoring of the peptide helices was indicated by FRET measurements between NBD-peptides 2-10 and Lissamine™ Rhodamine B labeled vesicles (Fig. 4). In addition, concentration dependent FRET was performed to analyze peptide-peptide interaction driven by specific nucleobase recognition. In solution NBD-peptides were titrated with TAMRA-labeled peptides. With increasing molar fractions of the TAMRA-labeled peptide, a FRET-signal was observed for the combination of peptide sequences CTG and CAG (7 and 9, Fig. 5a), whereas minor indication for a stable interaction was obtained for the ATT/AAT recognition motif (3 and 6, Fig. 5b). The mismatch combination of peptide sequences CTG/ATT (7 and 6, Fig. 5c) serves as reference for lacking interaction. Depending on the nucleobase sequence the peptide helices can be specifically aggregated at least in duplexes. Further studies will investigate the specific peptide helix aggregation of membrane bound oligomers.

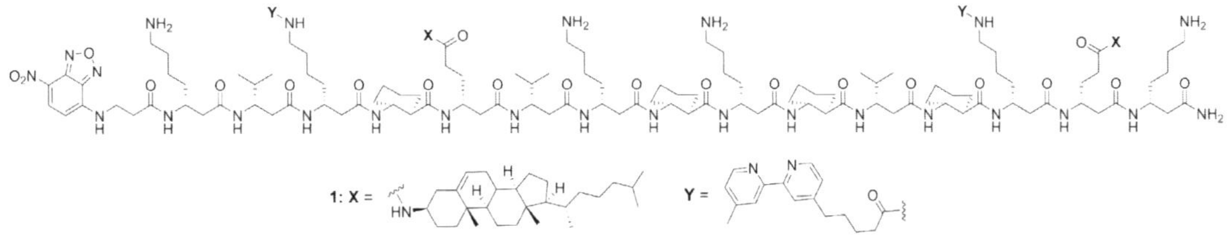

Scheme 1

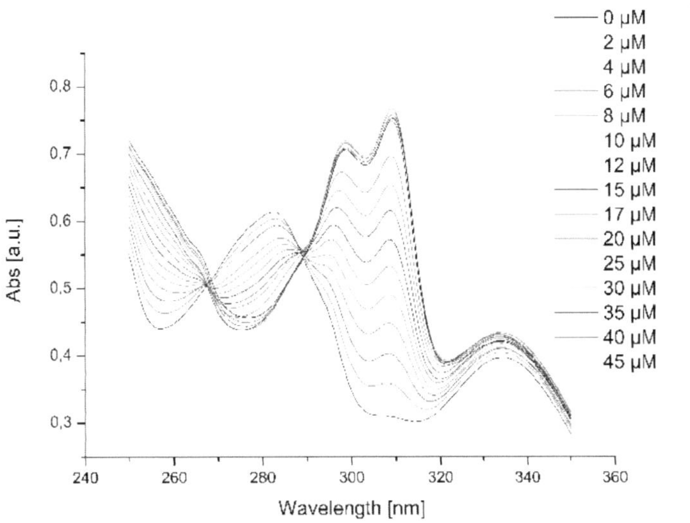

Figure 2: UV/Vis titration of b-peptide 1 (c = 30 µm in MeOH) with CuCl2.

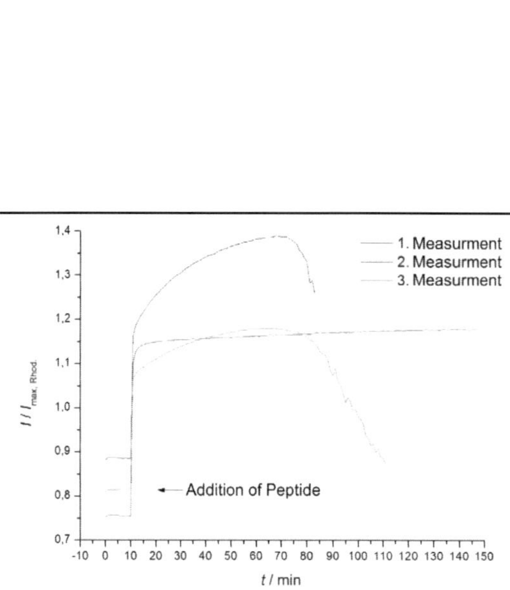

Figure 3: Normalized FRET spectra of NBD labeled peptide 1 and DOPC-Rhodamine vesicles in a 1:300 ratio.

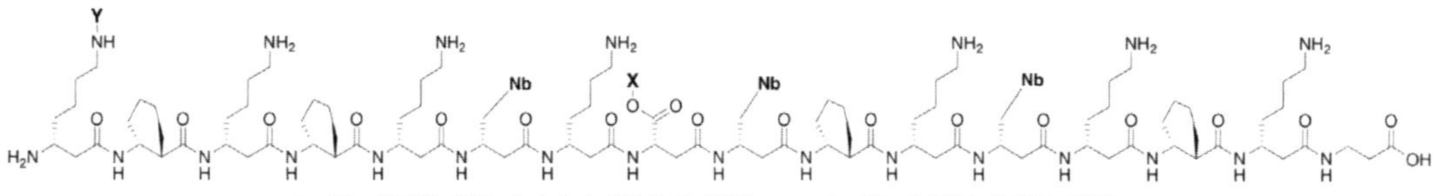

2: **Nb**= ATT **X**= C14 alkyl chain (C14) **Y**= NBD
3: **Nb**= ATT **X**= methyl group (Me) **Y**= NBD
4: **Nb**= ATT X= Me Y= TAMRA
5: **Nb**= **A**TT X= **Me** Y= **acetyl group (ac)**

6: **Nb**= AAT **X**= C14 **Y**= NBD
7: **Nb**= AAT **X**= Me **Y**= NBD
8: **Nb**= CTG **X**= Me **Y**= TAMRA
9: **Nb**= CTG **X**= Me **Y**= Ac
10: **Nb**= CAG **X**= Me **Y**= NBD

Scheme 2

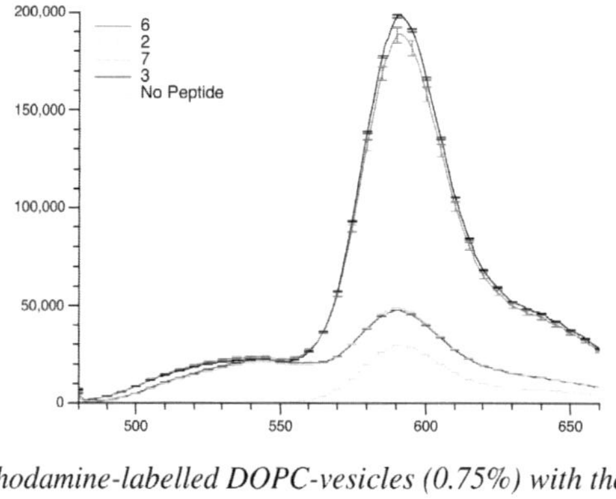

Figure 4: FRET spectra of rhodamine-labelled DOPC-vesicles (0.75%) with the indicated b-peptides (5 μm) in Tris-HCl 10 mm pH 7.5.

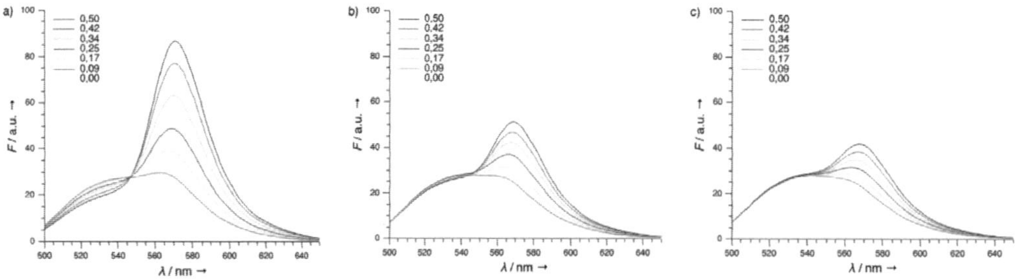

Figure 5: Concentration-dependent FRET experiments in Tris-HCl 10 mm pH 7.5 at 10°C with a total peptide concentration of 8 μm and the indicated molar fractions of TAMRA-labelled peptides. (a) Fluorescence emission spectra of b-peptides 4, 5 and 7. (ATT-AAT) (b) Fluorescence emission spectra of b-peptides 8, 9 and 7. (CTG-AAT) (c) Fluorescence emission spectra of b-peptides 8, 9 and 10 (CTG-CAG).

References

[1] J. A. Kritzer et al., J. Am. Chem. Soc. 2005, 127, 167-178, DOI: 10.1021/ja0459375.

[2] J. Bonifacino, J. Lippincott-Schwartz, Nat. Rev. Mol. Cell. Biol. 2003, 4, 409-414, DOI: 10.1038/nrm1099.

[3] U. Koert, Angew. Chem. 1997, 109, 1922-1923, DOI: 10.1002/ange.19971091708.

[4] G. J. Doherty, H. T. McMahon, Annu. Rev. Biophys. 2008, 37, 65-95, DOI: 10.1146/annurev.biophys.37.032807.125912

[5] J. Saarikangas, H. Zhao, P. Lappalainen, Physiol. Rev. 2010, 1, 259-289, DOI: 10.1152/physrev.00036.2009

[6] E. E. Ambroggio et al., Biochim. Biophys. Acta 2004, 1664, 31–37, DOI: 10.1016/j.bbamem.2004.03.013.

[7] a) A. M. Brückner, P. Chakraborty, S. H. Gellman, U. Diederichsen, Angew. Chem. Int. Ed. 2003, 42, 4395–4399, DOI: 10.1002/anie.200351871;

b) M. Sadek, D. Berndt, D. Milovanovic, R. Jahn, U. Diederichsen, ChemBioChem, 2016, 17, 479-485, DOI: 10.1002/cbic.201500517.

[8] W. S. Horne, S. H. Gellman, Acc. Chem. Res. 2008, 41, 1399–1408, DOI: 10.1021/ar800009n.

[9] D. Seebach, A. K. Beck, D. J. Bierbaum, Chem. Biodiversity. 2004, 1, 1111-1239, DOI: 10.1002/cbdv.200490087.

[10] G. R. Newkome, A. K. Patri, E. Holder, U. S. Schubert, Eur. J. Org. Chem. 2004, 2, 235–254, DOI: 10.1002/ejoc.200300399

Depsipeptide cyclisation: Lactonization vs. lactamisation

Sacha Stadelmann, Stéphane Mayor, Roger Marti, Marc Mathieu

HESSO, Switzerland

https://doi.org/10.17952/35EPS.2018.166

The therapy of life-threatening infections significantly weakened by the global spread of antibiotic resistance has prompted an urgent need for the development of novel, effective, and safe antibiotics. Novel antibacterial agents with unprecedented mechanisms of action, which are devoid of pre-existing cross-resistances, are therefore necessary [1].

Peptides constitute one of the most promising platforms for drug development due to their biocompatibility, chemical diversity, and resemblance to proteins. Novel design approaches along with efficient and economic peptide synthesis have contributed to revitalize peptide-based drugs in the current pharmaceutical market [2]. Linear peptides are not very promising therapeutic agents owing to their low stability towards proteolysis, consequently reducing their feasibility and profitability for the pharmaceutical industry. Diverse chemical modification protocols that have evolved to diminish the mentioned drawbacks; these include cyclization, N-methylation, incorporation of non-natural amino acids and other structural constraints.

Our literature searches brought us to consider natural products(actinomycetes), which have served as a promising source of new structural leads in the antibiotic resistance area and comply with these structural enhancing features (listed below), in particular the cyclodepsipeptides (CDP) [2]. CDP's show promising pharmacological activities and meet all criteria for good solubility and permeability. CDP's are capable like natural peptides, to interact with numerous proteins and show a large panel of activities (antitumor, anthelmintic, insecticidal, antibiotic, antifungal, immunosuppressant, anti-inflammatory and antimalarial).

However, further research of their medical applications is necessary and natural CDP's are challenging lead structures from a chemical and synthetic viewpoint [3], this triggered our attention. But, several Structure-Activity Relationship (SAR) and medicinal chemistry studies enabled strong post-evolution of the natural scaffold and enabled us to design novel structures considering known issues. Limiting chemical factors described in the literature to include: (a) the lactone core can be hydrolyzed in basic and acidic aqueous media; (b) the acylated serine hydroxy group eliminated readily under non-aqueous basic conditions; (c) the conjugated triene was sensitive to temperature and light (cyclization and aromatization reactions) (defined in Figure 1 as the R group); (d) solubility was not sufficient for parenteral application.

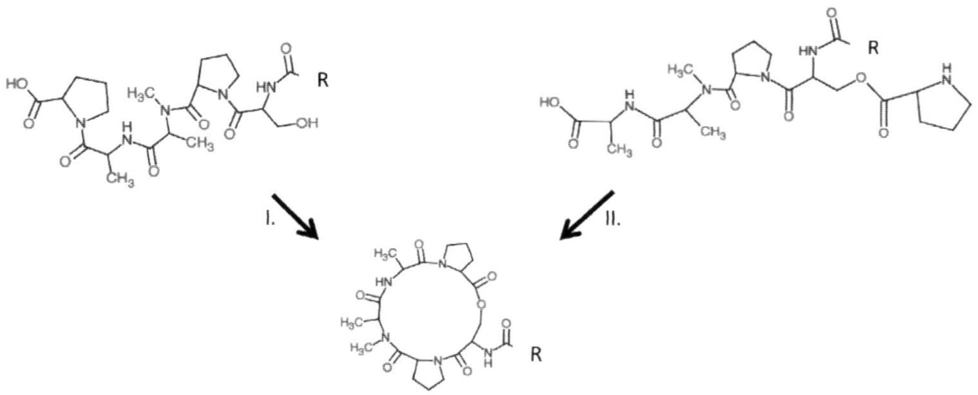

Figure 1: Enopeptin A cyclisation strategies: (I) Lactonization vs. (II) Lactamisation. R=variable biocompatible functionality.

Our solid-phase methodology contribution focuses on the evaluation of two distinct strategies for the cyclisation process preserving the ester functionality: (I) lactonization and (II) lactamisation are evaluated in order to produce Enopeptin A analogs (R = variable functionality), as presented in the Figure 1. Lactonisation[(I)] with our sequences did not give satisfying results, regardless of the chosen route: (i) 2 eq HATU; 2.5 eq DIEA; 1h at 0°C followed by 24 h at 25°C and (ii) using Yamaguchi reagents [4] (data not shown).

Lactamisation[(II)] in opposition gave the target products, although in poor yield. Different concentrations were tested, in order to minimize polymerization but no visible improvement could be achieved. Other cyclisation methods for the lactam formation are currently under investigation.

References

1. Hinzen B. et al., Med. Chem., 2006, 1, pp. 689 - 693, DOI:

2. Tapeinou A. et al., Pep. Sci., 2015, 5, vol. 104,

3. Jordan D. Goodreid, J. Nat. Prod. 2014, 77, pp. 2170–2181 DOI: 10.1021/np500158q

4. Hikot M. et al., THL, 1990, 31, pp. 6367-6370 DOI:

Trifluoromethylated proline surrogates as part of 'Pro-Pro' turn-inducing templates for the design of β-hairpin mimetics

Charlène Gadais[1], Kevin Van Holsbeeck[1], Samuel Moors[2], Krisztina Fehér[3], Julien Pytkowicz[4], José Martins[3], Frank De Proft[2], Dirk Tourwé[1], Thierry Brigaud[4], Charlotte Martin[1], Grégory Chaume[4], Steven Ballet [1]

[1]ORGC - Vrije Universiteit Brussel, Belgium
[2]ALGC - Vrije Universiteit Brussel, Belgium
[3]NMR department- Universiteit Gent, Belgium
[4]LCB - Université de Cergy-Pontoise, France

https://doi.org/10.17952/35EPS.2018.168

Proline is often found as a turn inducer in peptides or proteins. Exploitation of its propensity to induce structuration led to the development of a d-Pro-Pro(pP) 1 segment as a 'templating' unit, frequently used in the design of β-hairpin peptidomimetics.[1] Following this well-established strategy, our lab recently published the design of a series of cyclic peptides mimicking the β-hairpin structure of the CDR3 loop of Nb80.[2] Although one of the reported structures could adopt a conformation featuring a high overlap with the backbone structure of Nb80 CDR3 loop (MD Cα RMSD = 1.4 Å), NMR analysis revealed several sets of resonances arising from the existence of several conformers, and complex NOE patterns and secondary chemical shifts indicated random coil conformations for all of the residues. Those results illustrate well the limitations of the diproline template: it is known that the conformational stability of the type II' β-turn adopted by d-Pro-Pro(and herein required for a stable β-hairpin) is highly compromised by the cis-trans isomerization of the prolyl amide bond in larger ring systems. Use of constrained prolines, such as CF_3-pseudoprolines, demonstrated the possibility of stabilizing the backbone geometry by favouring the cis or inversely the trans population of a series of pseudo tetrapeptide sequences.[3]

In view of finding alternative templates with a stronger capacity to fix the β-hairpin conformation, we investigated different fluorinated analogues of the well-established pP segment as a β-turn promoter.

An in silico conformational study was performed on a set of 12 variants of the pP sequence, incorporating α-trifluoromethyl-proline (TfmPro) and α-trifluoromethyl-oxazolidine (TfmOxa),[4] and capped with acetyl (Ac) and N-methyl amide (NHMe) end groups, providing therefore the required H-bond donor and acceptor for a type II' β-turn (Figure 1A). For each peptide, all rotameric combinations were generated with Open Babel software, using the MMFF94 force field, and followed by a geometry optimization of the lowest energy trans and cis conformers. Out of the investigated combinations only TfmPro-Pro 2 and d-Pro-(R)-TfmOxa 3a exhibited a strongly stabilized β-turn conformation, relative to the 'parent' pP (Figure 1, panel B). For those two dipeptides, the rotational barrier energy between the trans-cis and trans-trans conformers was indeed increased by 3.1 to 4.6 kcal/mol, compared to the d-Pro-Pro template and still maintained H-bond formation (Figure 1C). It is also noteworthy that the favourable effect of (R)-TfmOxa was totally lost for diastereoisomer 3b bearing (S)-TfmOxa. Considering synthetic accessibility, d-Pro-TfmOxa was selected for the experimental incorporation into the model and the cyclic Nb80 CDR3 loop peptidomimetics.

A.

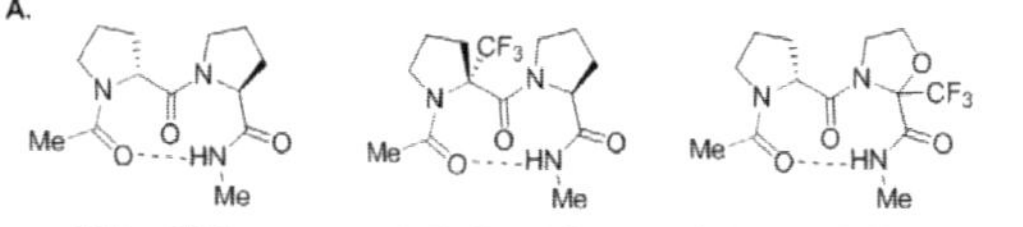

(*R*)-Pro-(*S*)-Pro **1** | (*S*)-TfmPro-(*S*)-Pro **2** | **3a** (*R*)-Pro-(*R*)-TfmOxa, **3b** (*R*)-Pro-(*S*)-TfmOxa

B. Energies relative to the lowest energy *trans-trans* conformer (calculated with the MMFF94 force field, OpenBabel®, in kcal/mol)
H-bond: lowest energy conformer with a H-bond between *O*-acetyl and *N*-methyl-NH

	D-Pro-Pro	(*S*)-TfmPro-Pro	D-Pro-(*R*)-TfmOxa	D-Pro-(*S*)-TfmOxa
trans-trans	0	0	0	0
trans-cis	3.6	6.7	8.2	2.4
cis-trans	2.2	3.8	5.6	3.0
cis-cis	7	12.2	15.2	8.7
H-bond	0.3	0	0	2.3

C.

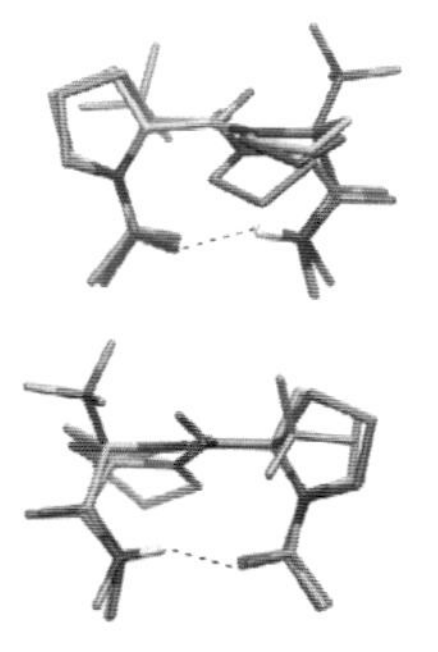

Figure 1: A. Diproline templates. B. Table of the calculated lowest energy conformers. C. Overlay between 1 (green), 2 (orange) and 3a (grey) in front (panel top) and back view (panel bottom).

A strong deactivation of the secondary amine's nucleophilicity and the steric hindrance generated by the CF3 group in TfmOxa prevent any direct application in SPPS. We therefore chose to 'cap' the TfmOxa with the adequate amino acids, by synthesizing in solution the tripeptideFmoc-d-Pro-TfmOxa-Val-OH 8 for incorporation *via* SPPS. Through a reported procedure,[4] a racemic mixture of the ethyl ester of TfmOxa was first saponified and then engaged in a coupling with the bench stable Fmoc-d-Pro-Cl 6. To our knowledge, such an activation is the only method reported to date, permitting coupling with amines of α trifluoromethylated amino acids. Eventually, the desired dipeptide 7 could be isolated and purified in good yield once an intermediate esterification and extraction of the unreacted Fmoc-d-Pro-Cl 6 was performed. Then, access to the tripeptide by C-terminal coupling of TfmOxa turned out to be challenging: preliminary synthetic screening efforts with standard coupling reagents performed on H-TfmOxa-OHall failed in giving the targeted amide bond. This led us to investigate the Mukaiyama reagent. Due to their poor solubility in most organic solvents, pyridinium iodide compounds are not preferred reagents in peptide synthesis. It proved however to be highly efficient in TfmOxa's C-terminal coupling. After a thorough optimization study, examining stoichiometry, time and temperature conditions, the two diastereoisomers of Fmoc-d-Pro-TfmOxa-Val-OtBu were obtained in good yield, and ultimately in situ deprotected and separated by chromatography to give 8a and 8b (Scheme 1). Crystallization and X-Ray analysis were eventually carried out to attribute the stereochemistry of each diastereoisomers.

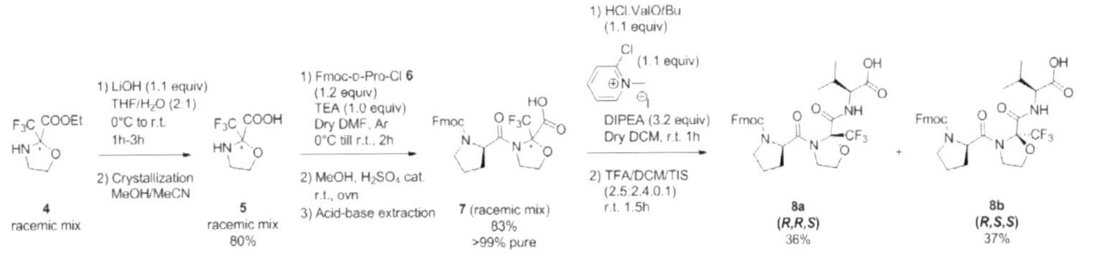

Scheme 1: Synthetic pathways toward tripeptides bearing TfmOxa

Next, efforts were dedicated to the incorporation of the Fmoc-d-Pro-TfmOxa-Val-OH fragments 8a and 8b into the Nb80 CDR3 loop sequence. Interestingly, coupling of both tripeptides and subsequent peptide elongation on 2-chlorotrityl resin and cleavage went smoothly in standard conditions. However, due to low solubility of the fully protected linear peptides in CH_2Cl_2, head-to-tail cyclisation required the use of trifluoroethanol(TFE) as a co-solvent. Associated TFE ester formation could not be fully prevented, but it did not hamper the successful formation and isolation of the cyclic peptidomimetics 9a-(R,R) and 9b-(R,S) (Figure 2).

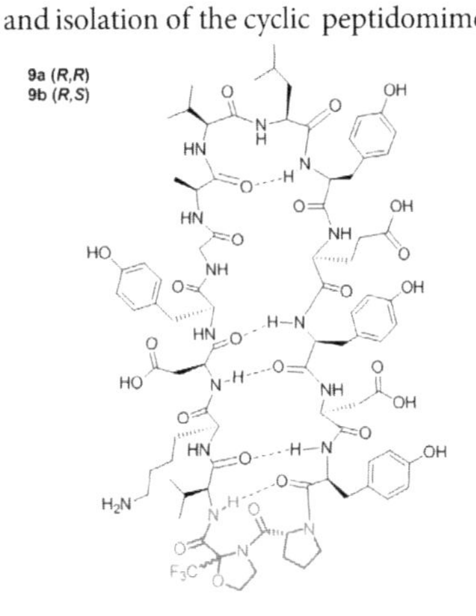

Figure 2: CDR3 Nb80 Cyclic peptidomimetics 9a & b bearing TfmOxa

Conformational NMR analysis of both mimetics 9a-(R,R) and 9b-(R,S) was first initiated in a mixture of water/acetonitrile 80:20 (this percentage allowed full solubility of the mimetics). In such conditions, no clear structuration could be identified. However, an assignment of the signals, achieved unambiguously up to 90%, showed for each compound the existence of only one conformer; an improvement knowing that the d-Pro-Pro bearing cyclic peptidomimetic exhibited 4 conformers and multiple overlapping resonances between major and minor conformers. These early findings demonstrate the beta-hairpin stabilizing properties of the trifluoromethylated diproline analogues. The influence of NMR solvents and temperature changes are currently inspected. In particular, focus on the trans conformation of the prolyl amide bonds and the existence of intra-strand H-bond are examined to further verify the hypothetical beta-turn hairpin inducing feature of d-Pro-TfmOxa template.

References

[1] a) R. Rai, S. Aravinda, K. Kanagarajadurai, S. Raghothama, R. Shamala, P. Balaram. Journal of the American Chemical Society, 2006, 128 (24), 7916-7928. DOI: 10.1021/ja060674v b) J. A. Robinson, Accounts of Chemical Research, 2008, 41 (10), 1278-1288. DOI: 10.1021/ar700259k

[2] C. Martin, S. L. C. Moors, M. Danielsen, C. Betti, C. Fabris, D. Sejer Pedersen, E. Pardon, M. Peyressatre, K. Fehér, J. C. Martins, J. Mosolff Mathiesen, M. C. Morris, N. Devoogdt, V. Caveliers, F. DeProft, J. Steyaert, S. Ballet, Chem. Eur. J.2017, 23, 9632. DOI: 10.1002/chem.201701321

[3] G. Chaume, D. Feytens, G. Chassaing, S. Lavielle, T. Brigaud, E. Miclet. New Journal of Chemistry, 2013, 37, 1336-1342. DOI: 10.1039/c3nj41084f

[4] J. Simon, J. Pytkowicz, N. Lensen, G. Chaume, T. Brigaud. Journal of Organic Chemistry, 2016, 81, 5381–5392. DOI : 10.1021/acs.joc.6b00704

Design of new cyclic plasmin inhibitors

Yuko Tsuda[1], Koushi Hidaka[1], Keiko Hojo[1], Keigo Gohda[2], Naoki Teno[3], Keiko Wanaka[4]

[1]Kobe Gakuin University, Japan
[2]Computer-Aided Molecular Modeling Research Center, Kansai, Japan
[3]Hiroshima International University, Japan
[4]Research Projects on Thrombosis and Haemostasis, Japan

https://doi.org/10.17952/35EPS.2018.171

Plasmin (Plm), a trypsin like protease, is involved in the fibrinolysis pathway. Additionally, the localized activation of plasminogen (Plg) and action of Plm at the specific regions of the extracellular milieu lead to the digestion of many proteins at the localized cell surface that results in inducing cell inversion and metastasis and alternating the expression of cytokines. A growing body of data suggests that a Plm inhibitor is a potential candidate as an anti-inflammatory and anti-cancer agent as well as for hemostatics. The development of selective Plm inhibitors is required to study the precise role of Plm and its relationship in treating several diseases.

We previously revealed the binding mode of the YO-2 (Plm inhibitor)[1]- ≠ Plm complex with the crystal structure analysis [2]. In the complex, the aminomethylcyclohexyl moiety of YO-2 extends into the S1 subsite. At the S1' subsite, the Tyr moiety forms a hydrogen bond with K607, and the pyridine moiety forms an imperfect π-π stacking. In contrast, the hydrophobic octyl moiety does not have extensive interaction with the S2' subsite; it points away from the protease surface. In this work, we deal with the structure-based design of Plm inhibitors by focusing on the P' residues; namely, we paid attention to the enhanced interaction at the S' pocket. Incorporation of a ring segment could create extensive interactions to make the inhibitor fit better in the S' subsites. We cyclized YO-2 *via* ring closing metathesis (RCM), or through lactam formation between the side chains of P2' and P3' residues.

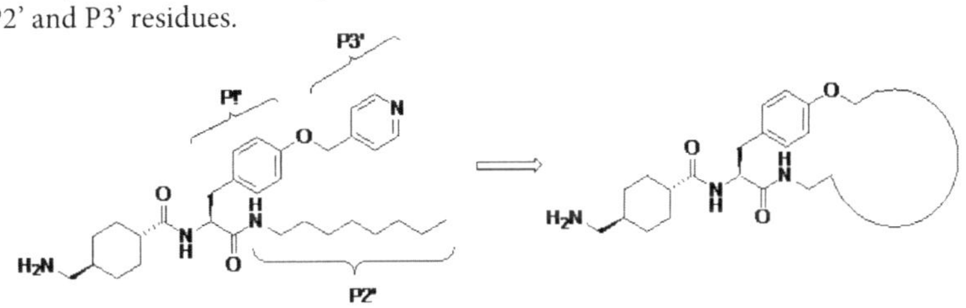

Figure 1: The structure of YO-2 and design of new cyclic derivatives.

Scheme 1 summarizes the synthetic route for the preparation of cyclic derivatives 3-5. The synthesis started from H-Tyr-OMe, which was derived to the dialkene intermediates, which were treated with TFA giving the linear derivatives. On the other hand, the intermediates were cyclized with RCM using Grubbs catalyst, followed by treatment of TFA, giving the cyclic derivatives 3-5. Final products were identified by analytical HPLC and MALDI (ESI)-TOF mass spectrometry. The Plm inhibitory activity of newly synthesized compounds is summarized in Table 1. An addition of benzene ring into the hydroxy group of a Tyr residue dramatically enhanced the inhibitory activity (1, 2 vs 3, 4, 5), which proved that the π-π interaction at the S3' subsite played a critical role. Among the compounds, the cyclic compound possessing an olefin with 20-membered ring had the strongest Plm inhibition (4: IC_{50} = 3.68 ≠ M), which, however, was seven-times weaker than YO-2, the parent compound.

In conclusion, the cyclization performed between the side chains of P2' and P3' residues and effectively enhanced the interaction at S' subsites; however, the other elements, such as hydrophobic interaction or their combination are necessary to optimize the interaction at the contact surface. Further efforts to improve Plm inhibition are continuing.

This research was supported in part by KAKEN (16K08333).

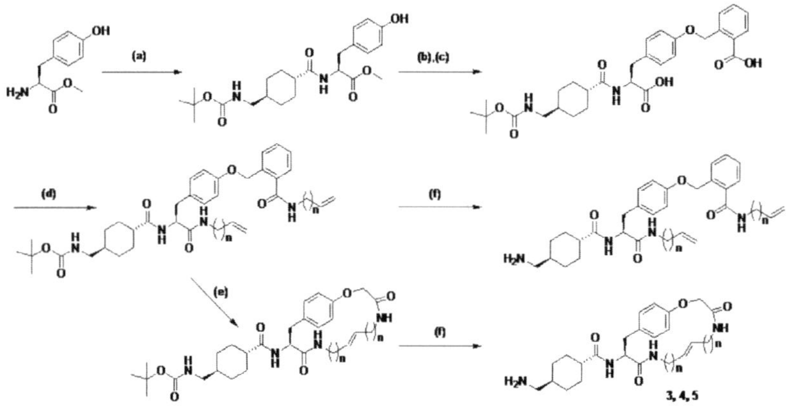

Scheme 1: Preparation of cyclic derivatives 3-5

Reagents and conditions: (a) TEA, Boc-Tra-OH, HOBt·H2O, EDC·HCl (1.2 equiv); (b) K2CO3 (4.0 equiv), Methyl 2-(chloromethyl)benzoate (2.0 equiv)/DMF, MeOH; (c) 1M NaOH/DMF; (d)Unsaturatedamines (2.5 equiv), HOBt·H2O, EDC·HCl (2.2 equiv); (e) Grubbs catalyst (0.25 equiv)/DCE at 60oC under N2; (f) TFA-anisole for 1h.

Table 1: Plm inhibition of newly synthesized derivatives

compd.	cyclic			linear		Index:
	structure	ring size	IC_{50} (μM)	structure	IC_{50} (μM)	IC_{50}(linear)/IC_{50}(cyclic)
YO-2				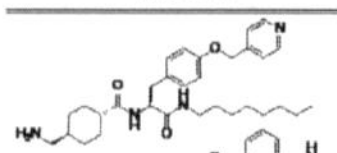	0.53	/
PSI-325				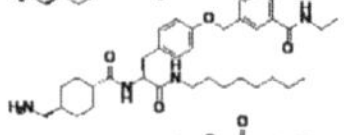	0.20	/
1	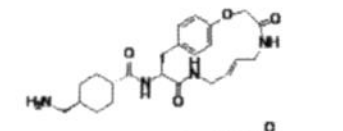	16	25.0% inhibition at 100 μM	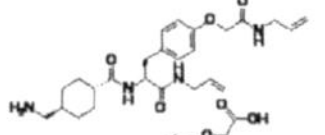	20.9	<0.21
2	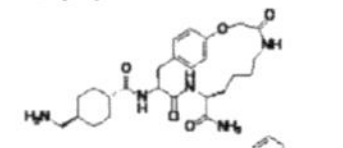	17	42.6% inhibition at 100 μM	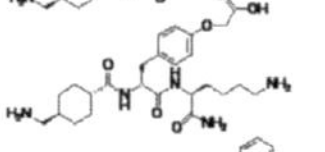	39.9% inhibition at 100 μM	1.06
3	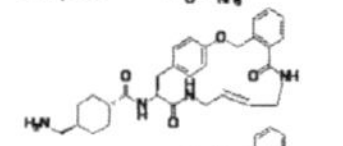	18	5.88	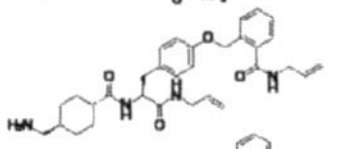	27.6	4.69
4	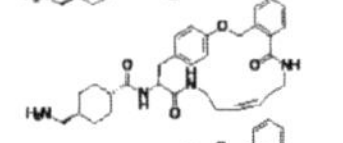	20	3.68	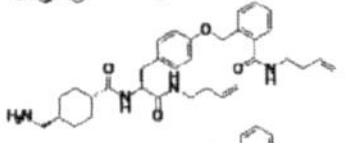	16.8	4.56
5	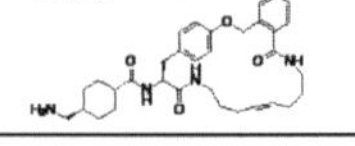	22	7.14	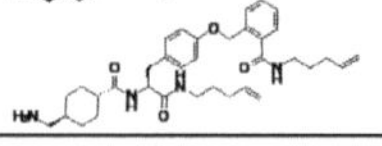	8.97	1.25

References

1. Okada, Y., Tsuda, Y., Tada, M., Wanaka, K., Hijikata-Okunomiya, A., Okamoto, S. Chem. Pharm. Bull., 48, 1964-1972 (2000).

2. Law, R. H. P., Wu, G., Leung, E. W. W., Hidaka, K., Quek, A. J., Caradoc-Davies, T. T., Jeevarajah, D., Conroy, P. J., Kirby, N. M., Norton, R. S., Tsuda, Y., Whisstock, J. C. Blood Adv. 1, 766-771 (2017).

ON/OFF repetitive control of protease activity using peptidomimetic ligands through biotin-streptavidin affinity

Koushi Hidaka, Keiko Hojo, Yuko Tsuda

Kobe Gakuin University, Japan

https://doi.org/10.17952/35EPS.2018.173

Introduction

Peptidomimetic protease inhibitors targeting the enzyme active site are useful as research tools and drugs. However, such small compounds are difficult to remove from the proteases once they bound, requiring large dilution or protein denaturation procedures. Our research group developed "removable inhibitors" that was designed with a direct conjugation of biotin and a strippable property, through extremely strong affinity to streptavidin (SA) (1) even after the binding to the target protease. This inhibitor stripping action was caused by affinity competition between target protease and SA, named as ISAAC. To extend the utility of removable inhibitors, we studied the ON/OFF repetitive control of enzymatic activity derived from affinity-purified proteases.

Results and Discussion

Previously, we reported a removable inhibitor, bPI-11 which was synthesized from a HIV protease inhibitor, a derivative of KNI-10006 containing an amino group by coupling with biotin using mixed anhydride (2). Enzymatic activity of HIV-1 protease was inhibited by bPI-11, which was sufficiently recovered by addition of SA, and then inhibited again by extra addition of the inhibitor. The result showed ON/OFF/ON/OFF repetitive control of the enzymatic activity of HIV protease (Figure 1).

Based on this result, we performed the activity detection of the affinity purified protease. Recombinant HIV-1 protease was mixed with human serum. The target protease was purified using magnetic beads conjugated with the inhibitor with a long aminocaproyl spacer. The bound protease was eluted with a solution containing bPI-11 and the eluate was examined for the activity detection using FRET substrates by ISAAC. As we expected, the substrate cleavage was confirmed by addition of SA and suppressed by following addition of clinical protease inhibitors such as darunavir, lopinavir and atazanavir (Figure 2). Similar experiments using a mutated protease, A17m5, which is lopinavir-resistant, were performed. We detected the lower enzymatic activity and slightly different inhibition by the clinical drugs from that of the wild-type protease. Some biotinylated derivatives of pepstatin A, were also utilized to detect cathepsin D activity purified from HCT116 cells and the activity was inhibited by addition of pepstatin A.

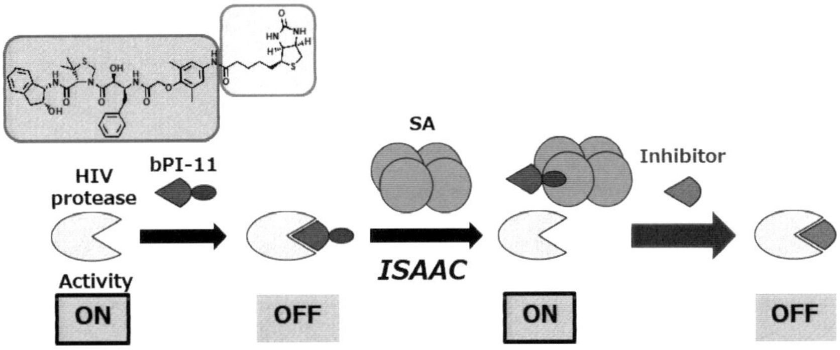

Figure 1: Repetitive control of protease activity using bPI-11.

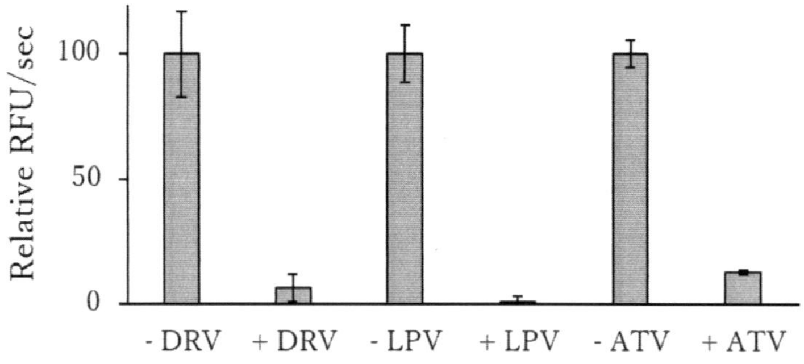

Figure 2: Activity detection of wild-type HIV protease mixed in with human serum.

In conclusion, we confirmed that a directly biotinylated protease inhibitor, bPI-11, repetitively controlled HIV protease activity. We also succeeded in detecting the enzymatic activity after the affinity purification from human serum and inhibited by clinical drugs. These results suggest that ON/OFF repetitive control of protease activity using removable inhibitors could be applied for testing susceptibility and drug resistance before the treatment with protease inhibitor drugs.

Acknowledgements

This research was supported by a grant from JSPS KAKENHI, 18K05344.

References

1. Howarth, M., et al., Nat. Methods, 3, 267-273 (2006).

2. Beck-Sickinger, A. G. and Diederichsen, U. (Eds.) Proceedings of the 34th European Peptide Symposium, The European Peptide Society, Leipzig, 2017, p. 178.

Polyethylene glycol (PEG)-based peptidomimetics (Pegtides)

Marc Devocelle, Alan Hibbitts, Aoife O'Connor, Sally-Ann Cryan, Deirdre Fitzgerald-Hughes

RCSI, Ireland

https://doi.org/10.17952/35EPS.2018.175

Peptide and protein drug discovery generates potent leads in most therapeutic areas.[1] Their optimisation and development may involve peptidomimetic conversion, or conjugation to biocompatible polymers. The features of these two techniques are combined in polymer-based peptidomimetics. For instance, functionalised poly(norbornene), poly(acrylic) and nylon-3 backbones have been described as synthetic mimics of Antimicrobial Peptides (AMPs).[2, 3] Polyethylene Glycol (PEG), one of the most established biocompatible polymers for improving the pharmacokinetic and pharmacodynamic properties of polyamide-based therapeutics, has not been exploited as a backbone for peptidomimetics on the other hand.[4]

Co- and post-polymerisation functionalisation of substituted ethylene oxides can produce PEG-based backbones with branching chains separated by an equal number of bonds than side chains in peptides (Fig. 1). These hybrid structures of PEG and peptides, termed 'Pegtides', are different from PEGtide dendrons which alternate monodisperse non-functionalised PEG chains and dipeptide motifs.[5]

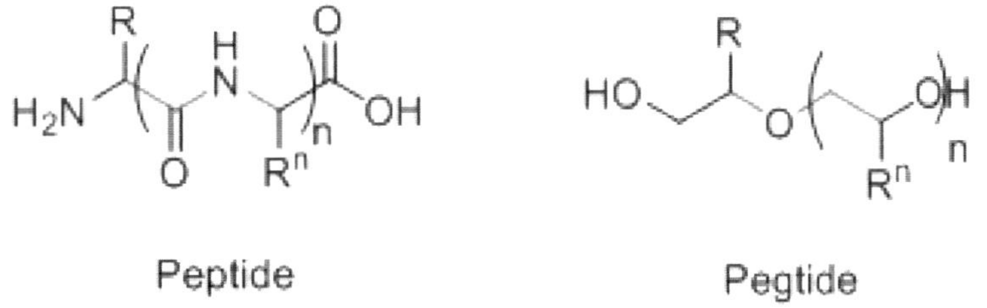

Figure 1: General structures of peptides and PEG-based peptidomimetics 'pegtides'.

Homopolymerisation and copolymerisation of substituted oxiranes, followed by possible functional group interconversion, can produce PEG-based peptidomimetics of poly(α-amino acids) and of sequences enriched in two types of amino acids, respectively. Hence, copolymers with hydrophobic and positively charged repeating units can mimic amphipathic peptides such as AMPs. Their sequences are indeed characterised by a net excess of positively charged residues and a hydrophobic amino acid content approximating 30-50%.[6] Moreover, the main activity determinants of these peptides are their overall amino acid content and sequence pattern, rather than sequence-dependent arrangement. These characteristic features can be mimicked with statistical copolymers. Consequently, an AMP pegtide was prepared by anionic ring-opening copolymerisation of tert-butyl N-(2-oxiranylmethyl)carbamate and 1,2-epoxyhexane, using sodium benzyloxide as the initiator, followed by Boc removal by acidolytic cleavage and guanylation of the resulting amines with N,N'-di-Boc-1H-pyrazole-1-carboxamidine. Final deprotection by treatment with trifluoroacetic acid yielded poly(hexene oxide-co-glycidylguanidine)s (Fig. 2, left), which were purified by Size-Exclusion Chromatography (SEC) or precipitation and analysed by ^{1}H NMR and MALDI-TOF Mass Spectrometry. Short copolymers isolated by SEC (m = 1-2, n = 2-3) were tested in preliminary susceptibility and antibiofilm assays. The former indicated Minimum Inhibitory Concentrations (MICs) of 150-300 $\neq$ M against the model Gram-positive (*Staphylococcus aureus*) and Gram-negative (Escherichia coli) organisms. These pegtides also showed some growth inhibitory properties in a Staphylococcus epidermidis biofilm prevention assay (Fig. 3). Longer copolymers with 4 glycidylguanidine (m = 4) and 10 hexene oxide repeating units (n = 10) were also obtained by this approach.

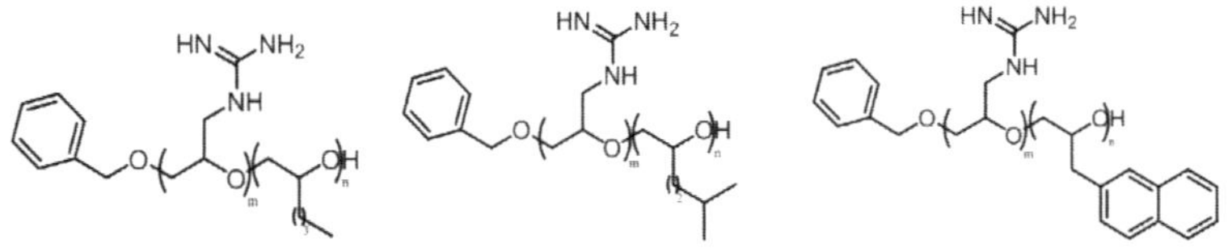

Figure 2: Pegtides of (arginine, norleucine)-, (arginine, homoleucine)- and (arginine, tryptophan)-rich AMPs.

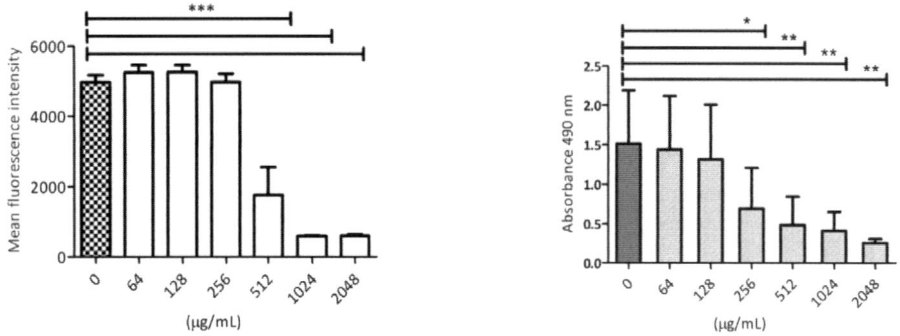

*Figure 3: Prevention of S. epidermidis biofilm formation by short Arg-Nle mimetics; (left) viability of cultures determined by a Resazurin-conversion assay; (right) biofilm formation quantified by the crystal violet staining method. Presented are means±SD of the independent experiments. Statistical significance was determined by 2-way ANOVA with Dunnett's multiple comparison test, * $P < 0.05$ ** $P < 0.005$ & ***$P<0.001$.*

Two other AMP pegtides, differing by their hydrophobic contents, were prepared by using epoxide monomers mimicking branched or aromatic amino acids. They were 2-(3-methylbutyl)oxirane for homoleucine (Fig. 2, centre) and 2-(naphtalen-2-ylmethyl)oxirane for 3-(2-naphtyl)alanine, itself a substitute of tryptophan (Fig. 2, right). Copolymers were obtained with degrees of polymerization of 10 on average. Preliminary susceptibility testing indicated that peptidomimetics based on branched lipophilic units displayed greater antimicrobial activities than their linear counterparts and that the Arg-Trp pegtides displayed the lowest MICS, one dilution away from those of their parent decapeptide, against both Gram-negative and Gram-positive bacteria.

Acknowledgements

This material is based upon works supported by the Science FoundationIreland underGrant No. 13/TIDA/B2657. The authors would like to thank Dr Andrea Molero Bondia and Mr Daniel O'Neill for the synthesis of the peptidomimetics, Drs Deirdre Fitzgerald-Hughes, Eoghan O'Neill and Marta Zapotoczna for the susceptibility testing.

References

1 Henninot A, Collins JC, Nuss JM. The CurrentState of Peptide Drug Discovery: Back to the Future? J Med Chem. 2018, 61(4):1382-1414.

2 Lienkamp K, Tew GN. Synthetic mimics of antimicrobial peptides–a versatile ring-opening metathesis polymerization based platform for the synthesis of selective antibacterial and cell-penetrating polymers. Chemistry. 2009, 15(44):11784-11800.

3 DornerF, Lienkamp K, in Polymeric Materials with Antimicrobial Activity: From Synthesis to Applications, ed. A. Munoz-Bonilla, M. Cerrada, M. Fernández-García, Royal Society of Chemistry, Cambridge, UK, 1st edition, 2013, chapter 5, 97-138.

4 Herzberger J, Niederer K, Pohlit H, Seiwert J, Worm M, Wurm FR, Frey H. Polymerization of Ethylene Oxide,

Propylene Oxide, and Other Alkylene Oxides: Synthesis, Novel Polymer Architectures, and Bioconjugation. Chem Rev. 2016, 116(4):2170-2243.

5 Gao J, Chen P, Singh Y, Zhang X, Szekely Z, Stein S, Sinko PJ. Novel monodisperse PEGtide dendrons: design, fabrication, and evaluation of mannose receptor-mediated macrophage targeting. Bioconjug Chem. 2013, 24(8):1332-1344.

6 Zasloff M. Antimicrobial peptides of multicellular organisms. Nature. 2002, 415(6870):389-395.

1,5-Benzothiazepinone and arylazepinone dipeptide mimetics as local constraints in peptidomimetic design

Robin Van Den Hauwe[1], Olivier Van der Poorten[1], Steven Ballet[1], Emilie Eiselt[2], Phillipe Sarret[2], Louis Gendron[2], Dirk Tourwé[1]

[1]Vrije Universiteit Brussel, Belgium
[2]Université de Sherbrooke, Canada

https://doi.org/10.17952/35EPS.2018.178

Introduction

Herein, the application of novel conformationally constrained amino acids, 4-amino-8-bromo-2-benzazepin-3-one (8-Br-Aba), 3-amino-3-4-dihydroquinolin-2-one(Dhq), and regioisomeric 4-amino-naphthoazepinones (1- and 2-Ana), is described. Introduction of these constricted scaffolds into the N-terminal tetrapeptide of dermorphin(i.e., H-Tyr-D-Ala-Phe-Gly-NH_2) induced significant shifts in binding affinity, selectivity, and *in vitro* activity at the μ- and δ-opioid receptors (MOP and DOP, respectively). A reported constrained μ-/δ-opioid lead tetrapeptide H-Dmt-D-Arg-Aba-Gly-NH_2 was modified through application of various constrained building blocks to identify optimal spatial orientations in view of activity at the opioid receptors. Interestingly, when the aromatic moieties were turned towards the C-terminus of the peptide sequences, (partial) (ant)agonism at MOP and weak (ant)agonism at DOP were noticed, whereas the incorporation of the 1-Ana residue (with the aromatic moiety shifted towards the N-terminus) led towards balanced low nanomolar MOP/DOP binding and *in vitro* agonism.[1]

Additionally, a 3-step methodology for the synthesis of 1,5-benzothiazepin-4(5H)-one dipeptidomimetics has been elaborated *via* an Ugi-4CR followed by a S-trityl deprotection and an intramolecular Cu(I)-catalyzed Ullmann condensation with moderate to good yields. In silico and NMR conformational studies showed that the lowest energy conformers stabilize γ- and β-turn structures.[2]

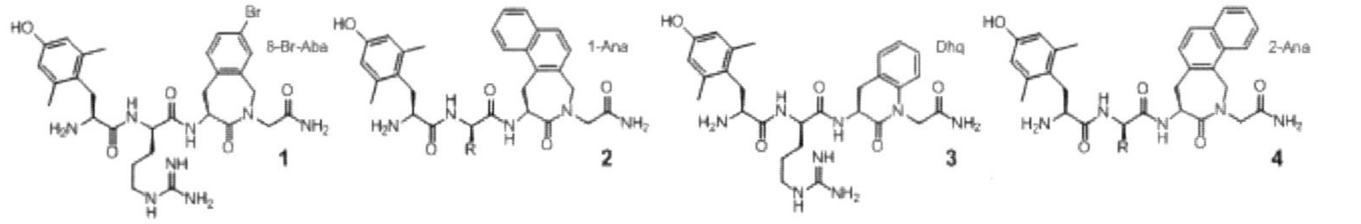

Figure 1: 8-Br-Aba, Dhq, and 1/2-Ana scaffolds incorporated in the N-terminal tetrapeptide of dermorphin

Results and Discussion

The insertion of constricted 4-amino-8-bromo-2-benzazepin-3-one(8-Br-Aba), 3-amino-3-4-dihydroquinolin-2-one (Dhq), and regioisomeric 4-aminonaphthoazepinones7a-c (1/2-Ana) (1-4, Figure 1) dipeptidomimetics into an optimized dermorphin tetrapeptide lead sequence has resulted in compact and high affinity MOP/DOP opioid receptor ligands. Whereas insertion of the conformationally constrained 8-Br-Aba-Gly, Dhq-Gly, or 2-Ana-Gly dipeptides resulted in decreased DOP recognition, application of the regioisomeric 1-Ana-Gly building block led to excellent low nanomolar MOP and DOP binding affinities and *in vitro* functional activities (Table 1). The described cyclic constrained aromatic amino acids can be regarded as additional tools to modulate receptor selectivity and activity, but they also provide a way to enhance proteolytic stability and bioavailability of lead peptides. Bulky naphthylalanine residues often provide a powerful means of modulating receptor selectivity. The current building blocks present the additional feature to control χ dihedral angles, thus presenting specific ligand topologies. The present study clearly showcases the advantages of such constrained residues for improving ligand potency and opioid receptor selectivity. It is expected that these building blocks will rapidly find application in other biologically active peptides.

Table 1: In vitro opioid receptor affinity and activity of dermorphin tetrapeptides

Compound	GPI[a] (IC_{50} nM)	MVD[a] (IC_{50} nM)	MOP[b] (IC_{50} nM)	DOP[b] (IC_{50} nM)	Selectivity IC_{50} DOP / IC_{50} MOP
H-Dmt-D-Arg-[8-Br-Aba-Gly]-NH_2	22.0±2.4[c]	8.51±0.70	0.76±0.49	60.7±13.7	80
H-Dmt-D-Arg-[Dhq-Gly]-NH_2	33.7±3.6[d]	90.5±14.4	1.62±0.03	630±65.0	389
H-Dmt-D-Arg-[1-Ana-Gly]-NH_2	0.252±0.014	1.42±0.18	0.33±0.01	1.00±0.45	3
H-Dmt-D-Ala-[1-Ana-Gly]-NH_2	1.13±0.13	0.86±0.16	0.39±0.07	2.07±0.30	5
H-Dmt-D-Arg-[2-Ana-Gly]-NH_2	21.4±1.9	114±7.0	2.19±0.70	153±12.0	70
H-Dmt-D-Ala-[2-Ana-Gly]-NH_2 major	2.88±0.10	306±12.0	1.78±0.40	167±38.0	94
H-Dmt-*N*Me-D-Ala-[1-Ana-Gly]-NH_2	0.43±0.06	1.09±0.04	0.35±0.04	1.47±0.04	4

Additionally, a short synthetic strategy towards constrained thiazepinones was developed which comprised an Ugi-4CR and an intramolecular Cu(I)-catalyzed C-S cyclization reaction as the two key synthetic steps (Scheme 1). Our goal was to generate a diverse set of substituted 1,5-benzothiazepin-4(5H)-one-based dipeptide mimetics 10, wherein the Cα-position of the exocyclic amino acid was substituted with different alkyl and aryl substituents. Starting from commercially available Boc-L-Cys(Trt)-OH 5, different aldehydes 6, o-iodoaniline 7, and tert-butylisocyanide 8, linear Ugi dipeptides of type 9 were constructed. Orthogonal S-trityl deprotection, was followed by a CuI-catalyzed C-S Ullmann condensation, yielding the desired (R)-1,5-benzothiazepin-4(5H)-one-containing dipeptide mimetics of type 10.

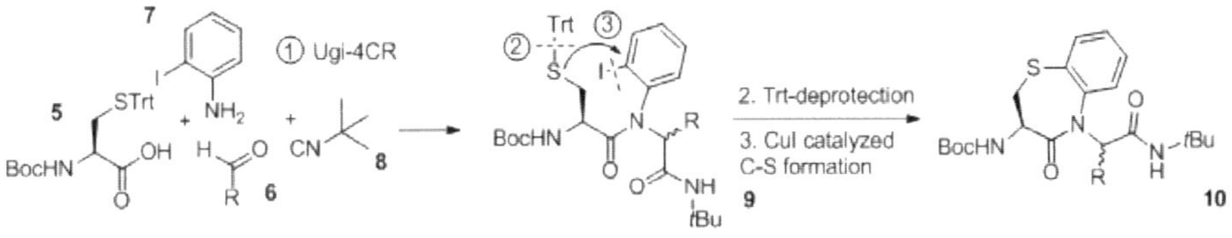

Scheme 1: Optimized synthethic pathway towards (R)-1,5-benzothiazepin-4(5H)-one dipeptide mimetics via subsequent Ugi-4CR, S-trityl deprotection and CuI-catalyzed C-S ullman condensation

The effect of the second stereocenter on the overall turn-inducing properties of the dipeptidomimetics was explored *via* structural in silico calculations. Molecular modeling indicated that all lowest energy conformers of (R,S)- and (R,R)-benzothiazepinones possessed a hydrogen bond-stabilized γ-turn conformation. However, the expected β-turn conformations were also observed within a range of 1.95 to 5.60 kcal mol^{-1} above the energetically lowest conformer. The conformational preferences and overall turn-inducing properties of these dipeptidomimetics can potentially be influenced by the steric bulk of the C-terminal tert-butylamide moiety. Therefore, structural calculations were repeated for and both (R,S)- and (R,R) with less sterically hindering primary and secondary amides (i.e. $-CONH_2$, –CONHMe, –CONHiPr), and indeed, significantly lower energy differences were recorded between the lowest energy γ-turn conformers and the first β-turn structures if the bulky C-terminal tert-butylamide group was replaced.

Acknowledgments

RVDH and SB are grateful to the NSE Doctoral School and the OZR of VUB for the financial support and acknowledge the organizing committee of 35th EPS for the registration bursary.

References

1. Van der Poorten, O.; Van Den Hauwe, R.; Eiselt, E.; Betti, C.; Guillemyn, K.; Chung, N. N.; Hallé, F.; Bihel, F.; Schiller, P. W.; Tourwé, D. ACS Medicinal Chemistry Letters 2017, 8, 1177-1182. DOI: 10.1021/acsmedchemlett.7b00347

2. Van der Poorten, O.; Van Den Hauwe, R.; Hollanders, K.; Maes, B. U.; Tourwe, D.; Mouhamad, J.; Ballet, S. Organic & Biomolecular Chemistry 2018. DOI:10.1039/C7OB03094K

Impact of bulky dehydroamino acids on the structure and proteolytic stability of peptides

Ankur Jalan, David W Kastner, Kei G I Webber, Steven Castle

Brigham Young University, United States

https://doi.org/10.17952/35EPS.2018.180

The great potential of peptides as therapeutic agents has stimulated numerous efforts to increase their stability to proteolytic cleavage. Stammer and co-workers previously established that α,β-dehydroamino acids (ΔAAs) such as ΔPhe and ΔLeu are able to impart rigidity and proteolytic stability to peptides that contain them.[1–4] Presumably, the aryl or alkyl substituent attached to the β-carbon of these residues introduces $A_{1,3}$ strain into the peptide that destabilizes many of its conformations. This decrease in low-energy conformations can lead to an enhanced preference for rigid folded states over flexible random coil conformations. Intrigued by the presence of ΔVal and ΔIle in the anticancer peptide yaku'amide A,[5] we reasoned that these bulky dehydroamino acids containing two β-substituents would exhibit high levels of $A_{1,3}$ strain. Accordingly, they should have a greater impact on peptide rigidity and proteolytic stability than ΔAAs such as ΔPhe and ΔLeu that possess just one β-substituent (Figure 1).

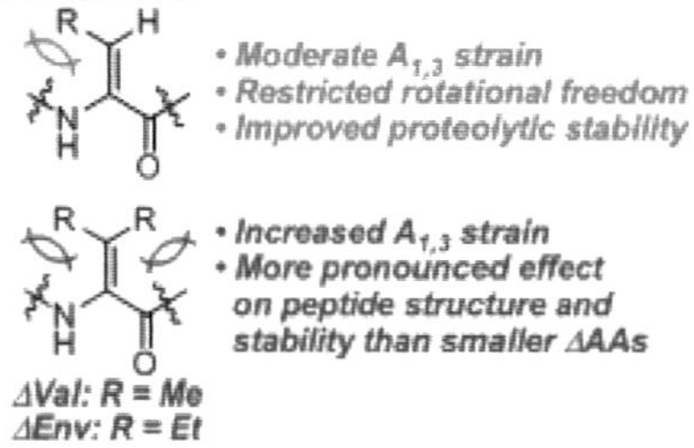

Figure 1: Normal Versus Bulky ΔAAs

Based on previously published studies of ΔVal by others,[6,7] we hypothesized that placing this residue or its slightly larger analogue dehydroethylnorvaline (ΔEnv) at the (i + 1) position of a β-turn would result in increased proteolytic stability without compromising the secondary structure. We selected the β-hairpin designed by Waters and co-workers[8] (NG, Figure 2) as a system for evaluating this concept. We synthesized two variants of NG (ΔValG and ΔEnvG, Figure 2) *via* solid-phase peptide synthesis by utilizing azlactone ring-opening chemistry developed previously in our laboratory.[9] ^{1}H NMR and NOE data for ΔValG and ΔEnvG indicated that these peptides retained the β-hairpin motif characteristic of NG. NOE-restrained structural calculations using molecular dynamics showed that NG and ΔValG exhibited very similar solution structures (Figure 3). RMSD calculations suggested that ΔValG is more rigid than NG, an assertion that is supported by qualitative examination of the structural ensembles of the two peptides.

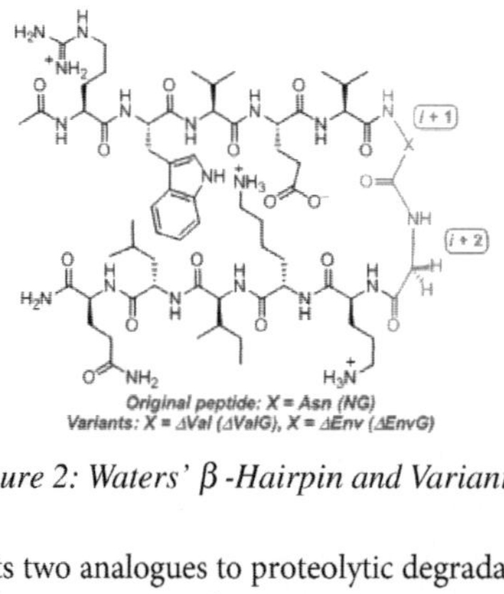

Figure 2: Waters' β-Hairpin and Variants

We evaluated the stability of NG and its two analogues to proteolytic degradation by treating them with Pronase, a cocktail of proteases that cleaves peptides in a nonspecific manner. Disappearance of the peptides was monitored by HPLC, and half-lives were calculated for each β-hairpin. We were pleased to discover that replacing the (i + 1)

Asn residue with a bulky ΔAA led to a substantial increase in proteolytic stability, as both ΔValG and ΔEnvG exhibited half-lives that were six- to sevenfold longer than the half-life of NG (Figure 4).[10] This exciting result caused us to wonder if the beneficial impact of bulky ΔAAs was confined to the (i + 1) position, or if insertion of ΔVal or ΔEnv at the (i + 2) position would also protect β-hairpins from proteolysis. Accordingly, two other variants of NG in which the Gly residue is replaced by a bulky ΔAA (NΔVal and NΔEnv) were synthesized and evaluated in the proteolysis assay. While these analogues were more stable than NG, the ca. threefold increase in half-life was not as dramatic as the effect seen with bulky ΔAAs at the (i + 1) position.[10] Nonetheless, this measurable improvement in proteolytic stability was substantial enough to justify the preparation of β-hairpins pairing a stabilizing D-Pro residue at (i + 1) with a bulky ΔAA at (i + 2) (pΔVal and pΔEnv). Each of these peptides adopted a β-hairpin structure as evidenced by NMR spectroscopy. Subjection of them to the Pronase assay showed that the half-life of pΔEnv was ca.1.5 times longer than that of the β-hairpin pG, whereas the half-life of pΔVal was ca. 2.5 times longer than that of the control peptide.[10] Thus, we established that a D-amino acid and a bulky ΔAA can synergistically increase the stability of a β-hairpin to proteolysis without disrupting its secondary structure.

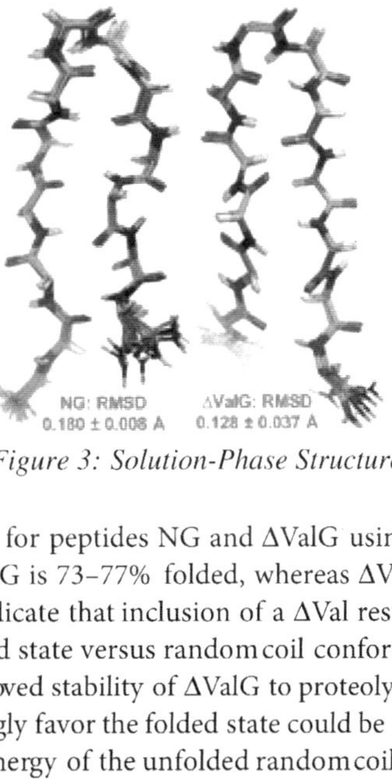

Figure 3: Solution-Phase Structures

We calculated fraction folded values for peptides NG and ΔValG using two different methods based on 1H NMR spectroscopy. We found that NG is 73–77% folded, whereas ΔValG is folded to a substantially greater degree (90–94%).[10] These data indicate that inclusion of a ΔVal residue at the (i + 1) position of a β-turn enhances the preference for the folded state versus random coil conformations by ca. 0.6–1.0 kcal/mol. This can at least partially explain the improved stability of ΔValG to proteolysis relative to NG. The ΔVal-promoted shift of the equilibrium to more strongly favor the folded state could be achieved by either lowering the energy of the folded state or by raising the energy of the unfolded random coil conformations. While further studies would be necessary to distinguish between these two possibilities, we suggest that ΔVal (and by extension ΔEnv) destabilizes numerous random coil conformations of the model peptides due the high levels of $A_{1,3}$ strain inherent in its tetrasubstituted alkene moiety.

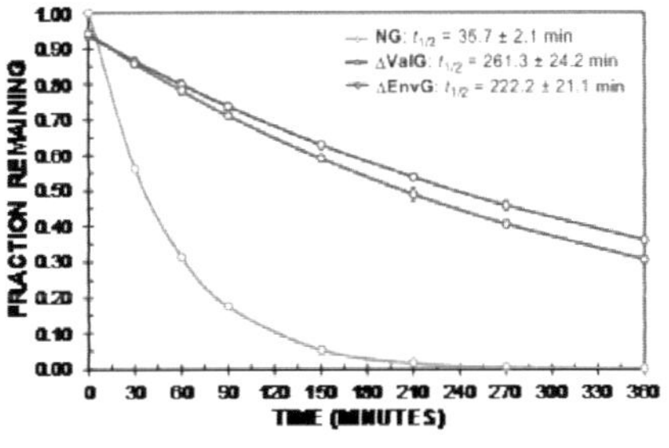

Figure 4: Proteolysis of NG, ΔValG, and ΔEnvG by Pronase

In summary, we have discovered that the bulky dehydroamino acids dehydrovaline and dehydroethylnorvaline can substantially increase the proteolytic stability of β-hairpins without negatively impacting the secondary structures of these peptides. The stabilizing effect is more pronounced at the (i +1) position than at the (i +2) position of the β-turn region. We also found that a bulky ΔAA can work synergistically with a D-amino acid to render β-hairpins highly resistant to proteolysis. Future studies are focused on the impact of bulky ΔAAs on other secondary structures as well as their ability to protect bioactive peptides from proteolysis without decreasing their potency.

References

1. English, M. L.; Stammer, C. H. Biochem. Biophys. Res. Commun. 1978, 83, 1464.

2. English, M. L.; Stammer, C. H. Biochem. Biophys. Res. Commun. 1978, 85, 780.

3. Shimohigashi, Y.; Chen, H.-C.; Stammer, C. H. Peptides 1982, 3, 985.

4. Shimohigashi, Y.; Stammer, C. H. J. Chem. Soc., Perkin Trans. 1 1983, 803.

5. Ueoka, R.; Ise, Y.; Ohtsuka, S.; Okada, S.; Yamori, T.; Matsunaga, S. J. Am. Chem. Soc. 2010, 132, 17692.

6. Vijayaraghavan, R.; Kumar, P.; Dey, S.; Singh, T. P. Acta Crystallogr., Sect. C 2001, 57, 1220.

7. Siodłak, D.; Rzeszotarska, B.; Broda, M. A.; Kozioł, A. E.; Kołodziejczyk, E. Acta Biochim. Pol. 2004, 51, 145.

8. Tatko, C. D.; Waters, M. L. J. Am. Chem. Soc. 2004, 126, 2028.

9. Jiang, J.; Luo, S.; Castle, S. L. TetrahedronLett. 2015, 56, 3311.

10. Jalan, A.; Kastner, D. W.; Webber, K. G. I.; Smith, M. S.; Price, J. L. Org. Lett. 2017, 19, 5190.

Study of β-sheet helix interactions based on conformational properties of synthetic D,L-alternating decapeptides

Emma Fenude, Michele Saviano

ICB-CNR, Departmentof Chemical Science and Material Technology- Italy, Italy

https://doi.org/10.17952/35EPS.2018.183

Abstract

De novo peptide design is based on the ability to construct peptide sequences with predictable folding patterns. Helices and β-sheet have been the focus of considerable synthetic attention in attempts to assemble mimics for helical bundles and all β-motifs. In contrast, strategies for construction of β-sheet helix have been less widely explored. The β-sheet helical architecture is constructed from polypeptides that are coiled into a large helix, formed by stacks of β-sheets separated by loops. β-Sheet helices are present in the fibrous form of transyretin, that play an important role in bovine spongiform encephalopathy (BSE) and form the crucial structural elements in insect antifreeze proteins. We report on synthetic analogues able to form β-sheet helices and on the structural comparison with β-helices formed by D,L-alternating peptides.

Introduction

The β-sheet helical architecture is constructed from polypeptides that are coiled into a large helix, formed by stacks of β-sheets separated by loops. β-Sheet helix is also known to occur in Nature, for example, form the crucial structural elements in insect antifreeze proteins. Antifreeze protein consists of an irregular β-sheet helix with a triangular cross-section and a long α-helix that runs parallel on one side of the β-sheet helix, both domains are stabilized by hydrophobic interactions. Several studies have indicated that the ion-conducting transmembrane channels formed by gramicidin A in lipid bilayer originated througha dimerization process, and Urry [1] has proposed that the channels are helical dimers consisting of two single stranded β-helices (6,3 residues per turn) connected head-to-head (formyl-end-to-formyl-end). Urry's proposal is consistent with a number of experimental observations, and the ability of single- and double-stranded β-helices to give head-to-head dimers has been demonstrated either with gramicidin A [1,2] as with synthetic N-formyl oligopeptides[3-4]. A previous work on elastin fragments has clearly revealed the presence of type I and type II β-turns. Based on these results, Urry proposed that elastin adopts a β-spiral structure that contains approximately three pentamers (VPGVG) per turn and the repeating type II Pro-Gly β-turns function as spacer between turns of the β-spiral, utilizing dominantly the Val⟵⟶Pro hydrophobic interactions. Proline is the most frequent residue that occurs at the i+1 position of type I and type II β-turnin proteins. Here we investigated the role played from proline in the conformational behavior of β-turn-containing peptides.

Method

Hydrogen-bonding vs Hydrophobicinteractions: there are many occasions in the literature where conformational preference (propensity) is varied through hydrophobic interactions, even at the expense of hydrogen-bonds. The balance between hydrogen bonds and hydrophobic interactions is addressed to differing extents by existing molecular modeling packages. The relative importance of hydrophobic and polar interactions is especially important when attempting to predict the orientation of conformational behavior in the absence of experimental observations. The conformational behavior of D,L-alternating peptides clearly show the conformational changes due to feeble shift of the interactions equilibrium. Oligonorleucines, that on theoretical ground would prefer to form double stranded β -helix, form instead a extended chain structure that is slightly arched (or has a shape of large incomplete ring). This contrast is due to the prevalence of hydrophobic interaction of the linear side chain (interdigitation)that stabilize the extended structure and is prevalent respect to the effect of hydrogen bonds in stabilize the helical structure. Boc-(D-Nle-L-Nle)5-OMe (X) form aggregates which are unsoluble in common organic solvents even at moderate chain lengths Fig.1[3]. An N-methylated residue at the n-3 position reduce the maximum numberof H-bondsrealizable by Boc-(D-Nle-L-Nle-)3-D-NMeNle-L-Nle-D-Nle-L-Nle-OMe that forms a single β-helix in solution Fig.2 [4].

Insertion of a Leu residue in central position of decapeptide backbone reduce hydrophobic interaction between norleucine side chains Boc-(D-Nle-L-Nle)3-D-Leu-L-Nle-(D-Nle-L-Nle)-OMe forms double stranded β-helix in solution [5].

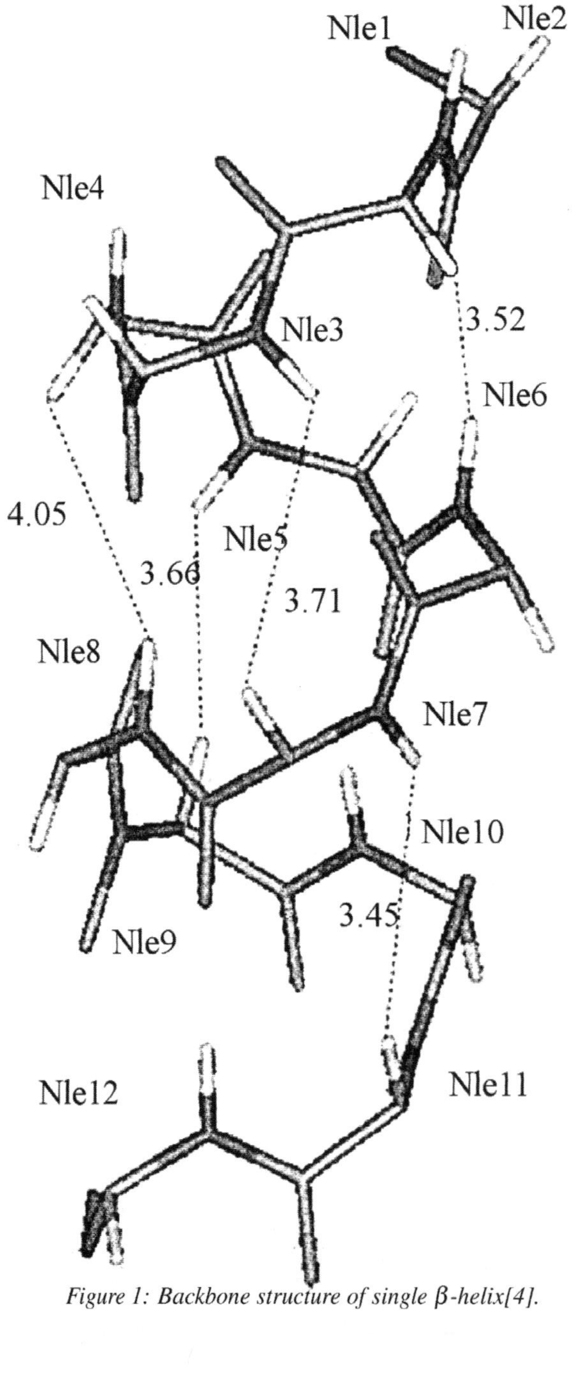

Figure 1: Backbone structure of single β-helix[4].

Results. Here we report about conformational study of synthetic analogue of β-casomorphin-8; Boc-Tyr1(tBu)-Pro2-Phe3-Pro4-Gly5-Pro6-Ile7-Pro8-OMe(I). NMR spectra of peptide I show a equilibrium of five different conformations. Internuclear distances between not vicinal nuclei have been obtained from spectral data, by measuring correlation peak's volumes that are proportional to the inverse of the internuclear distance's sixth potency (Fig.4). The results show the presence of a conformation having all the amide bonds Xxx-Pro trans (64% events). After deprotection of the phenolic group, NMR spectra show the presence of 3 H-bond(Fig.): (Y1)C=O—-H-N(F3), 1.820Å; (F3)C=O—-H-N(G5) 1.821 Å (with formation of a γ-turn in both cases); (Y1)O-H—-O=C(G5) 1.454. The contribution of steric/hydrophobic interactions to the stability of pseudocyclic structure of β CM-8 (Fig.6) can be well described with homochiral sequences Boc-(Xxx-Pro)2-OMe with Xxx=bulky, hydrophobic residue. The results suggest that, like occur in β -spiral, the balance between hydrogen bonding and hydrophobic interaction is particularly important for the stability of pseudocyclic structure. All peptide bonds Xxx-Pro are trans.

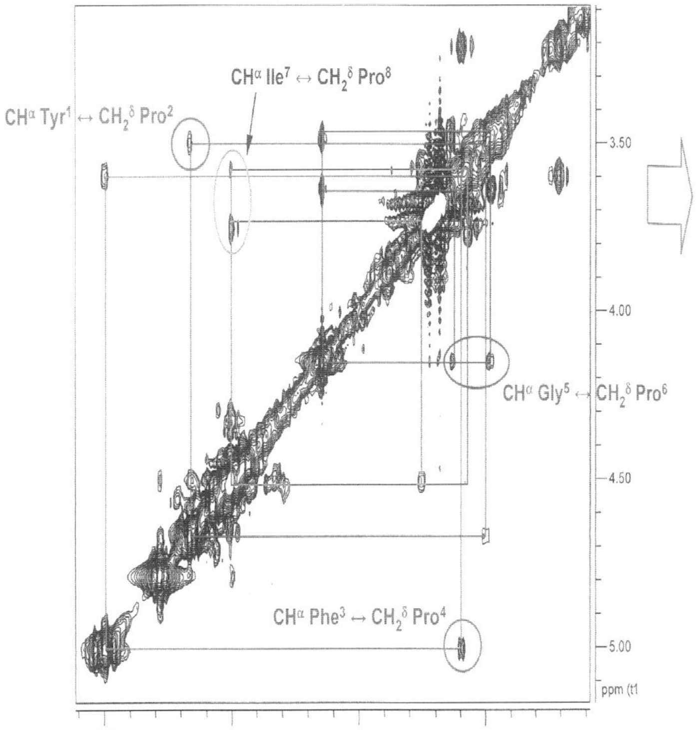

Figure 2: ROESY (CDCl3) spectrum of Boc-Tyr(tBu)-Pro-Phe-Pro-Gly-Pro-Ile-Pro-OMe.

Dihedral angles H-N-C-H(Karplus eq.)

F3 139.5° (calc.) vs 151.7° (exp.)

G5 120.9° (calc.) vs 117.4° (exp.)

I7 146.7° (calc.) vs 153.7° (exp.)

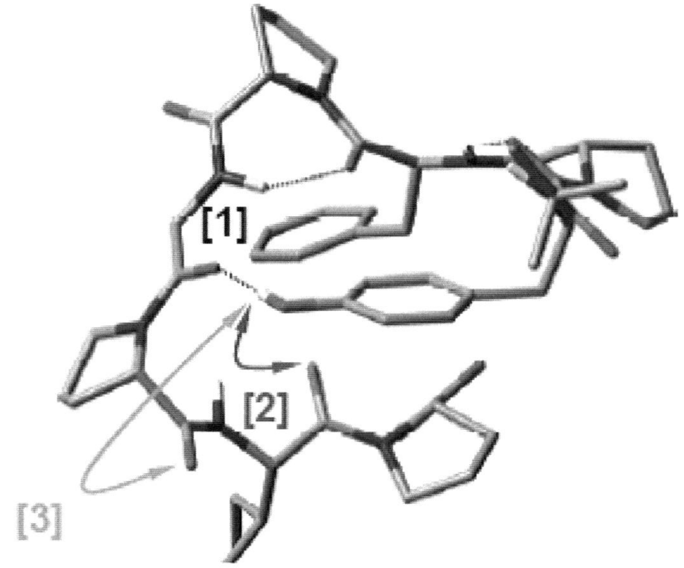

Figure 3: Molecular Dynamic (Y1)C=O—-H-N(F3); (F3)C=O—–H-N(G5)

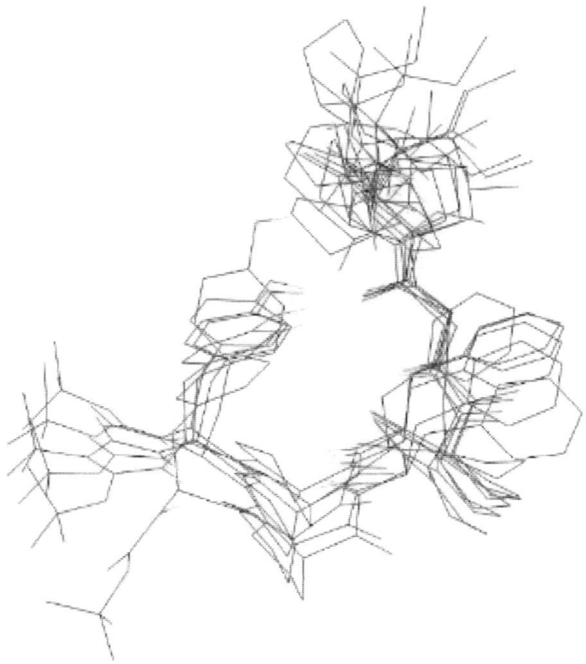

Figure 4: Molecular Dynamic (G5) CO—HO(Y1).

NMR spectral data of Boc-Gly-Tyr-Tyr-Pro-OMe and Boc-Gly-Tyr-Tyr-Pro-Thr-OMe reported in this work show a predominant equilibrium between two β-turn structures and a very low presence of the cis/trans isomers. The gluten exorphins A4 and A5 are δ-selective ligands; these two peptides had not affinity for κ-receptors. The gluten exorphin A4 is found 21 times in the repetitive domains of glutenin and structural prediction studies suggest the presence of β-turns (fig. 5).

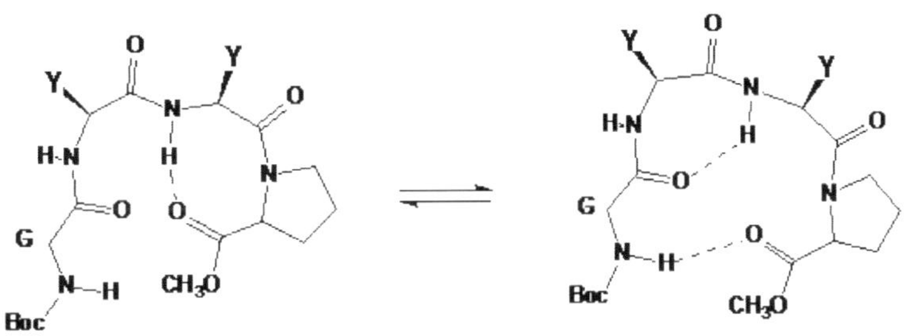

Figure 5: Boc-Gly1-Tyr2-Tyr3-Pro4-OMe equilibrium of structures

Conclusions

The results presented above provided a possible method for relating the propensity of amino acid residues with the structural features of single stranded β-helix (formed by D,L-alternating peptides) and of β-sheet helix (formed by all-L-peptides). The data of conformational analysis of peptides here reported suggest that important determinants of β-sheet helix formation are amino acid residues with high β propensity, well-defined turns, and inter- and intra-strand side chain interactions between hydrophobic residues. Our interest in this peptide is from the standpoint of deducing the structure and dynamics of homopeptides that assume preferentially a β-sheet helical structure. These oligopeptides can be considered as synthetic analogs of naturally occurring β-sheet helices. The cooperative behavior of the hydrogen-bonded network in these oligopeptides allows their architecture and thus their properties to be finely tuned. The availability of a large number of natural and unnatural amino acids opens the possibility to design and synthesize a wide array of well-defined β -helical polymers and block copolymers.

References

[1] Urry D.W,. Proc. Natl. Acad. Science U.S.A. (1983) 68, 672

[2] Urry D.W., Phil. Trans. R. Soc. London. B (2002) 357, 169-184

[3] Fenude Schoch E., Römer U.D., Lorenzi G.P., Int. J. Peptide Protein Res.(1994) 44, 10-18

[4] Navarro E., Fenude E., Celda B., Biopolymers (2002) 64, 198-209

[5] Unpublished results

[6] Lorenzi G.P., Gerber C., Jackle H., Macromolecules (1985) 18, 154-159

Modulation of the properties of elastin-like polypeptides by structure variations

Katerina Sancova[1], Lucie Bednarova[2], Romana Sulakova[1], Sergej Karel[1], Barbora Brtkova[3], Martin Flegel[4], Vladimír Velebný[1]

[1]Contipro a.s., Czechia
[2]Institute of Organic Chemistry and Biochemistry of the CAS, Czechia
[3]Contipro a.s., Slovakia
[4]University of Chemistry and Technology Prague, Czechia

https://doi.org/10.17952/35EPS.2018.188

Introduction

Elastin-like polypeptides (ELPs) with VPGXG repeating sequence are artificial biomacromolecules derived from hydrophobic domain of tropoelastin.[1] Their characteristic feature is coacervation (self-association) characterized by transition temperature T_t. ELPs are soluble in water or buffer below its T_t and they form β-structures and aggregates above its T_t. Up to date, the reversibility of the coacervation at specific temperature T_t has been studied in many research areas, e.g. in drug delivery, protein purification and tissue engineering.[2] Moreover, the transition temperature can be modulated by both extrinsic (pH, ionic strength) and intrinsic factors (primary sequence and chain length).[3]

The aim of our study was to characterize secondary structure behavior and to define T_t with respect to primary sequence of ELPs with general formula $H(VPGLG)_2(VPGXG)_2(VPGLG)_6$-OH, where we used the hydrophobic (Gly, Ala, Leu), hydrophilic (Gln, Asn, Tyr), acidic (Glu) and basic amino acids (His, Lys) as the variable element X.

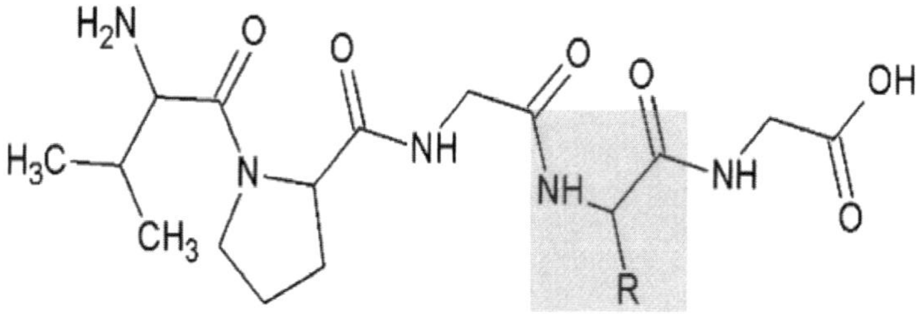

Scheme 1: Chemical structure of elastin-like polypeptide subunit VPGXG

Table 1: Properties of prepared ELPs

peptide	sequence	MW (Da)	pI*	pI (X)
ELP 1	H-(VPGLG)$_2$(VPG**G**G)$_2$(VPGLG)$_6$-OH	4301.47	6.09	5.97
ELP 2	H-(VPGLG)$_2$(VPG**A**G)$_2$(VPGLG)$_6$-OH	4169.33	6.09	6.0
ELP 3	H-(VPGLG)$_2$(VPG**L**G)$_2$(VPGLG)$_6$-OH	4253.51	6.09	5.98
ELP 4	H-(VPGLG)$_2$(VPG**K**G)$_2$(VPGLG)$_6$-OH	4283.53	10.72	9.74
ELP 5	H-(VPGLG)$_2$(VPG**H**G)$_2$(VPGLG)$_6$-OH	4301.47	8.13	7.59
ELP 6	H-(VPGLG)$_2$(VPG**Y**G)$_2$(VPGLG)$_6$-OH	4353.53	5.96	5.66
ELP 7	H-(VPGLG)$_2$(VPG**E**G)$_2$(VPGLG)$_6$-OH	4285.41	3.03	3.22
ELP 8	H-(VPGLG)$_2$(VPG**Q**G)$_2$(VPGLG)$_6$-OH	4283.45	6.09	5.65
ELP 9	H-(VPGLG)$_2$(VPG**N**G)$_2$(VPGLG)$_6$-OH	4255.39	6.09	5.41

defined by Peptide Property Calculator.[4]

Solid Phase Peptide Synthesis and HPLC-MS purification

Peptides were synthesized using SPPS protocol by Fmoc/tBu strategy on Wang resin (L= 1.12 mmol/g, mesh 100 200). The coupling steps were performed using DIC/oxyma/AA in molar ratio 3/3/3. Subsequent Fmoc deprotection was provided by 20 % piperidine in DMF for 5 and 15 min. After synthesis, all peptides were cleaved by cleavage cocktail consisting of DMC/anisole/thioanisole/TIS/TFA (5/2/1/2/90).

Crude peptides were purified by RP-HPLC on Shimadzu L CMS-2020 system equipped with a splitter and ESI-MS and PDA detection at 210 nm. The mobile phases were 0.1 % formic acid in distilled water and acetonitrile. For separation, C18 column (Jupiter 4 μm Proteo 90 Å AXIA, 250 × 10 mm) was used. The flow rate was 2.5 ml/min in the suitable slow gradient profiles.

Temperature measurements

ELPs were dissolved in phosphate buffer (pH 7.5) to concentration 0.1 mg/ml. Peptides structural changes were studied by circular dichroism (CD) spectroscopy (Figure 1). Temperature dependence of CD spectra were measured in 1 mm quartz cell in spectral range from 190 nm to 280 nm with temperature step 10 °C, scanning speed 10 nm/min and time response 8 sec. To compare molar ellipticity changes at two different wavelengths (Figure 2A), peptide sample was heated with gradient 1 °C/min and for each measurement the temperature equilibration was 30 sec. Temperature interval was 5 °C to 90 °C.

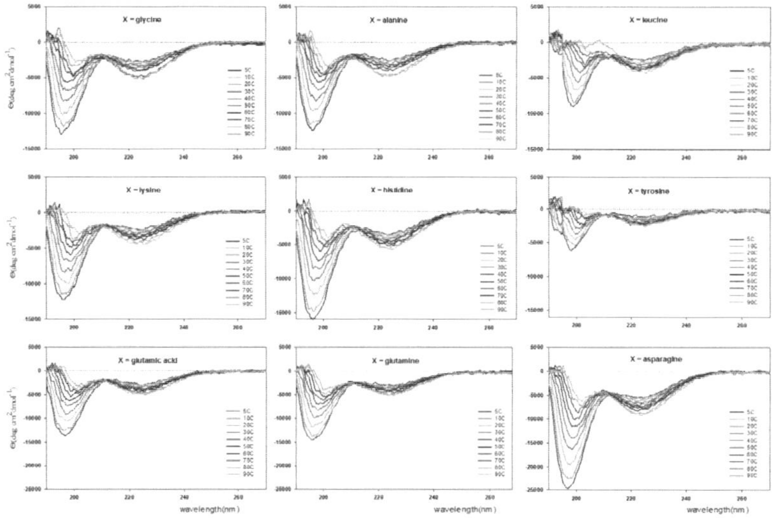

Figure 1: CD spectra of elastin like polypeptide ELP1-9. The CD spectra at low temperature are characterized by high proportion of random coil (distinct minimum at 197 nm) and low proportion type II β -turns (less-pronounced minimum at 224 nm). In general, the secondary structure distribution is gradually inverted with increasing temperature. The differences in intensity of CD spectra for discrete peptide is caused by presence of β -turn structure already presented at low temperature.

Turbidity profiles and transition temperatures (Figure 2B) were estimated using UV-Vis spectroscopy (OD at 400nm). The sample (concentration 1 mg/ml in phosphate buffer pH 7.5) was placed in 1 cm quartz cell and continually stirred and heated with temperature gradient 1 °C/min in the temperature range from 10 °C to 70 °C.

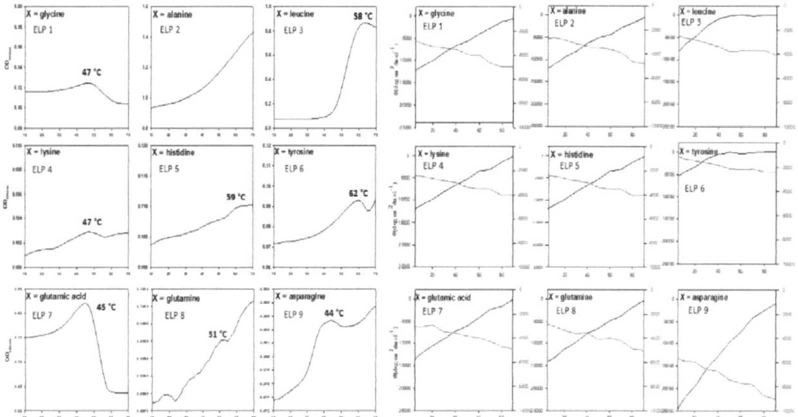

A B

Figure 2: Temperature measurements: A) ELPs coacervation profiles and B) ELPs molar ellipticity (Θ) changes at 197 nm (black) and 224 nm (red)

Conclusion

Turbidity profile and transition temperature for each prepared ELP were estimated with exception of the peptide containing alanine, where aggregation was observed immediately after dissolving in PBS buffer. Pronounced coarcervation was observed for ELPs with hydrophobic (L) or polar (Q and N) and negative (E and T) side chain.

To obtain the information of secondary structure dependency on temperature changes, CD spectroscopy was employed. For all studied ELPs, we observed higher portion of unordered conformation in combination with type

II β-turns at low temperature (5 °C). The distribution of type II β-turns increased with rising temperature and simultaneously the distribution of unordered structures decreased. However, the presence of β-turn structures was observed for ELPs containing bulky side chains (T, N and L) or negative charge (E) at low temperature. ELPs with polar (Q and N) and negative (E) side chain showed pronounced structural transition from unordered structure to type II β -turns. On the other hand, saturation of β-turns if any was observed for peptides containing X with bulky side chain (L, K, H and T). For ELPs, the reversibility of their secondary structure changes with increasing and afterwards decreasing temperature was confirmed, but on the different time scale for discrete ELPs.

References

1. Foster, J. A.; Bruenger, E.; Gray, W. R.; Sandberg, L. B. J. Biol. Chem.1973, 248, 2876–2879

2. Aladini, F.; Araman, C.; & Becker, C. F. W. J. Pept. Sci. 2016, 22(5), 334–342.

3. Chilkoti, A.; Christensen, T.; & Mackay, J. Curr. Opin. Chem. Biol. 2006, 10(6), 652–657.

4. Peptide Property Calculator - Ver 3.1.; 2018, on-line

New insights into the structural and dynamics properties of collagen model peptides using CD spectroscopy, MD simulations and innovative NMR approaches

Emeric Miclet[1], Anas Terrien[1], Claire Loison[2], Luminita Duma[3], Thierry Brigaud[4], Nathalie Lensen[4], Keyvan Rahgoshay[4], Grégory Chaume[4]

[1]Sorbonne Université, France
[2]Université Lyon 1, France
[3]UTC, France
[4]Université de Cergy Pontoise, France

https://doi.org/10.17952/35EPS.2018.192

Collagen, the most abundant protein in the animal kingdom, is also an important biomaterial owing to its great mechanical strength and its excellent thermal stability. This last decade, both the industry and the academia have been involved in the development of new collagen-related biomaterials.[1] One common strategy consists in using Collagen Model Peptides (CMP) as collagen surrogates. Numerous studies have demonstrated that short sequences, based on the repetition of the Pro-Hyp-Gly triplet, can be used to mimic the primary sequence of collagen.[2]

Using such peptides, we have developed an approach combining CD and NMR spectroscopies in order to characterize the formation of the triple helix (TH) and its structural features in solution. MD simulations and NMR spectroscopy have allowed us to gain insights into the conformational ensembles explored by the monomeric CMP peptides and their conversion into larger oligomeric species. Innovative 13C, 15N strategies and NMR developments have been carried out to obtain structural information on these assemblies at the atomic scale. In addition, we have been able to introduce 19F nuclei in model CMP using previously reported strategies.[3-6] Kinetic and structural effects of this insertion onto the collagen triple helix have been discussed using the developed biophysical tools and MD simulations.

Structural requirements for the stability of the collagen Triple Helix

The primary structure of collagen has a high content of (Pro-Hyp-Gly) units. The secondary structure adopts a PPII conformation: all peptide bonds are trans, α and ψ dihedral angles are -70° and 150°, respectively. The stability of the right-handed triple helix (TH) is ensured by inter-strand hydrogen bonds and the down-up alternation in puckering of pyrrolidine rings (figure 1). [2] Glycine residues are regularly distributed along the peptide sequence and are thus valuable probes of the TH structure.

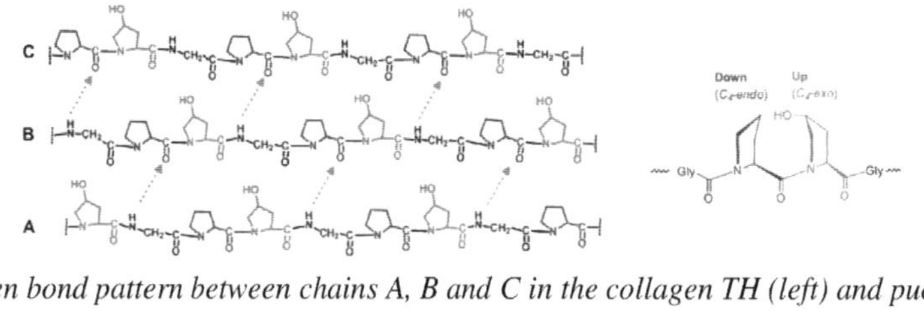

Figure 1: Hydrogen bond pattern between chains A, B and C in the collagen TH (left) and puckering alternation in Pro-Hyp pairs

Use of the CH2-TROSY NMR spectroscopy to observe methylene groups in TH models

The CH2-TROSY pulse scheme is based on the Rance-Kay ^{1}H-^{13}C HSQC experiment. A spin state selective element has been introduced prior to the t_1 evolution making it possible to select the ^{13}C downfield component. Then, by removing both the ^{1}H decoupling during the ^{13}C evolution and the ^{13}C decoupling during the acquisition and by adjusting the transfer delays in the double INEPT, it is possible to suppress the $^{2}J_{HH}$ and to accurately measure small $^{3}J_{HH}$ couplings. This strategy was applied on specifically ^{13}C,^{15}N Gly-labelled CMPs and allowed the measurements of $^{3}J_{H\alpha HN}$ along the peptide sequence (figure 2). Several conformers of the (Pro-Hyp-Gly)$_7$ peptide were characterized using this approach and the $^{3}J_{H\alpha HN}$ values measured at each Gly position are used to constrain the α angles in MD simulations.

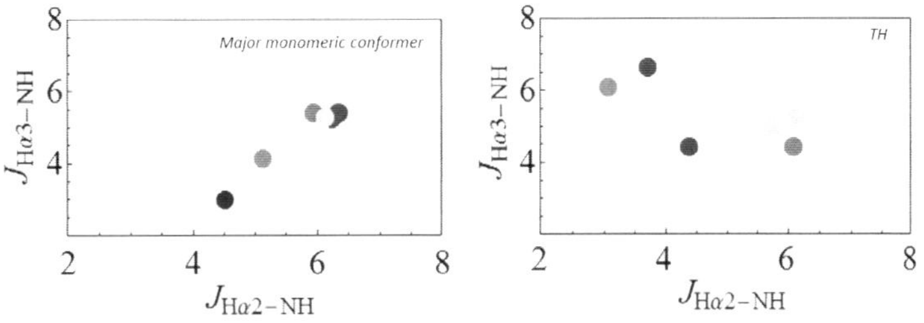

Figure 2: 3JHaHN couplings measured along the (Pro-Hyp-Gly)7 peptide sequence using the CH2-TROSY-pulse scheme. The two diastereotopic protons Ha2 and Ha3 of the Gly residues provide a measurement of the f-dihedral angle for each (Pro-Hyp-Gly)7 conformer.

TH formation in fluorinated CMP

FluorinatedCMPs have been synthetized by replacing Pro and/orHyp residues by CF_3-substituted pseudoprolines.[3-6] The stability of the TH on these new CMP was monitored by both CD spectra and NMR spectroscopy (figure 3). In CD spectra, the dependence of the ellipticity at 225 nm as a function of the temperature gave an estimation of the TH melting temperature (Tm) since sigmoidal curves were obtained and the Tm value corresponds to the inflexion point. By NMR, the integration of a Hδ Pro which belongs to the TH provided independent measurements of the Tm values. In addition, the quantitative analysis of the 2D NMR spectra allowed us to evaluate at each temperature the population of each CMP conformers (TH and up to six monomeric states).

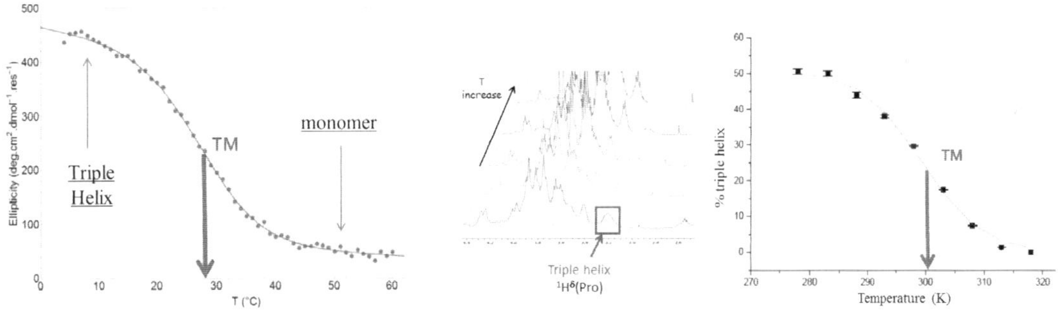

Figure 3: Monitoring the TH as a function of the temperature. CD spectra were recorded between 4°C and 60°C and the [q]225 values were plotted as a function of the temperature (left panel). 1H NMR spectra obtained with increasing temperatures (middle). Integration of the Hd signal of the Pro residues within the TH as a function of the temperature.

^{19}F NMR DOSY were recorded to show evidence of the TH formation in Fluorinated CMPs. A diffusion coefficient ratio of 1 .58 was found between the monomer and the triple helix. MD simulations have been performed to gain insights into the structural features of these new TH. It was found that the C^{δ} atoms in CF_3-pseudoprolines must have specific configurations to engender the required ring puckerings and that the puckering alternation was essential to stabilize the fluorinated TH.

In summary, the results obtained showed evidence that CF_3-substituted pseudoprolines can be introduced in collagen model THs although slightly less stable than their (Pro-Hyp-Gly)$_n$ homologues. It was found that the alternation in puckering of the 5-membered cyclic residues was a key point to promote the TH formation.

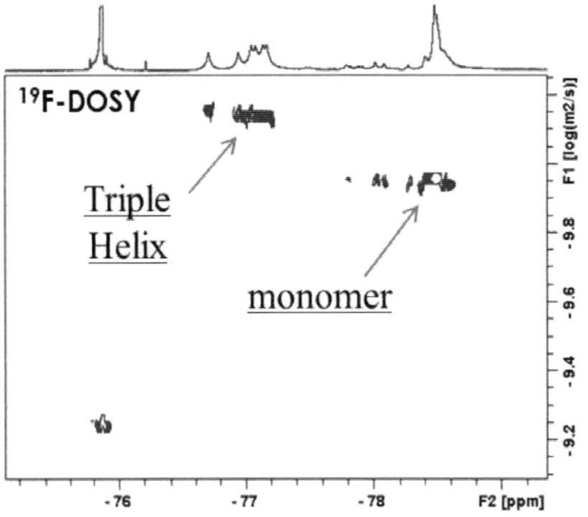

Figure 4: 19F NMR DOSY spectra recorded on a Fluorinated CMPs.

References

[1] Koide, T., Philos. Trans. R. Soc. B. 362, 1281-1291 (2007).

[2] Shoulders, M. D., Raines, R. T., Annu. Rev. Biochem. 78, 929-958 (2009).

[3] Feytens, D., Chaume, G., Chassaing, G., Lavielle, S., Brigaud, T., Byun, B. J., Kang, Y. K., Miclet, E. J. Phys. Chem. B 116, 4069-4079 (2012).

[4] Chaume, G., Simon, J., Caupene, C., Lensen, N., Miclet, E., Brigaud, T., J. Org. Chem. 78, 10144-10153 (2013).

[5] Chaume, G., Feytens, D., Chassaing, G., Lavielle, S., Brigaud, T., Miclet, E. New J. Chem. 37, 1336-38 (2013).

[6] Chaume, G., Simon, J., Lensen, N., Pytkowicz, J., Brigaud, T., Miclet, E. J. Org. Chem. 24, 13602-13608 (2017).

[7] Guichard G., Violette A., Chassaing G., Miclet E. Magn. Reson. Chem., 46, 918-24 (2008).

Synthesis and conformation of short peptides modeled after peptide LL-37

Barbara Biondi[1], Silvia Millan[2], Fernando Formaggio[2], Alessandra Semenzato[3], Cristina Peggion[2]

[1]ICB, Padova Unit, CNR, Department of Chemistry, University of Padova, Italy, Italy
[2]Department of Chemistry, University of Padova, Italy
[3]Department of Pharmaceutical Sciences, University of Padova, Italy

https://doi.org/10.17952/35EPS.2018.195

Introduction

Numerous tissues and cells of our body (e.g., epithelial or innate immune cells) are able to produce the 37-residue peptide cathelicidin LL-37 [1]. LL-37 plays important roles in the innate immune system as it acts as a natural antibiotic against Gram-positive and Gram-negative bacteria, fungi, and even viruses. In addition, it may inhibit the viral replication, promote wound healing [1,4] and kill tumor cells [1,2,3,4,5]. LL-37 is an amphipathic, α-helical peptide with an overall positive charge (+6). The helical structure promotes cell killing by binding to the negatively-charged bacterial membrane. Its possible mechanisms of action, based on the disruption of cell membranes, were recently reviewed [3,6].

Alteration of LL-37 were observed to occur in some pathologies, as for instance psoriasis (overproduction of LL-37) and topic dermatitis (inability to produce LL-37).

Results and Discussion

To make LL-37 industrially appealing as a drug, we started a program aimed at identifying shorter segments with the same potency of LL-37. We synthesized short, helical peptides stabilized by one or two helix-promoting Aib (α-aminoisobutyric acid) residues:

1 Pal-RKSKEKIG-NH_2 (Pal = palmitoyl)

2 Pal-KSKEKIG-NH_2

3 Ac-RKUKEKIG-NH_2 (Ac = acetyl; U = Aib)

4 Pal-RKUKEKIG-NH_2

5 Ac-KRUVQRUKDFLR-NH_2

6 Pal-KRUVQRUKDFLR-NH_2

Peptides 1 and 2 correspond to segments 7-14 and 8-14, respectively, of LL-37. To protect them from the enzymatic degradation we added at the N-terminus a palmitoyl moiety, able also to promote membrane insertion. In peptides 3 and 4 the helix-promoting and hydrophobic Aib residue replaces a Ser9. We chose this substitution after having observed that the hydrophilic Ser^9 is located on the hydrophobic face of the α-helical LL-37. With peptides 5 and 6 we focused on the central part of LL-37, segment 18-29, known to be quite active. Also in this case we aimed at assessing the possible beneficial effects of palmitoylation and Aib insertion.

All peptides were synthesized by the solid phase approach and purified by reverse-phase HPLC. Their 3D-structures were assessed by means of circular dichroism, 2D-NMR and FT-IR absorption. For all peptides we observed α-/ 3_{10}-helix equilibria, strongly dependent on the solvent used (see Figure 1 as an example). The substitution of Ser with Aib led to rather stable helices in membrane-mimetic environments (SDS and TFE). Interestingly, peptide 1 adopts a helical conformation although lacking any helix-promoting Aib residue. The presence of the N-terminal palmitic chain does not appear to affect the helical conformation.

Preliminary experiments of interactions with model membranes reveal for peptide 4 an interesting selectivity towards eukaryotic membrane mimics (Figure 2). Peptide 1, having the same sequence as 4 but with Ser instead of Aib, has no interaction with both type of vesicles (Figure 2). Therefore, the insertion of Aib appears to be a promising tool to modulate membrane selectivity and, hopefully, bioactivity.

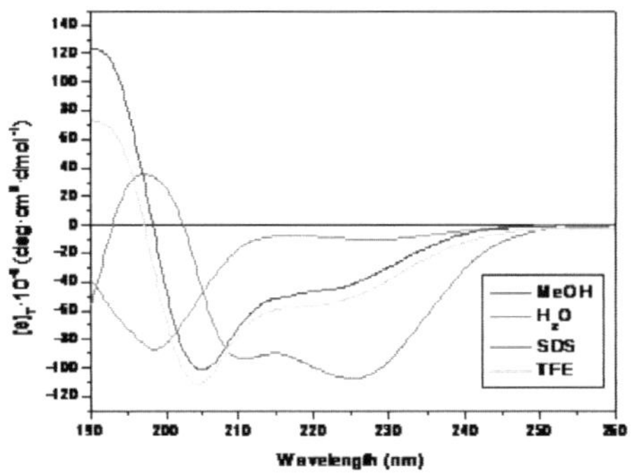

Figure 1: Circular dichroism of peptide 3, Ac-RKUKEKIG-NH_2, in different solvents.

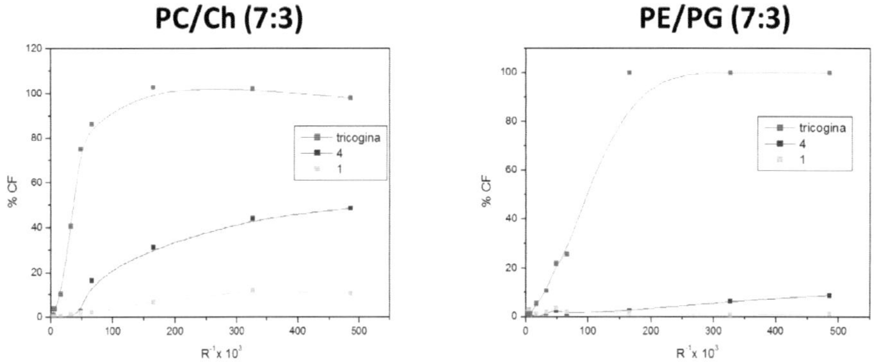

Figure 2: Peptide-induced carboxyfluorescein leakage from artificial small unilamellar vesicles (SUV), as a function peptide concentration. Left: SUV mimicking eukaryotic cells, (7:3) egg phosphatidylcholine/cholesterol; right: SUV mimicking bacterial cells, (7:3) phosphatidylethanolamine/phosphatidylglycerol. The fractional releases were determined as a function of the ratio R = [peptide]/[lipid](x10-3) 20 minutes after peptide addition.

References

[1] P. Chieosilapatham, S. Ikeda, H. Ogawa, F. Niyonsaba, Cur. Pharm. Design 2018, 24, 1079-1091. DOI: 10.2174/1381612824666180327113418

[2] M. E. Pachón-Ibánez, Y. Smani, Jerónimo Pachón, J. Sánchez-Céspedes, FEMS Microbiol. Rev. 2017, 41, 323-342. DOI: 10.1093/femsre/fux012

[3] K. Zeth, E. Sancho-Vaello, Front. Chem. 2017, 5, art. 86. DOI: 10.3389/fchem.2017.00086

[4] A. A. Baxter, F. T. Lay, I. K. H. Poon, M. Kvansakul, M. D. Hulett, Cell. Mol. Life Sci. 2017, 74, 3809-3825. DOI: 10.1007/s00018-017-2604-z

[5] E. Ogawa, Y. Sato, A. Minagawa, R. Okuyama, J. Dermat. 2018, 45, 264–272. DOI: 10.1111/1346-8138.14139

[6] Z. Khurshid, M. Naseem, F. Y. I. Asiri, M. Mali, R. S. Khan, H. A. Sahibzada, M. S. Zafar, S. F. Moin, E. Khan, Biomolecules 2017, 7, art. 80. DOI: 10.3390/biom7040080

Synthesis and conformational investigation of hetero-chiral sequential oligopeptides based on the (αMe)Aze/Ala dyad

Bruno Drouillat[1], Cristina Peggion[2], Barbara Biondi[2], Karen Wright[1], François Couty[1], Marco Crisma[2], Fernando Formaggio[2], Claudio Toniolo[2]

[1]Institute Lavoisier de Versailles, UMR CNRS 8180, Université de Versailles, France
[2]ICB, Padova Unit, CNR, Department of Chemistry, University of Padova, Italy

https://doi.org/10.17952/35EPS.2018.197

Introduction

Methylation at the Cα-position of a Pro residue is known to lock the preceding tertiary amide (ω) torsion angle of the resulting (αMe)Pro (Fig. 1) residue to the trans disposition and to restrict the α,ψ surface mainly to the region where the 310-/α-helices are found [1]. Moreover, the known high propensity for β-turn formation of the Pro residue is further enhanced in peptides based on its Cα-methylated derivative. A few years ago some of us developed a practical, multigram scale synthesis of both (αMe)Aze (Fig. 1) enantiomers [2]. This residue is characterized by a ring contraction (from five to four atoms) with respect to (αMe)Pro. Herein, we discuss synthesis and conformational analysis of the following set of hetero-chiral sequential oligopeptides based on the -(S)-Ala-(R)-(αMe)Aze- dyad:

(a) Boc-[(S)-Ala-(R)-(αMe)Aze]n-OMe (n = 1-3)

(b) Boc-[(S)-Ala-(R)-(αMe)Aze]n-(S)-Ala-OMe (n = 1,2)

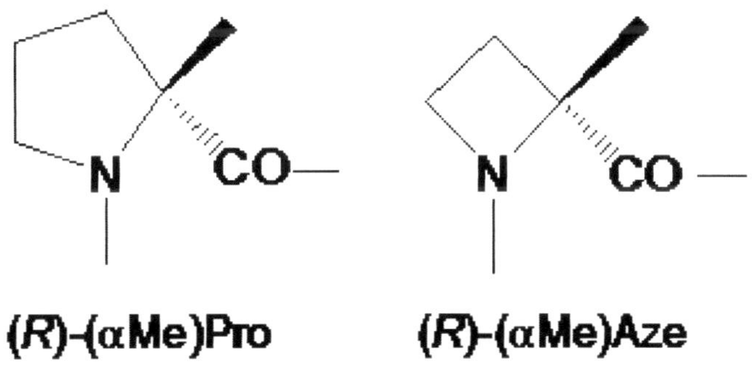

Figure 1: Chemical structures of (R)-(αMe)Pro, left, and (R)-(αMe)Aze, right.

Results and Discussion

Peptide syntheses were achieved in acceptable yields by use of the HATU coupling reagent.

Results from our combined X-ray diffraction and solution studies of terminally protected, hetero-chiral sequential oligopeptides based on the -(S)-Ala-(R)-(αMe)Aze- dyad support the view that this type of oligopeptide sequence strongly favors folding of the main chain into turns. In particular, the occurrence of a type-II β-turnhas been unambiguously demonstrated for the -(S)-Ala-(R)-(αMe)Aze- sequence in the crystal state (Fig. 2). This finding matches that published for the strictly related -(S)-Ala-(R)-(αMe)Pro- sequence [1] (Fig. 2).

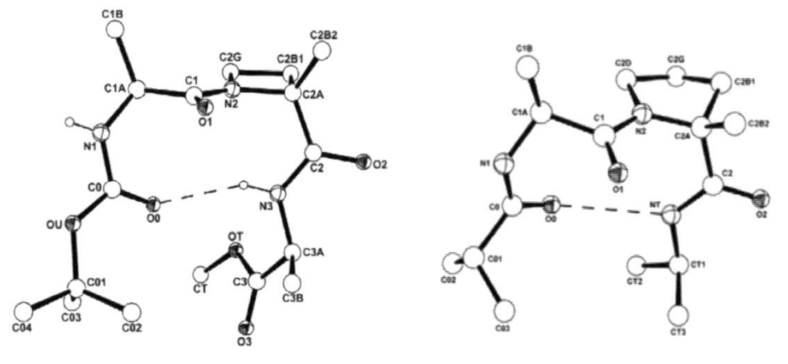

Figure 2: -Ray diffraction structures of the tripeptide ester Boc-(S)-Ala-(R)-(aMe)Aze-(S)-Ala-OMe (left) and the dipeptide alkylamide iPr-CO-(S)-Ala-(R)-(aMe)Pro-NHiPr (right) [1]. In both peptides the tertiary amide (w) torsion angle is trans.

However, FT-IR absorption and 2D-NMR data provide evidence that in CDCl3 solution the tripeptide ester is mainly folded in a γ-turn conformation. As for its higher homologs, in solution they seem to be involved in conformational equilibria to which both β- and γ-turn conformations may contribute. It is worth recalling here that the γ-turn promoting ability of (αMe)Aze was previously documented by us with the crystal structure of the homo-chiral tetrapeptide Boc-[(S)-MeAze-(S)-Ala]2-OMe [3]. In any case, even if β-turns may prevail over γ-turns in the longest hetero-chiral oligomers, the resulting overall structure would be a "flat" β-turn ribbon [4], distinct from the (helical) β-turn ribbon spiral typical of sequential oligopeptides based on the -Aib-Pro- dyad [5].

References

[1] Moretto, A.; Terrenzani, F.; Crisma, M.; Formaggio, F.; Kaptein, B.; Broxterman, Q. B.; Toniolo, C. Biopolymers 2008, 89, 465. DOI: 10.1002/bip.20839

[2] Drouillat, B.; Wright, K.; Marrot, J.; Couty, F. Tetrahedron: Asymmetry 2012, 23, 690. DOI: 10.1016/j.tetasy.2012.05.006

[3] Crisma, M.; De Zotti, M.; Moretto, A.; Peggion, C.; Drouillat, B.; Wright, K.; Couty, F.; Toniolo, C. New J. Chem. 2015, 39, 3208. DOI: 10.1039/C4NJ01564A

[4] Hayward, S.; Leader, D. P.; Al-Shubailly, F.; Milner-White, E. J. Proteins: Struct., Funct., Bioinform. 2014, 82, 230. DOI: /10.1002/prot.24357

[5] Di Blasio, B.; Pavone, V.; Saviano, M.; Lombardi, A.; Nastri, F.; Pedone, C.; Benedetti, E.; Crisma, M.; Anzolin, M.; Toniolo, C. J. Am. Chem. Soc. 1992, 114, 6273. DOI: 10.1021/ja00042a001

The 1,3-diyne linker as a tunable tool for peptide secondary structure stabilization

Steven Verlinden[1], Niels Geudens[2], José Martins[2], Steven Ballet[1], Guido Verniest[1]

[1]Vrije Universiteit Brussel, Belgium
[2]Ghent University, Belgium

https://doi.org/10.17952/35EPS.2018.199

Introduction

Protein-protein interactions (PPIs) control essential cellular processes and influence several biological functions.[1] However, despite their fundamental roles, PPIs are considered to be rather cumbersome targets for drug design due to their large and often flat contact surfaces, which cannot easily be addressed by small molecules. Nonetheless, the modulation of PPIs is important for the elucidation of biological processes and it represents a promising strategy towards next-generation therapeutics. One rational approach for the design of PPI inhibitors consists of a structure-based approach, wherein the topology of protein epitopes can serve as a starting point.[2] These epitopes are defined by the secondary structure of the underlying amino acid chain.[1] Among different successful strategies, the stabilization of α-helical peptide conformations can be accomplished by introducing covalent links between amino acid side chains at selected positions.[3] As a result, these 'stapled' peptides are often characterized by an improved binding affinity, metabolic stability and enhanced cell penetration, as compared to their linear peptide counterparts.[4,5] When aiming at α-helix mimetics, in particular, a macrocyclization involving the amino acids side chains at the i,i+4 or i,i+7 positions is often applied to bridge one or two helical turns, respectively.[1,6] Whereas a vast amount of work has been done on i,i+4 cyclizations [7], the current report focusses on constraints bridging two loops *via* an i,i+7 cyclization.

Results and Discussion

In light of the increasing interest for alkyne-alkyne coupling reactions in peptide chemistry,[8,9] our group verified whether O-propargylated serine residues were able to cross-link two helical turns in a i,i+7 fashion. Additionally, we evaluated the influence of the O-propargylated serines' configuration on the α-helical stabilization by means of circular dichroism (CD).

In search for a suited model system, stapled peptide analogues able to inhibit MDM2 and MDMX proteins, important regulators of the transcriptional activity and stability of p53, were selected. In this context, Sawyer and co-workers designed an α-helical peptide (ATSP-7041) bearing an all-hydrocarbon staple, which proved to be a highly potent dual inhibitor of both MDM2 and MDMX.10 We employed this 12-mer peptide as a model system and replaced the α-methyl-α-alkenyl amino acids in ATSP-7041 with O-propargylated serines for the evaluation of the 1,3-diyne linker as a rigid staple.

Previously, our group cyclized small peptidic macrocycles *via* a Glaser-Hay coupling of terminal alkynes.[8] We used the optimized coupling conditions to cyclize the envisaged linear peptide Ac-Leu-Thr-Phe-Ser(O-propargyl)-Ala-Tyr-Trp-Ala-Gln-Leu-Ser(O-propargyl)-Ser-NH_2 in solution, as well as on solid support. However, in both cases, the seemingly straightforward alkyne alkyne coupling failed to provide the cyclized peptide sequence. Therefore, the cyclization step was switched to a lactamisation, whereby the 1,3-diyne staple is introduced *via* insertion of a diyne-bearing dipeptide (4) by means of regular SPPS.

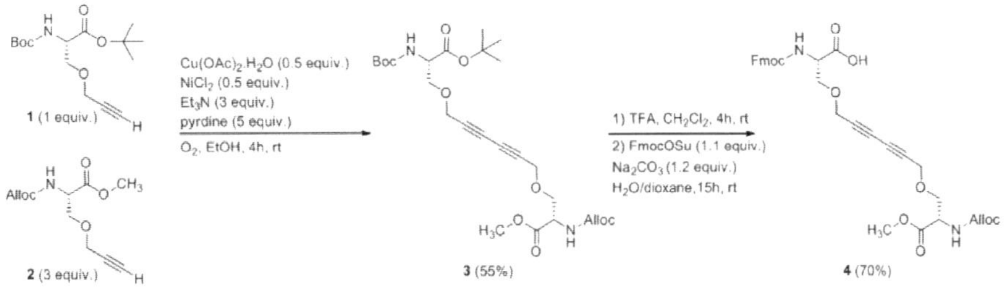

Scheme 1: Synthesis of dipeptide diyne building block 4.

Using 1 equivalent of 1 and 3 equivalents of 2 in the presence of $Cu(OAc)_2.H2O$, $NiCl_2$, Et_3N and pyridine in EtOH after 4 h at room temperature under an oxygen balloon, resulted in the formation of 3 in 55% isolated yield. Under these reaction conditions, homodimerization of terminal alkynes 1 and 2 was also observed, but both types of homodiynes were easily separated from heterodiyne 3 *via* silica gel chromatography in 6% and 17% yield, respectively. Simultaneous Boc- and tert-butyl deprotection by TFA treatment and subsequent Fmoc-protection led to the formation of dipeptide 4.

Dipeptide 4 was then used as a building block in regular Fmoc-SPPS on a Rink amide resin. After assembly of the first subsequence 5, saponification of the methyl ester on resin was performed using 10 equivalents of LiOH in DMF/H_2O. In a next step, lactamization was achieved after 15 h by adding 3 equivalents of PyBOP and 6 equivalents of DIPEA in DMF, and followed by Alloc-deprotection with $Pd(PPh_3)_4$ and morpholine as nucleophilic scavenger. Finally, the final 3 amino acid residues were added, followed by acetylation of the terminal amine, cleavage from the resin and purification *via* preparative RP-HPLC, yielding cyclic peptide 7. To evaluate the influence of the configuration of the O-propargylated serine residues on the stabilization of helicity, 3 additional diyne dipeptides with different configurations (i.e., D-L, L-D and D-D) were synthesized in a similar fashion, to assess the 'stapling efficiency' of the cyclic peptides. The linear reference counterparts were also synthesized.

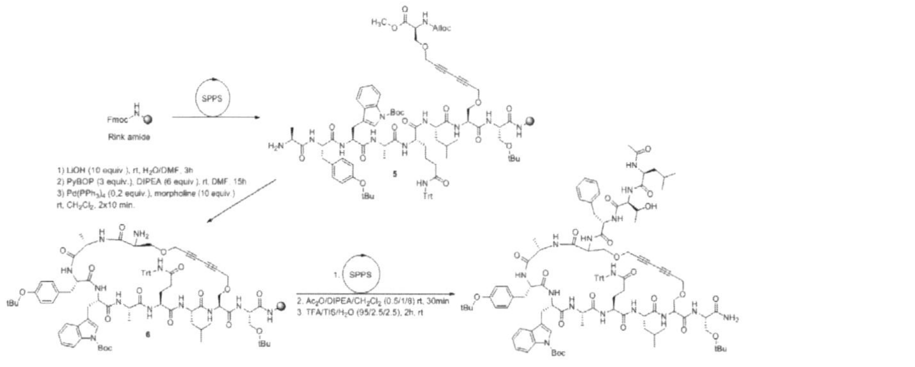

7: Ac-Leu-Thr-Phe-[L-Ser(O-prop.)^4-Ala-Phe-Trp-Ala-Gln-Leu-L-Ser(O-prop.)^{11}]-Ser-NH_2
8: Ac-Leu-Thr-Phe-[D-Ser(O-prop.)^4-Ala-Phe-Trp-Ala-Gln-Leu-L-Ser(O-prop.)^{11}]-Ser-NH_2
9: Ac-Leu-Thr-Phe-[L-Ser(O-prop.)^4-Ala-Phe-Trp-Ala-Gln-Leu-D-Ser(O-prop.)^{11}]-Ser-NH_2
10: Ac-Leu-Thr-Phe-[D-Ser(O-prop.)^4-Ala-Phe-Trp-Ala-Gln-Leu-D-Ser(O-prop.)^{11}]-Ser-NH_2

Scheme 2: Representative synthesis of macrocyclic peptde 7.

Next, the helical character of stapled peptides 7-10 was analyzed with circular dichroism (CD) spectroscopy. The peptides were dissolved in water and water/TFE mixtures with increasing concentrations of TFE to mimic a hydrophobic environment. Peptides (L,L)-7 and (D,L)-8 show a more pronounced helical character in H2O/TFE mixtures and TFE. On the contrary, peptides (L,D)-9 and (D,D)-10 display only a slightly helical conformation in H2O/TFE mixtures. Comparing the CD measurements of the stapled peptides 7 and 8 with their linear counterparts in H2O/TFE mixtures (50/50) shows that cyclic peptide 7 has less helical character than his linear counterpart, whereas stapled peptide 8 a similar helical profile has compared to the corresponding acyclic peptide.

Conclusion

In conclusion, we showed that the 1,3-diyne linker can be used as a rigid staple to stabilize short α-helical peptides by crosslinking residues in positions i, i+7. CD analysis confirmed the helical secondary structure of the prepared peptides. A full secondary structure characterization by NMR is envisaged to confirm the α-helical propensity of the macrocyclized peptides.

References

1. M. Pelay-Gimeno, A. Glas, O. Koch, T. N. Grossmann, Angew. Chem. Int. Ed. 2015, 54, 8896–8927. (doi: 10.1002/anie.201412070)

2. B. N. Bullock, A. L. Jochim, P. S. Arora, J. Am. Chem. Soc. 2011, 133, 14220–14223. (doi: 10.1021/ja206074j)

3. A. Gross, C. Hashimoto, H. Sticht, J. Eichler, Front. Bioeng. Biotechnol. 2016, 3, 1–16. (doi: 10.3389/fbioe.2015.00211)

4. C. J. White, A. K. Yudin, Nat. Chem. 2011, 3, 509–524. (doi: 10.1038/nchem.1062)

5. P. G. Dougherty, Z. Qian, D. Pei, Biochem. J. 2017, 474, 1109–1125. (doi: 10.1042/BCJ20160619)

6. A. J. Wilson, Chem. Soc. Rev. 2008, 38, 3289–3300. (DOI: 10.1039/b807197g)

7. Y. H. Lau, P. de Andrade, Y. Wu, D. R. Spring, Chem. Soc. Rev. 2015, 44, 91. (doi: 10.1039/C4CS00246F)

8. S. Verlinden, N. Geudens, J. C. Martins, D. Tourwé, S. Ballet, G. Verniest, Org. Biomol. Chem. 2015, 13, 9398–9404. (doi: 10.1039/C5OB01153A)

9. P. A. Cistrone, A. P. Silvestri, J. C. J. Hintzen, P. E. Dawson, ChemBioChem 2018, 19, 1031–1035. (doi: 10.1002/cbic.201800121)

10. Y. S. Chang, B. Graves, V. Guerlavais, K. Tovar, K. Packman, K.-H. To, K. A. Olsen, K. Kesavan, P. Gangurde, et al., Proc. Natl. Acad. Sci. U. S. A. 2013, 110, 3445–3454. (doi: 10.1073/pnas.1303002110)

Conformation and domain movement analysis of Human Matrix Metalloprotease-2: The role of associated Zn^{2+} and Ca^{2+} ions

Leah Voit-Ostricki[1], Sándor Lovas[2], Charles Watts[1]

[1] Mayo Clinic Health System, United States
[2] Creighton University, United States

https://doi.org/10.17952/35EPS.2018.202

Introduction

Matrix Metalloproteinase-2 (MMP-2) is a 550 amino acid residue extracellular Zn^{2+} proteinase that recognizes and degrades type I and IV collagens.[1] MMP-2 expression is associated with several inflammatory, degenerative, and malignant diseases. MMP-2 has three domains catalytic (Cat), fibronectin (Fib), and hemopexin (Hpx) and five crystallographically assigned divalent cations (two Zn^{2+} and three Ca^{2+}). The Cat domain (Tyr^{110} through Tyr^{445}) is a conserved matrix fold consisting of five β-sheets and three α-helices. The Fib domain (Glu^{217} through Gln^{393}) is inserted within the catalytic domain between the β 5-sheet and α2-helix and contains three type II fibronectin fingers consisting of two antiparallel β-sheets connected by a short α-helix. The Hpx domain (Leu^{461} -Cys^{660}) is a four bladed propeller fold that is partially oriented towards the Cat domain. The Cat and Hpx domains are connected by a proline rich flexible link (Lnk) region (Gly^{446} -Thr^{460}). In the present study we examine the domain movements within MMP-2 and evaluate the structural stability of the bound divalent ions using 1.0 μs NPT MD simulations.

Simulations were performed using GROMACS 5.1.2[2] with the CHARMM36m[3] force field, modified TIP3Pm water model, and a divalent metal ions model.[4] The system was solvated in a truncated dodecahedron with 45105 water molecules neutralized with Na^+ and Cl^- ions with a final NaCl concentration of 150 mM. Protein-ion distances were measured, ion hydration evaluated, the sampled conformational space analyzed with dihedral Principle Component Analysis (dPCA)[5] and Dynamic Cross-Correlation Matrix (DCCM) analysis[6]. Energy of binding of the Zn^{2+} and Ca^{2+} ions to the protein was calculated using the MM-PBSA method.[7]

Results and Discussion

A Cα-trace overlay of the centroids representing the lowest energy conformations for each cluster (Fig. 1, Left) demonstrates that the most conformationally stable regions are the Cat and Hpx domains. The RMSDs of Cat and Hpx are 0.50 ± 0.03 nm and 0.34 ± 0.04 nm, respectively, while that of Fib is 3.69 ± 0.03 nm. The Fib domain is not disordered and the most significant flexibility occurs within the loops connecting the three type II fibronectin fingers. The four bladed anti-parallel β-sheet configuration the Hpx domain lends to its considerable conformational stability. The Cα-trace overlay of the first five principle components of the domain cross correlation analysis are shown in Fig. 1 (Right). The most significant domain movements are accounted for by the first two principle components which represent 86.2% of the correlated and anti-correlated motions. The most significant contributions are an opening and closing of the center-of-mass distance between the Fib and Hpx domains mediated by the flexible Lnk domain

Energetically most favorable interactions occur between Zn^{2+} ion 1, Ca^{2+} ion 1 and Ca^{2+} ion 3 and the protein (Table 1). Zn^{2+} ion 1 is bound by three His residues arising from the α1-helix and the Ω-loop and is associated with a single hydrating water molecule. Ca^{2+} ion 1 is bound within a pocket created by the distal portion of the S-loop and the interim loop connecting the Cat and Fib domains. Ca^{2+} ion 3 is closely associated with the Asp side chains surrounding the central cavity of the Hpx domain. Zn^{2+} ion 2 does not maintain its crystallographically defined binding site position but does stay in close proximity to the β-COO- group of Asp^{180}. Ca^{2+} ion 2 freely dissociates from its binding site and diffuses through the hydration box during the simulation.

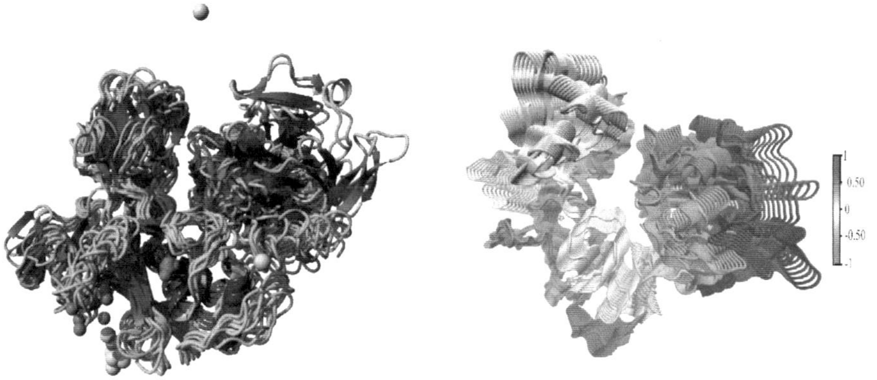

Figure 1: Overlay of 10 representative lowest energy conformations identified from k-means clustering of the first two components of the dihedral principal components (Left). Red arrows representing β-sheet, blue α-helix, green tubes turn, aqua tubes coil, pink Zn^{2+}, and Yellow Ca^{2+}. Cα-trace overlay of the first five principal components obtained from the DCCM analysis (Right). Correlated domain movements are indicated in blue and anti-correlated are in red.

Table 1: Binding energies between MMP-2 and the associated divalent cations as determined by the MMPBSA methods with its associated components.

Ion(s)	ΔE_{vdw}	ΔE_{elec}	ΔG_{polar}	$\Delta G_{non\text{-}polar}$	$\Delta E_{binding}$
All Ions	95.04 ± 2.92	-9264.45 ± 194.34	3142.58 ± 81.6	-3.95 ± 0.13	-6020.73 ± 116.75
Zn^{2+} Ion 1	25.41 ± 0.77	-2427.66 ± 72.19	1291.3 ± 42.75	-0.88 ± 0.04	-1114.16 ± 27.12
Zn^{2+} Ion 2	13.79 ± 0.96	-799.35 ± 63.45	188.73 ± 35.14	-0.47 ± 0.04	-598.61 ± 29.09
Ca^{2+} Ion 1	38.15 ± 1.90	-1796.42 ± 89.21	605.84 ± 30.55	-1.14 ± 0.07	-1154.73 ± 56.65
Ca^{2+} Ion 2	6.52 ± 0.58	-367.21 ± 36.96	73.71 ± 9.94	-0.35 ± 0.05	-289.24 ± 27.33
Ca^{2+} Ion 3	14.94 ± 0.57	-1480.59 ± 42.61	320.38 ± 11.31	-1.32 ± 0.05	-1147.96 ± 30.78

Only minor structural fluctuations occur within the Cat and Hpx domains. Flexibility within the Fib domain is accounted for by the loops connecting the three type II fibronectin fingers. Flexibility within this region may play an important role in collagen binding and unravelling necessary for substrate binding to the Cat domain. Both dPCA and DCCM analysis identify movement and change in orientation of the Hpx domain in relation to the Cat and Fib domains as mediated by the flexible Lnk region. This Hpx-to-Cat and Fib domain movement comprises the vast majority of correlated and anti-correlated motions within the protein and may allow for the enzyme to slide along the collagen triple helix to the next substrate binding site. The catalytic Zn^{2+} ion 1 has the greatest degree of conformational stability staying closely associated with the sidechains of His^{403}, His^{407}, and His^{413} and an associated water molecule. Zn^{2+} ion 2, Ca^2+ ion1, and Ca^{2+} ion 3 are conformationally less stable but do stay within close proximity to the crystallographically identified binding sites.

References

1. Morgunova E. *et al.* Science 1999; 284: 1667-1670.

2. Abraham M. J., *et al.* SoftwareX 2015; 1-2, 19-25.

3. Huang J. *et al.* Nature Methods 2017; 14: 71-73.

4. Li, P. *et al.* Theory Comput. 2013; 9, 2733-2748.

5. Mu Y. *et al.* Proteins. 2005; 58: 45-52.

6. Romanowska J. *et al.* J Chem Theory Comput. 2012; 8: 2588-2599.

7. Kumari R. *et al.* J Chem Inf Model. 2014; 54: 1951-1962.

Inhibitors of the Neuropilin-1 and Vascular Endothelial Growth Factor 165 interaction

Bartlomiej Fedorczyk[1], Dagmara Tymecka[1], Karolina Grabowska[1], Anna Puszko[1], Piotr Lipinski[2], Gerard Y. Perret[3], Aleksandra Misicka[1]

[1]Faculty of Chemistry, University of Warsaw, Poland
[2]Mossakowski Medical Research Center, PAS, Poland
[3]Université Paris 13, Sorbonne Paris Cité, France

https://doi.org/10.17952/35EPS.2018.204

Introduction

Neuropilin-1 (NRP-1) is a protein expressed by a number of types of somatic cells [1]. Moreover, it has been found to be overexpressed in several kinds of malignant tumors e.g. breast cancer [2] and it is postulated that in pathological angiogenesis its interaction with the Vascular Endothelial Growth Factor 165 (VEGF165) leads to progression of tumor vascularization and growth [3]. Additionally, this signaling complex is also involved in suppression of immune response against tumor cells [4].

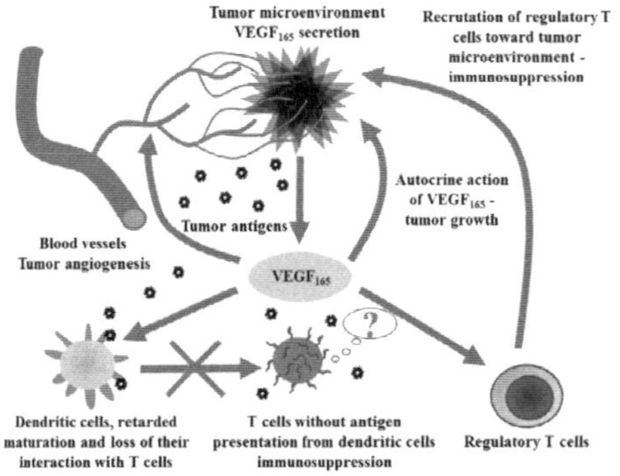

Figure 1: Scheme of signaling pathways mediated by NRP-1/VEGF165 involved in tumor development and its escape from immune surveillance

Results and Discussion

Inhibitors of the NRP-1/VEGF165 complex are very attractive compounds to block several processes regulating tumor growth and metastasis. In the frame of presented project efforts have been made to design strong inhibitors of this interaction. All compounds were synthesized and characterized with HPLC and mass spectrometry. By the means of ELISA-based test and molecular dynamics simulations, the inhibitory activity toward NRP-1/VEGF165 complex has been characterized. Lead compounds were tested in an preliminary assay toward proteolytic stability in human serum.

The presented paper is a review of our recent four publications on peptidic or peptidomimetic inhibitors of NRP-1/VEGF165 complex.

Conformational latitude – inhibitoryactivity relationshipof KPPR tetrapeptide analogues (KxxR)[5]

Several analogues (KxxR) with various conformational latitudes have been synthesized and found as inhibitors of NRP-1. Detailed insight provided by molecular dynamics simulation allowed forming a clear relationship between “flexibility” of the -xx- part (positions 2 and 3) of the molecules and their inhibitory activity. Introduction of amino acids with greater conformational latitude decreases inhibitory activity in a non-equal manner. The most important sites of the molecule are N-terminal K and C-terminal R, which directly interact with the NRP-1. The -xx- part allows the remaining part of the molecule to find interaction partners on the NRP-1 surface and by this greatly affects the inhibitory activity. For sufficient inhibitory activity, proline with low conformational latitude in position 3 should be maintained (KxPR).

Structure – inhibitory activity relationship study of tetrapeptide inhibitors [6]

Here we present a structure-activity relationship study of the systematic optimization of amino acid residues in positions 1–3 (Kxx). Substitution of N-terminal K in position 1 to homoarginine (Har) increases the inhibitory activity. The HarPPR analogue has been used as the next starting compound for further modifications. Position 2 is the best site to incorporate basic amino acid such as Dap or Dab. Analogue with Dab is the strongest NRP 1 inhibitor with sub-micromolar IC_{50} (0.8 μM) in this generation.

Branched pentapeptides and cyclic peptides as potent inhibitors [7], [8]

Our findings suggest that branching at position 1 with additional homoarginine (Har) residue, to obtain K(Har)PPR, allows more effective inhibition. Increasing the conformational latitude of the middle part of molecule, in particular with simultaneous introduction of basic amino acids at the second or third position, pro-duced lead compounds K(Har)DabPR and K(Har)PDabR (both with IC_{50} 0.2 μM). These two linear and branched pentapeptides exhibit the same inhibitory activity, however the first one is more stable toward proteolysis in human serum (half-lifes are 41 h and 44 min).

Moreover, simple side chain – side chain cyclization of EPKR peptide resulted in obtaining a very potent inhibitor c[EPK]R (A) with IC_{50} = 0.16 μM. Also dimeric forms of such compound possess inhibitory activity (B) IC_{50} = 0.89 μM

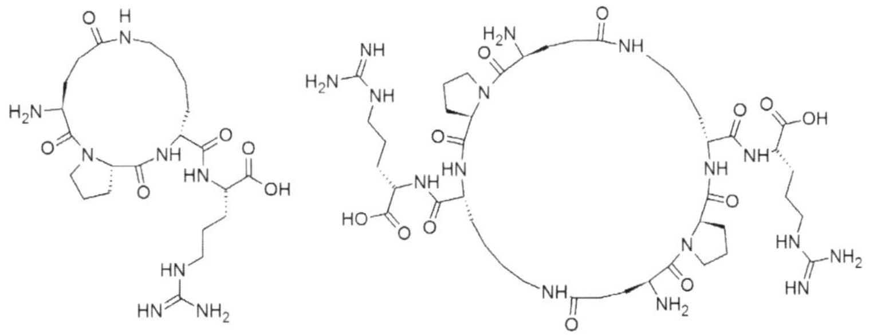

Figure 2: Structures of two exemplary cyclic inhibitors of NRP-1/VEGF165 complex

In the recent years novel peptide therapeutics have been approved for clinical use. Therefore the field of peptide based inhibitors of NRP-1 could be a promising direction for development of new antitumor drugs.

Acknowledgements

This work was supported by NCN grant no N 204 350940. Project was carried out with the use of CePT infrastructure financed by the European Union – the European Regional Development Fund within the Operational Programme "Innovative economy" for 2007-13.

References

[1]. F. Nakamura and Y. Goshima, Neuropilin: From Nervous System to Vascular and Tumor Biology, D. Bagnard, Ed. Boston, MA: Springer US, pp. 55–69 (2002).

[2]. A. Jubb M et al., J. Pathol., vol. 226, no. 1, pp. 50–60, (2011). .

[3]. H. L. Goel and A. M. Mercurio, Nat. Rev. Cancer, vol. 13, p. 871, (2013).

[4]. M. Terme et. al., Oncoimmunology, vol. 2, no. 8, (2013).

[5]. B. Fedorczyk et al., J. Pept. Sci., vol. 23, no. 6, pp. 445–454, (2017).

[6]. D. Tymecka et al., Peptides, vol. 94, pp. 25–32, (2017).

[7]. D. Tymecka et al., Eur. J. Med. Chem., vol. 158, pp. 453–462, (2018).

[8]. K. Grabowska et. al., Bioorg. Med. Chem., vol. 25(2), pp. 597-602 (2017).

Proceedings of the 35th European Peptide Symposium
Patrick B. Timmons, Chandralal M. Hewage, Michal Lebl (Editors)
European Peptide Society & PSP, 2018

Use of trifluoromethylated pseudoprolines for the design of collagen triple helix containing unusual C(5)-substituted proline surrogates

Keyvan Rahgoshay[1], Anaïs Terrien[2], Nathalie Lensen[1], Thierry Brigaud[1], Emeric Miclet[2], Grégory Chaume[1]

[1]Université de Cergy-Pontoise, France
[2]Sorbonne Université, France

https://doi.org/10.17952/35EPS.2018.206

Introduction

Collagen is the most abundant protein of mammals. This structural protein is omnipresent and represents the major component of the extracellular matrix. Its tertiary structure consists of three individual left-handed polyproline II helices folded into a right-handed triple helix which is stabilized by inter-strand hydrogen bonds with a single-residue offset.[1] Each strand comprises the repeat of the (Xaa-Yaa-Gly primary sequence where the proline (Pro, P and the (4R-4-hydroxyproline (Hyp, O are the most prevalent at Xaa and Yaa positions, respectively.[2] Crystal structures of short-length Collagen Model Peptides (CMPs show that Pro and Hyp residues exhibit preferentially C(4-endo and C(4-exo ring pucker, respectively.[3] This alternation has been proposed to be a prerequisite for the formation of the triple helix (TH since it preorganizes the individual strands in a suitable conformation, thereby decreasing the entropic cost.[4] Following these principles, numerous CMPs have been designed using C(4- and C(3-substituted proline derivatives. It was shown that these latter allowed the formation of the TH, provided that they satisfactorily mimic the conformational properties of Pro and Hyp and do not interfere with the interstrand H-bonds.[1] However, due to the packed nature of the collagen TH, substituents in the close vicinity of the backbone could be very deleterious for the TH stability and, to our knowledge, no work in the literature reports on TH containing C(5-substituted proline derivatives. Our group has developed the synthesis of the CF_3-pseudoprolines (CF_3-ψPro)and their incorporation into peptides.[5] We previously demonstrated that these proline analogues allow one to locally tune the peptide bond geometry, the puckering of the 5-membered ring as well as the Φ and Ψ dihedral angles.[6] It was found that the diastereoisomers (S)-CF_3-ψPro (tfmψP)and (R)-CF_3-ψPro (TfmψP) act as good mimics of the Pro and Hyp residues, respectively, by adopting the prerequisite puckering necessary for the formation and stabilization of the collagen PPII triple helix. However, we showed that the replacement of Hyp by TfmψP at the Yaa position of the Ac-Xaa-Yaa-Gly-NHMe collagen triplet model strongly affects the PPII-like conformation and perturbs the puckering alternation.[7] This result was confirmed with the fact that TfmψP^{11}- $[POG]_7$ CMP 2 displays a thermal denaturation lowered by 30°C compared to the $[POG]_7$ 1.[8] In other hand, we demonstrated that the incorporation of both tfmψP and TfmψP at respectively the Xaa and Yaa position of the Ac-Xaa-Yaa-Gly-NHMe collagen triplet model favors the TH conformational requirements.[7] Finally, predictive molecular modeling realized on the [(Pro-Hyp-Gly)$_3$-(tfmψP-TfmψP -Gly)-(Pro-Hyp-Gly)$_3$] CMP shows that the CF_3 groups point outside of the TH and do not cause interstrand steric penalties. Here, we propose to assess and compare the stability of a collagen TH bearing both tfmψP and TfmψP in the middle of a $[POG]_7$ sequence.

Results and Discussion

Because of the lack of nucleophilicity of the amino group and the steric bulkiness of the vicinal CF_3-group, the direct incorporation of the CF_3-ψPro residues into the $[POG]_7$ sequence *via* solid-phase peptide synthesis (SPPS)was excluded. Therefore, we planned to incorporate the two CF_3 -ψPro residues into the fluorinated host-guest CMP tfmψP^{10}-TfmψP^{11}-$[POG]_7$ 3 using a "ready to use" tripeptide building blocks for SPPS. At this stage, it was anticipated that the use of the Fmoc-tfmψP-TfmψP-Gly-OH building block would be extremely challenging since we previously reported that a complete epimerization of the C(5 stereocenter of the tfmψP residue occurred during peptide coupling reactions at its N-terminus, leading exclusively to the TfmψP moiety.[5b] However, we observed that N-acylation reactions using non-hindered acyl chlorides preserved the integrity of the (S)-configuration at the C(5)position of the tfmψP.[5a] Therefore, the triplet frame of the building block was shifted and the Gly residue at the N-terminal position was introduced sequentially starting from the bromoacetyl bromide acylating reagent (Scheme 1)As expected, the N-acylation of the tfmψP residue proceeded without epimerization of the C(5 stereocenter. Treatment with sodium azide followed by hydrogenation and Fmoc-protection of the corresponding amine afforded the Fmoc-Gly-tfmψP-OH dipeptide. The latter is then activated as acyl chloride and coupled with the TfmψP residue. Subsequent hydrogenolysis of the benzyl ester gave the Fmoc-Gly-tfmψP-TfmψP-OH building block as a single diastereomer in large scale.

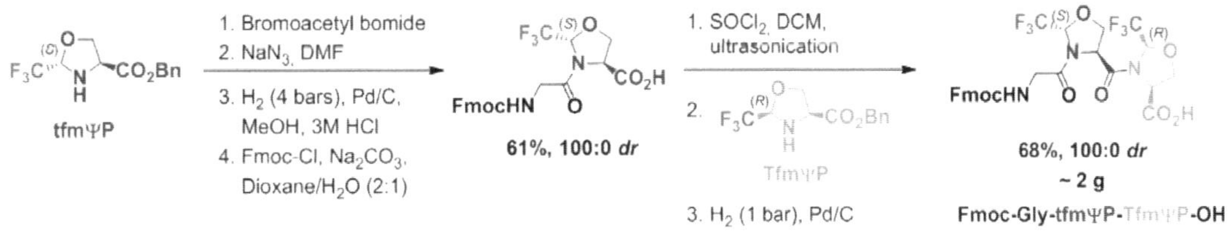

Scheme 1: Synthesis of the "ready to use" Fmoc-Gly-tfmψP-TfmψP-OH building block

The fluorinated host-guest CMP tfmψP^{10}-TfmψP^{11}-$[POG]_7$ 3 was prepared on Wang resin by automated SPPS under microwave activation using OxymaPure/DIC as a coupling reagent with similar yield (~20%) to those obtained for CMPs 1 and 2 (Scheme 2). This result demonstrates that our CF_3-ψPro residues are compatible with SPPS and TFA cleavage. Nonetheless, it should be stressed that the Fmoc removal step after the insertion of the fluorinated building block required the use of Morpholine:DMF (1:3) mixture in the presence of HOBt (0.1 M) to prevent the tfmψP-TfmψP amide bond cleavage.

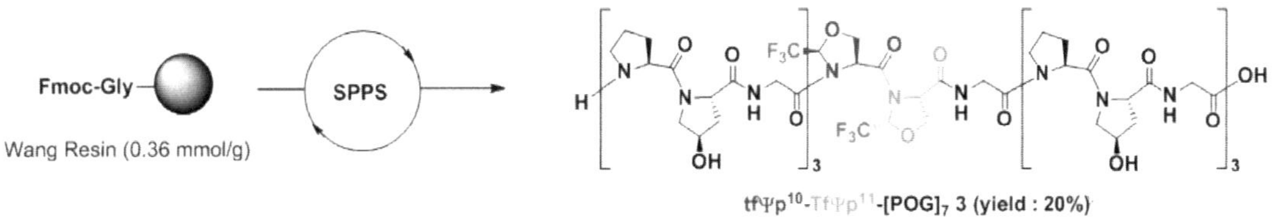

1. Piperidine (16 eq), MW*; 2. DIC (6 eq), Oxyma pure (6 eq), AAs (6 eq), MW; 3. TFA/TIS/H_2O (95:2.5:2.5)

* Deprotection conditions for the Fmoc-Gly-tfΨp-TfΨp moeity : Morpholine:DMF (1:3), HOBt (0.1 M)

Scheme 2: Synthesis of the fluorinated host-guest CMP 3

The stability of the triple helix of tfmψP^{10}-TfmψP^{11}-$[POG]_7$ 3 was then assessed. CD spectroscopy is the most common analytical method for the determination of collagen TH melting temperatures. At 10 mM, the fluorinated CMP 3 exhibits a PPII structuration with the characteristic positive band at 225 nm on the CD spectrum (Figure 1A). The comparison of the CD spectra of CMPs 1-3 shows that the non-fluorinated CMP 1 displays the higher $[\theta]$ maximum. Sigmoidal curves $[\theta] = f(T)$ were obtained by increasing the temperature as a result of the thermal denaturation of the collagen TH. CMPs 2 and 3 displayed lower melting temperatures (T_m) for the TH compared to the reference $[POG]_7$ 1 (Figure 1B). However, CMP 3 displays a 12°C higher T_m value compared to CMP 2, despite the fact that it bears a second CF_3-ψPro residue. As mentioned before, it was observed that when incorporated into the Ac-Xaa-Yaa-Gly-NHMe collagen triplet model, a single TfmψP at the Yaa position strongly affects the puckering of the Xaa=Pro residue, while when Xaa=tfmψP and Yaa=TfmψP the puckering alternation required in the TH model was maintained.[7] This result suggests that satisfying the C(4)-endo/C(4)-exo alternation allows a significant increase of the Tm value in TH models incorporating Pro and Hyp analogues. Therefore, the collagen TH can accommodate large CF_3 groups close to the peptide backbone provided that the puckering alternation is correct and that the CF_3 groups point outside of the TH.

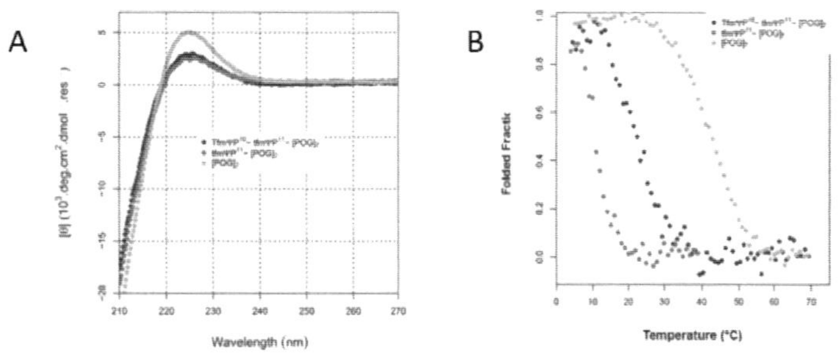

CMP	Tm (°C)
$[POG]_7$	41
$tf\Psi p^{10}$-$Tf\Psi p^{11}$-$[POG]_7$	23
$Tf\Psi p^{11}$-$[POG]_7$	11

Figure 1: A. CD spectra of CMPs 1-3 (10 mM, phosphate buffer (50 mM), 4°C) showing the maximum positive band at 225 nm; B. Sigmoidal curves obtained from the thermal denaturation of 1-3 at a heating range of 10°C/h.

References

[1] a) Shoulders, M. D.; Raines, R. T. Annu. Rev. Biochem. 2009, 78, 929; b) Bella, J. Biochem. J. 2016, 473, 1001.

[2] Ramshaw, J. A. M.; Shah, N. K.; Brodsky, B., J. Struct. Biol. 1998, 122, 86.

[3] a) Vitagliano, L.; Berisio, R.; Mazzarella, L.; Zagari, A. Biopolymers 2001, 58, 459; b) Okuyama, K.; Miyama, K.; Mizuno, K.; Bächinger, H. P. Biopolymers 2012, 97, 607.

[4] Shoulders, M. D.; Satyshur, K. A.; Forest, K. T.; Raines, R. T. Proc. Natl. Acad. Sci. U.S.A. 2010, 107, 559.

[5] a) Chaume, G.; Barbeau, O.; Lesot, P.; Brigaud, T. J. Org. Chem. 2010, 75, 4135; b) Chaume, G.; Simon, J.; Caupène, C.; Lensen, N.; Miclet,. E.; Brigaud, T. J. Org. Chem. 2013, 78, 10144; c) Chaume, G.; Simon, J.; Lensen, N.; Pytkowicz, J.; Brigaud, T.; Miclet, E. J. Org. Chem. 2017, 82, 13602.

[6] Feytens, D.; Chaume, G.; Chassaing, G.; Lavielle, S.; Brigaud, T.; Byun, B. J.; Kang, Y. K.; Miclet, E. J. Phys Chem B. 2012, 116, 4069.

[7] Chaume, G.; Terrien, A.; Renaglia, E.; Marquant, R.; Jacquot, Y.; Brigaud, T.; Miclet, E. Proceedings of the 33rd European Peptide Symposium. Naydenova, E.; Pajpanova, T.; Danalev, D. (Editors), European Peptide Society, 2014, pp 174-175.

[8] Terrien, A.; Rahgoshay, K.; Lensen, N.; Brigaud, T.; Chaume, G.; Miclet, E. Proceedings of the 34th European Peptide Symposium. Beck-Sickinger, A.; Mörl, K.; Bellmann-Sickert, K.; Els-Heindl, S. (Editors), European Peptide Society, 2016, pp 22-23.

Molecular dynamics simulations studies of Gonadotropin Releasing Hormone (GnRH) and GnRH Receptor

Haralambos Tzoupis[1], Agathi Nteli[1], Maria-Eleni Androutsou[2], Theodore Tselios[1]

[1]University of Patras, Greece
[2]Vianex S.A., Greece

https://doi.org/10.17952/35EPS.2018.209

Abstract

Sexual maturation of human cells in ovaries and prostate is closely linked to the secretion of Gonadotropin Releasing Hormone (GnRH). The hormone binds to receptors (GnRH receptors) in the surface of the cells and instigates an extensive biochemical cascade that leads to production and release of luteinizing hormone (LH) and the follicle-stimulating hormone(FSH). Sequence analysis data classify GnRH receptors (GnRHR) as part of the G protein-coupled receptors (GPCRs). Structural data from GPCRs show the presence of a seven trans-membrane helical bundle. The lack of any structural data about GnRHR, impedes the rational design of agonist or antagonist GnRH peptides or non-peptide mimetics for use in the treatment of infertility and hormone dependent cancer. In this study, we employ these common structural and sequence characteristics for the construction of the homology model of GnRH receptor. Structural information from the humanB2-adrenergic receptor as well as rhodopsin has been used in order to create a theoretical model of GnRHR. Furthermore, molecular dynamics (MD) simulations have been used to refine the proposed model and investigate the impact of the bilayer membrane in GnRHR conformation. Additionally, MD simulations were implemented for the analysis of the conformations of GnRH in aqueous solution. The results support the proposed model in the literature regarding the bent conformationof the hormone. The presence of a β-turn in the conformation of the peptide has been shown to be paramount in the binding of GnRH to the receptor and its subsequent activation.

Results

GnRH Receptor Homology Model.

The initial model for the receptor was based on the crystal structures of human β 2-adrenergic receptor (PDB ID: 2rh1), bovine (PDB ID: 3dqb) and squid (PDB ID: 3ayn) rhodopsins. The model of GnRHR was superimposed on the Cα, N and C backbone atoms with homology models based on other rhodopsin/GPCRstructures. The comparison between the proposed model in this study with those reported in the GPCR database, reveals the structural similarities, despite the differences in the templates employed. The comparison of the models is also employed in order to verify the positioning of conserved residues Asn$^{2.50(87)}$, Asp$^{2.61(98)}$, Asn$^{2.65(102)}$, Lys$^{3.32(121)}$ and Asp$^{7.49(319)}$. Our model positions the specific residues with orientation inside the helical bundle as reported in the literature [1]. Residues Asn$^{2.50(87)}$ and Asp$^{7.49(319)}$ are oriented so that their side chains point towards each other inside the helical bundle. The particular orientation has been proposed to stabilize the conformation of GnRHR *via* the formation of a water mediated bridge between the residues.

ECL2 loop refining and modeling.

ECL2 is highly variable in the different members of the family and flanks the trams membrane domains TM4 and TM5. ECL2 has been implicated in ligand binding and it has been proposed that regulates ligand entry into the receptor [2]. The disulfide bond between Cys^{114} and Cys^{196} has been proposed to drive the proper folding and activation of GPCRs.

GnRH molecular dynamics (MD).

GnRH has been isolated as a decapeptide from hypothalamic cells. Post translational modification leads to the modification of the C– and N–termini residues [3]. The MD simulations of GnRH, performed in this study in aqueous environment, have shown the presence of a bent conformation for the hormone at positions 5-8. The analysis of our simulations shows the presence of three distinct clusters during the MD simulation (Figure 1). The data derived from the MD production run are in accordance with studies suggesting that the presence of a β-turn between residues 5 to 8 is paramount for the function of GnRH (Figure 1). Furthermore, our simulations have shown that the bent conformation of the hormone is retained through the simulation time from the moment of its formation.

Conclusions

GnRH has been implicated in various regulatory pathways in the human organism [4]. Any imbalances in the production of GnRH or its binding to GnRH receptor leads to the development of various cancers [5]. The conformational analysis of GnRH provides the relevant structural information for the design of analogues that further stabilize the bent conformation around residues 5-8 of the hormone. In combination with the development of a reliable model for GnRH receptor we can implement MD simulations to further analyze the interactions in the GnRH-GnRH receptor complex. Thus, it is possible to provide valuable information for the rational design of peptide analogues and non-peptide mimetics with improved binding affinity and greater stability [6,7].

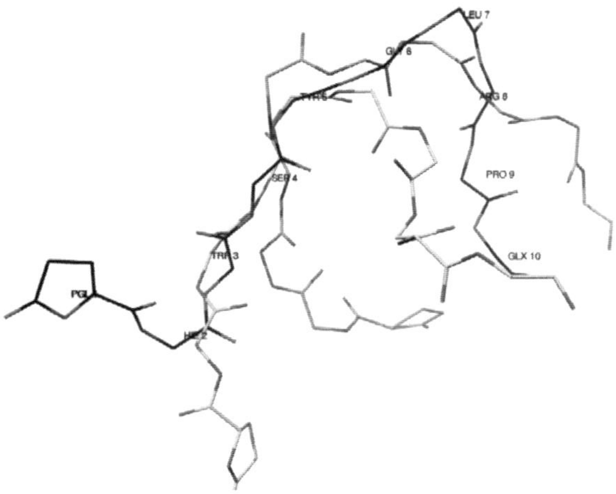

Figure 1: Superimposition of the backbone of the three representative GnRH clusters reported in the MD simulation. The bend is appeared between the residues 5-8.

Methods

Homology model. The homology modeling process was performed with MODELLERv9.17 software.

Loop modeling. Loops ICL1-3, ECL1 and ECL3 were not considered for loop modeling/refinement. The SuperLooper2 web server (http://proteinformatics.charite.de/ngl-tools/sl2/start.php) was implemented for the modeling of ECL2.

Molecular Dynamics (MD). All MD simulations were performed using the AMBER14 software package.

Acknowledgements

The research is funded by the States Scholarships Foundation (IKY) through the "Post-doctoral Financial Support action" part of the "Development of Human Resources, Education and Life-long learning action" co-financed by the European Social Fund (http://ec.europa.eu/esf/)and the Hellenic Republic.

References

(1) Forfar, R.; Lu, Z.L. The Journal of Biological Chemistry 2011, 286, 34617.

(2) DeVree, B. T.; Mahoney, J.P.; Velez-Ruiz, G.A.; Rasmussen, S.G.; Kuszak, A.J.; Edwald, E.; Fung, J.J.; Manglik, A.; Masureel, M.; Du, Y.; Matt, R.A.; Pardon, E.; Steyaert, J.; Kobilka, B.K.; Sunahara, R.K. Nature 2016, 535, 182.

(3) Czieselsky, K.; Prescott, M.; Porteous, R.; Campos, P.; Clarkson, J.; Steyn, F.J.; Campbell, R. E.; Herbison, A.E. Endocrinology 2016, 157, 4794.

(4) Maggi, R.; Cariboni, A.M.; Marelli, M.M.; Moretti, R M.; Andre, V.; Marzagalli, M.; Limonta, P. Human Reproduction Update 2016, 22.

(5) Montagnani Marelli, M.; Manea, M.; Moretti, R.M.; Marzagalli, M.; Limonta, P. International Journal of Oncology 2015, 46, 243.

(6) Laimou, D.; Katsila, T.; Matsoukas, J.; Schally, A.; Gkountelias, K.; Liapakis, G.; Tamvakopoulos, C.; Tselios, T. European Journal of Medicinal Chemistry 2012, 58, 237.

(7) Laimou, D. K.; Katsara, M.; Matsoukas, M. T.; Apostolopoulos, V.; Troganis, A. N.; Tselios, T.V. Amino Acids 2010, 39, 1147.

(8) Webb, B.; Sali, A. Current Protocols in Bioinformatics 2014, 47, 561.

(9) Case, D.A.; Cheatham, III, T.E.; Simmerling, C.L.; Wang, J.; Duke, R.E.; Luo, R.C.W.; Zhang, W.; Merz, K.M.; Roberts, B.; Hayik, S.; Roitberg, A.; Seabra, G.; Swails, J.; Kolossváry, I.; Wong, K.F.; Paesani, F.; Vanicek, J.; Wolf, R.M.; Liu, J.; Wu, X.; Steinbrecher, T.; Gohlke, H.; Cai, Q.; Ye, X.; Wang, J.; Hsieh, M.J.; Cui, G.; Mathews, D.H.; Seetin, M.G.; Salomon-Ferrer, R.; Sagui, C.; Babin, V.T; Luchko, S.G.; Kovalenko, A.; Kollman, P.A.University of California 2012.

On-bead analysis of substrate specificity of caspases using peptide modified by quaternary ammonium group as ionization enhancers

Remigiusz Bąchor[1], Aneta Paluch[1], Wioletta Rut[2], Marcin Drag[1], Zbigniew Szewczuk[1]

[1]Faculty of Chemistry, Wrocław University, Poland
[2]Departmentof Bioorganic Chemistry, Faculty of Chemistry, Wroclaw University of Technology, Poland

https://doi.org/10.17952/35EPS.2018.212

Caspases are proteolytic enzymes at the heart of networks that govern apoptosis and inflammation. Therefore the investigation of new substrates for monitoring of their activity is of great importance. The one-bead-one-compound (OBOC) peptide libraries, allow for obtaining and screening a wide range of compounds in short time, are commonly used in such kind of investigation and electrospray mass spectrometry (ESI-MS) is currently the method of choice for the identification of compounds on single beads. However, up to now, the analysis of on-bead peptide proteolysis and cleavage site have not been investigated by MS due to the necessity of analysis of trace amount of compound obtained from a single resin bead (femtomole level, 10^{-15} mole) which may be insufficient for reliable sequence analysis. Previously we invented the application of quaternary ammonium (QA) group as ionization tags for ultrasensitive sequencing of peptide obtained from a single resin beads of OBOC peptide library by tandem mass spectrometry [1, 2].

The aim of this work was to analyze the on-bead peptide proteolysis by using resin-bound peptide modified by ionization tag and LC-ESI-MS/MS technique (Fig. 1). Ionization tag in the form of quaternary ammonium (QA) group (4-aza-1-azoniabicyclo[2.2.2]octylammonium acetyl group), increasing the ionization efficiency during ESI-MS analysis, was located at the N-terminus of the synthesized peptides. The on-resin peptide bond cleavage by caspase, occurring even with low yield, results in the formation of QA-peptide present in supernatant which may be analyzed by ESI-MS.

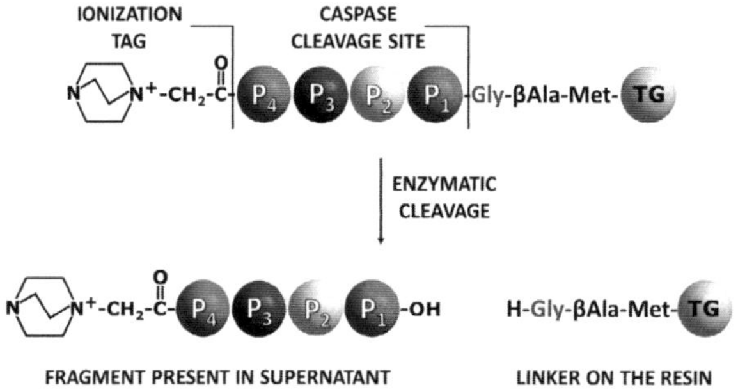

Figure 1: Schematic presentation of the idea of on-bead peptide proteolysis analysis using resin-bound peptide modified by ionization tag.

Model peptide sequences bearing QA group at the N-terminus were synthesized on the TentaGel HL-NH_2 resin where the peptide substrates were connected to the resin by the linker containing methionine residue, allowing peptide cleavage by CNBr, β Ala as a spacer and Gly as the best residue for caspases at P_1' position (Fig. 1). To obtain high proteolytic activity the DEVD/G sequence was selected as the optimal motif for the executioner caspases, caspase 3 and 7 (Fig. 2). It was also presented that the caspase 3 and 7 cleave the DEVE/G sequence however very poorly [3]. Therefore this motif was used to analyze the sensitivity of the proposed here analytical tool for the on-bead monitoring of caspase activity. The DEVA/G sequence was synthesized as a control due to its stability towards caspases.

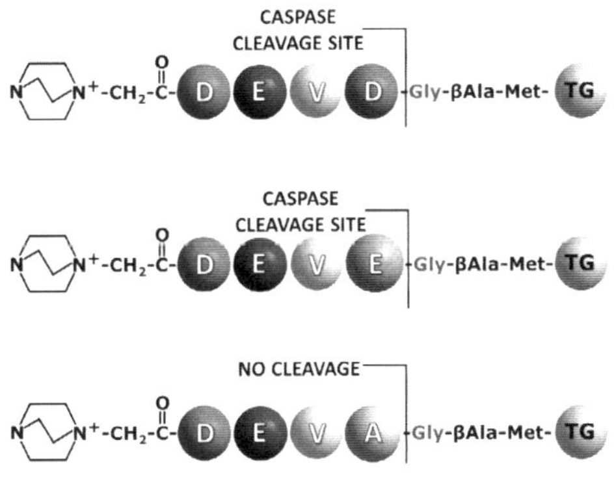

Figure 2: Schematic presentation of the synthesized model sequences containing QA group for the on-bead analysis of peptide bond hydrolysis by caspases.

After the peptide synthesis the fixed charge tag in the form of 4-aza-1-azoniabicyclo[2.2.2]octylammonium acetyl group was synthesized at the α-amino group of the N-terminal amino acid residue according to the procedure described by us previously [4]. Then the side chain protecting groups were removed under acidic conditions and 3 mg of the obtained QA-modified peptidyl resins were incubated with a caspases 3 and 7 for 5, 15, 30, 60 and 180 minutes. The enzymatic reaction was terminated by the addition of formic acid (5 μl) and the presence of fixed charge tagged peptides released to the supernatant after on-bead enzymatic digestion was analyzed by LC-ESI-MS/MS. The following transitions were investigated: for QA-DEVD-OH: 629.4 $\rightarrow$ 268.1 (b

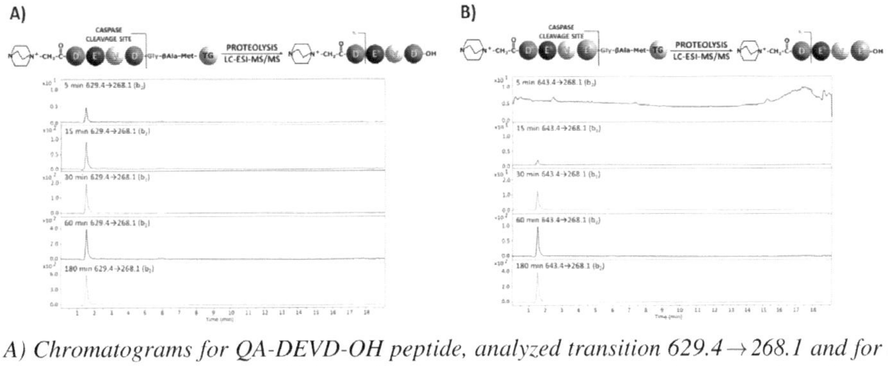

Figure 3: A) Chromatograms for QA-DEVD-OH peptide, analyzed transition 629.4$\rightarrow$268.1 and for B) QA-DEVE-OH peptide (analyzed transition 643.4 $\rightarrow$268.1) investigated in supernatant after incubation with caspase 3.

It was found that in the case of QA-DEVD/G sequence located on the resin the proteolysis product was present in the supernatant even after 5 minutes of enzymatic reaction. The hydrolysis of QA-DEVE/G sequence was also confirmed using our method however the proteolysis product was found in supernatant after 15 minutes of incubation with caspase. It is commonly known that the DEVE/G sequence is poorly recognized by caspases 3 and 7. Therefore the amount of compound released to the supernatant may be very small however the presence of QA tag increases the ionization efficiency of peptide allowing for the ultrasensitive analysis of compounds in complex mixture (supernatant). The signal corresponding to the hydrolysis of QA-DEVA/G sequence was not observed which was expected. The proposed method confirmed the known specificity of caspase 3 and 7 – cleavage after Asp residue, poor cleavage after Glu residue and no cleavage after Ala residue. The solution presented here is a new, rapid and straightforward method of the on-bead caspases activity analysis by LC-ESI-MS/MS technique. The application of fixed charge tag in the form of QA group allows the ultrasensitive analysis of peptide released even from a single resin bead. The proposed method may be used as a tool for the analysis of substrate specificity of enzymes.

This work was supported by a grant No. UMO-2015/17/D/ST5/01329 from the National Science Centre, Poland.

References

[1] R. Bąchor, M. Cydzik, M. Rudowska, A. Kluczyk, P. Stefanowicz, Z. Szewczuk. Mol. Divers., 16, 613 (2012).

[2] R. Bąchor, A. Kluczyk, P. Stefanowicz, Z. Szewczuk. Mol. Divers., 17, 605 (2013).

[3] M. Poręba, A. Szalek, W. Rut, P. Kasperkiewicz, I. Rutkowska-Wlodarczyk, S.J. Snipas, Y. Itoh, D. Turk, B. Turk, Ch.M. Overall, L. Kaczmarek, G.S. Salvesen, M. Drag. Sci. Rep., 23, 7, 43135 (2017).

[4] M. Cydzik, M. Rudowska, P. Stefanowicz, Z. Szewczuk. J. Pept. Sci., 17, 445 (2011).

Modulation of CYP3A4 by the RGD- and Neurotensin (8-13)-analogues

Kaloyan Georgiev[1], Maya Radeva[1], Anelia Balacheva[2], Sylvia Michaylova[1], Tamara Pajpanova[2]

[1]Medical University of Varna, Bulgaria
[2]Bulgarian Academy of Sciences, Bulgaria

https://doi.org/10.17952/35EPS.2018.215

Introduction

RGD- and Neurotensin (8-13)-based drug design is a hopeful perspective for drug development [1, 2]. There is a great deal of data of the observed therapeutic effects of the both sequences (e.g. anticancer, analgesic etc.), but little is known about the pharmacokinetic of these peptides and the possible drug interactions in which they can be involved. Because of that, we decided to check whether these sequences are able to affect the activity of the most common cytochrome P450 enzyme CYP3A4, responsible for the most drug interactions.

Materials and Methods

Peptide synthesis and analysis

Peptides in this study were synthesized by manual solid-phase procedures using techniques for Fmoc-protected amino acids on Wang or MBHA Rink-Amide peptide resins respectively. 20% Piperidine in DMF was used for deprotection of Fmoc-groups and DIC and HBTU were employed as a coupling agent. Simultaneous deprotection and cleavage from the resin was accomplished by treatment with TFA/TIS/ water (95:2.5:2.5) for 3 h at room temperature. Crude peptides were purified by preparative TLC and their purity was checked by analytical HPLC.

Measurement of CYP3A4 activity *in vitro*

The test compound was mixed with a master pre-mix comprising CYP450 BACULOSOMES® reagent and regeneration system, which contained glucose-6-phosphate and glucose-6-phosphate dehydrogenase. The mixture was incubated at room temperature for 20 min. Following incubation, CYP enzyme-specific substrate (Vivid DBOMF for CYP3A4) and NADP+ were added and the mixture was incubated at room temperature for 30 min. CYP activity was evaluated by measuring the fluorescence of the fluorescent metabolite generated from CYP3A4 enzyme-specific substrate. The fluorescence was measured using a microplate reader BioTek Synergy.

Results and Discussion

We used two RGD-analogues modified in the first position with nonproteinogenic amino acids – Agb (S-2-amino-3-guanidino-butyric acid) (1, AgbGD) and Agp (S-2-amino-3-guanidino-propionic acid) (2, AgpGD), Neurotensin (8-13) (3, Arg-Arg-Pro-Tyr-Ile-Leu) and modified at position eight and nine with Lys-Cav (4, Lys-Cav-Pro-Tyr-Ile-Leu). We used the fourth peptides at concentrations of 25 to 100 μM.

Both RGD-analogues (1 and 2) showed minor modification in the activity of CYP3A4 (Figure 1). Only with the largest used concentration, 100 μM, the effect was pronounced with inhibition of 50% of the activity of enzyme.

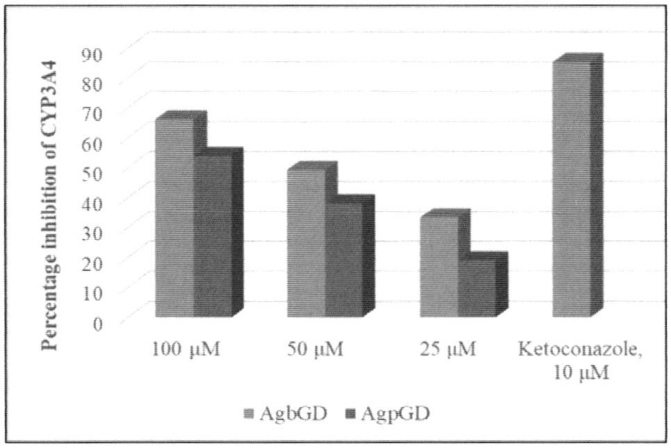

Figure 1: Modulation of CYP3A4 by the RGD analogues - AgbGD and AgpGD.

Neurotensin(8-13) (3) and its analogue (4) did not show activity on CYP3A4 (Figure 2). All three concentrations modulate insignificantly CYP3A4 enzyme.

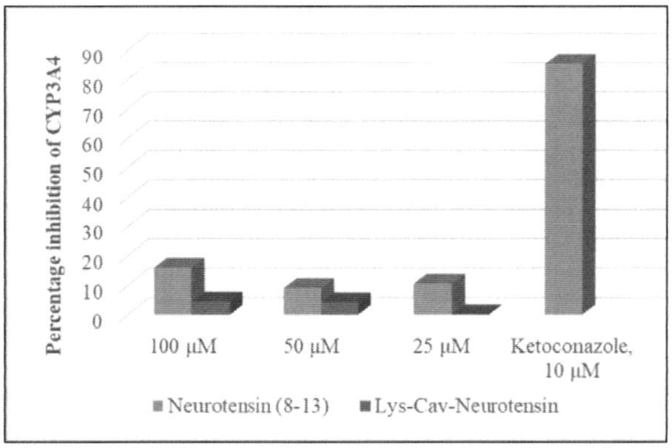

Figure 2: Modulation of CYP3A4 by the Neurotensin (8-13) and its analogue Lys-Cav-Neurpotensin

Conclusion

The results have shown that the use of the RGD-analogues (1 and 2) should be monitored for possible drug interactions, while using Neurotensin(8-13)-sequences does not endanger drug interactions at the cytochrome P450 3A4 level.

Acknowledgements

The study was supported by National Science Fund, Ministry of Education and Science, Bulgaria; Project: DN19/17 and by Fund Science, at the Medical University of Varna/Bulgaria; Project: 16021

References

1. Georgiev, K., Balacheva, A., Peycheva, E., Georgieva, M., Dzimbova, T., Georgiev, L., Iliev, I., Detcheva, R., Miloshev, G., Pajpanova T. In: Kokotos, G., Constantinou-Kokotou, V., Matsoukas, G. (eds) Peptides 2012. Proceedings of the 32st European Peptide Symposium, Athens, Greece, p. 368-369. ISBN: 978-960-466-121-3.

2. Dzimbova, T., Stoeva, S., Tancheva, L., Georgieva, A., Kalfin, R., Pajpanova, T. In: Naydenova, E., Danalev, D, Pajpanova, T (eds) Peptides 2014. Proceedings of the 33rd European Peptide Symposium, Sofia, Bulgaria, p 260–261. ISBN: 978-619-90427-1-7.

Synergy of short antimicrobial peptides with β-lactam antibiotics against MRSA resides in the degradation of peptidoglycan barrier

Ondřej Nešuta, Jakub Kyznar, Andrea Volejníková, Václav Čeřovský
Institute of Organic Chemistry and Biochemistry of the Czech Academy of Sciences, Czechia

https://doi.org/10.17952/35EPS.2018.217

Introduction

The emerging resistance of bacteria to currently used antibiotics is becoming a significant global problem that requires searching for alternative antimicrobial agents. One of the most frequently reported pathogens, methicillin-resistant *Staphylococcus aureus* (MRSA), which causes infections of wounds, bones, implants, bloodstream, skin, pneumonia, etc., is a typical example of Gram-positive bacterium that possess resistance to several antibiotics, including β-lactams. Therapeutic options are becoming limited, and vancomycin remains as the last resort for treating MRSA, but not for long.

Antimicrobial peptides (AMPs) have long been considered as a possible new class of anti-infective agents that could be used as a supplement to, or substitute for, conventional antibiotics in the fight against multi-drug-resistant bacteria. Their potential advantage resides in a unique mechanism of action that involves interacting with the negatively charged phospholipid bilayer of bacterial cell membrane causing its disruption *via* pore formation or detergent-like disintegration, thereby leading to cell death. However, in Gram-positive bacteria, AMPs have to first pass through the cell wall that consists prevalently of a peptidoglycan layer, before reaching its target - the cytoplasmic membrane.

Several reports have shown a significant synergistic antimicrobial effect when AMPs were applied in combination with common antibiotics. In this work, we examined a combination of short linear AMPs, previously invented in our laboratory, with β-lactam antibiotics against a reference strain of MRSA (ATCC 43300) in order to shed light on the mechanism of the synergy between AMPs and β-lactam antibiotics.

Results and Discussion

In this study we used four AMPs (I: GKWMKLLKKILK-NH_2, VI: GKWVKLLKKILK-NH_2, VIII: GKWMKLL-KKILK-NH_2, and IX: KWMKLLKKILK-NH_2) which were derived from the natural AMP halictine-2 [1] and were characterized recently [2]. These peptides showed potent anti-staphylococcal activity in concentrations ranging from 8 to 20 μM (Table 1). MIC values for both peptides and antibiotics were noticeably higher when they were measured in hypertonic LB 20 medium (LB medium supplemented with 20% [wt/vol] sucrose). All tested peptides acted synergistically in combination with β-lactam antibiotics - amoxicillin, ceftazidime, cefuroxime, and meropenem (FIC < 0.5, Table 2). On the other hand, the peptides showed only an additive effect with vancomycin ($0.5 <$ FIC ≤ 2). Synergy between short α-helical peptides and β-lactam was previously described [3]. Some authors proposed a mechanism hypothesizing that the cell wall breakdown caused by β-lactams resulted in the enhanced access of AMPs to the cytoplasmic membrane [3, 4].

Table 1: Minimum inhibitory concentrations (MIC) of peptides and antibiotics against MRSA ATCC 43300 and its protoplasts measured in two different media.

Peptide Antibiotic	MIC (μM)		
	MRSA[a)]	Intact MRSA[b)]	MRSA protoplasts[b, c)]
Peptide I	8.1 ± 0.3	14.8 ± 1.7	3.8 ± 0.6
Peptide VI	15.5 ± 2.0	24.0 ± 2.0	6.0 ± 0.5
Peptide VIII	20.0 ± 1.1	48.0 ± 4.4	6.2 ± 1.0
Peptide IX	15.8 ± 0.6	27.0 ± 3.0	6.6 ± 0.6
Amoxicillin	20.7 ± 1.4	106.6 ± 8.0	>500
Ceftazidime	57.3 ± 5.0	466.7 ± 27.2	>500
Cefuroxime	9.4 ± 0.6	>500	>500
Meropenem	6.2 ± 0.6	26.9 ± 5.8	>500
Vancomycin	0.9 ± 0.1	0.8 ± 0.1	4.0 ± 2.1

a) Determined in LB medium; b) determined in LB20 medium; c) protoplast were prepared by digesting the peptidoglycan layer of bacteria with lysostaphin

To validate this theory, we measured MIC values for all AMPs against MRSA cells which were previously transformed into their protoplasts – cells that lack a cell wall. Interestingly, they exhibited a significant four-to eight-fold reduction. This was in contrast to all the antibiotics which had either higher, or at least same, MIC values against the protoplast compared to intact MRSA (Table 1). The protoplasts became resistant to the antibiotics since β-lactams and vancomycin are known to inhibit cell wall synthesis but do not act on the bacterial cell membrane.

Moreover, AMPs, as cationic molecules, can interact not only with the phospholipids of bacterial membranes but also with peptidoglycan, which contains components carrying a negative charge. In fact, such a study describing the direct interaction of AMPs with peptidoglycan was recently published [5].

Assuming that the peptide may bind to the peptidoglycan barrier before reaching the cytoplasmic membrane, we studied its interaction with peptidoglycan isolated from *Staphylococcus aureus* (InvivoGen, France). The study, based on RP-HPLC methodology, showed that when 10 μM of a peptide was mixed with peptidoglycan (0.02 mg in 1 mL), up to 25% of the peptide was almost immediately trapped into peptidoglycan. Under the same conditions, approximately the same amount of peptide was also absorbed into MRSA cells without killing them immediately, but the cells were killed during next two hours as determined by CFU counting.

Table 2: Fractional inhibitory concentration (FIC) indexes for the combination of studied peptides with selected antibiotics against MRSA ATCC 43300.

Peptides	Antibiotics				
	amoxicillin	ceftazidime	cefuroxime	meropenem	vancomycin
Peptide I	**0.30 ± 0.01**	**0.42 ± 0.03**	**0.33 ± 0.03**	**0.24 ± 0.04**	0.67 ± 0.07
Peptide VI	**0.25 ± 0.00**	**0.44 ± 0.05**	**0.35 ± 0.02**	**0.25 ± 0.00**	0.83 ± 0.14
Peptide VIII	**0.31 ± 0.03**	**0.38 ± 0.06**	**0.33 ± 0.07**	**0.21 ± 0.02**	0.86 ± 0.15
Peptide IX	**0.25 ± 0.00**	**0.38 ± 0.00**	**0.35 ± 0.06**	**0.20 ± 0.02**	1.07 ± 0.02

The FIC indexes were interpreted as follows: FIC ≤ 0.5 is synergistic (in bold), $0.5 < \text{FIC} \leq 2$ is additive, and FIC > 2 is antagonistic effect.

Our results indicate that the binding of AMP to peptidoglycan (and thus the bacterial cell wall) causes considerable interference during AMP-mediated killing of the bacteria due to a decreased AMP concentration that is available for the disruption of the cytoplasmic membrane. We hypothesize that in the course of the combination treatment, β-lactams cause the degradation of the peptidoglycan layer, resulting in a weakening of the cell wall, thus allowing AMP easier access to the cytoplasmic membrane and its subsequent disruption. Consequently, lower concentrations of both antibiotics and AMPs are required to kill the bacteria that finally results in a synergistic effect. This would also explain the different MIC values of AMPs against intact MRSA and the protoplasts. Our conclusion is in accordance with previously published data showing that treatment of *S. aureus* with β-lactam antibiotics can result in its conversion to protoplasts [6]. In addition, vancomycin, which is known to inhibit cell wall synthesis, did not show any synergy with the peptides. The increase in its MIC against the protoplast, compared to intact MRSA, was only four-fold, thus, indicating that vancomycin probably possesses other direct effects on the bacterial cytoplasmic membrane other than cell wall inhibition, as was proposed by Hancock and Fitz-James [7]. This could explain the additive effect of combining vancomycin with AMPs, as they also disintegrate the membrane.

Acknowledgments

This work was supported by Ministry of Health of the Czech Republic, Grant Number 16-27726A and by research project RVO 61388963 of the Institute of Organic Chemistry and Biochemistry of the Czech Academy of Sciences. We thank our technical assistant Mrs. L. Borovičková for peptide synthesis.

References

1. Monincová, L., Buděšínský, M., Slaninová, J., Hovorka, O., *et al.* Amino Acids 39, 763-75 (2010) doi:10.1007/s00726-010-0519-1.

2. Melicherčík, P., Nešuta, O., Čeřovský, V. Pharmaceuticals 11, 20 (2018) doi: 10.3390/ph110100200.

3. Giacometti, A., Cirioni, O., Del Prete, M.S., Paggi, A.M., *et al.* Peptides 21, 1155-1160 (2000) doi:10.1016/S0196-9781(00)00254-0.

4. Li, B., Kang, W., Liu, H., Wang, Y. *et al.* Arch. Pharm. Res. 39, 978–988 (2016) doi:10.1007/s12272-016-0769-x.

5. Neelay, Om P., Peterson, C.A., Snavely, M.E, Brown, T.C., *et al.* J. Mol. Stuct. 1146, 329-336 (2017) doi:10.1016/j.molstruc.2017.06.018.

6. Cushnie, T.P.T, O'Driscoll, N.H., Lamb, A.J. Cell. Mol. Life. Sci 73, 4471-4492 (2016) doi:10.1007/s00018-016-2302-2.

7. Hancock R, Fitz-James PC. J. Bacteriol. 87,1044-1050 (1964).

Structural concept of bitterness of peptides derived from food proteins using chemometric techniques

Anna Iwaniak, Monika Hrynkiewicz, Justyna Bucholska, Małgorzata Darewicz, Piotr Minkiewicz

Warmia & Mazury University, Faculty of Food Science, Chair of Food Biochemistry, Pl. Cieszynski 1, Olsztyn, Poland

https://doi.org/10.17952/35EPS.2018.220

Introduction

Apart from the variety of biological functions, peptides derived from food proteins may affect the all five taste sensations i.e. bitter, salty, sour, sweet, and umami. Peptides with bitter taste are the most dominant as well as the most extensively studied comparing to the other tastes of these molecules. It is also reflected when searching the literature as well as biomolecule databases to find the information on peptides as tastants [1]. Bitter peptides are often produced by hydrolysis of proteins and the generated taste may be considered as an off-flavor and thus be found as the one of the limitations when producing biologically, chemically, and functionally active foods. It specially concerns the situation when a tastant peptide shows other bioactivities [1]. For example, many food protein-derived bitter peptides act also as ACE inhibitors i.e. cause blood pressure reduction. Although they are health-beneficial, their off-flavor taste may discourage the food technologists to use them as food additives [2].

Some data about the molecules of interest, including peptides, are available in the specific databases of biological and chemical information. It is the result of the popularity and continuous development of information technologies, which contributed to the increase of data repositories (databases) as well as programs helpful for biomolecules analyses. One of the scientific disciplines involving the statistics, mathematics, and formal logics to acquire the maximum information and knowledge about chemical systems is called a chemometrics. Chemometric techniques with combination of data provided in biological and chemical information are useful for multivariate analyses aiming to study the relationships between structure and function (activity) of molecules e.g. peptides. Thus, the aim of the study was to apply some multivariate chemometric techniques (see below) to find out which structural attributes (i.e. descriptors) may decide about the bitterness of peptides [3].

Methods

Sequences of bitter di- and tripeptides were acquired from the BIOPEP-UWM database of sensory peptides and amino acids [4]. The experimental measure of their bitterness (necessary for regression analyses) was Rcaf. i.e. bitterness relative to that of 1 mM caffeine solution ($R_{caf.}$ = 1.0) [5]. These values were also found in the BIOPEP-UWM database. The variables (i.e. attributes, descriptors) were the numerical parameters describing the selected properties of each amino acid forming bitter di- and/or tripeptide sequence. They were: molecular weight, bulkiness, polarity, hydrophobicity, the number of carbon atoms, and the number of hydrogen atoms. First four descriptors were derived from ProtScale [6], while the other two from the Biological Magnetic Resonance Data Bank [7]. Finally well-conditioned matrices were obtained for 51 dipeptides and 12 variables and 51 tripeptides and 18 variables. These matrices were the foundation to perform PCA and MLR using Statistica13.1®.

Results and Discussion

The results of PCA and MLR calculations are shown in Table 1. The number of principal components (Comps.) in di- and tripeptide models was distinguished based on few criteria (data not shown). One of them was the percentage of cumulative variance explained, which usually should be at least 70 % and in some cases 80 %. Based on the above-mentioned criterion, 4 and 5 Comps. were found statistically significant for di- and tripeptide datasets, respectively (see Table 1). They explained over 87 % and 82 % of cumulative variance, respectively. It was discovered that one of the attributes affecting the bitterness of both peptide datasets was molecular weight. Positive correlations were found for the residue located at the C-terminus of a di- and tripeptide chain, which was revealed in Comp. 1 (dipeptides) and Comp. 2 (tripeptides). This attribute was also characteristic for Comp. 2 (tripeptides) but describing middle amino acid. Such observation can be also related to the increasing number of carbon and hydrogen atoms in a peptide chain. The impact of the relatively high molecular weight of a residue on the bitter taste of peptides was confirmed by other authors. For example, Kim and Li-Chan [5] reported that high molecular weight of amino acids forming the peptide sequence affected its bitterness and also influenced its hydrophobicity. When looking at both datasets analyzed, peptides comprised mostly of F, Y and/or P (data not shown). According to the literature F and/or Y were reported to be bitter [5]. MLR analysis revealed the impact of a presence of N-terminal bulky residue as well as the number of carbon atoms of C-terminal amino acid in dipeptide chain. No statistically significant attributes were found for MLR tripeptide model. According to the literature, it was found that peptides bitter taste was depended on the presence of hydrophobic/bulky (C-terminus) and basic/hydrophilic (N-terminus)of a peptide sequence. Despite the fact that results of tripeptides MLR were not statistically significant, our results obtained for di- and tripeptides were in line with the idea, according to which, bitterness of di-/tripeptidewas related to the physicochemical property of a specific amino acid present in a whole peptide sequence [8]. The majority of di- and tripeptides of our datasets were composed of hydrophobic C-terminal amino acid with a bulky chain like, e.g., V, L, and/or I which was indicated by variables revealed both by PCA and MLR. The detailed discussion on structure-bitterness of peptides using proposed approach was discussed in papers [9] and [10].

To conclude, when comparing both chemometric techniques applied, PCA (quite old and well-known chemometric method) revealed more attributes explaining their impact on peptide bitter taste than MLR. However, our approach was consistent with the results of other authors studying the relationships between structure and bitterness of peptides. Moreover, our applied chemometric techniques are universal and suitable for analysis structure-function relationships for peptides representing other bioactivities.

Table 1. Summary of PCA and MLR results calculated for bitter di- and tripeptide datasets.

Table 1: Summary of PCA and MLR results calculated for bitter di- and tripeptide datasets.

		Principal Component Analysis	
		Dipeptides	
% cumulative variance explained	*Number of statistically significant components*	*Statistically significant variables forming the specific component (Comp.); values in brackets represent positive/negative correlation*	*Location of amino acid in a peptide chain assigned to the variable present in a component*
~ 87.2	4	Comp. 1: molecular weight(-), bulkiness(-), number of carbon (-) and hydrogen atoms (-),molecular weight(+), bulkiness(+), number of carbon (+) and hydrogen atoms (+)	}N-terminal }C-terminal
		Comp. 2: hydrophobicity(+)	}C-terminal
		Comp. 3: polarity(+), hydrophobicity (-)	}N-terminal
		Comp. 4: polarity(-)	}C-terminal
		Tripeptides	
~82.5	5	Comp. 1: molecular weight(+), bulkiness(+), number of carbon (+) and hydrogen atoms (+)	}middle
		Comp. 2: molecular weight(+), bulkiness(+), number of carbon (+) and hydrogen atoms (+)	}C-terminal
		Comp. 3: molecular weight(-), bulkiness(-), number of carbon (-) and hydrogen atoms (-)	}N-terminal
		Comp. 4: hydrophobicity(+)	}N-terminal
		Comp. 5: polarity(+)	}N-terminal
		Multilinear regression	
		Dipeptides	
R^2	α	*Statistically significant variables affecting the bitterness of peptides; values in brackets represent positive/negative correlation*	*Location of amino acid in a peptide chain assigned to the variable*
~0.4	0.05	bulkiness(-) number of carbon atoms(+)	}N-terminal }C-terminal
		Tripeptides	
~0.7	0.05	-	-

Conclusions

Bitterness of di- and tripeptides derived from food proteins depended on the presence of bulky, branched side residue or a ring (e.g. L, I, V, Y, F) in their sequences. Hydrophobicity of amino acids present in peptide sequences was one of the important attributes affecting their bitterness. Our approach involving some chemometric techniques to analyse structure-function of peptides can be useful for studying peptides with other biological functions.

References

[1] Iwaniak A., Minkiewicz P., Darewicz M., Hrynkiewicz M. 2016. Food protein-originating peptides as tastants - physiological, technological, sensory, and bioinformatic approaches. Food Res. Int., 89, 27–38;

[2] Li-Chan E. C. Y., 2015. Bioactive peptides and protein hydrolysates: research and challenges for application as nutraceuticals and functional food ingredients. Curr. Op. Food Sci., 1, 28-37;

[3] Iwaniak A., Minkiewicz P., Darewicz M., Protasiewicz M., Mogut D., 2015. Chemometrics and cheminformatics in the analysis of biologically active peptides from food sources. J. Funct. Foods, 16, 334-351;

[4] Iwaniak A., Minkiewicz P., Darewicz M., Sieniawski K., Starowicz P., 2016. BIOPEP database of sensory peptides and amino acids. Food Res. Int., 85, 155-161;

[5] Kim H., Li-Chan E. C. Y., 2006. Application of Fourier Transform Raman spectroscopy for prediction of bitterness of peptides. Appl. Spectroscopy 60(11), 1297-1306;

[6] Gasteiger E., Hoogland C., Gattiker A., Duvaud S., Wilkins M. R., Appel R. D., Bairoch A., 2005. Protein Identification and Analysis Tools on the ExPASy Server. In: J. M. Walker (Ed.)

[7] Ulrich E. L., Akutsu H., Doreleijers J. F., Harano Y., Ioannidis Y. E., Lin J., Markley J. L., 2008. BioMagResBank. Nucl .Acids Res. 36, D402-D408;

[8] Nongonierma A. B., FitzGerald R. J., 2016 Learnings from quantitative structure-activity relationship (QSAR) studies with respect to food protein-derived bioactive peptides: A review. RSC Adv., 6, 75400-75413;

[9] Iwaniak A., Hrynkiewicz M., Bucholska J., Minkiewicz P., Darewicz M., 2018. Understanding the nature of bitter-taste di- and tripeptides derived from food proteins based on chemometric analysis. J. Food Biochem., 2018e:12500;

[10] Iwaniak A., Hrynkiewicz M., Bucholska J., Darewicz M., Minkiewicz P., 2018. Structural characteristics of food protein-originating di- and tripeptides using principal component analysis. Eur. Food Res. Technol., 244, 10, 1751-1758.

Acknowledgements

This work was financed by the Warmia and Mazury University in Olsztyn (project number: 17.610.014-300)

The effects of glutathione analogues on Na,K-ATPase activity in the kidneys of the wildtype and the Wfs1 gene mutated mice *in vitro* and *in vivo*

Ceslava Kairane, Riina Mahlapuu, Rando Porosk, Mihkel Zilmer, Ursel Soomets

University of Tartu, Estonia

https://doi.org/10.17952/35EPS.2018.223

Protein S-glutathionylation, the reversible formation of mixed disulfides between glutathione and cysteinyl residues of proteins, is a potential mechanism for dynamic, post-translational regulation of a variety of regulatory, structural, and metabolic proteins in cells.[1]

Glutathione(GSH) is the most abundant low-molecular weight thiol compound that occurs in millimolar ranges in the mammalian cells. The compensatory administration of GSH is not effective because of its rapid degradation. Various chemical modifications of the GSH molecule have been carried out to improve its stability and cellular uptake. We have designed and synthesized a small library of tetrapeptidic GSH analogues (UPF peptides). Glutathione analogue UPF1 is a tetrapeptide (o-methyl-L-tyrosine to the N-terminus of GSH). It is shown that this change increased the hydroxyl radical scavenging ability 60-fold compared to GSH itself *in vitro*. UPF17 is a UPF1 with α-glutamate bond instead of γ-glutamate bond. The change of γ-glutamate to the α-glutamate improves the hydroxyl radical scavenging properties.[2]

Na,K-ATPase is a crucial plasma membrane enzyme in kidneys composed of α subunit, β subunit and a regulatory subunit belonging to the FXYD protein family, which all contain free cysteine residues. An α subunit has 23 free cysteine residues, the β has only one and FXYD protein two of them. The glutathionylation of the Na,K-ATPase causes a decrease in the Na,K-ATPase activity.[3,4]

The WFS1 gene encodes wolframin, an endoplasmic reticulum membrane glycoprotein, which regulates protein folding and secretion, cellular transport and calcium homeostasis. Mutations in WFS1 gene lead to changes in Na,K-ATPase activity and in the expression of the and β subunits.[5]

The aim of this study was to investigate the effect of GSH and its analogues UPF1 and UPF17 on Na,K-ATPase activity in two months old wildtype and WFS1 knockout mice kidney *in vitro* and *in vivo* .

The activity of the Na,K-ATPase was determined by Innova Biosciences kit as the ability to hydrolize ATP and expressed as μmol Pi/mg min. *In vitro* the samples were pre-incubated with GSH and UPF peptides for 10 and 45 minutes at room temperature. *In vivo* glutathione analogues were administrated i.p. 1 mg/kg for 5 days.

The activity of Na,K-ATPase in two months old wildtype and WFS1 knockout mice kidney was measured (Figure 1). The study showed that glutathione and UPF17 in concentrations 10^{-3} M inhibit the activity of Na,K-ATPase in wildtype and WFS1 knockout mice kidney in a time dependent manner(Figure 2). UPF1 in concentrations 10-3 M at pre-incubation time 10 min inhibits the activity of Na,K-ATPase in wildtype, but activates in WFS1 knockout mice kidney (Figure 3). Same concentration of UPF1 at pre-incubation time 45 min did not have any effect in both genotypes *in vitro* . Intraperitoneal administration of UPF1 and UPF17 inhibit the activity of Na,K-ATPase in wildtype mice kidney, but did not have any effect in WFS1 knockout mice *in vivo* (Figure 4).

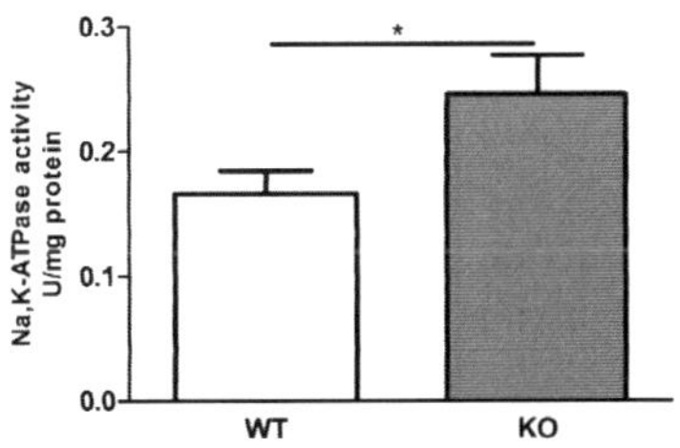

*Figure 1: The activity of Na,K-ATPase in two months old wildtype and WFS1 knockout mice kidney. * $p < 0.05$; $n = 8$.*

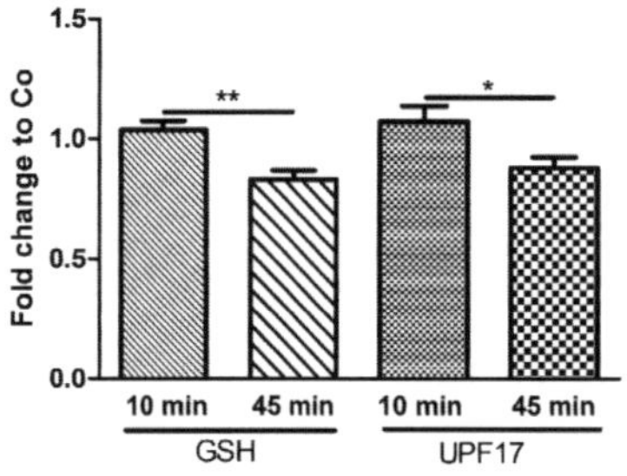

Figure 2: The modulation of Na,K-ATPase activity by GSH and UPF17 (1 mM) in two months old wildtype mice kidney in vitro. Pre-incubation time 10 and 45 minutes. The basal Na,K-ATPase activity is 1.0. Values are represented as mean ± SEM; n = 8.

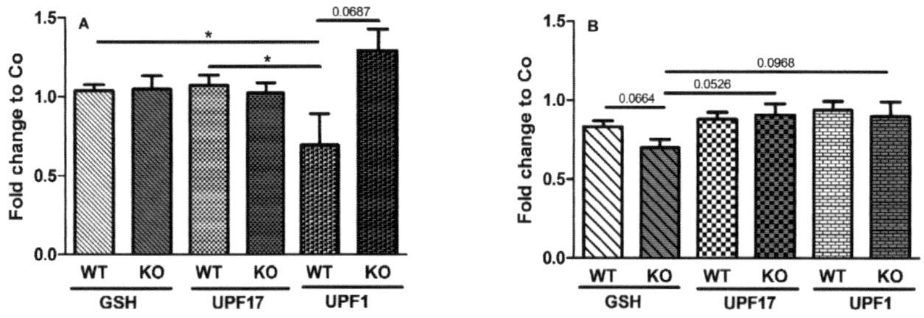

*Figure 3: The modulation of Na,K-ATPase activity by GSH, UPF17 and UPF1 (1 mM) in two months old wildtype and WFS1 knockout mice kidney in vitro. Pre-incubation time 10 (A) and 45 (B) minutes. The basal Na,K-ATPase activity is 1.0. Values are represented as mean ± SEM. *p < 0.05; n = 8.*

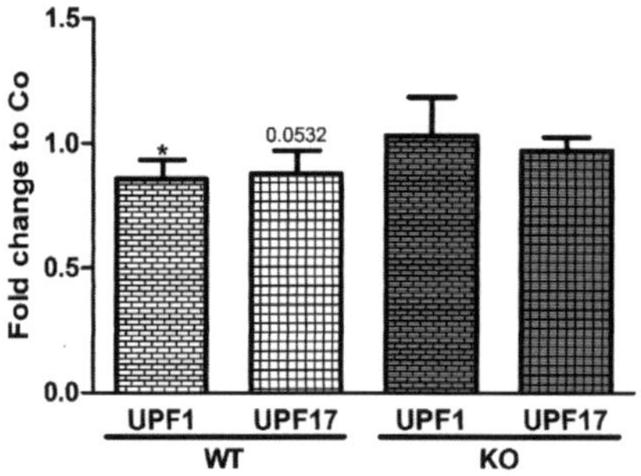

*Figure 4: The activity of Na,K-ATPase in two months old wildtype and WFS1 knockout mice kidney after intraperitoneal administration of UPF1 and UPF17 (1 mg/kg, 5 days). Values are represented as mean ± SEM. * p < 0.05; The basal Na,K-ATPase activity is 1.0; n = 8.*

The various effects of glutathionylation on the activity of Na,K-ATPase may be caused by structural differences between the GSH and UPF peptides (the addition of a methylated tyrosine moiety to the GSH backbone and the change of the peptide bond type). Substitution of the γ-glutamyl with α-glutamyl moiety drastically improved the effect on Na,K-ATPase activity of UPF peptides. As well as by conformational changes in the enzyme of WFS1 knockout mice (different access and affinities of thiol groups of the enzyme).

The present study was supportedby Institutional Research Funding No IUT20-42 from the Estonian Ministry of Education.

References

1. Dalle-Donne, I., Rossi, R., Colombo, G., Giustarini, D., and Milzani, A. (2009). Protein S-glutathionylation: A regulatory device from bacteria to humans. TrendsBiochem.Sci. 34,85-96.

2. Ehrlich, K., Viirlaid, S., Mahlapuu, R., Saar, K., Kullisaar, T., Zilmer, M., Soomets, U. (2007). Design, synthesis and properties of novel powerful antioxidants, glutathione analogues. Free Radical Research, 41(7), 779–87.

3. Bogdanova, A., Petrushanko, I. Y., Hernansanz-Agustín, P., & Martínez-Ruiz, A. (2016). "Oxygen sensing" by Na,K-ATPase: These miraculous thiols. Frontiers in Physiology, 7(AUG), 1–21.

4. Kairane, Č., Mahlapuu, R., Ehrlich, K., Zilmer, M. and Soomets, U. (2014). The effects of different antioxidants on the activity of cerebrocortical MnSOD and Na,K-ATPase from post mortem Alzheimer's disease and age-matched normal brains. Current Alzheimer Research, 11, 79-85.

5. Zatyka, M., Ricketts, C., da Silva Xavier, G., Minton, J., Fenton, S., Hofmann-Thiel, S., Rutter, GA., Barrett, TG. (2008). "Sodium-Potassium ATPase 1 Subunit Is a Molecular Partner of Wolframin, an Endoplasmic Reticulum Protein Involved in ER Stress." Human Molecular Genetics 17 (2). Oxford University Press: 190–200.

Serpin A1 derivatives as collagen turnover modulators for cosmeceutical uses

Fosca Errante[1], Simona Pascarella[2], Caterina Cipriani[2], Lisa Giovannelli[2], Anna Maria Papini[2], Paolo Rovero[2]

[1]University of Florence and Espikem srl, Prato, Italy
[2]University of Florence, Italy

https://doi.org/10.17952/35EPS.2018.226

Introduction

Collagen type I is the archetype above all collagens and it is the main component of connective tissue. Alterations of the delicate balance of its synthesis and degradation lead to various pathophysiological conditions, such as osteogenesis imperfecta, Ehlers-Danlos Syndrome, inflammatory conditions and skin aging. An increased degradation of collagen type I has been shown in chronic wounds and photo-aged skin. Serine proteases, together with matrix metalloproteinases (MMPs), are the main enzymes responsible for collagen degradation within the extracellular matrix (ECM). The physiological inhibitors of these two protein families are serpins and TIMPs, respectively. Serpin A1 is one of the components of the serpins family and is involved in modulating wound healing, cell proliferation, and procollagen production. Among the reported activities, many, such as the wound healing ability, reside in the C-terminal portion of the protein, i.e. region 393-418, characterized by a β-turn structure.

Starting from these evidences, we previously performed a structure-activity relationship study on Serpin A1 C-terminal portion and identified a 10-mer sequence, named SA1-III, which is active in increasing collagen concentration in cultured human dermal fibroblasts. We subsequently decided to investigate whether a shorter sequence of the SA1-III peptide could maintain this promising activity, in order to understand its role in collagen turnover modulation. For this reason, we synthetized four overlapping peptides derived from SA1-III sequence and measured collagen type I concentration in treated fibroblasts by Western Blot analysis, keeping into account both soluble collagen secreted by fibroblasts in the culture medium and the newly-synthesized collagen forms present inside the cells.

Results and Discussion

We prepared SA1-III and its overlapping peptides by solid-phase peptide synthesis (SPPS) with the Fmoc/tBu approach on a 0.48 mmol/g Rink-Amide AM resin. All the sequences were synthetized with C-terminal amidation and N-terminal acetylation to mimic a peptide bond. Crude peptides were purified by semi-preparative HPLC, using a reverse-phase C18 column. After lyophilization, all the sequences were characterized by analytical HPLC and electro-spray ionization mass spectrometry. All peptides were used for *in vitro* studies on cell cultures derived from neonatal human dermal fibroblasts (NHDFs) at passages from 6 to 9, in order to determine the concentrations of collagen type I produced in the presence of each peptide in comparison with the basal concentration of collagen, measured in untreated fibroblasts. We used TGF-β 1 as a positive control for its well-known ECM remodeling properties.

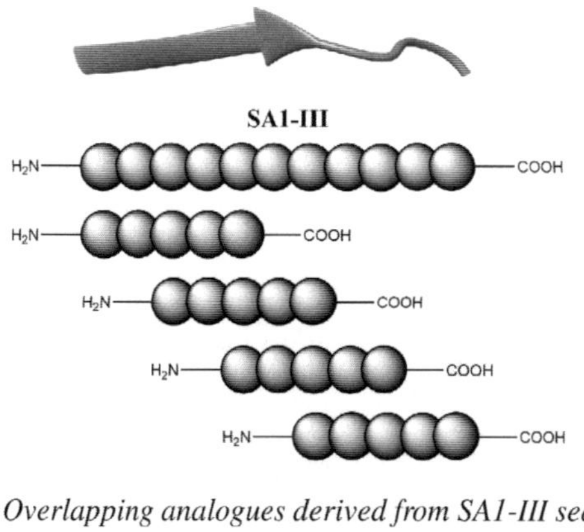

Figure 1: Overlapping analogues derived from SA1-III sequence.

The evaluation of collagen type I concentrations was obtained by Western Blot analysis on culture medium conditioned by fibroblasts treated with the peptides in study for 48 hours, and on cell lysates as well. This immunoenzymatic technique allows to distinguish between different collagen type I forms in the membranes obtained after gel blotting, in particular those related to procollagen and to collagen degraded by the proteases released by fibroblasts in the culture medium. Our experiments show that all the shorter sequences tested display an activity comparable to that of their precursor SA1-III. In fact, all peptides induce an increase in procollagen concentration(170-150 kDa) both in cell lysates and in the media, and, most importantly, we observe in all cases a decrease of the bands related to collagen degradation fragments (100-75-60 kDa). We observed a significant (ANOVA, $p<0.05$) activity of SA1-III and of one of its analogues in cell lysates. These data suggest that the mechanism of action of SA1-III and of the derived peptides is the modulation of collagen degradation mediated by endogenous enzymes. To confirm this result, we also tested MMP-2 and MMP-9 activity in presence and in absence of SA1-III with zymography studies; these preliminary data show enzyme activity reduction.

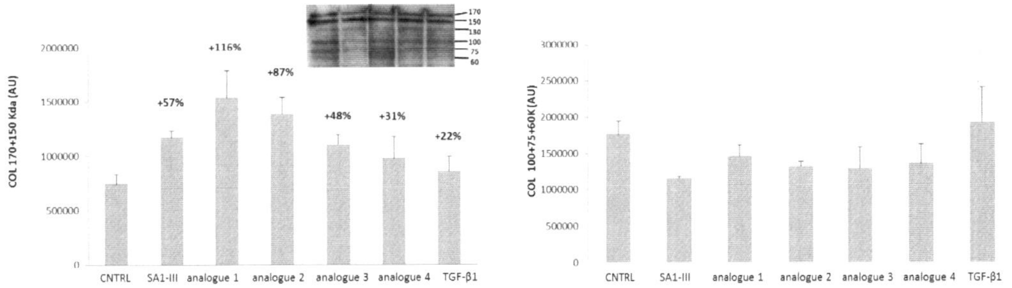

Figure 2: Increased procollagen concentration (on the left) and decreased degradation fragments (on the right) observed for SA1-III and its analogues in fibroblasts culture medium.

SA1-III was then used to formulate a cosmeceutical face cream that might be useful in all those conditions in which collagen degradation is increased, i.e. as an anti-aging factor. The claimed efficacy of this cream was demonstrated with an ultrasonographic evaluation of the cheek skin density in a small cohort of volunteers, before and after a 21-day treatment with the cream. The results clearly indicate a statistically significant increase of skin density after the treatment, thus confirming the efficacy of the cream.

References

[1] Caley M.P. et al., "Metalloproteinases and Wound Healing," Adv Wound Care (New Rochelle), vol. 4, pp. 225-234, 2015.

[2] Pillai S. et al., "Ultraviolet radiation and skin aging: roles of reactive oxygen species, inflammation and protease activation, and strategies for prevention of inflammation-induced matrix degradation - a review," Int J 34 of Cosmet Sci, vol. 27, pp. 17-34, 2005.

[3] Dabbagh K. et al., "Alpha-1-antitrypsin stimulates fibroblast proliferation and procollagen production and activates classical MAP kinase signalling pathways," J Cell Physio, vol. 186, no. 1, pp. 73-81, 2001.

[4] Congote L.F. et al., "Comparison of the effects of serpin A1, a recombinant serpin A1-IGF chimera and serpin A1 C-terminal peptide on wound healing," Peptides, vol. 29, no. 1, pp. 39-46, 2008.

[5] Pascarella S. et al., "Serpin A1 C-Terminal Peptides as Collagen Turnover Modulators," ChemMedChem, vol. 11, no. 16, pp. 1850-1855, 2016.

[6] Cipriani C. et al., "Serpin A1 and the modulation of type I collagen turnover: Effect of the C-terminal peptide 409-418 (SA1-III) in human dermal fibroblasts," Cell Biol Int, vol. 42, no. 10, pp. 1340-1348, 2018.

How to control *Pseudomonas aeruginosa*-induced pneumonia and keratitis? A lesson from the amphibian skin-derived peptide Esculentin(1-21) and its diastereomer

Maria Luisa Mangoni[1], Floriana Cappiello[1], Bruno Casciaro[1], Chen Chen[2], Debarun Dutta[3], Alison McDermott[4], Mark Dp. Willcox[3], Peter Di[2]

[1]Sapienza University of Rome, Italy
[2]Pittsburgh University, United States
[3]UNSW, Sydney, Australia
[4]NorthumbriaUniversity Newcastle upon Tyne, United Kingdom

https://doi.org/10.17952/35EPS.2018.228

The alarming emergence of "superbugs" that are resistant to commercially available antibiotics represents an increasingly serious threat to global public health.

The Gram-negative bacterium *Pseudomonas aeruginosa*is among microbial pathogens for which new anti-infective agents are urgently needed. It causes a large variety of infections including pneumonia especially in cystic fibrosis (CF) sufferers and keratitis in contact lens wearers.

Naturally occurring antimicrobial peptides (AMPs) hold promise as novel therapeutics (1). They are gene-encoded molecules produced by almost all living organisms as key components of the innate immune system. Amphibian skin secretions are one of the richest sources for AMPs. We recently discovered that the frog skin-derived cationic AMP Esc(1-21) has rapid killing kinetics against both free-living and biofilm forms of *P. aeruginosa*, with membrane-perturbing activity as a plausible mode of action which was shown to limit the emergence of resistance (2). This was demonstrated by the invariant minimal inhibitory concentration(MIC) of Esc(1-21) after multiple cycles of treatmentof *P. aeruginosa* in contrast with what found for the currently-used colistin or other traditional antibiotics, e.g. ciprofloxacin, tobramycin and aztreonam, whose MICs significantly increased already after 15 cycles of treatment.

Beside displaying antimicrobial activity, Esc(1-21) can also stimulate migration of bronchial epithelial cells, stably expressing either a functional or mutated (ΔF508) form of the CF transductance regulator (CFTR). Indeed, Esc(1-21) was found to promote the closure of a pseudo-wound of epithelial cells with 500 ≠ m width, produced in a monolayer of these cells within 24 h, at the optimal concentration of 10 ≠ M (3). This suggests that this peptide can accelerate the healing of injuries that are produced in the lung tissue, especially in CF patients, where the wound healing mechanisms are extremely slow favoring the establishment of microbial infections.

Our preliminary data have also indicated that the peptide-inducedmigration of bronchial epithelial cells involves the epidermal growth factor receptor (EGFR) mediated signaling pathway (3).

Among other non-antimicrobial properties, Esc(1-21) was found to display an anti-endotoxin activity, by preventing the secretion of the pro-inflammatory cytokine TNF-α from lipopolysaccharide (LPS)-activated macrophages (4). This is likely due to the peptide-induced disruption of LPS aggregates into smaller-sized particles. Nevertheless, we cannot ignore the limitations in developing AMPs as future therapeutics, namely: (i) the cytotoxicity; (ii) the low biostability and (iii) limited diffusion and delivery to the target site. Over the past years we discovered that substitution of two amino acids in Esc(1-21), i.e. Leu^{14} and Ser^{17} with the correspondingD-enantiomers (to obtain the diastereomer Esc(1-21)-1c) is sufficient to drastically reduce the cytotoxicity of the peptide against a large number of mammalian cells. Furthermore,these two D-amino acids substitution is sufficient to significantly increase the peptide stability not only in biological fluids, but also in the presence of proteolytic enzymes that are abundant in the lungs of CF patients (e.g. elastase from human neutrophils). In addition, the presence of these two D-amino acids can confer the peptide a higher antibiofilm activity and improved wound healing activity, with an optimal concentration of 1 ≠ M compared to 10 ≠ M for the all-L peptide (3). Importantly when tested *in vivo* in murine models of acute lung infection, a single intra-tracheal instillation of the diastereomer at a very low dosage (0.1 mg/kg) was found to cause 2-log reduction in the lung bacterial burden, 24 h after infection without provoking an inflammatory response, similarly to colistin (5). However, in contrast with Esc(1-21)-1c, colistin is active only against Gram-negative bacteria, it induces antibiotic resistance and does not have any wound healing activity.

As mentioned above, *P. aeruginosa* is also responsible for keratitis, especially in contact lenses (CLs) wearers, because of the ability of this pathogen to strongly adhere to the silicone hydrogel CLs forming a biofilm. Among the most feasible approaches to reduce the risk of CL-associated keratitis there is the eradication of bacterial biofilm formed on CLs and the prevention of bacterial colonization of this medical device. We recently found out that the diastereomer can rapidly kill *Pseudomonas* biofilm formed on both sides of CLs. Most importantly, we demonstrated that covalent binding of the peptides to silicone hydrogel soft CL e.g. etafilcon A, represents and effective strategy to achieve an antimicrobial medical device (2 $\neq$ g peptide per lens). More precisely, metacrylic acid present in the lens matrix was activated by 1-Ethyl-3-(3-dimethylaminopropyl) carbodiimide and functionalized with the peptide *via* carbodiimide coupling. The resulting peptide-immobilized CLs were able to cause 4-log decrease in the number of bacterial cells in 20 min. Furthermore, they were able to reduce bacterial adhesion to the lens surface. This is expected to hamper the bacterial transfer from the CL to the cornea, and therefore the establishment of infection.

Remarkably, such peptide conjugation did not make CLs toxic to mammalian cells and did not alter the lens parameters, e.g. diameter, center thickness whose values were similar to those of control lenses; and no detectable abnormalities were obtained for the peptide-conjugated CLs when examined under a Nikon profile projector.

We also proved that Esc(1-21) can retain its bactericidal activity in the presence of human basal tears and reduce the level of infection in murine models of keratitis after three application per day for 3 days at a concentration of 40 $\neq$ M.

Overall, these peptides represent attractive alternatives to control *Pseudomonas*-associated pneumonia and keratitis (Fig.1).

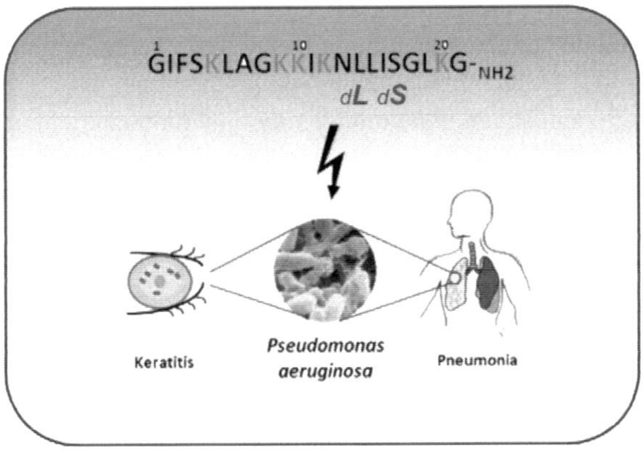

Figure 1: Schematic representation for the potential application of Esc peptides against Pseudomonas aeruginosa-induced infection. dL and dS indicate amino acids in the D-configuration. Basic residues are marked in red

Acknowledgments

This work was supported by grants from Sapienza University of Rome; the Italian Cystic Fibrosis Research Foundation (Project FFC#11/2014; FFC#15/2017); NIH (R01 HL-125128 and AI-133361), NHMRC development grant (APP107620) and an ARC discovery grant (DP160101664) from the Australian Government

References

1. Zasloff M. Nature (2002) 415: 389–3952. doi: 10.1038/415389a
2. Luca V *et al.* Cell Mol Life Sci. (2013) 70:2773-2786. doi: 10.1007/s00018-013-1291-7.
3. Cappiello F *et al.* Antimicr Agents Chemother (2016) 60:7252-7262. doi: 10.1128/AAC.00904-16
4. Di Grazia *et al.* Aminoacids (2015) 47:2505-19. doi:10.1007/s00726-015-2041-y
5. Chen C *et al.* Sci Rep. (2017) 7(1):8548. doi: 10.1038/s41598-017-08361-8.

Development of an IgG/IgY sandwich-ELISA for the bioactive polypeptide Prothymosin alpha

Chrysoula-Evangelia Karachaliou[1], Persefoni Klimetzou[1], Maria Paravatou-Petsotas[1], Ourania Tsitsilonis[2], Wolfgang Voelter[3], Hubert Kalbacher[3], Christos Zikos[1], Evangelia Livaniou[1]

[1]National Centre for Scientific Research "Demokritos", INRASTES, Greece
[2]Departmentof Biology, University of Athens, Greece
[3]Interfaculty Institute of Biochemistry, University of Tuebingen, Germany

https://doi.org/10.17952/35EPS.2018.230

Introduction

Prothymosin alpha (ProTα) is a highly acidic polypeptide expressed in all mammalian cells and consisting of 109 amino acids in humans. The polypeptide exerts important biological functions acting both intracellularly, as a chromatin-remodeling, proliferation-enhancing and anti-apoptotic factor, and extracellularly, as a pleiotropically-operating immunostimulating mediator [1]. Under certain cellular conditions ProTα may be enzymatically cleaved resulting in biologically active peptidyl fragments, such as the N-terminal region ProTα[1-28] (also known as thymosin alpha 1, Tα1) and the C-terminal decapeptide ProTα[100-109]. More specifically, it has been reported that the ProTα[1-28] fragment may be formed after proteolytic cleavage of ProTα by a lysosomal asparaginyl endopeptidase [2], while ProTα[100-109] may be produced after ProTα's cleavage by caspace-3 [3]. Therefore, immunoassays distinguishing between intact parental ProTα and the above mentioned active fragments thereof in specific biological samples are considered valuable laboratory tools, especially for elucidating the precise mechanism(s) of action of the polypeptide.

Experimental/Results

In the current work we have employed rabbit antibodies G (IgGs) and chicken antibodies Y (IgYs), which were previously raised by our teams, to develop a prototype sandwich-ELISA (Figure 1-A). This assay has proved to be highly specific for the intact ProTα molecule. Briefly, the assay steps are as follows: IgG-antibodies raised in rabbits against a conjugate of synthetic ProTα[100-109] with keyhole limpet hemocyanin (KLH) have been indirectly immobilized on the ELISA microwells. After blocking the remaining binding sites, ProTα solutions of known concentration (1 μg/mL - 1 ng/mL), prepared by dissolving commercially available ProTα in various cell culture media, were transferred to the plate for incubation. IgY-antibodies raised in chickens against a ProTα/KLH conjugate which show high specificity for the intact polypeptide [4, 5] have been subsequently incubated with the microwells. Finally, commercially available enzyme-labelled secondary anti-IgY antibodies along with the proper chromogenic substrate have been used for signal development. The ProTα sandwich-ELISA developed shows high reproducibility (inter-assay CV < 5%) and a detection limit of < 2 ng/mL (Figure 1-B). Moreover, assay specificity for the intact ProTα molecule is very high as verified by cross-reactivity studies with synthetic ProTα[100-109] (Figure 1-B) and ProTα[1-28] as well as synthetic peptides of the beta-thymosin family (data not shown here).

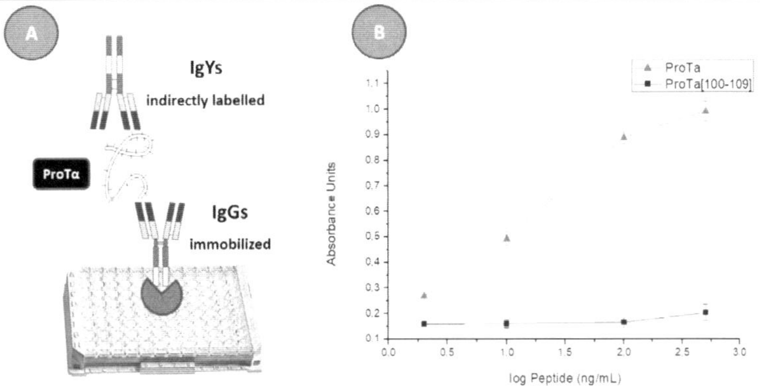

Figure 1: A. Schematic representation of the IgG/IgY-based sandwich-ELISA setup for prothymosin alpha. B. Standard curve of ProTα (pink triangle) in the sandwich-ELISA setup over a concentration range of 1 ng/mL - 500 ng/mL. The corresponding curve of synthetic ProTα[100-109] (blue square) verifies no cross-reactivity of this fragment with the ELISA system.

Discussion/Future perspectives

ProTα is a highly conserved among mammalian organisms polypeptide that in humans may serve as a specific biomarker and/or a target for developing new therapies, especially in cancer and immune diseases. In this work, we have exploited antibodies G and antibodies Y previously raised by our groups against ProTα[100-109]/KLH and ProTα/KLH, respectively, to develop a prototype sandwich-ELISA with high specificity for intact ProTα, which exhibits practically no cross-reaction with bioactive fragments of the parental molecule. Moreover, the assay is sensitive and highly reproducible. If necessary, assay sensitivity may be further increased, e.g. using different enzyme substrate systems such as chemiluminescence-based systems. The developed sandwich-ELISA is currently being applied to specific biological samples, i.e. the supernatants of human cells and cells of cancer cell lines, cultured in the presence of various apoptotic/necroticfactors, in order to verify the literature scenario, according to which human cells may release intact ProTα as a damage-associated molecular pattern (DAMP)/alarmin, under certain survival-threatening circumstances [1]. Potentially, other biological samples, e.g. human sera, may be also analyzed, after slightly modifying the assay procedure. The results obtained after analyzing suitably selected biological samples through the above sandwich ELISA will contribute to the elucidation of the mechanism(s) of action of ProTα under certain conditions, which is a prerequisite for any further clinical exploitation of the polypeptide.

Acknowledgements

This research is implemented through IKY scholarships program and co-financed by the European Union (European Social Fund - ESF) and Greek national funds through the action entitled "Reinforcement of Postdoc-toral Researchers", in the framework of the Operational Program "Human Resources Development Program, Education and Lifelong Learning" of the National Strategic Reference Framework (NSRF) 2014 - 2020. We also acknowledge partial support of this work by the project "NCSRD - INRASTES research activities in the framework of the national RIS3" (MIS 5002559) which is implemented under the "Action for the Strategic Development on the Research and Technological Sector", funded by the Operational Program "Competitiveness, Entrepreneurship and Innovation" (NSRF 2014-2020) and co-financed by Greece and the European Union (European Regional Development Fund).

References

[1] Samara P, Karachaliou CE, Ioannou K et al., Prothymosin alpha: An alarmin and more..., Current Medicinal Chemistry. 2017;24(17):1747-60; doi: 10.2174/0929867324666170518110033

[2] Sarandeses CS et al., Prothymosin alpha is processed to thymosin alpha 1 and thymosin alpha 11 by a

lysosomal asparaginyl endopeptidase, Journal of Biological Chemistry. 2003 Apr 11;278(15):13286-93; doi: 10.1074/jbc.M213005200

[3] Evstafieva AG et al., Prothymosin alpha fragmentation in apoptosis, FEBS Letters. 2000 Feb 11;467(2-3):150-4; doi.org/10.1016/S0014-5793(00)01139-X

[4] Klimentzou P. et al., Development and immunochemical evaluation of antibodies Y for the poorly immunogenic polypeptide prothymosin alpha, Peptides. 2006 Jan;27(1):183-93; doi: 10.1016/j.peptides.2005.07.002

[5] Karachaliou CE et al., Highly specific polyclonal antibodies Y for intact prothymosin alpha, a human polypeptide with diagnostic and/or therapeutic potential, Journal of Biotechnology. 2018; 280S:S70; doi.org/10.1016/j.jbiotec.2018.06.228

Lipopeptides derived from BP100 containing a D-amino acid or a His residue

Angel Oliveras, Esther Badosa, Emilio Montesinos, Lidia Feliu, Marta Planas

University of Girona, Spain

https://doi.org/10.17952/35EPS.2018.233

Introduction

Plant diseases caused by bacteria and fungi are currently one of the major problems in agriculture, producing important economic losses. The solution to this problem relies on the use of copper, antibiotics or fungicides. These compounds are efficient; however, they are regarded as serious environmental contaminants. Moreover, antibiotics are not allowed in many countries and their use is hampered by the appearance of resistant strains [1].

Antimicrobial peptides have emerged as a good alternative to traditional pesticides. They display a broad spectrum of activity and do not easily facilitate the development of microbial resistance [2]. In this context, we have identified the peptide H-Lys-Lys-Leu-Phe-Lys-Lys-Ile-Leu-Lys-Tyr-Leu-NH_2 (BP100), which displays high antibacterial activity *in vitro* and low hemolysis [3].

The acylation of antimicrobial peptides is a strategy to increase their biological activity and their stability to protease degradation [4]. Based on these considerations, a collection of lipopeptides derived from BP100, incorporating an acyl chain (butanoyl, hexanoyl or lauroyl group) at the N-terminus or at the side-chain of each residue of the sequence, was synthesized [5]. Lipopeptides with an interesting biological activity profile were identified.

On the other hand, the incorporation of one or more D-amino acids into the BP100 sequence was evaluated. BP143, that incorporates a D-Phe at position four, showed a similar antibacterial activity to that of BP100 but it was less hemolytic and more stable to protease degradation [6].

Herein, with the aim of improving the hemolysis and the stability to protease degradation of the lipopeptides derived from BP100, 21 new lipopeptides containing one or two D-amino acids or a histidine residue were designed, synthesized and evaluated for their biological activity.

Results and Discussion

Lipopeptides were synthesized on solid phase as C-terminal amides following a standard 9-fluorenyl-methoxycarbonyl (Fmoc)/tert-butyl (tBu) strategy using Fmoc-Rink-MBHA as solid support. To obtain the lipopeptides incorporating the acyl group at the side-chain of a Lys residue, this amino acid was incorporated as Fmoc-Lys(ivDde)-OH. All lipopeptides were purified, characterized by HPLC and HRMS and obtained in excellent purities (>99%).

Peptides were tested for their *in vitro* antibacterial activity. The minimum inhibitory concentration(MIC) was determined. In general, the lipopeptides containing one D-amino acid at position four were as active as the corresponding parent lipopeptide against all the phytopathogens assayed. On the other hand, lipopeptides containing a His residue or two D-amino acids were less active than their corresponding counterparts.

The toxicity of antimicrobial peptides targeting the bacterial membrane can be assessed with animal cell model systems, being erythrocytes the most commonly used. Hemolysis was determined from erythrocyte suspensions of horse blood at a lipopeptide concentration of 250 μM. The introduction of one or two D-amino acids or a His residue in lipopeptides containing a butanoyl or a hexanoyl group resulted in a decrease of the hemolysis. In contrast, no improvement was observed for peptides incorporating a lauroyl group.

The phytotoxicity was evaluated for their effect upon infiltration on tobacco leaves as described previously [7]. All lipopeptides containing a D-amino acid or a His residue were less hemolytic than melittin, which was used as reference product. However, no general trend was observed when comparing the phytotoxicity of these lipopeptides with that of the parent lipopeptides.

Lipopeptides with the best biological activity profile were D-BP387 and D-BP389 that displayed MIC values between 0.7 and 12.5 μM against all eight phytopathogens, they were not hemolytic and were less phytotoxic than melittin (Table 1).

Table 1: MIC values, hemolysis and phytotoxicity of the lipopeptides with the best biological activity profile

Lipopeptide	MIC (μM)								H[b]	Phyt[c]
	Ea[a]	*Pss*[a]	*Xap*[a]	*Xf*[a]	*Psa*[a]	*Xav*[a]	*Pe*[a]	*Fo*[a]	(%)	(%)
BP387	3.1-6.2	3-1-6.2	3.1-6.2	1.6-3.1	3.1-6.2-	1.6-3.1	6.2-12.5	0.8-1.6	14	9
D-BP387	6.2-12.5	3.1-6.2	3.1-6.2	6.2-12.5	1.6-3.1	3.1-6.2	3.1-6.2	0.8-1.6	0	4
BP389	3.1-6.2	6.2-12.5	0.8-1.6	1.6-3.1	3.1-6.2	0.8-1.6	6.2-12.5	1.6-3.1	22	9
D-BP389	3.1-6.2	3.1-6.2	0.8-1.6	1.6-3.1	1.6-3.1	1.6-3.1	3.1-6.2	0.8-1.6	0	5

a) Ea, *Erwinia amylovora*; Pss, *Pseudomonas syringae pv. syringae*; Xap, *Xanthomonas arboricola pv. pruni*; Xf, *Xanthomonas fragariae*; Psa, *Pseudomonas syringae pv. actinidiae*; Xav, *Xanthomonas axonopodis pv. vesicatoria*; Pe, *Penicillium expansum*; Fo, *Fusarium oxysporum*.
b) H, Percentage of hemolysis at 250 μM.
c) Phyt, Percentage of phytotoxicity at 250 μM

References

1 Agrios, GN. Plant Pathology, 5th ed. Academic Press: San Diego, California, 2005.

2 Baltzer, SA; Brown, MH. Antimicrobial Peptides – Promising Alternatives to Conventional Antibiotics. J. Mol. Microbiol. Biotechnol. 2011; 20: 228–235; DOI:10.1159/000331009.

3 Badosa, E; Ferre, R; Planas, M; Feliu, L; Besalú, E; Cabrefiga, J; Bardají, E; Montesinos, E. A Library of Linear Undecapeptides with Bactericidal Activity against Phytopathogenic Bacteria. Peptides 2007; 28: 2276–2285; DOI:10.1016/j.peptides.2007.09.010.

4 Vilà, S; Badosa, E; Montesinos, E; Planas, M; Feliu, L. Synthetic Cyclolipopeptides Selecti ve against Microbial, Plant and Animal Cell Targets by Incorporationof D-Amino Acids or Histidine. PLoS One 2016; 11: e0151639; DOI:10.1371/journal.pone.0151639.

5 Oliveras, À; Baro, A; Montesinos, L; Badosa, E; Montesinos, E; Feliu, L; Planas, M. Antimicrobial Activity of Linear Lipopeptides Derived from BP100 towards Plant Pathogens. PLoS One 2018; 13: 1–17; DOI:10.1371/journal.pone.0201571.

6 Güell, I; Cabrefiga, J; Badosa, E; Ferre, R; Talleda, M; Bardají, E; Planas, M; Feliu, L; Montesinos, E. Improvement of the Efficacy of Linear Undecapeptides against Plant-Pathogenic Bacteria by Incorporationof D-Amino Acids. Appl. Environ. Microbiol. 2011; 77: 2667–2675; DOI:10.1128/AEM.02759-10.

7 Badosa, E; Moiset, G; Montesinos, L; Talleda, M; Bardají, E; Feliu, L; Planas, M; Montesinos, E. Derivatives of the Antimicrobial Peptide BP100 for Expression in Plant Systems. PLoS One 2013; 8: 1–12; DOI:10.1371/journal.pone.0085515.

Synthesis and antimicrobial activity of the bacteriocin pediocin PA-1 and analogs thereof

François Bédard[1], Hammami Riadh[2], Séverine Zirah[3], Sylvie Rebuffat[3], Ismail Fliss[2], Eric Biron[1]

[1]Faculty of Pharmacy and Institute of Nutrition and Functional Foods, Université Laval, Canada
[2]STELA Dairy Research Centre, Institute of Nutrition and Functional Foods, Université Laval, Canada
[3]Muséum national d'Histoire Naturelle, Sorbonne Universités, France

https://doi.org/10.17952/35EPS.2018.235

Introduction

Bacteriocins are a family of ribosomally-synthesized antimicrobial peptides produced by a wide variety of bacteria. With their attractive antimicrobial properties and activities, these peptides are promising alternatives to conventional antibiotics in the food and animal production industry as well as veterinary and human medicine. [1-3] Among the large number of bacteriocins, we were particularly interested in pediocin PA-1, a peptide of 44 amino acids containing two disulfide bonds produced by *Pedioccocus acidilactici* able to inhibit the growth of several clinically relevant pathogens such as *Listeria monocytogenes*.[4] Despite its great potential as an antimicrobial agent, the problems associated with its production continue to limit its applicability and delay regulatory approval. To pave the way for the future use of pediocin PA-1 in different food, veterinary and medical sectors, our main objective was to develop a convenient and efficient approach to prepare bioactive pediocin PA-1.

Results and Discussion

The chemical synthesis of pediocin PA-1 has been attempted in the past but with limited success (yields below 1%, use of multiple purification steps).[5] To overcome difficult couplings and prevent aggregation during peptide synthesis, PEG-based ChemMatrix resin bearing the HMPB linker was used and pseudoprolines introduced at the critical positions Val8-Thr9, Ala21-Thr22 and Ala34-Thr35 (Fig. 1). After cleavage from the resin, linear pediocin PA-1 1 was purified by HPLC and cyclized in presence of N-chlorosuccinimide to yield oxidized analogs 2a-c. After HPLC purification, the oxidized Met of 2c was selectively reduced to give pediocin PA-1 3c. Pediocin PA-1 oxidized at the Met31 position is 100X less active.6-8 This high sensitivity to aerobic oxidation strongly limits the production, storage and use of pediocin PA-1. For this reason, the Met31 was replaced by a Leu residue in analogs 4 and 5.[7,8] Finally, the Cys residues were replaced by Ala in analog 6 to determine if disulfide bonds are essential for activity. Pediocin PA-1 3c thus synthesized was obtained in 11% overall yield and showed similar strong inhibition of *L. monocytogenes* (MIC = 13.5 nM), similar to the bacteriocin produced naturally by *P. acidilactici* (Table 1).9 On the other hand, linear M31L analog 4 was obtained in 55% overall yield and showed an activity comparable to that of 3c with a MIC of 13.5 nM against *L. monocytogenes* (Table 1).

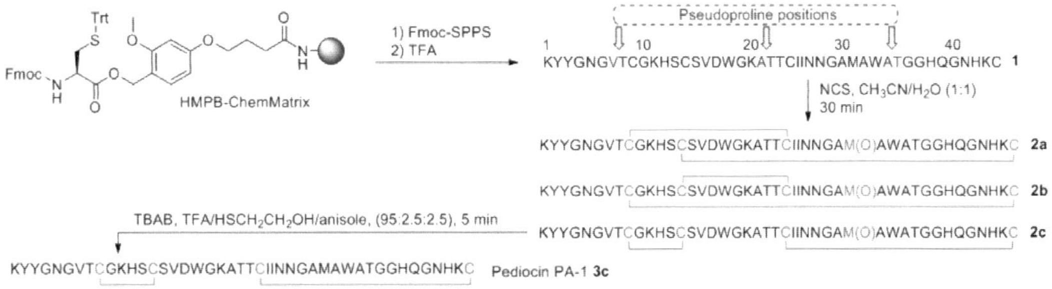

Figure 1: Synthesis of pediocin PA-1 3c and its analogs.

Table 1: Minimal inhibitory concentrations of synthetic pediocin PA-1 3c and its analogs

	Peptide	Minimal inhibitory concentration[a] (nM)	
		L. ivanovii HPB28	*L. monocytogenes* ATCC 19111
1	KYYGNGVTCGKHSCSVDWGKATTCIINNGAMAWATGGHQGNHKC	6.8	13.5
2c	KYYGNGVTCGKHSCSVDWGKATTCIINNGAM(O)AWATGGHQGNHKC	1562	25000
3a	KYYGNGVTCGKHSCSVDWGKATTCIINNGAMAWATGGHQGNHKC	27.0	27.0
3b	KYYGNGVTCGKHSCSVDWGKATTCIINNGAMAWATGGHQGNHKC	13.5	13.5
3c	KYYGNGVTCGKHSCSVDWGKATTCIINNGAMAWATGGHQGNHKC	6.8	13.5
4	KYYGNGVTCGKHSCSVDWGKATTCIINNGALAWATGGHQGNHKC	6.8	13.5
5	KYYGNGVTCGKHSCSVDWGKATTCIINNGALAWATGGHQGNHKC	1.7	13.5
6	KYYGNGVTAGKHSASVDWGKATTAIINNGAMAWATGGHQGNHKA	*n.a.*[b]	*n.a.*[b]

[a]Determined from micro-dilution assay. [b]No activity detected at concentrations up to 100μM

Table 2: Spectrum of activity of linear pediocin PA-1 M31L analog 4

Strain	Reference	MIC[a] (nM)	Diffusion assay[b] (mm)
Carnobacterium divergens	ATCC 35677	1.9	35
Leuconostoc mesenteroides	ATCC 23386	1.9	33
Listeria seeligeri	ATCC 35967	4.7	32
Clostridium perfringens	AAC 1-222	37.8	25
Clostridium perfringens	AAC 1-223	75.7	22

[a]Determined from micro-dilution assay. [b]Inhibition zone diameter

As expected, the oxidized Met31 analog 2c showed an important decrease in inhibition while analog 6 was inactive, demonstrating that disulfide bonds are important for pediocin PA-1 activity. Analogs 3a and 3b with incorrect disulfide bond pairings were 2-4 times less active (Table 1), suggesting the presence of an equilibrium in disulfide bond pairing that ultimately leads to the most stable and active conformation of pediocin PA-1 for a portion of the peptide. The equivalent antimicrobial activities observed for pediocin PA-1 3c and its linear analogs 1 and 4 suggest that the disulfide bonds are formed in situ in the bioassay medium since they are essential for activity. Based on the bacterial target of pediocin PA-1, a protein Blast with the sequence of manPTS IID from *L. monocytogenes* (WP_003721724.1) allowed us to identify several potential sensitive strains.9 Among the tested strains with peptide 4, the strongest activity was observed against *C. divergens*, *L. mesenteroides* and *L. seeligeri* with MICs ranging from 1.9 to 4.7 nM (Table 2). The most interesting result from this focused screening was the strong activity against *Clostridium perfringens* with MICs of 37.8-75.7 nM (Table 2).9 This important food-borne pathogen is one of the most common causes of food poisoning in North America.

In summary, a combination of solid- and solution-phase strategies was used to overcome synthetic pitfalls and produce pediocin PA-1 in good yields. Replacement of the sensitive methionine at position 31 with leucine did not affect antimicrobial activity and improved peptide stability. The linear analogs 1 and 4 were equally potent as their bicyclic counterparts3c and 5 with low nanomolar MIC against *L. monocytogenes*. With higher yields, increased stability and strong antimicrobial activities against major food-borne pathogens such as *Listeria* spp. and *Clostridium perfringens*, the linear M31L pediocin PA-1 analog 4 has great potential in the prevention and treatment of infections in the food and animal production industry as well as in human and veterinary medicine.

Acknowledgements

This work was supported by the Fonds de recherche du Québec-Nature et Technologie (FRQ-NT). F. Bédard thanks the FRQ-NT and Fondation de l'Université Laval for postgraduate scholarships.

References

1. Cotter, P. D., Ross, R. P. & Hill, C. Nat. Rev. Microbiol. 11, 95-105 (2013) DOI 10.1038/nrmicro2937

2. Mathur, H., Field, D., Rea, M.C., Cotter, P.D. Hill, C., Ross, R.P. Front. Microbiol. 8, 1205 (2017) DOI 10.3389/fmicb.2017.01205

3. Ahmad, V., Khan, M.S., Jamal, Q.M.S., Alzohairy, M.A., Al Karaawi, M.A., Siddiqui, M.U. Int. J. Antimicrob. Agents 49, 1-11 (2017) DOI 10.1016/j.ijantimicag.2016.08.016

4. Henderson, J.T., Chopko, A.L., van Wassenaar, P.D. Arch. Biochem. Biophys. 295, 5-12 (1992) DOI 10.1016/0003-9861(92)90480-K

5. Bédard, F., Biron, E. Front. Microbiol. 9, 1048 (2018) DOI 10.3389/fmicb.2018.01048

6. Johnsen, L., Fimland, G., Eijsink, V. & Nissen-Meyer, J. Appl. Environ. Microbiol. 66, 4798-4802 (2000) DOI 10.1128/aem.66.11.4798-4802.2000

7. Derksen, D. J., Boudreau, M. A. & Vederas, J. C. ChemBioChem 9, 1898-1901 (2008) DOI 10.1002/cbic.200800272

8. Kaur, K., Andrew, L. C., Wishart, D. S. & Vederas, J. C. Biochemistry 43, 9009-9020 (2004) DOI 10.1021/bi036018e

9. Bédard, F., Hammami, R., Zirah, S., Rebuffat, S., Fliss, I., Biron, E. Sci. Rep. 8, 9029 (2018) DOI 10.1038/s41598-018-27225-3

Proceedings of the 35th European Peptide Symposium
Patrick B. Timmons, Chandralal M. Hewage, Michal Lebl (Editors)
European Peptide Society & PSP, 2018

Identification and characterization of cyclotides from Brazilian *Psychotria* species

Meri Emili Pinto, Mariana Rosa, Vanderlan Bolzani, Eduardo Cilli

Sao Paulo State University, Brazil

https://doi.org/10.17952/35EPS.2018.238

Psychotria genus below to tribe Psychotrieae (Rubiaceae) and comprises approximately 2,000 species[1]; some of them are reported as cyclotide-containing plant species[2,3]. Cyclotides are characterized by a head-to-tail cyclized backbone usually composed of 28-37 amino acids, and a unique knotted disulfide topology involving six conserved cysteine residues termed the cyclic cystine-knot (CCK) motif[4,5]. The CCK motif is responsible for their exceptional resistance to chemical, enzymatic and thermal degradation[6]. This topology confers them remarkable stability and wide range of biological and/or therapeutic applications, between them, anti-HIV, anthelmintic, insecticidal, molluscicidal, antimicrobial, uterotonic, cytotoxic, hemolytic, trypsin inhibitory, and immunosuppressive activities. This work deals with the identification and characterization of cyclotides accumulated in the Brazilian plant species: *Psychotria vellosiana* (stems and leaves) and *Psychotria leiocarpa* (aerial parts).

Dried and pulverized stems and leaves from *P. vellosiana* and aerial parts from *P. leiocarpa* were extracted with CH_3OH-H_2O (6:4, v/v) at room temperature 4 times over a period of 24 h. The extract was partitioned with CH_2Cl_2- CH_3OH -H_2O (1:1:1, v/v/v) (4 times) and the aqueous phases were separated and concentrated on a rotary evaporator prior to freeze drying, yielding what is further referred to as aqueous extract. The aqueous extracts were dissolved in CH_3CN-H_2O (1:9; v/v) and immediately used for solid-phase extraction (SPE). C_{18} SPE cartridge (Strata-Phenomenex C_{18} 55 μm, 70 , 500 mg) were activated with CH_3OH and subsequently equilibrated with aqueous 0.1% trifluoroacetic acid. After application of the extracts, the cartridge was washed with mixtures of buffer A (0.1% aqueous trifluoroacetic acid) in B (90% acetonitrile, 0.08% trifluoroacetic acid) in the proportions 8:2 (v/v) and 2:8 (v/v), respectively. The fractions eluted in buffer B 80%, named stems C_{18}_80% PV, leaves C_{18}_80% PV and C_{18}_80% PL showed the presence of masses in the range of 2900-3700 Da as analyzed by MALDI-TOF mass spectrometry (MS). To determine whether these fractions contained peptides with disulfide bonds, they were treated with the reducing agent 10 mM dithiothreitol and alkylation agent 100 mM iodoacetomide, and the mass increase of 348 Da were observed, suggesting that three disulfide bonds had been reduced.

The C_{18}_80% PL fraction was analyzed using LC-MS (HRESIMS) and was possible to find seven cyclotides; two of them are already described in the literature.

A

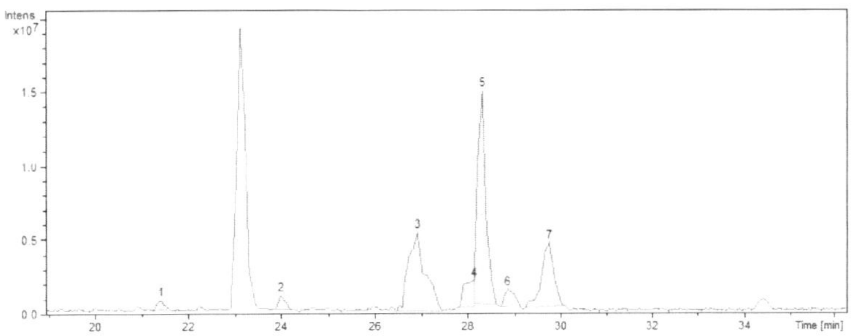

B

Mass	Retention time (min)	Molecular Mass (Da)*	Cyclotide	Reference
1	21.3	3044.01	-	
2	23.9	3026.88	-	
3	26.6	2972.97	-	
4	27.8	2948.13	Psyleio B	3
5	28.1	2988.24	Psyleio A	3
6	28.7	2982.03	-	
7	29.2	3048.09	-	

*Figure 1: Putative cyclotides identified by LC-MS in C_{18}_80% PL from Psychotria leiocarpa. A) Base peak chromatography. B) Table showing each mass related with its retention time. C_{18} Kromasil column (250 × 4.6 mm, 5 µm; flow rate: 1 mL per min), 300 Å, linear gradient of CH_3CN (25-65% B in 40 min), and UV 220 nm. Solvents: Buffer A (H_2O /0.1% HCOOH) and Buffer B (90% CH_3CN /0.1% HCOOH). *The masses shown are monoisotopic.*

The leaves and stems C_{18}_80% PV fractions were analyzed using LC-MS (HRESIMS) and were possible to find some cyclotides to both tissues.

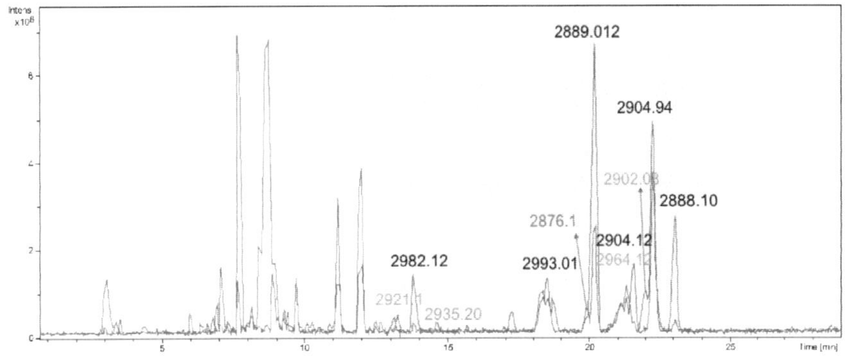

Figure 2: Putative cyclotides identified by LC-MS in leaves (red line) and stems (green line) C_{18}_80% PV fractions from Psychotria leiocarpa. Highlighted in gray, the region eluting putative cyclotides. The masses are represented in black when are common to both tissues, in green to stem-specific and in red to leave-specific.

From stems C_{18}_80% PV were isolated the möbius cyclotides, with m/z 2889 and 2905. The cyclotides sequencing were performed by reduction, alkylation and enzymatic digest with endoproteinases Glu-C and trypsin and followed by MS/MS analysis. The spectra were carefully examined and the sequences were proposed, based on the presence of both *b*- and *y*-series of ions (*N*- and *C*-terminal fragments). The Ile and Leu are assigned on basis of homology.

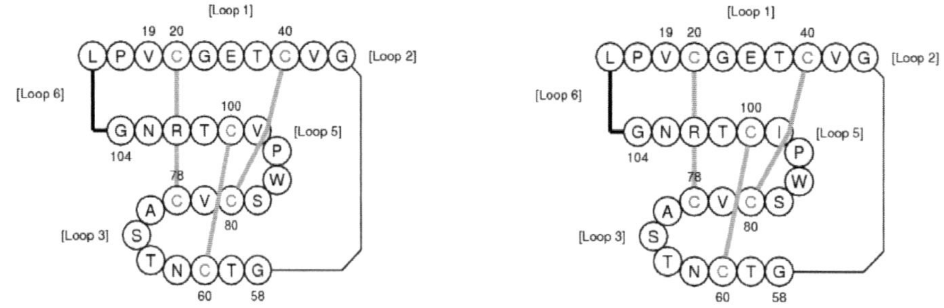

Figure 3: Cyclotides isolated from P. vellosiana. m/z of 2889 and 2905, respectively.

Here, some cyclotides are identify to *P. leiocarpa* and *P. vellosiana*. From this last species, were possible to isolated and to sequence two novel möbius cyclotides.

New 4-Aminopyridine derivatives containing peptide fragment designed for the treatment of Alzheimer disease and multiple sclerosis

Lyubomir Vezenkov[1], Borislav Anchev[1], Ivanka Kostadinova[2], Kristian Fidanov[1], Daniela S. Tsekova[1], Nikolay G. Vasilev[3], Nikolai D. Danchev[1]

[1]University of Chemical Technology and Metallurgy, Bulgaria
[2]Medical University, Sofia, Bulgaria
[3]Bulgarian Academy of Sciences, Bulgaria

https://doi.org/10.17952/35EPS.2018.241

Alzheimer's disease (AD) and Multiple sclerosis (MS) are neurodegenerative diseases. The AD process is associated with plaques and tangles in the brain. MS is a disease that causes inability of the CNS cells to communicate each to other. Neurodegeneration leads to problems with cognitive function (dementias) and / or movement. According to the literature data, no cure for AD and MS, but the treatments can help for the changing of the diseases progression. Among the most effective medicals used for both diseases treatment are 4-aminopyridine (4-AP) and galanthamine. 4-AP has very high toxicity, limiting its use in the treatment of neurodegenerative diseases, Based on the literature [1-3], we propose an approach to produce new hybrid molecules between 4-aminopyridine and different peptide fragments possessing β–secretase inhibitoty activity, thus we expect to combine important pharmacological effects to influence AD and MS: Inhibition of β-secretase and a significant reduction in toxicity due to peptide fragments; Blocking of potassium channels leads to an increase in the level of acetylcholine in the brain, an anti-inflammatory action as well as easier passage through BBB, due to 4-aminopyridine. Here we report the synthesis of three new hybrid compounds containing 4-AP and a peptide fragment AA- Asp(OBzl)-Leu-Ala, where AA is Val, Nva or tert-Leu (see Scheme 1).

Results and Discussion

Synthesis. The obtaining of the new compounds were performed according to Scheme 1. The synthesis of the final compounds 5-7 were carried out in solution by sequentially attachment of the protected amino acids Boc-Ala-OH, Boc-Leu-OH, Boc-Asp(OBzl)-OH and Boc-Val-OH/Boc-Nva-OH/Boc-tert-Leu-OH to 4-AP. The condensation was carried out by TBTU method, with the reagents being dissolved in a minimal amount of DMF. The resulting crude products were recrystallized to chromatographically pure products. The compounds were characterized by m.p. thin layer chromatography and NMR.

Toxicological tests. The new derivatives of 4-AP, containing peptide fragments were investigated for acute toxicity by OECD-425-FDA-USA method by intraperitoneal (IP) application on male mice line H, weight 20-25 g. [4] The data for LD50 are presented as mg/kg body weight in the Table 1. The newly compounds are considerably less toxic (70-80 times) than 4-AP that has toxicity LD50 = 19 mg/kg (highly toxic compound according to Hodge and Sterner classification [4]).

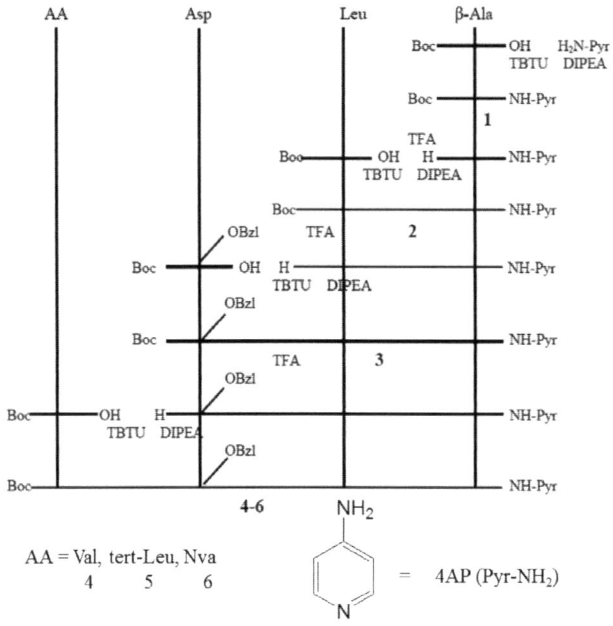

Scheme 1: Synthesis of 4-Aminopyridine derivatives containing a peptide fragment: compounds 4, 5 and 6

Table 1: Acute toxicity of final compounds

code	compound	LD50 (mg/kg)
4 (LK-6)	Boc-Val-Asp(OBzl)-Leu-β-Ala-4-AP	>1500
5 (LK-7)	Boc-tert-Leu-Asp(OBzl)-Leu-β-Ala-4-AP	>1500
6 (LK-4)	Boc-Nva-Asp(OBzl)-Leu-β-Ala-4-AP	>1500

Cytotoxic tests. Cytotoxicity tests were applied also on the the newly compounds. Cell toxicity screening was performed towards the following two types cell cultures: 1. Mice neuroblastoma cell Neuro 2a; 2. Human chronic myeloid leukemia - BV-173. The data are presented in the Fig. 1, Fig. 2 and in the Table 2.

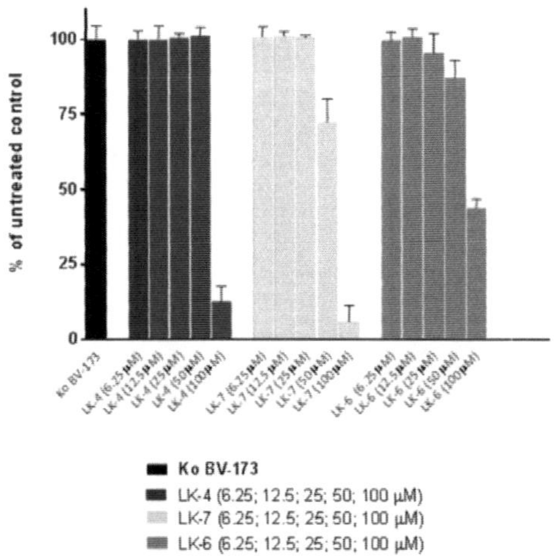

Figure 1: Viability of BV-173 cells following 72h exposure to LK-4, LK-7 and LK-6.

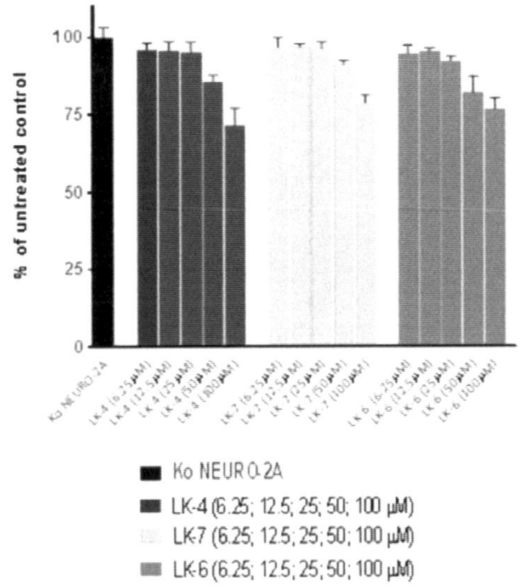

Figure 2: Viability of NEURO-2A cells following 72h exposure to LK-4, LK-7 and LK-6

Table 2: IC50 values of test compounds on the corresponding cell lines (μM).

code	compound	Cell line	
		BV-173	NEURO-2A
LK-4	Boc-Nva-Asp(OBzl)-Leu-β-Ala-4-AP	88.7 ± 23.1	> 100
LK-7	Boc-tert-Leu-Asp(OBzl)-Leu-β-Ala-4-AP	59.8 ± 4.1	> 100
LK-6	Boc-Val-Asp(OBzl)-Leu-β-Ala-4-AP	92.2 ± 6.3	> 100

Conclusions. A series of new 4-AP derivatives comprising peptide fragment were synthesized. Our study on acute toxicity shows that the investigated compounds are less toxic than 4-AP (about 80 times). Results from cytotoxicity studies are in a good correlation with results from acute toxicity test. Obviously, the used peptide fragments decrease significantly *in vivo* toxicity on mice and cytotoxicity on the studied cell cultures.

Acknowledgements

This study is supported by the Scientific Research Fund of Bulgaria through the project DH 03/8, "Galanthamine's and 4-aminopyridine's derivatives containing peptide motif with expected effect on the Alzheimer's disease and multiple sclerosis".

References

1.Niacin, an Old Drug, has New Effects on Central Nervous System Disease Jieli Chen and Michael Chopp The Open Drug Discovery Journal, 2010, 2: 181-186.

2. LT. Vezenkov, D I. Zaharieva, ND. Danchev, NG. Vasilev, AT. Bakalova, TchB. Ivanov. Proceedings of the 33rd EPS 31 Aug-5 Sep 2014; Sofia, Bulgaria, pp 200-201.

3. I. Kostadinova, N. Danchev, I. Nikolova, L. Vezenkov, Tch. Ivanov, Pharmacia, 2017,64(3): 22-39

4. Hodge, A. and Sterner, B. (2005) .Toxicity classes. In:Canadian center for occupational Health and safety. Retrieved from https://www.ccohs.ca/oshanswers/chemicals/ld50.html

5. OECD Guidelins for the testing of chemicals, Health effects. 16 oct. 2008

GLP-1 and GIP receptors co-agonists

Elise S.C. Bernard, Jacky Metcalfe, Isabelle Sermadiras, Andie Collinson, Jacqueline Naylor, David Hornigold, Maria A. Bednarek

MedImmune Ltd, United Kingdom

https://doi.org/10.17952/35EPS.2018.244

In 2030, it is predicted that about 15% of the United States population will be affected by diabetes [1]. Glucagon-like peptide 1 (GLP-1) and glucose-dependent insulinotropic peptide (GIP) are incretin hormones that regulate blood glucose. GLP-1 stimulates insulin secretion, regulates glucose levels and promotes reduction of food intake but also induces nausea. GIP stimulates insulin secretion and induces insulin resistance. Dual agonists with balanced activity at both GLP-1R and GIPR have been shown to control glucose levels better than the parent GLP-1 and GIP peptides; they also facilitate larger body weight loss and provide opportunity to minimize nausea through appropriate balance of GLP-1R and GIPR agonism.

GLP-1 based scaffolds for dual GLP-1R and GIPR agonists have been secured through screening of a recombinant peptide library [2]. One of the scaffold included the "AID" central motif (peptide g1681) which, in addition to other modifications to the GLP-1 sequence, allowed for the generation of dual agonists of various potency ratio at GLP-1 and GIP receptors. New peptides were evaluated *in vitro* for their ability to stimulate cAMP release from transfected CHO cells expressing humanGLP-1R and GIPR. Lipid scan of peptide g1681 showed preferential positions for the attachment of this half-life extension modality. The effect of lipid chain length on potency ratio of the dual potency agonist g1681 was also examined. Table 1 shows sequences of most potent dual agonists with a lipid attached in positions 10, 12, 13, 17, 18, 21 or 40. Activation of both GLP-1 and GIP receptors was stronger with the longer lipids attached to g1681: hexanoyl vs palmitoyl.

Table 1: Lipid scan of g1681, the GLP-1R and GIPR co-agonist

Alias	Sequence	hGLP-1R	hGIPR
		EC_{50} (pM)	
GLP-1	HAEGT FTSDV SSYLE GQAAK EFIAW LVKGR G	2	>23600
GIP	YAEGT FISDY SIAMD KIHQQ DFVNW LLAQK	>50700	35
g1681	Y-Aib-EGT FTSDL SILKE RQAID EFVNW LLKGG PSSGA PPPS	3	8
g1896	Y-Aib-EGT FTSD-**K(PEG$_4$-Pal)** SILKE RQAID EFVNW LLKGG PSSGA PPPS	0.4	1.3
g2017	Y-Aib-EGT FTSDL S-**K(PEG$_4$-Pal)**-LKE RQAID EFVNW LLKGG PSSGA PPPS	1.1	1.6
g1897	Y-Aib-EGT FTSDL SI-**K(PEG$_4$-Pal)**-KE RQAID EFVNW LLKGG PSSGA PPPS	0.6	2.1
g2021	Y-Aib-EGT FTSDL SILKE R-**K(PEG$_4$-Pal)**-AID EFVNW LLKGG PSSGA PPPS	1.0	2.0
g2022	Y-Aib-EGT FTSDL SILKE RQ-**K(PEG$_4$-Pal)**-ID EFVNW LLKGG PSSGA PPPS	2.8	2.4
g2024	Y-Aib-EGT FTSDL SILKE RQAID **K(PEG$_4$-Pal)**-FVNW LLKGG PSSGA PPPS	0.7	1.4
g1734	Y-Aib-EGT FTSDL SILKE RQAID EFVNW LLKGG PSSGA PPPS-**PEG$_4$-K(γE-Pal)**	1.0	2.0
g2068	Y-Aib-EGT FTSDL SI-**K(PEG$_4$-Hex)**-KE RQAID EFVNW LLKGG PSSGA PPPS	20.8	94.1
g1897	Y-Aib-EGT FTSDL SI-**K(PEG$_4$-Pal)**-KE RQAID EFVNW LLKGG PSSGA PPPS	**0.6**	**2.1**
g2065	Y-Aib-EGT FTSDL SI-**K(PEG$_4$-Stear)**-KE RQAID EFVNW LLKGG PSSGA PPPS	1.2	1.7

Like GLP-1 and GIP, dual agonist g1681 was prone to degradation by enzymes in circulation: trypsin, chymotrypsin, DPP-IV, and others. To enhance proteolytic stability of g1681, several unnatural amino acids were incorporated into the sequence and their effect on dual agonism was evaluated. g1726 with αMeLeu13 showed increase of potency on both receptors. G2008 with Nle6 was 15 times more potent on the GIP receptor and g2033 with Aib11 was 12 times more potent on the GLP-1 receptor than the parent co-agonist g1681.

Table 2: Stabilized non-lipidated analogues of g1681

Alias	Sequence	hGLP-1 R	hGIP R
		EC_{50} (pM)	
g1681	**Y**-Aib-EGT FTSDL S**I**LKE **R**QAID EF**VN**W L**L**KGG PSSGA PPPS	3	8
g1726	Y-Aib-EGT FTSDL SI-**αMeLeu**-KE RQAID EFVNW LLKGG PSSGA PPPS	0.5	1.4
g2008	Y-Aib-EGT **Nle**-TSDL SILKE RQAID EFVNW LLKGG PSSGA PPPS	28.1	**1.8**
g2033	Y-Aib-EGT FTSDL **Aib**-ILKE RQAID EFVNW LLKGG PSSGA PPPS	**2.7**	31.2

In the sequence of g1681, combination of chosen substitutions with unnatural amino acids and selective lipidation allowed to secure analogues of increased *in vivo* stability and high potency at both GLP-1R and GIPR.

Peptides g1682 and g1736 retained high potency on both receptors, while g2007 showed somewhat lower activity at GIP R. The long C18 lipid chain in position 13 substantially diminished potency at both receptors, in comparison to potency of g2007 with the shorter C16 lipid chain.

Table 3: Lipidated analogues of g1681 with unnatural amino acids in the sequence

Alias	Sequence	hGLP-1R	hGIPR
		EC_{50} (pM)	
g1681	**Y**-Aib-EGT FTSDL S**I**LKE **R**QAID EF**VN**W L**L**KGG PSSGA PPPS	3	8
g1682	Y-Aib-EGT FTSDL SILKE RQAID EFVN-**αMePhe** LLKGG PSSGA PPPS- **K(γE-Pal)**	7.1	6.3
g1736	Y-Aib-EGT FTSD- **K(γE-Pal)** SILKE RQAID E-**αMePhe**-VN-**αMePhe** LLKGG PSSGA PPPS	**2.2**	8.0
g2006	Y-Aib-EGT FTSDL SI-**K(γE-Stearyl)**-KE RQAID E-**αMePhe**-VN-**αMePhe** LLKGG	321.0	**90.0**
g2007	Y-Aib-EGT FTSDL SI-**K(PEG$_4$-Pal)**-KE RQAID E-**αMePhe**-VN-**αMePhe** LLKGG	**5.0**	11.0

Several dual agonist peptides were equipotent or more potent than GLP-1 at the insulinoma cell line (INS-1E). To isolate response of each receptor to the dual agonist peptides, these compounds were also tested at knock-out cell lines. Table 4 shows assay data confirmation that our novel peptides were balanced dual agonists of similar potency in WT and KO cell lines.

Table 4: INS1-E cell assay; EC50 (pM)

Alias	INS-1E	INS-1 832/3		
		WT CONTROL	GLP-1R KO	GIPR KO
GLP-1	33.3	31.3	-	30.4
GIP	718.6	614.0	387.0	42300
g1681	33.0	17.5	25.1	14.5
g1682	57.0	24.2	17.0	68.6
g1736	18.0	23.7	23.1	15.7

In this study, lipidated dual agonists at GLP-1R and GIPR were secured; they displayed high potency at both receptors and enhanced proteolytic stability. These new peptides could be used as lead compounds for the development of peptide therapeutics.

Potent and balanced GLP-1R and GIPR dual agonists are reported to be under scrutiny not only for the treatment of diabetes and obesity but also in search for novel treatments for Alzheimer disease, a condition believed to be stemmed from insulin de-sensitisation in the brain [3].

References

1. Rowley WR, Bezold C, Arikan Y, Byrne E, Krohe S. Popul. Health Manag. 2017; 20(1): 6-12.

2. Sermadiras I, Ravn P, Bednarek MA, Suckow A, Papworth M, Bernard E. WO2016/131893A1.

3. Hölscher C. Neuropharmacology 2018; 136: 251-259.

Chemical synthesis and investigation of the native form and an improved gamma-core analogue of *Neosartorya fischeri* antifungal protein 2 (NFAP2)

Györgyi Váradi[1], Liliána Tóth[2], Ákos Vendrinszky[1], László Galgóczy[2], Gyula Batta[3], Attila Borics[4], Zoltán Kele[1], Gábor K. Tóth[1]

[1]Departmentof Medical Chemistry, University of Szeged, Hungary
[2]Departmentof Microbiology, Faculty of Science and Informatics, University of Szeged, Hungary
[3]Department of Organic Chemistry, University of Debrecen, Hungary
[4]Institute of Biochemistry, Biological Research Centre, Hungarian Academy of Sciences, Hungary

https://doi.org/10.17952/35EPS.2018.246

Introduction

NFAP2 is a novel cysteine-rich antifungal protein from *Neosartorya fischeri* N RRR 181 [1]. The Minimum Inhibitory Concentration(MIC) values on clinically relevant *Candida* species vary between 0.391 and 1.563 μg/mL. NFAP2 is a cysteine-rich, cationic 52-mer protein stabilized by three disulfide bridges. Based on homology to other cysteine-rich antifungal proteins, the *abcabc* topology is feasible (Fig. 1).

IATSPYYACNCPNNCKHKKGSGCKYHSGPSDKSKVISGKCEWQGGQLNCIAT

Figure 1: Sequence and supposed disulfide bond pattern of NFAP2

The aims of this study were the chemical synthesis and functional mapping of NFAP2. In order to improve the antifungal activity, the evolutionary conserved γ-core motif was modified, and an analogue of NFAP2 having a more active γ-core was synthesized. All these efforts can lead to a better understanding of the importance of γ-core and other parts of the protein on the antifungal effect.

Results and Discussion

NFAP2 was prepared by native chemical ligation of the N-terminal 22-mer thioester fragment and the C-terminal 30-mer part [2]. The thioester was synthesized on the previously described Cys-SH resin [3]. Both peptides were prepared on the solid-phase using Boc chemistry and DCC/HOBt coupling. Native chemical ligation of the purified fragments was conducted in a pH 7.5 ammonium acetate buffer in the presence of 3% thiophenol for 3 hours. Oxidation of the thiols of cysteines was carried out in a glutathione redox buffer containing 1mM GSH and 1 mM GSSG at pH 7.5 for 24 hours. Synthetic NFAP2 (sNFAP2) was compared to the native protein isolated from the ferment of *Neosartorya fischeri* N RRR 181 (nNFAP2) and to recombinant NFAP2 which was prepared by heterologous expression in a *Penicillium chrysogenum* based system (rNFAP2). RP-HPLC analysis, NMR and ECD spectroscopies, and antifungal susceptibility tests revealed the identity of synthetic, native and recombinant NFAP2.

For functional mapping, small fragments of NFAP2 were synthesized by solid-phase synthesis applying Fmoc chemistry and DCC/HOBt coupling. Surprisingly, not the highly conserved γ-core motif, but the mid-N-terminal part of the protein found to be functionally active. In contrary to the γ-core motif which is almost neutral (estimated charge at pH 7 is -0.2) and slightly hydrophilic (GRAVY value is -0.993), the mid-N-terminal fragment is positively charged (estimated charge at pH 7 is +3.1) and hydrophilic (GRAVY value is -1.682). Shuffling of amino acid sequence of the most effective fragment did not have effect on the antifungal activity. It proved the importance of physico-chemical properties and not the sequence on the antifungal effect.

The γ-core motif is a unifying structural signature in all classes of cysteine-rich antimicrobial peptides [4]. It is an evolutionary conserved region both in its amino acid sequence and its structure: two antiparallel β-strands are connected by a short turn region.

To examine the effect of γ-core on the antifungal activity, a 14-mer peptide (NFAP2γ) and an improved analogue of it (NFAP2impγ) were synthesized. While NFAP2γ did not inhibit the growth of the examined fungi (*Saccharomyces cerevisiae* SZMC 0644, *Candida albicans* ATCC 10231 and *Candida parapsilosis* CBS 604), the more positively-charged improved analogue (NFAP2impγ) showed some antifungal effect.

An analogue of NFAP2 protein containing the peptide NFAP2impγ as γ-core was synthesized by native chemical ligation. Two methods were tried for the formation of disulfide bridges: air oxidation and the above mentioned glutathione redox buffer. According to our experience, either an oxidizing agent alone (e.g. O2 of air) or a redox system (e.g. GSH-GSSG buffer) can form the right pairing of cysteines in a protein. Surprisingly, both methods led to the same disulfide bond pattern in this case (Fig. 2). The major product was isolated from the reaction mixtures and was subjected to structural and biological investigations. NMR and ECD conformational studies showed an unordered structure for the protein, and microdilution test revealed weaker antifungal activity of the improved γ-core analogue than that of the native NFAP2 protein.

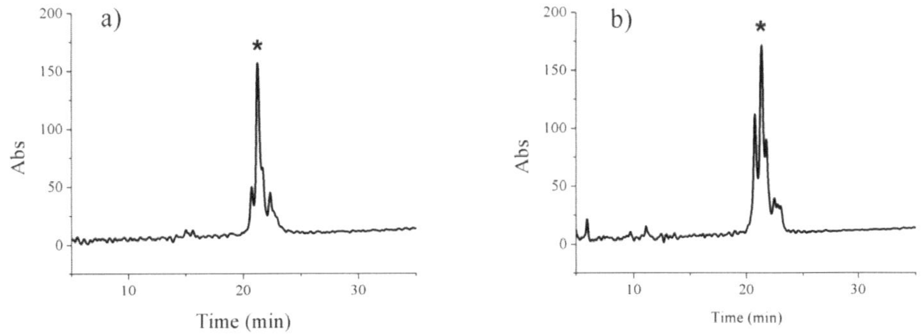

Figure 2: HPLC profiles of disulfide bond formation of the improved γ-core analogue of NFAP2 protein applying air oxidation (a) and a glutathione redox buffer (b).

In conclusion, native chemical ligation followed by oxidation led to a protein that was proved to be identical with native NFAP2. Functional mapping revealed the mid-N-terminal part and not the γ-core as the most active fragment. Antifungal activity of the γ-core could be improved by increasing the number of positive charges and the hydrophilic character. It demonstrated the importance of physico-chemical properties on the antifungal effect. In contrary to our expectations, the improved γ-core-containing analogue of NFAP2 possessed lower antifungal activity than the native protein.

Acknowledgements

This work was supported by the EU and co-financed by the European Regional Development Fund under the project GINOP-2.3.2-15-2016-00014.

References

[1] Tóth, L., Kele, Z., Borics, A., Nagy, G. L., Váradi, G., Virágh, M., Takó, M., Vágvölgyi, C., Galgóczy, L., AMB Express, 2016, 6:75, doi:10.1186/s13568-016-0250-8.

[2] Dawson, P. E., Muir, T. W., Clarklewis, I., Kent, S. B. H., Science, 1994, 266, 776-779, doi:10.1126/science.7973629.

[3] Váradi, G., Tóth, G. K., Kele, Z., Galgóczy, L., Fizil, Á., Batta, G., Chemistry – A European Journal, 2013, 19, 12684-12692, doi:10.1002/chem.201301098.

[4] Yount, N.Y. Yeaman, M.R., Proc. Natl. Acad. Sci. U.S.A., 2004, 101, 7363-7368, doi: 10.1073/pnas.0401567101

Novel peptide-based control measures against the rice fungal pathogen *Pyricularia oryzae*

Marta De Zotti[1], Luca Sella [1], A. Quarantin[1], Carla Castiglioni[1], Angela Bolzonello [1], Francesco Favaron[1], A.M. Picco[2], R. Govind[1], Silvio Tundo[1], Silvana Odorizzi[1], Le T. Do[3], Vu V. Van[4], Nguyen M. Hung[3]

[1]University of Padova, Italy
[2]University of Pavia, Italy
[3]Duy Tan University, Viet Nam
[4]Nguyen Tat Thanh University, Viet Nam

https://doi.org/10.17952/35EPS.2018.248

Introduction

The filamentous fungus *Pyricularia oryzae* is the main causal agent of the rice blast disease, which accounts for 10-30% yield losses per year globally (Talbot, 2003). The disease is worldwide distributed in more than 85 countries, including Italy and Vietnam (Fig. 1; Talbot, 2003), and causes typical leaf spot eye shaped symptoms characterized by large, spreading lesions with a necrotic center and a chlorotic margin (Fig. 1; Wilson and Talbot, 2009).

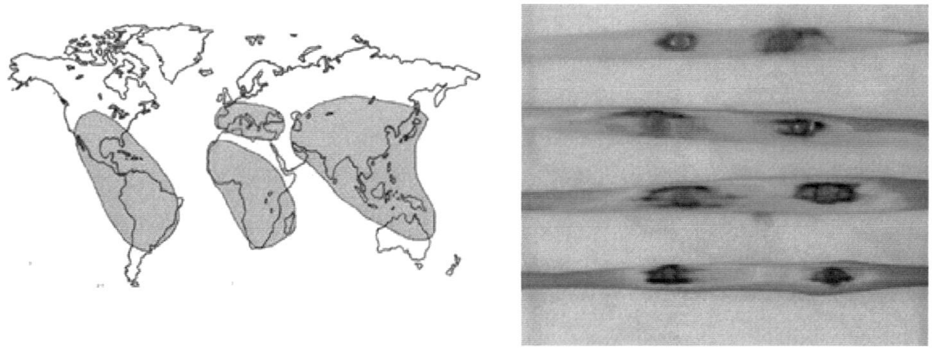

Figure 1: Left: global distribution of rice blast disease. Right: typical blast disease symptoms on rice leaves.

P. oryzae has been recently ranked as the most important fungal pathogen of crops (Fig. 2; Dean et al., 2012).

Rank	Fungal pathogen
1	*Magnaporthe oryzae*
2	*Botrytis cinerea*
3	*Puccinia* spp.
4	*Fusarium graminearum*
5	*Fusarium oxysporum*

Figure 2: Most important fungal pathogens of crops worldwide.

New antimicrobial peptides, analogs of the natural peptaibol trichogin GA IV (tric), have been synthesized (De Zotti et al., 2009; De Zotti et al., 2012) and tested *in vitro* against several *P. oryzae* strains from different geographic origin. Trichogin GA IV is the main component of the mixture of peptide congeners produced by *Trichoderma longibrachiatum* as part of its defense mechanism against other microorganisms (Battaglia et al., 2013). As all peptaibols, its sequence is characterized by the presence of the non-coded residue Aib (α-aminoisobutyric acid), a C-terminal 1,2-aminoalcohol (Leucinol, Lol) and an acylated N-terminus(1-Octanoyl). All peptide sequences are reported in Table 1.

Table 1: Primary structure of tested trichogin analogs.

Abbreviation	Sequence
tric	1-octanoyl-Aib-Gly-Leu-Aib-Gly-Gly-Leu-Aib-Gly-Ile-Lol
K2	1-octanoyl-Aib-Lys-Leu-Aib-Gly-Gly-Leu-Aib-Gly-Ile-Lol
K259G6	1-octanoyl-Aib-Lys-Leu-Aib-Lys-Gly-Leu-Aib-Lys-Ile-Lol
K56	1-octanoyl-Aib-Gly-Leu-Aib-Lys-Lys-Leu-Aib-Gly-Ile-Lol
K5	1-octanoyl-Aib-Gly-Leu-Aib-Lys-Gly-Leu-Aib-Gly-Ile-Lol
K6	1-octanoyl-Aib-Gly-Leu-Aib-Gly-Lys-Leu-Aib-Gly-Ile-Lol

Results & Discussion

Synthesis

All analogs were synthesized by SPPS using a cost-effective strategy with Oxyma Pure / DIC (diisopropylcarbodiimide) as activating agents. All peptides were obtained in high yield and crude purities.

Conformational analysis

Several spectroscopic techniques were applied to shed light on the 3D-structure preferably adopted by the synthesized analogs in solution. Our Circular Dichroism (CD) analysis showed that all peptides mainly assume a right-handed310-helical conformation in a membrane-mimicking environment (Fig. 3).

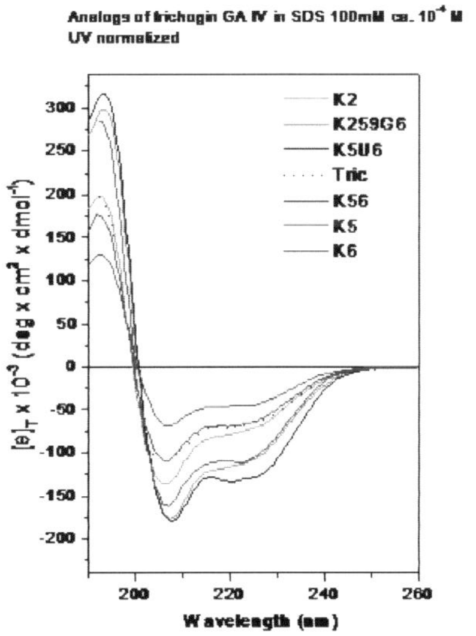

Figure 3: CD spectra of selected analogs in aqueous SDS 100mM (peptide concentration: 0.1 mM).

Antifungal activity

The synthesized peptides have been screened *in vitro* against two different strains (AP and Guy11) of the rice fungal pathogen *P. oryzae*. *In vitro* germination and growth tests conducted in microtiter plates inoculating conidia in potato dextrose medium in presence of peptides showed significant differences in the biocidal activity among the synthesized peptaibols. Although the assay revealed some differences in efficacy between strains, all analogs showed a strong inhibitory effect against *P. oryzae* at 50 μM concentration.

Microscopy

The effect of the most active peptides on fungal conidia was also verified by optical microscopy analysis. In particular, the most effective peptides resulted able to inhibit spore germination and caused in fungal cells morphological modifications probably involving vacuolation and membrane damages.

Conclusions

We exploited a cost-effective, solid-phase synthetic strategy to afford trichogin GA IV analogs with high yield and purity. The conformational studies showed that the analogs adopt a 310-helical structure regardless the sequence modifications. Preliminary tests have highlighted that these peptaibol analogs are non-toxic against plant cells.

Our *in vitro* screening and microscopic analysis have allowed to identify several peptide sequences very effective in inhibiting *P. oryzae* spore germination and fungal growth. The most effective sequences are currently being tested *in vivo* on rice leaves to verify their possible efficacy in protecting rice from the blast disease.

References

Battaglia D. et al., Mol. Plant-Microbe Interact. 2013, 26: 1249. DOI: 10.1094/MPMI-02-13-0059-R

Dean R. et al., Mol. Plant Pathol. 2012, 13: 414–430. DOI: 10.1111/j.1364-3703.2011.00783.x

De Zotti M. et al., J. Pept. Sci. 2009, 15: 615. DOI: 10.1002/psc.1135

De Zotti M. et al., Amino Acids, 2012, 43: 1761. DOI: 10.1007/s00726-012-1261-7

Talbot N.J., Annu. Rev. Microbiol. 2003, 57: 177–202. DOI: 10.1146/annurev.micro.57.030502

Wilson R.A. and Talbot N.J., 2009, Nature Rev. Microbiol. 2009, 7: 185-195. DOI: 10.1038/nrmicro2032

Acknowledgments

This work is part of the Scientific and Technological Cooperation Agreement between the Italian Ministry of Foreign Affairs and International Cooperation and the Department of International Cooperation of the Ministry of Science and Technology of Vietnam.

Novel oligopeptides with angiotensin I-converting enzyme inhibitory activity found in an elastase-treated hydrolysate of porcine aortic elastin

Toshiya Hatakenaka[1], Tamaki Kato[1], Kouji Okamoto[2]

[1]Graduate School of Life Science and Systems Engineering, Kyushu Institute of Technology, Japan
[2]Vital Resources Applied Laboratory, Inc., Japan

https://doi.org/10.17952/35EPS.2018.251

Introduction

Hypertension is one of the main risk factors causing cerebro and cardiovascular diseases. Angiotensin I-converting enzyme (ACE) catalyzes the production of a vasoconstrictive peptide, angiotensin II from angiotensin I. Therefore, the inhibition of ACE activity is an essential target for antihypertension. Elastin is an important macromolecular protein which exists widely in elastic tissues such as arteries, ligaments, lung, skin, etc. and exerts its elasticity to such elastic tissues. It was recently shown that elastin is the protein with ACE inhibitory activity. The peptides with ACE inhibitory activity were found in a thermolysin-treated hydrolysate of bovine ligamentum nuchae elastin [1] and an elastase-treated hydrolysate of porcine aortic elastin [2]. However, ACE inhibitory activities of these peptides were relatively weak. Therefore we are particularly interested in novel ACE inhibitory peptides derived from porcine aortic elastin. Here, we investigated to isolate the ACE inhibitory peptides from a pancreatic elastase-treated hydrolysate of porcine aortic elastin.

Methods

Porcine aortic elastin was treated with pancreatic elastase and its hydrolysate was separated into eleven crude fractions (F1~F11) and analyzed by ACE inhibition assay. The mixture of F2 and F3 with ACE inhibitory activity was separated into five refractions (Ref1~Ref5) (Fig. 1) and analyzed by ACE inhibition assay. The separation into crude fractions (F1~F11) or into refractions (Ref1~Ref5) was carried out using Inertsil ODS-3 column (250 mm x 20 mm) by reversed-phase high-performance liquid chromatography (HPLC). The ACE inhibition assay was performed according to the method of Cushman and Cheung [3] as modified by Yamamoto *et al.* [4]. The potent ACE inhibitory refraction, Ref5 was separated and identified by means of LC/MS/MS. Seven peptides identified were synthesized by a conventional solid-phase method using Fmoc strategy, and the molecular weights and purities of these synthetic peptides were confirmed by MALDI-TOF-MS and by HPLC measurement, respectively. ACE inhibitory activity of synthetic peptides was examined by ACE inhibition assay.

Results

The novel ACE inhibitory peptides derived from pancreatic elastase-treated hydrolysate of porcine aortic elastin were oligopeptides in the crosslinked regions as shown in Table 1. The most potent ACE inhibitory oligopeptide was exhibited to be a Leu-Ala-Ala.

Conclusion

The resulting ACE inhibitory peptides may be beneficial as ingredients of functional foods for preventing hypertension.

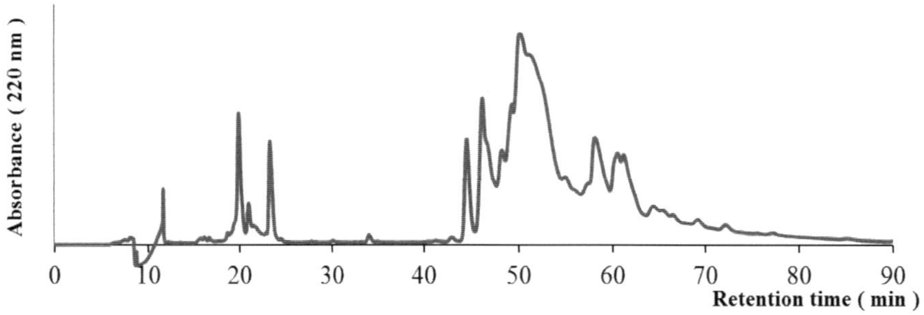

Figure 1: HPLC Chromatogram of Mixture of F2 and F3

The separation was carried out under the following conditions.

Solvent A: CH3CN 5% + 0.1% TFA Solvent B: CH3CN 95% + 0.1% TFA

Flow rate: 10 mL/min

Gradient system for refractionation:

Solvent B 0% (0 - 3 min), 0 - 3.5% (3 - 10 min), 3.5 - 10% (10 - 80 min)

The mixture of F2 and F3 was further fractionated into 5 refractions as follows:

Ref1 (18.3 - 24.0 min), Ref2 (43.5 - 47.5 min), Ref3 (47.5 - 55.3 min),

Ref4 (55.3 - 60.0 min), Ref5 (60.0 - 74.0 min).

*Table 1: IC50 of Oligopeptides Found in Porcine Aortic Elastin * Tropoelastin is a precursor protein of elastin*

Oligopeptides	The number of oligopeptide which is present in tropoelastin*	IC_{50} (μM)
Ala-Ala-Ala	33	214
Leu-Ala-Ala	1	6.1
Ser-Ala-Ala	1	162
Gln-Ala-Ala	3	255
Gly-Ala-Ala	6	258
Pro-Ala-Ala	7	310
Asp-Ala-Ala	1	705

References

[1] Y.Sato,et al., Biosci. Biotechnol. Biochem., 76 (7), 1329–1333(2012).

[2] I.Maeda,et al., Current Enzyme Inhibitation, 14(1), 67-74(2018).

[3] D. W. Cushman and H.S.Cheung, Biochem. Pharmacol., 20, 1637–1648 (1971).

[4] S. Yamamoto,et al., Nihon Kyobu Shikkan Gakkai Zasshi (in Japanese), 18, 297–303 (1980).

The synthesis, oxidation and characterization of GLP-1 peptide receptor fragments

János Szolomájer[1], Zoltán Kele[1], Gábor K. Tóth[1], Pál Stráner[2], András Perczel[2]

[1]University of Szeged, Department of Medical Chemistry, Hungary
[2]Eötvös Loránd University, MTA-ELTE Protein Model. Res. Group and Laboratory of Structural Chemistry and Biology, Hungary

https://doi.org/10.17952/35EPS.2018.253

Glucagon is a 29-amino acid peptide hormone that is produced by the post-translational cleavage of proglucagon, which is a 160-amino acid precursor polypeptide expressed in pancreatic α-cells, intestinal L cells and brain cells. In the intestine and the brain, proglucagon cleavage is catalysed by the PC1/3 enzyme, leading to the formation of glucagon-like peptide-1 (GLP-1), glucagon-like peptide-2 (GLP-2), glicentin and oxintomodulin. GLP-1 is a 31-amino acid peptide hormone that has approximately 50% amino acid sequence homology with glucagon, which is secreted mainly by the intestinal L-cells in response to nutrient ingestion. Despite this high degree of sequence homology, glucagon and GLP-1 perform opposing actions in glucose homeostasis. [1] The main physiological role of glucagon is to protect the organism against the damaging effects of hypoglycaemia, especially in the central nervous system, which requires a continuous supply of glucose. The central action of glucagon is to increase the circulating levels of glucose by stimulating glycogenolysis and gluconeogenesis and, at the same time, to inhibit glycolysis and glycogenesis in the liver. These changes in glucose metabolism, which are induced by glucagon, in turn induce an increase in hepatic glucose output, to ensure an appropriate supply of glucose to the body and the brain. GLP-1 stimulates glucose-dependent insulin release from pancreatic β-cells, inhibits glucagon secretion and increases β-cell proliferation, which contributes to improving the control of the blood glucose levels. Furthermore, GLP-1 modulates satiety and reduces gastric emptying and is thus associated with weight loss. These multiple effects have generated a great deal of interest in the discovery of long-lasting agonists of the GLP-1 receptor (GLP-1R) in order to treat type 2 diabetes. The receptor for GLP-1 (GLP-1R) was first cloned from a cDNA library derived from rat pancreatic islets, and the following year, the human receptor was cloned. The cloning revealed a receptor sequence of 463 residues that resembled the receptors for secretin, parathyroid hormone and calcitonin. These receptors formed a new branch of the GPCR superfamily, named 'Family B' (or 'Class B', or 'secretin receptor-like'), which to date includes 15 members. These receptors possess a unique extracellular N-terminal domain (NTD) of 100–150 residues, which is connected to the integral membrane core domain (or J domain) that is typical of all GPCRs. Family B GPCRs bind their peptide ligands *via* a common mechanism known as the 'two-domain model' in which the NTD first binds to the C-terminal helical region of the ligand, thereby enabling a second interaction between the N-terminal region of the ligand and the core domain of the receptor. The latter interaction is essential for enabling agonist-induced receptor activation. [2]

Our aims were the chemical synthesis, oxidation and characterization of GLP-1 peptide receptor fragments and the investigation of the peptide-ligand (GLP-1, liraglutide, exendin) interaction using NMR spectroscopy.

Due to the difficulty and the length of the sequence we have decided to synthesize the 108 amino acid containing GLP-1 peptide receptor by native chemical ligation procedure. The designed fragments compatible with native chemical ligation was synthesized using solid phase peptide synthesis applying Fmoc/tBu strategy and the synthesis was carried out using a CEM® microwave assisted fully automated peptide synthesizer.

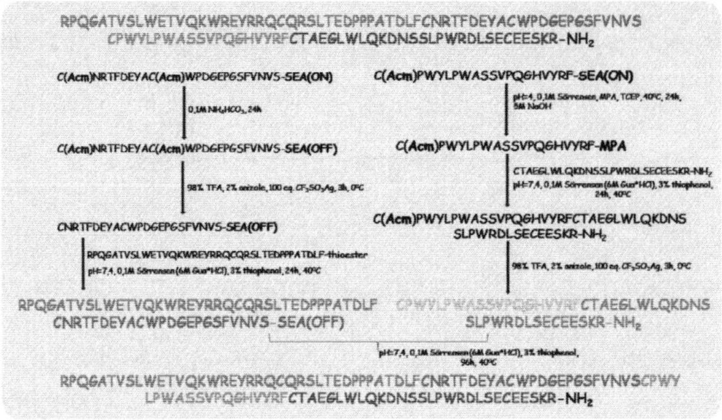

Scheme 1: The synthesis of GLP-1 receptor by native chemical ligation (NCL)

The synthesis of RPQGATVSLWETVQKWREYRRQCQRSL TEDPPPATDLF-thioester (fragment 1)

The synthesis of peptide thioester (fragment 1) was carried out using manual solid phase peptide synthesis applying Boc chemistry. First a Fmoc-Cys(Trt)-OH was attached to the MBHA resin using DCC/HOBT coupling. After that the Fmoc group was removed by using 20% piperidine/DMF and the resulting free amino group was acetylated using 30% acetic anhydride/dichloromethane. The trityl group was removed by the treatment of TFA. The first amino acid was attached to the free sulfhydryl group of cysteine by applying DCC/HOBT double coupling in the presence of DMAP.

The synthesis of CNRTFDEYACWPDGEPGSFVNVS-SEA(OFF) (fragment 2)

The peptide fragment 2 was synthesized using a CEM fully automated microvawe assisted peptide synthesizer applying Fmoc/tBu chemistry, using SEA resin, and Acm side chain protection for cysteins. The crude SEA-(ON) peptide was oxidized using 0,1M NH4HCO3 to obtain the crude SEA-(OFF) peptide. The Acm side chain protection was removed by using Ag(OTf) (50eq.) in TFA (10mg/ml) in the presence of anisole at 4°C for 4h.

The chemical ligation of peptide thioester (fragment1) and SEA-(OFF) peptide (fragment2) was carried out in the presence of thiophenol (3%) in 0,1M Sörrensen buffer, pH=7,4 (6M Gua*HCl), at 40°C for 24h.

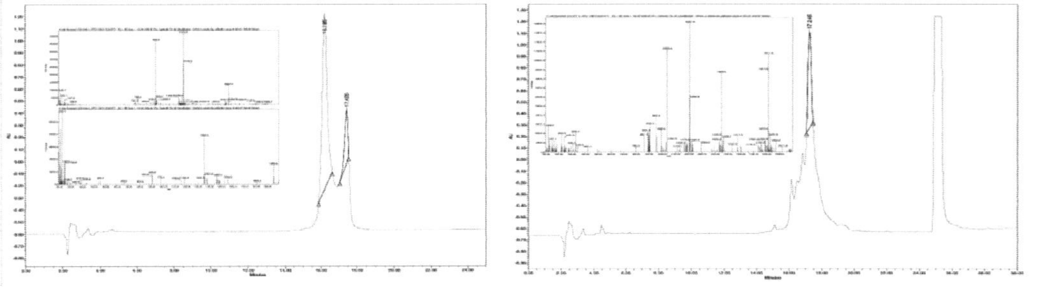

Scheme 2: The chemical ligation of peptide thioester (fragment 1) and SEA-(OFF) peptide (fragment 2), 0min (left side); The mass spectra and LC-chromatogram of ''N'' terminal GLP-1 receptor fragment obtained by chemical ligation (24h) (right side)

The synthesis of C(Acm)PWYLPWASSVPQGHVYRF-MP A (fragment 3)

The peptide fragment 3 was synthesized using a CEM fully automated microvawe assisted peptide synthesizer applying Fmoc/tBu chemistry, using SEA resin. The side chain of the "N" terminal cystein was protected with Acm protecting group. The crude Acm protected SEA-(ON) peptide was coverted into Acm protected peptide-MPA thioester by using MPA (mercaptoprqpionic acid), in the presence of TCEP (1000eq.) in 0,1M Sörrensen buffer (6M Gua*HCl) at 40°C, pH 4 for 24h.

The synthesis of CTAEGLWLQKDNSSLPWRDLSECEESKR-NH_2 (fragment 4)

The peptide amide (fragment 4) was synthesized using a CEM fully automated microvawe assisted peptide synthesizer applying Fmoc/tBu chemistry. The chemical ligation of the Acm protected peptide-MPA thioester (fragment3) and the peptide amide (fragment 4) was carried out in the presence of thiophenol (3%) in 0,1M Sörrensen buffer pH=7,4 (6M Gua*HCl), at 40°C for 24h.

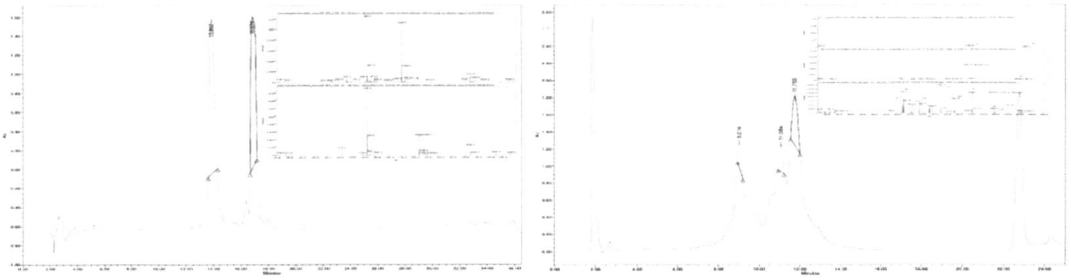

Scheme 3: The chemical ligation of Acm protected peptide-MPA thioester (fragment 3) and peptide amide (fragment 4), 0min (left side); The mass spectra and LC-chromatogram of Acm protected ''C'' terminal GLP-1 receptor fragment obtained by chemical ligation (24h) (right side)

The Acm protection of the "C" terminal peptide was removed by using Ag(OTf) (50eq.) in TFA (10mg/ml) in the presence of anisole at 4°C for 4h. The chemical ligation of "N" terminal SEA-(OFF) peptide and the Acm deprotected "C" terminal peptide amide was carried out in the presence of thiophenol (3%) in 0,1M Sörrensen buffer, pH=7,4, (6M Gua*HCl), 0,2M TCEP*HCl at 40°C for 96h.

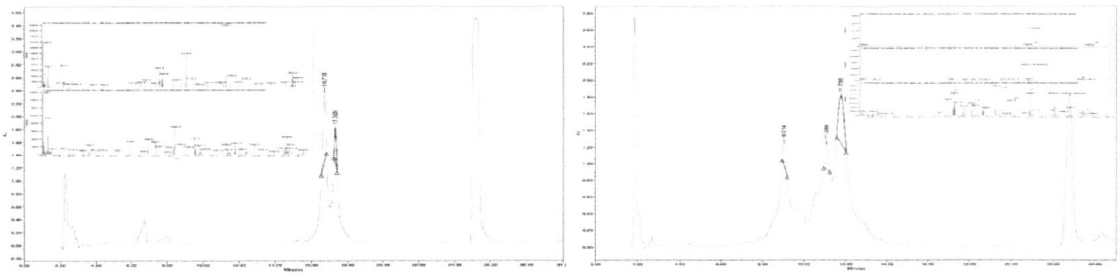

Scheme 4: The chemical ligation of of "N" terminal SEA-(OFF) peptide and the Acm deprotected "C" terminal peptide amide, 0min (left side); The mass spectra and LC-chromatogram of GLP-1 receptor obtained by chemical ligation (96h) (right side)

To obtain the desired disulfide bridges the purified 108 amino acid containing linear GLP-1R peptide was oxidized using different oxidation conditions. (see table 1.) Because of the presence of 6 cystein residues ($C_{23(A)}$, $C_{39(B)}$, $C_{48(C)}$, $C_{62(D)}$, $C_{81(E)}$, $C_{103(F)}$) the formation of three disulfide bonds was expected, but unfortunately none of the applied oxidation conditions were successful.

GLP-1R oxidation 1. No
pH=7.5, 0,1M ammonium acetate buffer
0,2mg/ml, air, 24-48h

GLP-1R oxidation 2. No
pH=7.5, 0,1M ammonium acetate buffer
1,55mM GSH, 0,155mM GSSG, (GSH:GSSG-10:1)
0,2mg/ml, 48-72h

GLP-1R oxidation 3. No
pH=7,5, 0,1M Sörrensen buffer/6M Gua*HCl
1,55mM GSH, 0,155mM GSSG, (GSH:GSSG-10:1)
0,2mg/ml, 48-72h

GLP-1R oxidation 4. No
pH=7,5, 0,1M Sörrensen buffer/6M Gua*HCl
10mM GSSG, 2mM GSH (GSSG:GSH-5:1)
0,4mg/ml, 48-72h

GLP-1R oxidation 5. No
pH=7.5, 0,1M ammonium acetate buffer
(6M Gua*HCl), 0,2mg/ml, air, 24-48h, cystein
(wrong disulfide bridge formation)

GLP-1R oxidation 6. No
pH=7.5, 0,1M ammonium acetate buffer
1mM GSH, 1mM GSSG, (GSH:GSSG-1:1), 0,2mg/ml, 48-72h

GLP-1R oxidation 7. No
pH=8, 0,1M ammonium-hydrogencarbonate/acetonitrile (1:1)
0,2mg/ml, 24-48 air

GLP-1R oxidation 8. No
pH=7.5, 0,1M ammonium acetate buffer /acetonitrile (1:1)
Clearox (4eq.)
0.2mg/ml, 2-4h, closed system

GLP-1R oxidation 9. No
pH=8.5, 0,1M ammonium acetete buffer 0,2mg/ml, 24-48h
air, cystein, 10°C

GLP-1R oxidation 10. No
pH=8.5, 0,1M Tris buffer
1mM GSH, 5mM GSSG, (GSH:GSSG-1:5), 1mM EDTA,
500mM L-Arginine,
0,2mg/ml, 48-72h closed system

GLP-1R oxidation 11. No
pH=8.0, 0,05M Tris buffer
1mM GSH, 1mM GSSG, (GSH:GSSG-1:1), 1mM EDTA, 1M L-Arginine, 150mM NaCl
0,2mg/ml, 48-72h closed system

Figure 1

The synthesis of a GLP-1R by gene expression using MBP

GSRPQGATVSLWETVQKWREYRRQC$_{(A)}$QRSLTEDPPPATDLFC$_{(B)}$NRTFDEYAC$_{(C)}$WPDGEPGSFVNVSC$_{(D)}$-PWYLPWASSVPQGHVYRFC$_{(E)}$TAEGLWLQKDNSSLPWRDLSEC$_{(F)}$EESKRG-NH_2

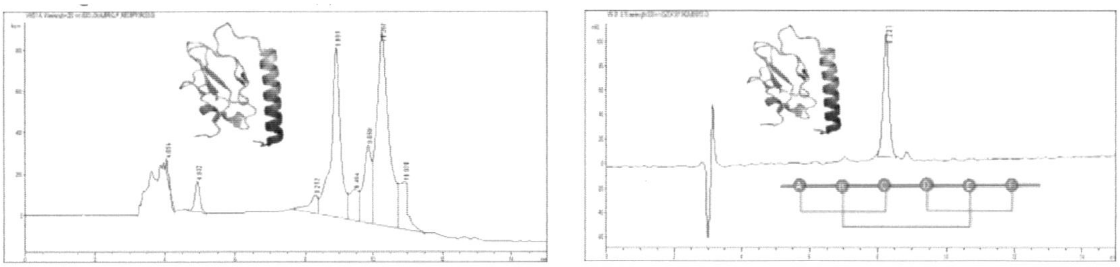

Scheme 5: The LC-chromatogram of oxidized GLP-1R containing protein mixture (right side); The LC-chromatogram of purified, oxidized GLP-1R having the desired disulfide bridges

The oxidation of linear GLP-1R obtained by native chemical ligation, and the investigation of the peptide ligand (GLP, exendine, liraglutide) interaction using NMR spectroscopic method are still in progress.

References

1. Insuela, D.B.R. and V.F. Carvalho, Glucagon and glucagon-like peptide-1 as novel anti-inflammatory and immunomodulatory compounds. European Journal of Pharmacology, 2017. 812: p. 64-72.

2. Donnelly, D., The structure and function of the glucagon-like peptide-1 receptor and its ligands. British Journal of Pharmacology, 2012. 166(1): p. 27-41.

Influence of arginine mimetics on the biological effects of NT(8-13) analogues

Silvia Michailova[1], Maya Georgieva[2], Thomas Bruckdorfer[3], Tamara Pajpanova[2]

[1]Medical University of Varna Prof. Dr. Paraskev Stoyanov, Faculty of Pharmacy, Bulgaria
[2]Roumen Tsanev Institute of Molecular Biology Bulgarian Academy of Sciences, Bulgaria
[3]Iris Biotech GmbH, Germany

https://doi.org/10.17952/35EPS.2018.257

Introduction

Neurotensin(NT) is a 13 amino acid containing neuropeptide expressed in both central nervous system (CNS) and periphery (gastrointestinal tract and the cardiovascular system). The biological effect is result from specific interactionsof the peptide with three different receptors (NTS1, NTS2 and NTS3/sortilin). The basic sequence for modifications was its C-terminal hexapeptide NT(8–13) (Arg^8-Arg^9-Pro^{10}-Tyr^{11}-Ile^{12}-Leu^{13}). Structure–activity studies on NT show that this fragment is equipotent or even more potent than the native NT(1–13) in binding to the NT receptor, while maintaining the same biological and pharmacological properties.

Besides their numerous central and peripheral functions it was reported that NTRs are over expressed in various human tumours. NT peptides are known to influence cell proliferation and pass through blood-brain barrier. Like many other neuropeptides, the main drawback in the use of NTs is extremely short half-life as a result of their rapid degradation by peptidases. To overcome this problem, various neurotensin analogues were synthesized.

Herein, we present the synthesis and biological properties of a small peptide library of the new neurotensin ligands 1–9 incorporating arginine mimetics (Figure 1).

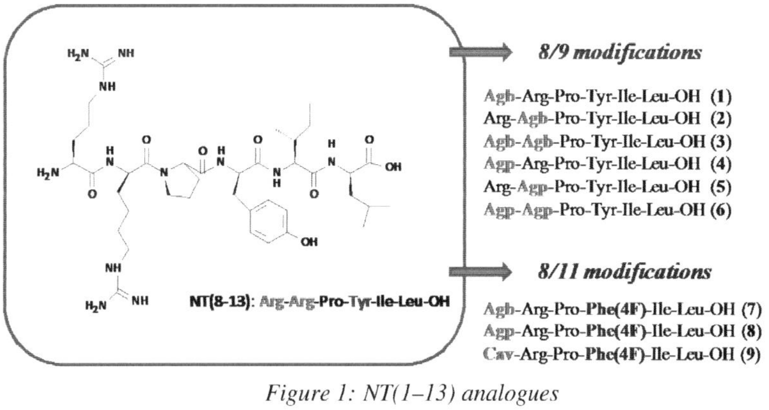

Figure 1: NT(1–13) analogues

Results and Discussion

Structurally, NT(8-13) exhibits the so-called "hot spots" that are suitable for further modification. Structure–activity studies on NT(8-13) shown that the last C-terminal residue of the peptide could not be removed or amidated without a complete loss of biological activity. The two guanidino groups of Arg^8-Arg^9, the hydroxyphenyl group of Tyr^{11}, the side chains of Ile^{12}-Leu^{13} and the carboxylic group of Leu^{13} are the functional groups of NT(8–13). The Arg^9 appeared to be more critical to the activity of NT(8-13) than was Arg^8. Furthermore, it was well known that the cleavage sites are located between Arg^8-Arg^9, Pro^{10}-Tyr^{11} and Tyr^{11}-Ile^{12}.

Previously, we replaced the terminal arginine units with its non-proteinogenic analogue canavanine (Cav) [1]. These modification resulted in the discovery of enzymatically and hydrolytic (in different values of pH, available physiologically) stable and more potent analogues [2, 3].

To learn more about the biological activity of NT(8-13), we envisioned the synthesis of target peptides 1-6 (Figure 1) incorporating shorter homologues of arginine - α-amino-4-guanidino-butyric acid (Agb), and α-amino-3-guanidino-propionic acid (Agp). Further replacement of the Tyr^{11} with Phe(pF) was also achieved (peptides 7-8). All peptides were synthesized using standard solid-phase protocol by the Fmoc/^{t}Bu strategy.

Elongation of the peptide chain was done by repetitive cycles starting from commercially available Leu-2-CT Resin. The crude peptides were precipitated into cold diethyl ether. Then, the precipitate was dissolved in 10% CH^3COOH and desalted by gel filtration on a Sephadex G25.

The yields of each product were over 60%. HPLC and CE (Figure 2) analysis revealed a purity of >95% for all peptides.

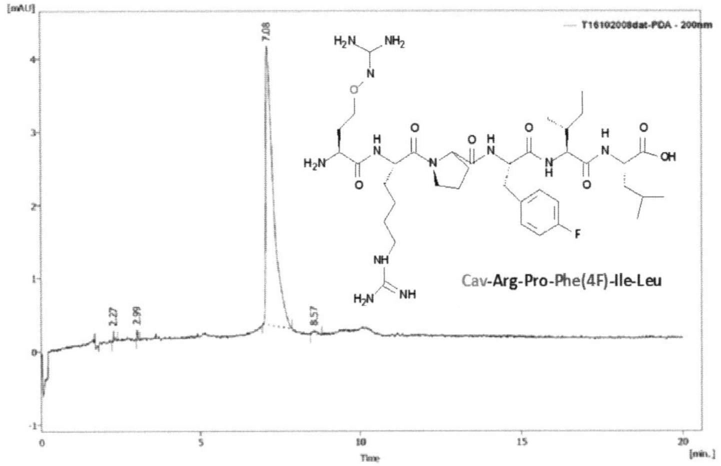

Figure 2: Electropherogram of NT(8-13), analogue 9; Chromatographic conditions: Effective capillary length: 340/407 mm, silica fused; BGE: 20 mMTris, 5mM H3PO4, 50 mM SDS, pH 8,6, U = 25KV, I = 25µA, T 23oC, UV 200 nm

The biological activity of the new NT(8-13) analogues is currently under investigation. Received preliminary data from MTT assay of the parent molecule tested in a wide concentration range (2 - 0.03 mM) on MCF-7 (breast cancer cells) and 3T3 (non-cancerous cells) after 72 hours, showed slight influence on viability over both 'normal' and cancer cell lines.

Acknowledgments

The study was supported by National Science Fund, Ministry of Education and Science, Bulgaria, Project: DN19/17, 20.12.2018 and by BAS, Project: DFNP-17-181.

References

1. Dzimbova, T. *et al.* In: Naydenova, E., Danalev, D,, Pajpanova, T (eds) Peptides 2014. Proceedings of the 33rd European Peptide Symposium, Sofia, Bulgaria, p 260–261. ISBN: 978-619-90427-1-7.

2. Michailova, S., *et al.* Bulg. Chem.Commun.,49(E), 113-117 (2017).

3. Lazarova, M., *et al.* J. Mol. Neurosci., DOI: 10.1007/s12031-018-1171-6 (2018).

In vitro assessment of the cytotoxic effects of novel RGD analogues and conjugates

Anelia Balacheva[1], Momchil Lambev[2], Roumyana Detcheva[1], Thomas Bruckdorfer[3], Tamara Paypanova[4]

[1]Roumen Tsanev Institute of Molecular Biology, Bulgarian Academy of Sciences, Bulgaria
[2]Medical University of Varna Prof. Dr. Paraskev Stoyanov, Faculty of Pharmacy, Bulgaria
[3]Iris Biotech GmbH, Germany
[4]Institute of Molecular Biology "Roumen Tsanev", Bulgarian Academy of Sciences, Bulgaria

https://doi.org/10.17952/35EPS.2018.259

Introduction

The amino acid sequence L-arginyl-glycyl-L-aspartic acid (RGD) plays role in interaction of many adhesion proteins by connecting with $\alpha v\beta$ 3 and $\alpha v\beta$ 5 integrin receptors that are overexpressed on various cancer and angiogenic endothelial cells [1]. Synthetic RGD peptides may affect adhesion, and tumour metastasis, or directly induce apoptosis by activating caspases [2].

Delivering peptides with antitumour action directly into cancer cells is one of the most important strategies in recent decades. Drug design based on the RGD structure may provide opportunity for targeted drug delivery and imaging, as well as a new chemotherapeutic treatment for cancer [3].

Herein, we present the synthesis of 9 novel RGD modified peptides and assessment of anti-proliferative activity of some of them.

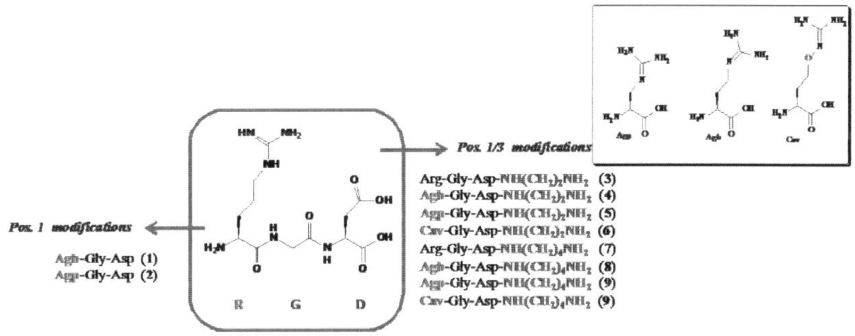

Figure 1: RGD analogues

Results and Discussion

In order to optimize physicochemical properties (stability, solubility), BBB permeability and cytotoxicity of the parent RGD molecule, the following steps we performed: First we prepared two analogues (1,2) (Fig. 1) containing the shorter homologues of Arg (α-amino-4-guanidino-butyric acid (Agb), and α-amino-3-guanidino-propionic acid (Agp); Next we synthesized a number of linear RGD analogues modified at 1st position with the arginine mimetics Agb, Agp and Cav, as well as at C-terminus with EDA and DAB (Fig. 1).

All peptides were synthesized using standard solid-phase protocol with the Fmoc/tBu strategy with the help of Activo-P14 synthesizer. For the synthesis of (3-9) commercially available EDA-2-CT Resin and DAB-2-CT Resin was used. The crude peptides were precipitated into cold diethyl ether; the precipitate was dissolved in 10% CH_3COOH and desalted by gel filtration on a Sephadex G25. The purity of peptides was analyzed by analytical RP- HPLC and CE. The yields of each product were over 60%. HPLC and CE analysis revealed a purity of >95% for the all peptides.

Next we analyzed the cytotoxic activity of RGD-analogues (1, 2) and the parent RGD molecule over two cancer cell lines by colorimetric assay (MTT) in a wide concentration range (2000 – 31.5 μM). The assay was performed on HT-29 and PC-3 cell lines, 72 h after the treatment. The cells were seeded in a concentration of 3x10^4 cells/ml. Optical density (OD) was determined at a wavelength of 550 nm and a reference wavelength of 630 nm (Fig.2).

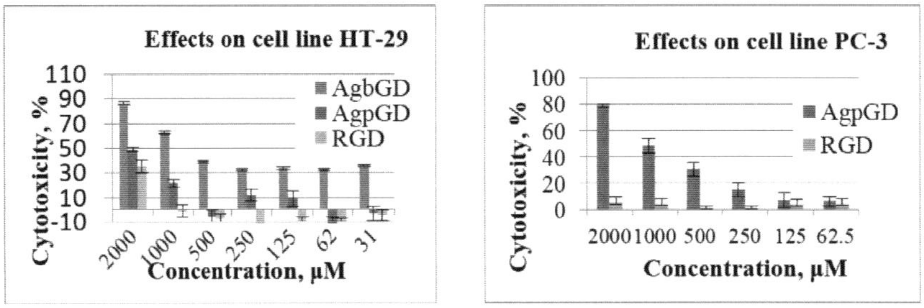

Figure 2: Comparative graph of cytotoxic activity of the tested peptides after 72h treatment

We found that the substitution of Arg with its analogues increased the cytotoxic effect of RGD-mimetics in comparison to parent RGD. The highest cytotoxic effect showed AgbGD on HT-29 cells. AgpGD showed significant increase of the cytotoxic effect on PC-3 cells in comparison to parent RGD molecule.

The combination of analgesia and antitumor activity with an absence of toxicity is highly appealing from clinical point of view, and broadens the therapeutic potential and application of RGD mimetics. Therefore, in order to obtain structures of dual-target peptide we synthesized the following endomorhin-2/RGD conjugate (Fig. 3):

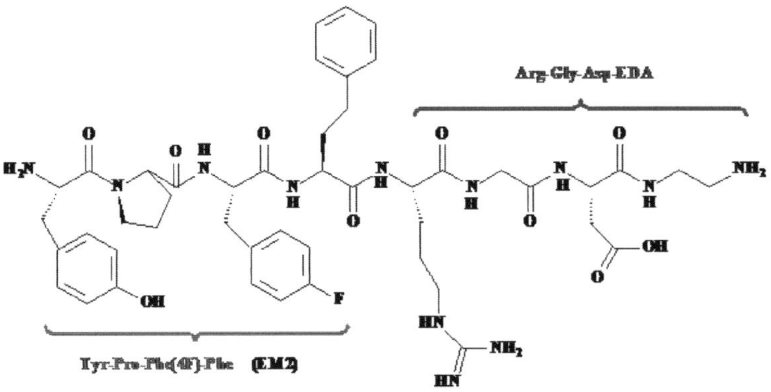

Figure 3: Endomorhin-2/RGD conjugate

The cytotoxic activity of the conjugate and the rest of the RGD analogues, as well as their analgesic activities are currently under investigation.

Acknowledgments

The study was supported by National Science Fund, Ministry of Education and Science, Bulgaria; Project, DN19/17, 20.12.2017.

References

1. Zamora, A., *et al.* Chem. Med. Chem. 2018, 10.1002/cmdc.201800282.

2. Danhier, F., *et al.* Mol. Pharmaceutics 2012, 9, 2961-2973.

3. Lee, M., Chem. Soc. Rev. 2018, 47, 28-52.

Development of murine model of fungal keratitis to test the antifungal efficacy of CPP conjugated Natamycin

Harsha Rohira[1], Sujithra Shankar[1], Sushmita G Shah[2], Shikha Yadav[3], Aastha Jain[4], Archana Chugh[1]

[1]Indian Institute of Technology Delhi, India
[2]Dr CM Shah Charitable Trust-Netra Mandir, Mumbai, India
[3]National Institute of Biologicals, Noida, India
[4]Karlsruhe Institute of Technology, Germany

https://doi.org/10.17952/35EPS.2018.261

Abstract

Corneal diseases including microbial keratitis are among the major causes of visual impairment and blindness worldwide. About 50% of the total microbial keratitis is caused by various fungal species. Fusarium and Aspergillus species are found to be the most common isolates in India. Natamycin is the only FDA approved drug which is used as a first line of treatment for fungal keratitis. Drawbacks associated with the use of natamycin is poor intraocular penetration and poor water solubility. Cell penetrating peptides (CPPs) are positively charged short peptides that can translocate across the cell membrane without damaging the cell. In our study, one of the CPP i.e. TAT is used as a nanocarrier to deliver an antifungal compound, Natamycin inside the corneal cells. Recently, we have successfully shown the increased uptake of TAT dimer conjugated natamycin by corneal cells *in vitro*. Also, conjugated natamycin showed increased water solubility as well as antifungal activity in comparison to natamycin alone. To further investigate the antifungal activity of CPP conjugated natamycin (TAT_2-Natamycin) *in vivo*, we have developed a murine model of *Fusarium keratitis*. Immunocompromised female BALB/c mice were inoculated with different concentrations of *Fusarium sp.* spores and were clinically graded (0 to 4 based on the severity of the disease) after three days of infection. Based on clinical grading and microbiological examination, 10^5 spores/5μl was found to be the optimum concentration for the establishment of fungal keratitis. We are currently testing TAT_2-Natamycin on this animal model for its antifungal activity and the results will be compared with the marketed formulation of natamycin (Natamycin 5% suspension).

Introduction

According to WHO, 285 million people are estimated to be visually impaired worldwide out of which 39 million are blind[1]. Among the various causes of blindness, most common causes include glaucoma, cataract and trachoma[2]. Other causes which also contribute to global blindness includes keratitis, conjunctivitis and endophthalmitis. Microbial keratitis is caused by variety of bacteria, fungi and viruses. Fungal keratitis accounts for 50% of the total microbial keratitis. In India, fungal keratitis is most commonly caused by *Fusarium sp.* and *Aspergillus sp.* [3–5]. To treat fungal keratitis caused by filamentous fungi, 5% sterile suspension of natamycin is used which is an FDA approved first line of treatment[6]. However, due to the poor intraocular penetration ability of natamycin, its bioavailability reduces in the corneal tissue. Hence, to increase the bioavailability, corneal epithelium is debrided[7]. Despite having so much of pharmacological developments, we still lack in good ocular drug delivery system. Currently available formulations and delivery strategies are based on the use of nanocarriers. However, precorneal (solution drainage, blinking, tear film, lacrimation etc) and anatomical factors (corneal epithelium) greatly limit the drug bioavailability to the anterior segment as well as posterior segment of the eye[8]. Advancements in genomics and biotechnology have led to the discovery of short chain peptides which can translocate cargos intracellularly. These short chain peptides, termed as Cell penetrating peptides (CPPs), are the delivery vectors used for transporting various hydrophilic drugs and nucleic acid across the cells[9].

Most known CPPs are positively charged sequences which can interact with negatively charged plasma membrane. By virtue of their electrostatic interaction with the plasma membrane, they are able to deliver cargo inside the cells. Some of the common examples include TAT_p, Penetratin, Polyarginines, MAP, Transportan etc. Although their mechanism of entry inside the cells remains highly debated, they are still used to deliver numerous pharmaceutical molecules inside various types of cells. In our previous study[10], we have successfully shown cellular uptake as well as enhanced antifungal activity of TAT_2 conjugated Natamycin in human corneal epithelial cell line. Now to further test the antifungal efficacy *in vivo*, we have developed a murine model of fungal keratitis using clinical isolate of *Fusarium dimerum*.

Materials and Methods

Fungal culture: The strain of fungus used in this study was *Fusarium dimerum*, a human clinical isolate obtained through corneal scraping. Fungus was grown on Potato Dextrose Agar (PDA) at 27°C for 5-7 days and the spores were harvested in potato dextrose broth (PDB). Spores were diluted in PDB to yield concentrations of 10^4 CFU/5μl, 10^5 CFU/5μl and 4X10^5 CFU/5μl.

Animals: Adult female inbred BALB/c mice of age 6-8 weeks were taken for the study. All animals were treated in accordance with the ARVO protocol for the Use of Animals in Ophthalmic and Vision Research. Animals were immunosuppressed with intraperitoneal injection of cyclophosphamide at 180mg/Kg body weight 5 days, 3 days and 1 day prior to inoculation. The mice were anaesthetized with isoflurane. Scarification of cornea of the right eye of mice was done with the help of hypodermic needle under dissecting microscope. Mice were divided into 3 groups according to the inoculum concentrations. 5μl of the inoculum was applied on the scarified eye followed by rubbing of eyelids for about 30 seconds. After 3 days of inoculation, establishment of fungal keratitis was confirmed by Slit lamp microscopy and quantitative fungal recovery.

Clinical grading: severity of keratitis in infected animals was graded visually on Day 4 with the help of Slit Lamp Microscope and Fluorescein staining of corneal epithelial layer. Clinical grades of 0-4 were assigned on the basis of epithelial defect and corneal infiltrate (Table 1).

Quantitative fungal recovery: Enucleated eyes were individually grinded in frosted glass grinder with 0.5ml of PBS. Homogenate was serially diluted till 4-fold in PDB containing 40μg/ml Chloramphenicol and incubated at 30°C under shaking conditions. After 3 days of incubation, the last blank in the dilution series with visible fungal growth was plated on PDA plates and the colonies were counted. End point was also confirmed by sub culturing and microscopic examination.

Results

Visual scoring of keratitis: Extent of keratitis was scored on the basis of epithelial defect, corneal clouding and the size of corneal infiltrate. Clinical grading was done with the help of slit lamp microscope and fluorescein strips (Table 2). Out of 3 concentrations, 10^4 CFU/5μl of inoculum (Group 1) was unable to establish keratitis successfully (only Grade 0 and 1). Concentrations of 10^5 CFU/5μl (Group 2) and 4X10^5 CFU/5μl (Group 3) resulted in the successful establishment of keratitis (Grade 2 and 3). However, presence of peripheral hemorrhage, corneal perforation and significant inflammation were seen in the group receiving highest concentration of inoculum. Epithelial defect was confirmed by visualizing fluorescein stained cornea under blue light.

Table 1: Criteria for clinical grading of fungal keratitis

Grade	Criteria
Grade 0	No epithelial defect or infection
Grade 1	Signs of epithelial defect +/- infiltrate
Grade 2	Corneal infiltrate covering 25-50% of corneal surface
Grade 3	Corneal infiltrate covering 50-75% of corneal surface
Grade 4	Corneal infiltrate covering >70% of corneal surface

Table 2: Severity of Fusarium keratitis with different inoculum size

Inoculum Size	Results		
	Grade	Observation	Image
1) 10^4 CFU/5µL	Grade 0, 1	No colony on PDA plate, mice were comparatively healthy with slight corneal epithelial defect.	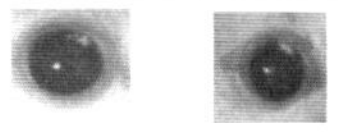 Grade 0 Grade 1
2) 10^5 CFU/5µL	Grade 2, 3	Fungal lawn on PDA plate, mice were weak with fungal infiltrate.	 Grade 2 Grade 3
3) $4x10^5$ CFU/5µL	Grade 2, 3	Fungal lawn on PDA plate, mice were weak with fungal infiltrate, corneal perforations and peripheral hemorrhage.	 Grade 2 Grade 3

Quantitative fungal recovery: *Fusarium sp.* infected eyes were enucleated and the homogenate was serially diluted in PDB containing chloramphenicol. After 3 days of inoculation at 30°C under shaking conditions, visible growth was seen in all three groups. No fungal colonies on PDA plates were seen in case of group 1 whereas fungal lawn was observed in both group 2 and group 3. Presence of fungal spores in broth was also confirmed microscopically (Figure 1).

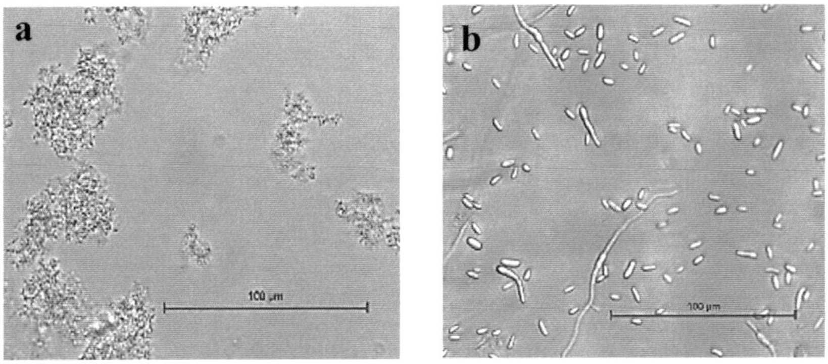

Figure 1: Microscopic images of wet mount prepared from serially diluted cultures (a) control eye (b) Fusarium sp. Spores and mycelia observed in culture obtained from the infected eye.

Conclusions

For murine model, successful establishment of infection was seen in concentrations 10^5 CFU/5μL and $4x10^5$ CFU/5μL after immunosuppression with cyclophosphamide. However, animals infected with highest concentration of fusarium spores showed corneal perforations resulting in peripheral hemorrhage. Establishment of infection was also confirmed by microscopic analysis of fungal spores in ocular homogenate. No fungal spore was detected in animals infected with lowest concentration of fusarium spores. Based on the above observations, concentration of 10^5 CFU/5μL was selected as the effective inoculum size to be used in further studies.

References

1. WHO | Visual impairment and blindness. at <http://www.who.int/mediacentre/factsheets/fs282/en/>

2. Roodhooft, J. M. J. Leading causes of blindness worldwide. Bull. Soc. belge Ophtalmol. 283, 19–25 (2002).

3. Tuft, S. J. Suppurative keratitis. Br. J. Ophthalmol. 87, 127 (2003).

4. Rajappa, S. A. & Shaji, J. A Clinical Study of Suppurative Keratitis. Int. J. Biomed. Res. 6, 869–873 (2015).

5. Gopinathan, U. et al. The epidemiological features and laboratory results of fungal keratitis: a 10-year review at a referral eye care center in South India. Cornea 21, 555–559 (2002).

6. Ansari, Z., Miller, D. & Galor, A. Current Thoughts in Fungal Keratitis: Diagnosis and Treatment. Curr.

Fungal Infect. Rep. 7, 209–218 (2013).

7. M. O' Day, D., Steven Head, W., D. Robinson, R. & A. Clanton, J. Corneal penetrationof topical amphotericin B and natamycin. Curr. Eye Res. 5, 877–882 (1986).

8. Gaudana, R., Ananthula, H. K., Parenky, A. & Mitra, A. K. Ocular drug delivery. AAPS J. 12, 348–360 (2010).

9. Stewart, K. M., Horton, K. L. & Kelley, S. O. Cell-penetrating peptides as delivery vehicles for biology and medicine. Org. Biomol. Chem. 6, 2242–2255 (2008).

10. Jain, A., Shah, S. G. & Chugh, A. Cell penetrating peptides as efficient nanocarriers for delivery of antifungal compound, natamycin for the treatment of fungal keratitis. Pharm. Res. 32, 1920–1930 (2015).

Fluorocarbon-peptideconjugates (FPC): new concept to increase the metabolic stability of peptides for therapeutic applications

Lucie Esteoulle[1], Adrien Flahault[2], Cendrine Seguin[3], Sylvie Fournel[1], Benot Frisch[1], Xavier Iturrioz[2], Catherine Llorens-Cortes[2], Dominique Bonnet[3]

[1]University of Strasbourg, France
[2]INSERM - Collège de France, France
[3]CNRS - University of Strasbourg, France

https://doi.org/10.17952/35EPS.2018.265

Introduction

During the last decade, peptides have gained a wide range of applications in medicine. However, they are often not directly suitable for use as convenient therapeutics because they have intrinsic weaknesses, including poor chemical and physical stability, and short *in vivo* half-life.[1] To increase their metabolic stability, various strategies have been developed that all present some limitations. Among them, the acylation of peptides with hydrocarbon chains (H-chains) such as a palmitoyl moiety has been developed to extend their half-life in the blood circulation as reported for liraglutide.[2] However, H-chains increase also the association with cell membranes as well as the hemolytic activity and the potential toxicity of modified peptides, such as observed for antimicrobial lipopeptides. As a consequence, the development of an alternative approach to increase the metabolic stability of peptides is needed.

While the grafting of H-chains onto peptides has been reported to enhance the circulation half-life of peptides, the incorporationof fluorocarbonchains (F-chains) within the sequence of potential therapeutic peptides has been sparsely described to date.[3, 4, 5] However, F-chains have unique properties, very different from those of H-chains. They are especially more stable, stiffer, lipophobic, and significantly more hydrophobic. In addition, F-chains are biologically inert molecules devoid of intrinsic pro-inflammatory, immuno-stimulatory activity or immunogenicity.[6] Considering those properties, we envisioned the grafting of F-chains onto peptides especially those targeting GPCRs that represent the target of more than 30% of drugs currently on the market. To test our novel approach called Fluorocarbon-Peptide Conjugation (FPC), the neuro-vasoactive peptide, apelin, previously reported for its low *in vivo* half-life (in the minute range) has been selected as a model. Apelin was isolated from bovine stomach extracts and identified as the natural ligand of the human orphan GPCR, APJ (ApelinR).[7] In the last decade, apelin and its receptors have emerged as playing a prominent role in controlling body fluid homeostasis and cardiovascular functions, with potential interesting therapeutic application in the treatment of cardiovascular diseases, water retention and/orhyponatremia.[8]

Results and Discussion

Herein, we describe the unprecedented FPC approach based on the grafting of F-chain onto apelin for improving its metabolic stability. To highlight the advantages of using F-chains over H-chains, fluoroapelin and lipoapelin were synthetized and their plasma stability, binding affinity for ApelinR, *in vitro* cytotoxicity and pro-inflammatory activities as well as their *in vivo* efficacy to decrease blood pressure in normotensive rats were carefully evaluated. Fluoroapelin were synthetized on solid-phase following a Fmoc/*t*Bu approach. Based on the molecular modeling of the three-dimensional structure of the humanApelinR, we decided to incorporate a F-chain ($CF_3(CF_2)_7(CH_2)_2CO^-$, F8-chain) on the *N*-terminal part of apelin. To highlight the advantages of using a F-chain over a H-chain, lipoapelin was obtained as control following the same approach.

The potential impact of F8-chain on the binding affinity of fluoroapelin for ApelinR was assessed by competition binding experiments using CHO cell membranes stably expressing the human ApelinR and $[^{125}I]$-pE13F as radioligand. Hence, as shown in Table 1, the affinity of fluoroapelin remained in the nanomolar range, similarly to that measured for apelin demonstrating the low impact of F8-chain on the binding properties of fluoroapelin. The plasma stability of fluoroapelin was then measured in the presence of different plasma species (mouse, rat and human)at 37 °C demonstrating that the grafting of F8-chain enabled to dramatically increase the half-life of apelin from few minutes to more than 24 h. Similar results were obtained with lipoapelin (Table 1).

Table 1: Structure, binding affinities and plasma stabilities of apelin, fluoroapelin and lipoapelin

R_1-KFRRQRPRLSHKGPMPF-OH

Peptides	R^1 group	ApelinR binding, K_i (nM)	Half-life in plasma (min)
Apelin	-	0.060 ± 0.010	4.6 ± 0.6
Fluoroapelin	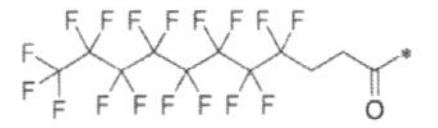	0.210 ± 0.027	>1440
Lipoapeline	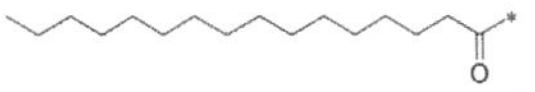	0.760 ± 0.130	>1440

To highlight the advantages of using the FPC approach vs the classical lipidation approach, the cell toxicity of fluoroapelin and lipoapelin was evaluated on RAW 264.7 macrophages by MTS-based assay (Figure 1a). Interestingly, fluoroapelin, as well as apelin, did not show any significant increase of cytotoxicity for concentrations up to 30 μM whereas lipoapelin displayed a high cytotoxicity even at low concentrations. In addition, lipoapelin exhibited higher hemolytic and pro-inflammatory effects than fluoroapelin (data not shown) demonstrating the higher biocompatibility of fluoropeptides as compared to lipopeptides. Finally, the efficacy of fluoroapelin was evaluated *in vivo* on arterial blood pressure (BP) in alert normotensive Sprague Dawley rats (Figure 1b). The intravenous injection of apelin (15 nmol/kg) did not significantly decrease arterial BP. In contrast, the intravenous injection of fluoroapelin and lipoapelin at the same dose (15 nmol/kg) maximally decreased arterial BP by 59 $\pm$ 3 and 38 $\pm$ 2 mmHg, respectively. Thereby, fluoroapelin was found 2-fold and 16-fold more efficient at decreasing arterial BP than lipoapelin and apelin, respectively. The high efficiency of fluoroapelin to decrease arterial BP compared to apelin is due both to an increase in the intensity and duration of the hypotensive response.

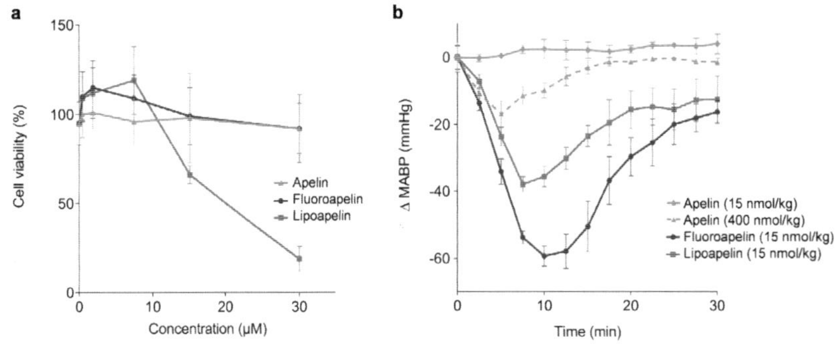

Figure 1: (a) Cytotoxicity and (b) in vivo effects on arterial blood pressure in alert normotensive rats of apelin, fluoroapelin and lipoapelin

Conclusion

As demonstrated for apelin, fluoropeptides are expected to decrease the hemolytic, cytotoxic and pro-inflammatory effects of lipopeptides and to improve their specificity of action both *in vitro* and *in vivo*. Compared to the classical lipidation of peptides, the FPC strategy offers therefore a more biocompatible, specific and efficient approach for enhancing the half-life of peptides and to facilitate their development as powerful pharmacological probes and/or drug candidates.

Acknowledgments

This work was supported by the Agence Nationale de la Recherche, the Centre National de la Recherche Scientifique, the Institut National de la Santé et de la Recherche Médicale (INSERM), the Université de Strasbourg and the Frontier Research in Chemistry (FRC).

References

1. Vlieghe P., Lisowski V., Martinez J. Drug Discov. Today 2010, 15, 40-56.

2. Buse J.B., *et al.* Lancet 2009, 374, 39-47.

3. Francis J.N. *et al.* Vaccine 2015, 33, 1071-1076.

4. Malik L., *et al.* Langmuir 2012, 28, 593-603.

5. Gerbier R., *et al.* FASEB J. 2017, 31, 687-700.

6. Krafft M.P. and Riess J.G. Chem. Rev. 2009, 109, 1714-1792.

7. Tatemoto K., *et al.* Biochem. Biophys. Res. Commun. 1998, 251, 471-476.

8. O'Carroll A.M., Lolait S.J., Harris L.E. and Pope G.R. J. Endocrinol. 2013, 219, R13-R35.

Electric field assisted inhibition of aggregation in amyloidogenic proteins

Gaurav Pandey, Vibin Ramakrishnan

Indian Institute of Technology Guwahati, Assam, India

https://doi.org/10.17952/35EPS.2018.268

Introduction

Tauopathies represent a group of progressive neurodegenerative disorders, associated with pathological aggregation of tau protein into neurofibrillary tangles. These neurodegenerative disorders are a significant unmet medical need because of their prevalence, severity, the cost associated with them, and lack of mechanism-based treatments. The failure of invasive molecular therapies has resulted in a shift towards developing relatively inexpensive and non-invasive approaches to impede the progression of tauopathies. The present study describes the application of the electric field for retarding the aggregation of core peptide segment of tau protein (Acetylated-VQIVYK-amide/PhF6) (Figure 1 (a),(b)). VQIVYK has been reported to be crucial for aggregation tau protein and is often used as model peptide system for developing aggregation inhibitors specific to tau protein. Our results utilising electric field as an aggregation inhibitor are in good agreement with the previously reported experimental [1-3] and theoretical [4,5] investigations on amyloid forming systems.

Method

Peptide synthesis using Fmoc chemistry -> Peptide -> Characterization by HPLC & MALDI -> Aggregation assay under ambient as well as electric field conditions -> Qualitative estimation of extent of aggregation using: Thioflavin T fluorescence (ThT) assay, static light scattering assay transmission; morphology examined by electron microscopy (TEM).

Results and Discussion

The model peptide PhF6 was synthesized by solid phase peptide synthesis and characterized for purity using HPLC and MALDI-TOF (Figure 1 (c), (d)). For testing our hypothesis, the PhF6 peptide sample was allowed to aggregate under the presence and absence of AC as well DC electric fields of varying strengths (Figure 1 (b)), for 16 hours. After 16 hours, the electric field was turned off. Both treated and untreated samples were analyzed for the extent of aggregation using ThT fluorescence and static light scattering assay. The degree of aggregation is said to be directly proportional to fluorescence and scattering intensities. We observed that samples allowed under AC/DC electric field showed reduced ThT fluorescence (Figure 1 (e)) and scattering intensity (Figure 1 (f)) compared to the untreated (0 Vcm^{-1}) sample. These observations indicate that electric mediated stress arrested the aggregation of the PhF6 peptide. The samples were examined under TEM microscopy, to investigate the effect of electric field induced modulation on the morphology of PhF6 aggregates. We observed that the untreated sample showed fibrillar structures in contrast the electric field treated samples displayed sphere-like structures.

Conclusions

We infer that electric field interceded stress can retard the aggregation and self-assembly of the core peptide segment of tau protein. Electric field mediated perturbationcan also modulate the morphology of aggregated structures and prevent the formation of toxic fibrillar assemblies. Initial results from this study showing the effect of electric field on the model peptide system are promising. Rigorous toxicity and side-effects related assessments are necessary to employ it as a cost-effective and non-invasive therapeutic against tauopathies.

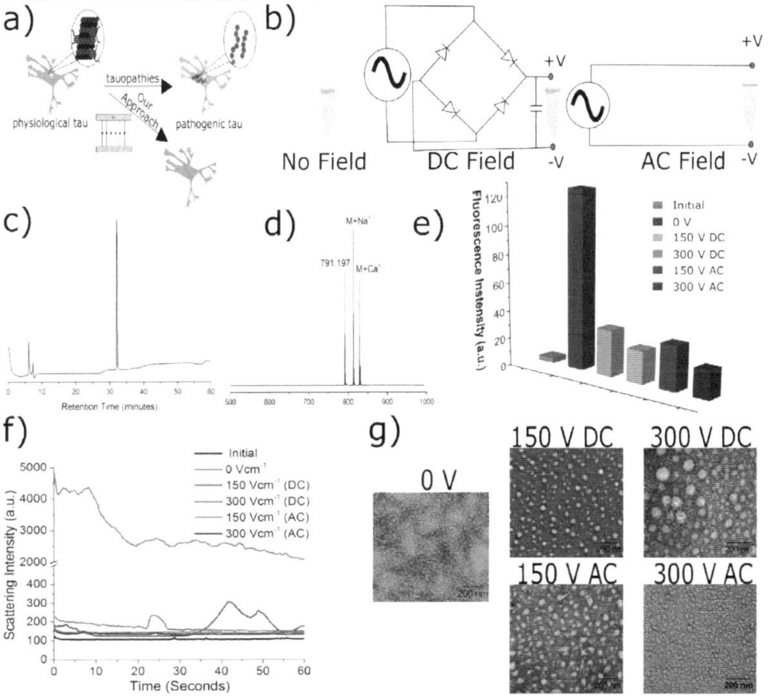

Figure 1: Schematic illustration (a) of our approach, (b) experimental set-up, charaterization of purity of synthesized peptide by (c) HPLC and (d) MALDI-TOF, quantification of extent of aggregation of peptide samples allowed to aggregate under electric field and ambient conditions by (e) ThT fluorescence assay and (f) static light scattering assay, (g) electron micrographs of electric field treated and untreated samples

Acknowledgments

The authors acknowledge Departmentof Biotechnology (grant No.BT/565/NE/U-Excel/2016) and Board of Research in Nuclear Sciences (grant no. 35/14/07/2017-BRNS/35090), Govt. of India, for financial assitance and Central Instruments Facility, IIT Guwahati for instrumentation support.

References

1. Pandey, N. K., Mitra, S., Chakraborty, M., Ghosh, S., Sen, S., Dasgupta, S., and Das Gupta, S., J. Phys. D., 2014, 47 (30), 305401. DOI:10.1088/0022-3727/47/30/305401

2. Pandey, G., Saikia, J., Sasidharan, S., Joshi, D. C., Thota, S., Nemade, H. B., Chaudhary, N., and Ramakrishnan, V., Sci. Rep., 2017, 7 (1), 2726. DOI:10.1038/s41598-017-02609-z

3. Ramakrishnan, V, Pandey, G., Nemade, H. B., Saikia, J., Sasidharan, S. and Chaudhary, N. Patent No. PCT/IN2018/050452.

4. Ojeda-May, P., and Garcia, M. E., Biophys. J., 2010, 99 (2), 595-599. DOI:10.1016/j.bpj.2010.04.040

5. Baumketner, A., J. Phys. Chem. B., 2014, 118 (50), 14578-14589. DOI: 10.1021/jp509213f

Triple-negative breast cancer: New potential therapeutics derived from SOCS3 protein

Sara La Manna, Daniela Marasco, Giancarlo Morelli

University of Naples Federico II, Italy

https://doi.org/10.17952/35EPS.2018.270

Suppressor of cytokine signaling (SOCS) proteins are a family of negative feedback regulators of the JAK-STAT pathway [1]. SOCS1,2,3 and CIS are the members of this family but, only SOCS1 and SOCS3, contain a kinase inhibitory region (KIR) essential for the inactivation of JAKs, bringing to the suppression of inflammatory cytokines [2]. SOCS proteins seem to play an important role in inflammatory diseases and in development and progression of cancers. Particularly, neonatal fatal inflammatory disease is associated with SOCS1 protein deficiency [3] while its over-expression in experimental autoimmune encephalitis (EAE) reduces negative effects of IFN-γ [4]; instead, the recombinant SOCS3 revealed able to inhibit triple negative breast cancer (TNBC) tumour growth and metastasis by suppressing inflammatory cytokines signaling while the proteolytic degradation of SOCS3 protein in TNBC subtype leads to the activation of inflammatory cytokines [5]. In last years, many studies have been carried out on the KIR region of SOCS1 providing both mimetic and antagonist compounds [6] [7].

Concerning SOCS3, structural and functional investigations showed that it simultaneously binds JAK2 and the cytokine receptor to which it is attached indicating the specificity of SOCS3 action and explaining why SOCS3 inhibits only a subset of cytokines. Recently, with the intention to develop new potential targets of JAK-STAT pathway, thanks to the X-ray structure of the protein complex among JAK2 kinase domain, mouse SOCS3 and gp130 phoshotyrosine peptide [8], a series of SOCS3'mimetic peptides derived from the N-terminal region of SOCS3, covering two domain, KIR and extended SH2 subdomain (ESS), located at the interface of the complex with JAK2, have been designed [9].

These peptides have been *in vitro* characterized by Circular Dichroism (CD) and Surface Plasmon Resonance (SPR) spectroscopies. In particular, the polypeptide formed by the union of KIR and ESS fragments, named KIRESS, showed in Figure 1, demonstrated to be enhanced in the helical content with respect to shorter protein fragments, KIR and ESS peptides. At the same time, in direct binding assays it revealed to have the highest affinity toward JAK2, showing a micromolar K_D.

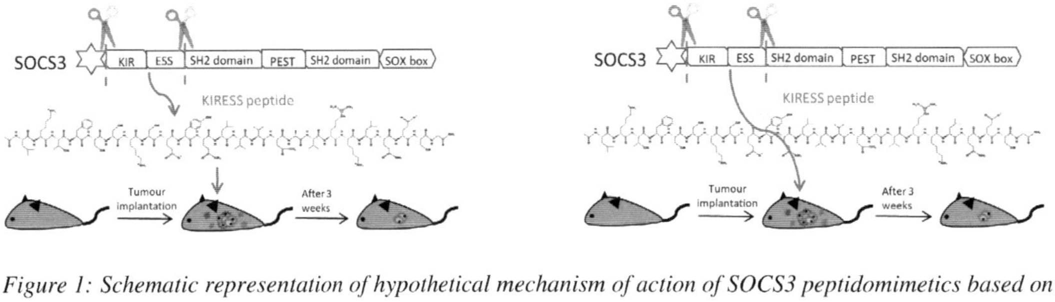

Figure 1: Schematic representation of hypothetical mechanism of action of SOCS3 peptidomimetics based on KIRESS domain.

The activity of KIRESS peptide was evaluated *in vivo* in mouse xenografts model of human TNBC subtype, MDA-MB-231-luci tumour. Interestingly, mice treated with KIRESS peptide showed a marked decrease in tumour growth. Moreover, it revealed able to completely eliminate pulmonary metastasis. At molecular level, STAT3 and NF-kB were identified as key intracellular targets of KIRESS peptide, since their phosphorylation was strongly reduced by its administration confirming that KIRESS is able to mimic the action of the entire SOCS3 protein. Fortunately, in histopathological examination of treated animals there was no observable toxicity caused by the peptide.

These overall data clearly indicate that KIRESS peptide can be considered a valid starting point to build, through structural and chemical modifications, compounds with high affinity and stability as potential therapeutics in TNBC.

References

[1] Carow B, Rottenberg ME. SOCS3, a Major Regulator of Infection and Inflammation. Frontiers in immunology 2014;5: 58.

[2] Yoshimura A, Yasukawa H. JAK's SOCS: a mechanism of inhibition. Immunity 2012;36: 157-9.

[3] Ahmed, C. M.; Dabelic, R.; Bedoya, S. K.; Larkin, J., 3rd; Johnson, H. M., A SOCS1/3 Antagonist Peptide Protects Mice Against Lethal Infection with Influenza A Virus. Front Immunol 2015, 6, 574.

[4] Balabanov, R.; Strand, K.; Goswami, R.; McMahon, E.; Begolka, W.; Miller, S. D.; Popko, B., Interferon-gamma-oligodendrocyte interactions in the regulation of experimental autoimmune encephalomyelitis. J Neurosci 2007, 27 (8), 2013-24.

[5] Kim G, Ouzounova M, Quraishi AA, Davis A, Tawakkol N, Clouthier SG, Malik F, Paulson AK, D'Angelo RC, Korkaya S, Baker TL, Esen ES, *et al.* SOCS3-mediated regulation of inflammatory cytokines in PTEN and p53 inactivated triple negative breast cancer model. Oncogene 2015;34: 671-80.

[6] Doti, N., Scognamiglio, P. L., Madonna, S., Scarponi, C., Ruvo, M., Perretta, G., Albanesi, C., and Marasco, D. (2012) New mimetic peptides of the kinase-inhibitory region (KIR) of SOCS1 through focused peptide libraries, The Biochemical journal 443, 231-240.

[7] La Manna, S.; Lopez-Sanz, L.; Leone, M.; Brandi, P.; Scognamiglio, P. L.; Morelli, G.; Novellino, E.; Gomez-Guerrero, C.; Marasco, D., Structure-activity studies of peptidomimetics based on kinase-inhibitory region of suppressors of cytokine signaling 1. Biopolymers 2017; e23082.

[8] Kershaw NJ, Murphy JM, Liau NP, Varghese LN, Laktyushin A, Whitlock EL, Lucet IS, Nicola NA, Babon JJ. SOCS3 binds specific receptor-JAK complexes to control cytokine signaling by direct kinase inhibition. Nature structural & molecular biology 2013;20: 469-76.

[9] La Manna S, Lee E, Ouzounova M, Di Natale C, Novellino E, Merlino A, Korkaya H, Marasco D. Mimetics of Suppressor of cytokine signalling 3: novel potential therapeutics in triple breast cancer. International Journal of Cancer 2018;143: 2177-2186.

Proceedings of the 35th European Peptide Symposium
Patrick B. Timmons, Chandralal M. Hewage, Michal Lebl (Editors)
272 *European Peptide Society & PSP, 2018*

Combinatorially screened peptide as targeted covalent binder

Shuta Uematsu[1], Yudai Tabuchi[1], Yuji Ito[2], Masumi Taki[1]

[1] The University of Electro-Communications, Japan
[2] Graduate School of Science and Engineering, Kagoshima University, Japan

https://doi.org/10.17952/35EPS.2018.272

Finding targeted covalent binders (also known as covalent drugs) is one of the cutting-edge disciplines such as biomedical sciences / chemical biology / pharmaceuticals fields. As shown in the figure below, we present here a novel concept for finding a peptide covalent binder from fluoroprobe-modified peptide library [1] on T7 bacteriophage constructed *via* the 10BASE $_d$-T [2], followed by a structure alteration into a reactive warhead [3]. Site- and position-specific conjugation toward the target protein and target selectivity were confirmed. This is the first demonstration for finding peptide-type targeted covalent binder by a combinatorial screening, instead of a rational designing. We believe the discovery technology would be generally applied to the discovery of novel covalent drugs for clinically-important proteins (e.g., cancer-related cell-surface proteins).

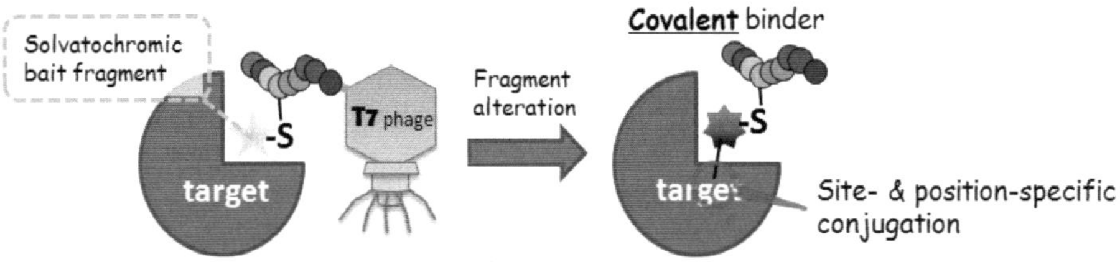

Figure 1

References

[1] Taki et al., Anal. Chem. 2016, 88, 1096; Uematsu et al., AIP Conf. Proc. 2017, 1807, 020028.

[2] Fukunaga et al., Chem. Commun. 2014, 50, 3921; inside cover article.

[3] Uematsu and Tabuchi et al., Bioconj. Chem. 2018, in press.

Hybrid peptidomimetics for the use in neuropathic pain

Ewa Witkowska[1], Jowita Osiejuk[1], Magda Godlewska[1], Sandra Gatarz[1], Beata Wilenska[1], Joanna Starnowska[2], Anna Piotrowska-Murzyn[2], Joanna Mika[2], Barbara Przewlocka[2], Aleksandra Misicka[1]

[1]University of Warsaw, Poland
[2]Institute of Pharmacology, Krakow, Poland

https://doi.org/10.17952/35EPS.2018.273

Introduction

Neuropathic pain is a type of chronic pain that results from the damage to the nerve tissue. The difficulty of treatment rely on the lack of suitably potent and long-acting drugs, because opioids successfully used in acute pain lose their analgesic efficacy in neuropathic pain. This effect is explained as a result of excessive pathological damage inducing activation of endogenous systems which cause pain intensification. The action of these systems is opposed to the analgesic effect of opioids, so the effect of these drugs is weakened. A number of such endogenous pain-enhancing systems have been described [1], among them the melanocortin system (MC), particularly the melanocortin 4 (MC4) receptor, is of great interest. Published data suggest that melanocortins are considered endogenous functional antagonists of opioids [2, 3]. Moreover, it was found that the administration of antagonists of MC4 receptor resulted in analgesia and intensify the effect of morphine [4]. Based on this and other studies on the MC4 receptor available in the literature [5], it has been hypothesized that the activation of opioid receptors while blocking MC4 receptors will improve the efficacy of opioids in the treatment of neuropathic pain. The simultaneous effect on both systems (opioid activation and melanocortin inhibition) in the same part (fragment) of nociceptive pathways may provide bifunctional hybrid compounds containing two components: the opioid receptor agonist (OP) and the MC4 receptor antagonist.

The purpose of this work was to design and synthesize hybrid peptidomimetics containing an opioid agonist and a melanocortin antagonist linked by various spacers, and to examine the analgesic activity of these compounds in neuropathic pain.

Results and Discussion

Nine new hybrid peptidomimetics containing two ligands: an enkephalin analogue (opioid agonist) [6] and SHU 9119 (an MC4 receptor antagonist) [4] linked by various spacers have been designed. Spacers in designed peptidomimetics can be classified as short (residues of D-Ala, β-Ala), flexible (residue of 6-aminohexanoic acid (Ahx)), rigid (residues of 4-aminomethylbenzoic acid (4AMB), 4-aminophenylacetic acid (4APhAc) and semi-rigid (dipeptidyl fragment of Pro-Gly [7]).

Tyr^{1}-D-Ala^{2}-Gly^{3}-Phe^{4} —-X—-Nle^{6}-c[Asp^{7}-His^{8}-D-Nal(2')^{9}-Arg^{10}-Trp^{11}-Lys^{12}]-NH_2 X= spacer

All hybrid peptides were synthesized on a MBHA resin (Bachem, 0.27 mmol/g) using the standard Boc strategy and carbodiimide (DIC) as the coupling reagent.

H_2N-**P** **P**= MBHA resin

↓ a, b, c, d, e

Boc-Nle-Asp-His(Bom)-D-Nal(2')-Arg(Tos)-Trp(For)-Lys-**P**

↓ f

Boc-Nle-c[Asp-His(Bom)-D-Nal(2')-Arg(Tos)-Trp(For)-Lys]-**P**

↓ b, g, b or e, d

Boc-Tyr(tBu)-D-Ala-Gly-Phe-X-Nle-c[Asp-His(Bom)-D-Nal(2')-Arg(Tos)-Trp(For)-Lys]-**P**

↓ b, h

H-Tyr-D-Ala-Gly-Phe-X-Nle-c[Asp-His-D-Nal(2')-Arg-Trp-Lys]-NH_2

↓

RP-HPLC purification and ESI MS analysis

Figure 1: Synthesis of hybrid peptidomimetics (a) Boc-AA, DIC, HOBt, (b) TFA, (c) DIEA, (d) steps a-c (e) piperidine, (f) DIC, HOBt, (g) Protected spacer: Boc-D-Ala or Boc-β -Ala or Fmoc-Ahx or Fmoc-4-AMB or Fmoc-4-APhAc or Boc-Gly, Boc-Pro, (h) HF

Table 1: Calculated ED50 values for effect of hybrids 1-9 and reference compounds: enkephalin analogue and SHU9119, in acute pain (tail-flick test) and neuropathic pain in CCI-exposed mice (von Frey and cold plate test). The experiments were performed on naive mice or 7-14 days after CCI procedure, all compounds were administered intrathecally (i.t.).

Compound	**ED_{50} (μg)**		
	Naive mice	**Mice subjected to CCI**	
	Tail-flick	**Von Frey (allodynia)**	**Cold plate (hyperalgesia)**
Tyr-D-Ala-Gly-Phe-NH_2 (parent opioid agonist)	0.03 (0.02 – 0.05)	0.16 (0.1 – 0.25)	9.3 (2.7 – 32)
SHU9119 (parent antagonist of MC4 receptor)	*	3.58 (0.01 – 803)	#
1 (X=D-Ala)	#	#	#
2 (X=Ahx)	#	0.0004 (0.0001 – 0.002)	0.008 (0.005 – 0.01)
3 (X= β-Ala)	104 (48 – 226)	0.08 (0.5 – 1.2)	251 (55 – 1157)
4 (X=Ahx-Ahx)	*	0.02 (0.00003 – 15)	0.2 (0.008 – 6.5)
5 (X=4AMB)	#	0.03 (0.01 – 0.1)	0.3 (0.2 – 0.4)
6 (X=4APhAc)	42 (21 – 85)	0,03 (0,0004 – 1,9)	0,04 (0,0009 – 2,2)
7 (X= Pro-Gly)	#	0,7 (0,5 – 1,1)	0,29 (0,13 – 0,63)
8 (X=Pro-Gly-Pro-Gly)	10 (0.8 – 121)	*	*
9[a] (X=Ahx)	#	0,005 (0,000005 – 6,35)	0,016 (0,01 – 0,02)

a- the residue of tyrosine was replaced by 2,6-dimethyl-L-tyrosine (DMT) residue, *- it was impossible to calculate ED50 due to lack of analgesic effect, # - it was impossible to calculate ED50 due to weak dose-related effect

References

1. Nickel F.T., Seifert F., Lanz S., Maihöfner C. (2012) Eur. Neuropsychopharmacol. 22, 81–91.

2. Contreras P.C., Takemori A.E.(1984) J. Pharmacol. Exp. Ther. 229, 21–26.

3. Székely J.I., Miglécz E., Dunai-Kovács Z., Tarnawa I., Rónai A.Z., Graf L., Bajusz S. (1979) Life Sci. 24, 1931–38.

4. Hruby V.J., Lu D., Sharma S.D., Castrucci A. de L., Kesterson R.A., Al.-Obeidi F.A., Hadley M.E., Cone R.D. (1995) J. Med. Chem. 38, 3454-61

5. a) Tao Y.-X. (2010) Endocr. Rev. 31, 506–543, b) Beltramo M., Campanella M., Tarozzo G., Fredduzzi S., Corradini L., Forlani A., Bertorelli R., Reggiani A. (2003) Mol. Brain Res. 118, 111–118, c) Starowicz K., Mousa S.A., Obara I., Chocyk A., Przewłocki R., Wedzony K., Machelska H., Przewłocka B. (2009) Pharmacol. Reports 61, 1086–95, d) Starowicz K., Obara I., Przewłocki R., Przewlocka B. (2005) Pain 117, 401–11.

6. McGregor W.H., Stein L., Belluzzi J.D. (1978) Life Sci. 23(13), 1371-76

7. Handl H.L., Sankaranarayanan R., Josan J.S., Vagner J., Mash E.A., Gillies, R. J. and Hruby V. J. (2007) Bioconjugates Chem. 18, 1101-09

Acknowledgements

This work was supported by National Science Centre grant No. 2012/06/A/NZ4/0002 (to B.P.)

Proceedings of the 35th European Peptide Symposium
Patrick B. Timmons, Chandralal M. Hewage, Michal Lebl (Editors)
276 *European Peptide Society & PSP, 2018*

A bifunctional biased Mu opioid agonist - Neuropeptide FF receptor antagonist as analgesic with improved acute and chronic side effects

Armand Drieu la Rochelle[1], Karel Guillemyn[2], Maria Dumitrascuta[3], Simon Gonzalez[2], Charlotte Martin[2], Valérie Utard[1], Raphaëlle Quillet[1], Séverine Schneider[4], François Daubeuf[4], Tom Willemse[2], Pieter Mampuys[5], Bert Maes[5], Nelly Frossard[4], Frédéric Bihel[4], Mariana Spetea[6], Frédéric Simonin[1], Steven Ballet[2]

[1]Biotechnologie et Signalisation Cellulaire, UMR 7242 CNRS, Université de Strasbourg, Illkirch, France
[2]Research Group of Organic Chemistry, Vrije Universiteit Brussel, Brussels, Belgium
[3]Opioid Research Group, Department of Pharmaceutical Chemistry, CMBI, University of Innsbruck, Innsbruck, Austria, Austria
[4]Laboratoire Innovation Thérapeutique, UMR 7200 CNRS, Université de Strasbourg, Illkirch, France
[5]Organic Synthesis, Department of Chemistry, University of Antwerp, Antwerp, Belgium
[6]Opioid Research Group, Department of Pharmaceutical Chemistry, CMBI, University of Innsbruck, Innsbruck, Austria

https://doi.org/10.17952/35EPS.2018.276

Introduction

Opioid receptors and their ligands are still considered as golden standards to alleviate moderate to severe pain. Nonetheless, this potent analgesic activity is usually paired with adverse side effects such as respiratory depression, hyperalgesia, analgesic tolerance or constipation.[1] Several studies have already demonstrated that other biological pathways and systems are involved in the development of such side effects, for instance, the recruitment of β-arrestin-2 at μ-opioid receptor (MOR) [2] and the neuropeptide FF receptors (NPFFRs).[3]

Previous work from our laboratories showed that the dermorphin-derived peptidomimetic KGOP01 (Table 1) was a high efficiency ligand at MOR, which lead to efficient pain relief *in vivo*.[4] Several studies have also demonstrated that NPFF ligands are recognized through their C-terminal part, corresponding to the dipeptide RF-NH_2.[5] Herein, we discuss the design as well as *in vitro* and *in vivo* evaluation of several hybrid peptidomimetics or multitarget drugs, resulting from the merger of a MOR agonist pharmacophore and a NPFF pharmacophore, to obtain improved side effect profiles (Figure 1).

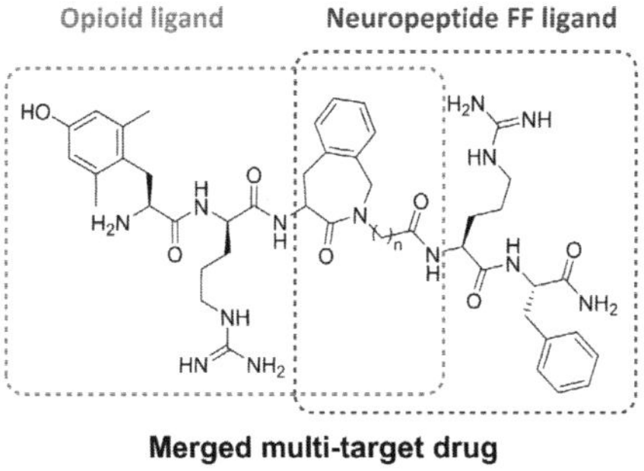

Figure 1: Example of an opioid-NPFF hybrid

Result and Discussion

In order to study the influence of changes imposed on each subunit at both MOR and the NPFFRs, several modifications were introduced in the peptide structure. Examples include the linker between the two pharmacophores (Gly or β Ala) and the basic residue within the NPFF subunit. For the latter feature, Orn and Arg derivatives were introduced based on previous studies6 and encompassed unnatural amino acids[7] (Figure 2). Finally, a merged multitarget drug design was attempted, in which the two pharmacophores were completely overlapping (e.g., H-Dmt-Arg-Phe-NH_2).

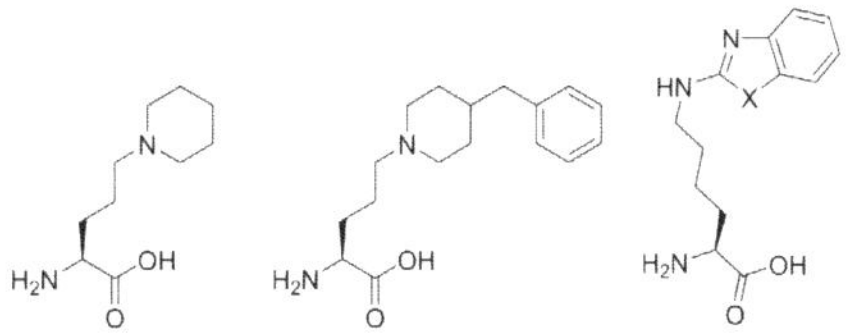

Apa Bpa X = N Lys(Bim) X = O Lys(Box) X = S Lys(Bth)

Figure 2: Structures of Orn and Arg derivatives

Although insertion of the Gly residue (KGFF01) yielded a slightly higher selectivity toward NPFF2R versus NPFF1R, as compared to the β Ala linker (KGFF03), the latter displayed a much better binding affinity at all 3 receptors (Table 1). Hence, this extra methylene group was kept for most other designed peptidomimetics. Regarding the NPFF pharmacophore's basic residue, substitution from Arg to an Orn residue (KGFF04) resulted in a reduced binding at both NPFFRs, particularly at NPFF1R. Conversely, other modifications only led to a slightly modified binding to NPFFRs and good affinity for MOR (Table 1).

Table 1: Binding affinities of selected opioid-NPFF peptidomimetics

Name	Sequence	K_i ± SEM (nM)		
		MOR	NPFF1R	NPFF2R
KGOP01	**H-Dmt-DArg-Aba-βAla-NH₂**	**0.12 ± 0.02**	**4,600 ± 1,300**	**4,500 ± 1,800**
KGFF01	H-Dmt-DArg-Aba-Gly-Arg-Phe-NH₂	0.59 ± 0.21	57 ± 10	1.2 ± 0.5
KGFF03	**H-Dmt-DArg-Aba-βAla-Arg-Phe-NH₂**	**0.24 ± 0.03**	**2.7 ± 0.3**	**0.077 ± 0.01**
KGFF04	H-Dmt-DArg-Aba-βAla-Orn-Phe-NH₂	0.67 ± 0.29	780 ± 90	45 ± 8
KGFF08	H-Dmt-DArg-Aba-βAla-Apa-Phe-NH₂	0.94 ± 0.19	136 ± 26	3.9 ± 0.1
KGFF09	**H-Dmt-DArg-Aba-βAla-Bpa-Phe-NH₂**	**2.43 ± 0.18**	**83 ± 21**	**3.2 ± 0.7**
KGFF14	H-Dmt-DArg-Aba-βAla-Lys(Bim)-Phe-NH₂	1.6 ± 0.3	4.54 ± 0.01	4.15 ± 1.35
KGFF15	H-Dmt-DArg-Aba-βAla-Lys(Box)-Phe-NH₂	1.8 ± 0.9	88 ± 8	9.6 ± 0.6
KGFF16	H-Dmt-DArg-Aba-βAla-Lys(Bth)-Phe-NH₂	2.6 ± 1.3	80 ± 13	10.4 ± 2.3

Further *in vitro* evaluation (data not shown) indicated that all designed peptidomimetics were agonists at MOR (EC50 ranging from 1.5 to 18.2 nM), with KGFF03 being the best agonist at the NPFFRs (EC50 of 84.8 and 11 nM at NPFF1R and NPFF2R, respectively). Interestingly, KGFF09 displayed a very weak activity at NPFFRs and was subsequently even characterized as an antagonist (pA2 values of 7.25 and 7.77). Evaluation of the β-arrestin recruitment *via* a functional test (BRET1 β-arrestin-2 recruitment)showed that both KGFF03 and KGFF09 were presenting a partially biased activity at MOR toward intracellular G protein signaling, conversely to the native opioid ligand KGOP01.

In order to assess their analgesic efficiency and investigate the influence of their interesting *in vitro* pharmacological features, the NPFF pharmacophore-devoid ligand KGOP01 as well as both hybrids KGFF03 and KGFF09 were evaluated *in vivo* by the tail-flick test on mice after s.c. administration. All 3 peptidomimetics induced a time- and dose-dependent antinociceptive effect with EC*50* of 0.53, 0.77 and 2.4 μmol/kg, respectively.

Chronic administration of KGOP01 and KGFF03 resulted in a significant and persistent decrease of the basal antinociceptive threshold, thus demonstrating the development of hyperalgesia (Figure 4A). This trend was not observed with the opioid agonist - NPFFR antagonist KGFF09. On the 8th day, mice treated with KGOP01 or KGFF03 showed a decreased maximal response and antinociception efficacy by 75% compared to day 1, whereas KGFF09 maintained potent analgesia (Figure 4A). These observations demonstrate that blocking the NPFFR system by an antagonist leads to a reduced analgesic tolerance development upon chronic administration.

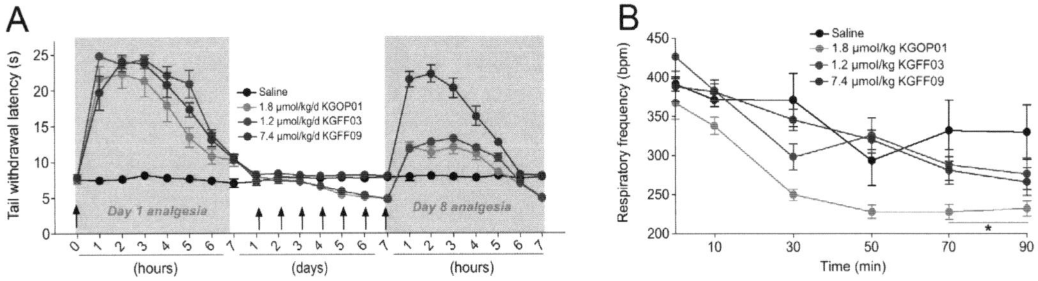

Figure 4: In vivo characterization of opioid-NPFF hybrids after chronic s.c. administration in mice. (A) Development of hyperalgesia and analgesic tolerance upon chronic treatment, (B) Effect of KGOP01, KGFF03, and KGFF09 on respiratory frequency after s.c. administration.

Finally, we studied respiratory depression, one of the major side effects that occurs upon acute opiate administration. The measurement of the respiratory frequency in mice revealed a significant respiratory depression induced by the non-biased opioid parent KGOP01, as compared to saline-treated animals (Figure 4B). This effect was not observed with either KGFF03 or KGFF09 which is in agreement with previous reports suggesting that respiratory depression is related to β-arrestin recruitment.[8]

Conclusion

Altogether, our data demonstrate the successful design of multitarget peptidomimetics merging, within a single molecule, opioid and NPFF pharmacophores. Indeed, the addition of the RF-NH_2 motif or its derivatives to our MOR ligand KGOP01 not only conferred a good NPFF1/2Rs affinity and preserved binding to MOR, but also induced a biased activity toward G protein signaling over the β-arrestin-2 recruitment. Interestingly, introducing variations of the basic residue within the NPFF pharmacophore led to two different profiles at the corresponding receptors i.e. an agonist, KGFF03, and an antagonist, KGFF09. *in vivo* experiments confirmed the benefits of one pharmacological profile, since KGFF09 displayed a potent antinociceptive activity as well as reduced respiratory depression, hyperalgesia, tolerance, and withdrawal syndrome (data not shown), when compared to the parent non-biased opioid ligand, KGOP01, or the dual opioid agonist - NPFFR agonist, KGFF03. Therefore, our data not only confirmed the relation between the NPFF system or β-arrestin-2 recruitment and opioid-inducedside effects, but also demonstrate the benefits to modulate these (signaling) systems in order to obtain safer pain treatments.

References

(1) Volkow, N. D.; McLellan, A. T., N. Engl. J. Med., 2016, 374, 1253-1263.

(2) Yekkirala, A. S.; Roberson, D. P.; Bean, B. P.; Woolf C. J., Nat. Rev. Drug. Discov., 2017, 16 (8), 545-564.

(3) Yang, H. Y.; Tao, T.; Iadarola, M. J., Neuropeptide, 2008, 42 (1), 1-18.

(4) Guillemyn, K.; Kleczkowska, P. et al., Eur. J. Med. Chem., 2015, 92, 64-77.

(5) Findeisen, M.; Rathmann, D.; Beck-Sickinger, A. G., ChemMedChem, 2011, 6, 1081-1093.

(6) Bihel, F.; Humbert, J. P. et al., ACS Chem. Neurosci., 2015, 6 (3), 438-445.

(7) Drieu la Rochelle, A.; Guillemyn, K. et al., PAIN, 2018, 159 (9), 1705-1718.

(8) Manglik, A. ; Lin, H. ; Aryal, D. K. ; McCorvy, J. D.; Dengler, D.; Corder, G.; Levit, A.; Kling, R. C.; Bernat, V.; Hübner, H.; Huand, X. P.; Sassano, M. F.; Giguère, P. M.; Löber, S.; Duan, D.; Scherrer, G.; Kobilka, B. K.; Gmeiner, P.; Roth, B. L.; Shoichet, B. K., Nature, 2016, 537, 185-190.

Design, synthesis and biological evaluation of stapled helical peptides targeting the myddosome

Victor Terrier[1], Jonas Verhoeven[2], Charlotte Martin[2], Karlijn Hollanders[2], Nicolas Triballeau[3], Patrick Mollat[3], Steven Van Der Plas[3], Gert De Wilde[3], Steven Ballet[2]

[1]Aelin Therapeutics, France
[2]Vrije Universiteit Brussel, Belgium
[3]Galapagos NV, Belgium

https://doi.org/10.17952/35EPS.2018.279

Introduction

Protein-protein interactions (PPI) are involved in the regulation of a large variety of biological processes and their malfunctioning may lead to the development of diseases. These targets, considered as largely undruggable less than a decade ago, are receiving an increased interest in the search for novel therapeutic agents.[1] An important challenge originates from the extended contact surfaces of interaction. Peptides as small protein segments can cover larger surface areas as compared to small molecules, while three-dimensional stability can be preserved *via* chemical modifications, such as macrocyclization or helical constraints.[2-4]

The target of this study is the intracellular adapter protein MyD88 which is involved in the signaling pathways of the Toll-like receptor (TLRs) and Interleukin-1 receptor (IL-1), both playing a crucial role in the innate immune and inflammatory response. The MyD88 protein consists of two domains; a TIR domain that interacts with TLRs and IL-1R, and a death domain which multimerizes with Interleukin-1 receptor-associated kinases (IRAKs) to form the myddosome.[5] A previously obtained X-ray structure of this challenging PPI target shows a left-handed helical oligomer that consists of several MyD88 proteins, IRAK4 and IRAK2 death domains, resulting in highly complex interaction networks.[6] Three types of interfaces (type I-III) are revealed when the multimeric protein is unfolded along the helical axis. The type I interface, composed by a seven amino acid long helical epitope (-[E]YLEIR[Q]-) from MyD88, is particularly interesting as it shows several interaction points with IRAK4 (highlighted in bold). Moreover, a non-occupied hydrophobic pocket on the IRAK4 side is found in front of one of the MyD88 epitope side chains. Supported by molecular modeling, rationally designed stapled peptides were synthesized as helix mimetics of the type I epitope to be evaluated as potential inhibitors of this PPI.

Results and Discussion

Short stapled peptide analogues of the MyD88 type I interface epitope (-[E]YLEIR[Q]-) have been designed using an i,i+4 macrocyclization strategy to induce a helical conformation and potentially increase *in vivo* stability and selectivity. Linear peptide substrates were synthesized *via* SPPS on Fmoc-Rink aminomethyl polystyrene resin (0.1 mmol/reactor scale) using a Liberty Lite peptide synthesizer, or were assembled through manual SPPS. Different cyclization strategies were evaluated using the linear peptide sequences on solid-support, including copper(I)-catalyzed alkyne/azide cycloaddition (CuAAC), lactamisation and ring-closing metathesis (Scheme 1). Robust reaction conditions have been obtained for the different macrocyclizations strategies, resulting in a series of different stapled peptides of which the most relevant examples are depicted in Scheme 1.

Surface plasmon resonance (SPR), used to determine the binding of the (macrocyclized) peptides to IRAK4 protein, qualitatively identified multiple compounds with binding affinity to the IRAK4 domain, with KD values estimated between 100 to 200 μM. Also, circular dichroism (CD) spectroscopy confirmed that macrocyclization successfully stabilizes alpha helicity in several of the short sequence peptides. By calculating the relative helicity of the different cyclic peptides[7], the highest helicity was observed for triazole macrocycle 2 (85%, Figure 1). According to the applied formula, this azide-alkyne cycloaddition product(2) showed a significant higher helicity compared to products 5-8 from lactamization and ring-closing metathesis cyclization strategies (helicity ranging from 42 to 58%). In attempts to further improve the potency and helicity of the cyclic CuAAC hit compound1 (Figure 1), the position of the staple was shifted to afford cyclic peptide 3, however, this was found to result in a loss of affinity compared to 1 which indicates the importance of the E1, R6 and Q7 residues for binding. With the same purpose, an Aib-scan (2-aminoisobutyric acid) was performed on the cyclic CuAAC hit compound1. This effort showed that the binding affinity could be preserved for analogue 4, although surprisingly a reduced helicity (58%) was calculated from the CD-spectrum compared to 1 (66%).

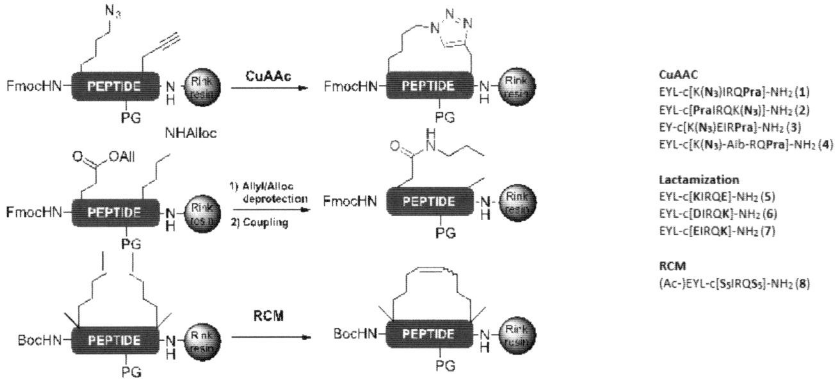

Scheme 1: Macrocyclization strategies via copper catalyzed-alkene/alkyne cycloaddition (CuAAC), lactamization and ring-closing metathesis (left) and prepared cyclic peptides (right)

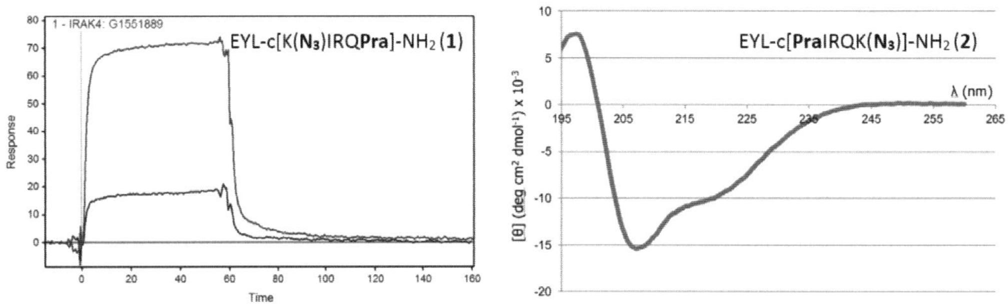

Figure 1: Surface plasmon resonance (SPR) spectrum of peptide 1, injected for 60 seconds at 20 μM (blue curve) and 200 μM (brown curve) displaying a high binding affinity to IRAK4 (left) and Circular Dichroism (CD) spectrum of 2 which showed the highest alpha helicity (right)

Conclusion

Different macrocyclization chemistries have been optimized on the type I epitope of MyD88 to design stapled, helix stabilizing peptide mimetics. Surface plasmon resonance (SPR) data showed that ten staples peptides were good binders to the IRAK4 target. Moreover, circular dichroism analysis demonstrated that the used cyclization chemistries successfully conferred helicity to the very short sequence peptide mimetics, as compared to the linear counterparts.

References

(1) Modell, A. E., Blosser, S. L., Arora, P. S., Trends Pharmacol. Sci., 2016, 37, 702.

(2) Petta, I., Lievens, S., Libert, C., Tavernier, J., De Bosscher, K., Mol. Ther. 2016, 24, 707.

(3) Nevola, L., Giralt, E., Chem. Commun., 2015, 51, 3302.

(4) Arora, P. S., J. Biomol. Struct. Dynam., 2013, 31, 120.

(5) Vyncke, L. Bovijn, C., Pauwels, E., Van Acker, T., Ruyssinck, E., Burg, E., Tavernier, J., Peelman, F. Structure, 2016, 24, 437.

(6) Lin, S. C., Lo, Y. C., Wu, H. Nature, 2010, 465, 885.

(7) Wang, D., Chen, K., Kulp, J. L., Arora, P. S. J. Am. Chem. Soc. 2006, 128, 9248.

Sequence-activity relationship analysis for peptide optimization using machine learning techniques

Hugo Villar[1], Jane Razumovskaya[2], Jason Hodges[2], Mark Hansen[2]

[1]University of California San Diego, United States
[2]Altoris, United States

https://doi.org/10.17952/35EPS.2018.281

A significant growth in the number of biologic drugs has led to noteworthy investments in research on bioactive peptides by the pharmaceutical industry. The number and diversity of peptides that are generated and evaluated in a project aimed to identify a clinical candidate has been rising, that created a need to manage ever larger data sets. The lack of adequate informatics tools to manage, analyze or visualize the data and sequences created has become evident to the point where tools used in the past, such as Excel result grossly inadequate for most projects and existing tools do not provide the analytical tools or an integrated platform to correlate sequences with trends in the data.

A few years back [1], SARvision|Biologics was developed to allow bench scientists relate a biopolymer sequence to other associated data, such as bioactivity or physical properties, regardless of the size of the data to be analyzed [1,2]. While developing the tool we were able to dissect the elements that are needed to maximize the information that can be extracted in retrospective studies of peptide series and to develop predictive models. We describe here several aspects that should be attended to create informative structure sequence relationships.

Monomer and Peptide Registration: A system to register monomers and the biopolymers is a precondition for any subsequent analysis. Chemical databases can be used to maintain a list of the monomers. In our case, monomer registration is handled *via* a flat file that contains each of the monomers, including their structure as SMILES strings, and physicochemical properties used to determine whole peptide properties, or to color the sequence according to different properties such as for example hydrophobicity. A database of over 700 amino acids that are used for research purposes is our default. The list could be significantly longer if modifications frequently made on the monomers are considered. Simply allowing for the different enantiomers would more than double the list. A second file that contains a list modifications could be used to avoid very large flat files, and operate on the residue list to generate the needed residue [1,2]. HELM notation [3] is used to register and read-in the biopolymers, but in many cases simple line notations suffice [2].

Sequence Alignments. Once sequences are read in, residues that serve equivalent functions in the sequence have to be identified, this is required to show what changes in the sequence result in changes in the peptide properties. The identification of equivalent residues in a sequence is usually done by analysis of the structure of the peptide or using sequence alignments. Clustal V [4] allows for multiple sequence alignments and is used as a default in our analysis but this approach is not suitable for unnatural amino acids. In those cases, the challenge is to create substitution matrices to appropriately score the alignments. After exploring a variety of possibilities substitution matrices based on similarity in computed physicochemical properties provide a good alternative. For short sequences global sequence alignments such as Needleman-Wunsch should be preferred to local sequence alignments such as FASTA [5].

Exploratory Data Analysis should then be carried out to inform the development of more complex statistical models. That is, patterns, trends, outliers, unexpected results in any existing data, using visual and quantitative methods to get a sense of clues that suggest logical next steps. This retrospective analysis of the data, requires the use of a variety of visualization techniques combined with means to identify trends in the data. SARvision|Biologics includes a number of techniques including Logo Plots, heat maps, and scatter plots, among others that can be correlated with the properties of the residues. Mutation cliffs and invariant maps [1] are part of the standard battery of tools when exploring the SAR of peptides. The exploratory data analysis is not a scripted process but one that is follow to build a construct of what is known so far about the peptides under scrutiny. This analysis is very important to develop a sense of models that can be built and their range of validity. Many times, exploring the existing data set with different visualization tools can provide insights that in themselves can suggest the right direction for the optimization of the peptides.

Predictive models [2] can then forecast different properties of interest. In many cases, molecular modeling techniques with analysis of three dimensional structures have been used. These methods confront the challenge of identifying the bioactive form of the peptide, i.e. the structure that actually is responsible for the biological activity, given the conformational flexibility of peptides, as well as the potential for different protonation states or tautomeric forms. Machine learning techniques offer an alternative, despite challenges of their own, that include the choice of molecular descriptors, to the selection of the most appropriate technique given the myriad of options available. One option is not to choose and build a very large number of models, using a variety of techniques, from linear regression to support vector machine. Hundreds of models can be generated in a few seconds on a regular personal computer that can be tested for their ability to fit the data and make predictions on data sets that were not part of the model training. The most predictive among those models can be combined and used to make recommendations as to what changes to make in the peptide sequence to optimize the property of interest. Indeed, models for multiple properties can be generated, such as potency, cell penetration, deamidation or half-life, among others. The recommendations can then be made on how to optimize multiple properties in parallel, by applying the different models. This is a significant departure from the typical SAR analysis where properties are optimized one at a time in an iterative cycle.

Our view is that these predictive techniques based on machine learning, when properly applied can greatly reduce the number of steps required to optimize peptides.

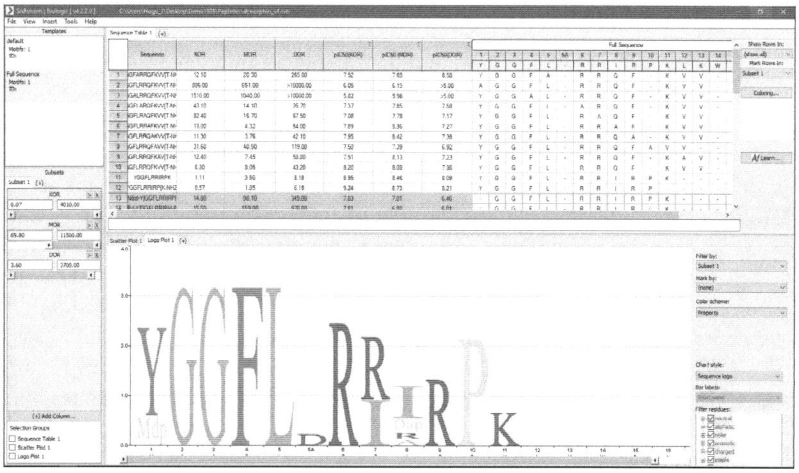

Figure 1: An example of Exploratory Data Analysis for a series of dynorphin analogs. Sequences are aligned with the data in the top right quadrant, while the bottom right shows a Logo Plot for the data. The panels on the right offer the possibility of subsetting the data. Other graphs are available including scatter plots, dendrograms, mutation cliff and invariant maps. All panels are responsive, since changes or selections in one affect the others.

References

[1] Hansen MR, Villar HO, Feyfant E., Development of an informatics platform for therapeutic protein and peptide analytics. J Chem Inf Model. 2013; 53:2774-2779.

[2] ChemApps Resources page, http://www.chemapps.com/resources (accessed: October 2018).

[3] Zhang T, Li H, Xi H, Stanton RV, Rotstein SH., HELM: a hierarchical notation language for complex biomolecule structure representation. J Chem Inf Model. 2012;52:2796-2806.

[4] Higgins DG, Bleasby AJ, Fuchs R. CLUSTAL V: improved software for multiple sequence alignment. Comput Appl Biosci. 1992; 8:189-191.

[5] Morgenstern, B. "Local versus Global Alignments." Sequence Alignment: Methods, Models, Concepts, and Strategies, edited by Michael S. Rosenberg, 1st ed., University of California Press, 2009, pp. 39–54.

Potent inhibition of CYP3A4 by the endomorphin-2 analogues

Kaloyan Georgiev[1], Maya Radeva[1], Tamara Pajpanova[2]

[1]Medical University of Varna, Bulgaria
[2]Bulgarian Academy of Sciences, Bulgaria

https://doi.org/10.17952/35EPS.2018.283

Introduction

Endomorphins(endomorphin-1,EM-1, Tyr-Pro-Trp-Phe-NH_2, endomorphin-2,EM-2, Tyr-Pro-Phe-Phe-NH_2) are endogenous peptides, which are very potent and highly selective μ-opioid receptor agonists. Moreover, they possess a potent analgesic activity comparable to that of morphine but are devoided of its' undesirable effects [1, 2]. However, their use as therapeutic agents is limited because of their pharmacokinetic features, such as stability and permeability *via* blood-brain barrier (BBB). In our previous developments, modifications in the structure of endomorphine-2have shown an improvement in stability and permeability through the membranes [3]. In the current study, the goal was to investigate another pharmacokinetic feature, namely the risk of drug interactions at the level of cytochrome enzymes. CYP3A4 is the most important drug metabolizing enzyme in humans. It is highly expressed in liver and gastrointestinal mucosa and is involved in the metabolism of more than 50% of the used drugs. Therefore, the risk of drug interactions is greatest in the drugs modifying the activity of CYP3A4.

Materials and Methods

Peptide synthesis and analysis

Peptides in this study were synthesized by manual solid-phase procedures using techniques for Fmoc-protected amino acids on Wang or MBHA Rink-Amide peptide resins respectively. 20% Piperidine in DMF was used for deprotection of Fmoc-groups and DIC and HBTU were employed as a coupling agent. Simultaneous deprotection and cleavage from the resin was accomplished by treatment with TFA/TIS/water (95:2.5:2.5) for 3 h at room temperature. Crude peptides were purified by preparative TLC and their purity was checked by analytical HPLC.

Measurement of CYP3A4 activity *in vitro*

The test compound was mixed with a master pre-mix comprising CYP450 BACULOSOMES ® reagent and regeneration system, which contained glucose-6-phosphate and glucose-6-phosphate dehydrogenase. The mixture was incubated at room temperature for 20 min. Following incubation, CYP enzyme-specific substrate (Vivid DBOMF for CYP3A4) and $NADP^+$ were added and the mixture was incubated at room temperature for 30 min. CYP activity was evaluated by measuring the fluorescence of the fluorescent metabolite generated from CYP3A4 enzyme-specific substrate. The fluorescence was measured using a microplate reader BioTek Synergy.

Results and Discussion

For the purpose of our study, we used four endomorphin-2analogues, two of them were modified at third position - Phe(pF) (1, Tyr-Pro-Phe(pF)-Phe-1,2-ethylenediamine) and conjugated at forth position with 1,2-ethylenediamine (2, Tyr-Pro-Phe-Phe-1,2-ethylenediamine). The other two, was conjugated at first position with deoxycholic acid and modified at third position - Phe(pF) (3, Deoxycholic-Tyr-Pro-Phe(pF)-Phe-OH)and Phe (pCl) (4, Deoxycholic-Tyr-Pro-Phe(pCl)-Phe-OH)(Figure 1).

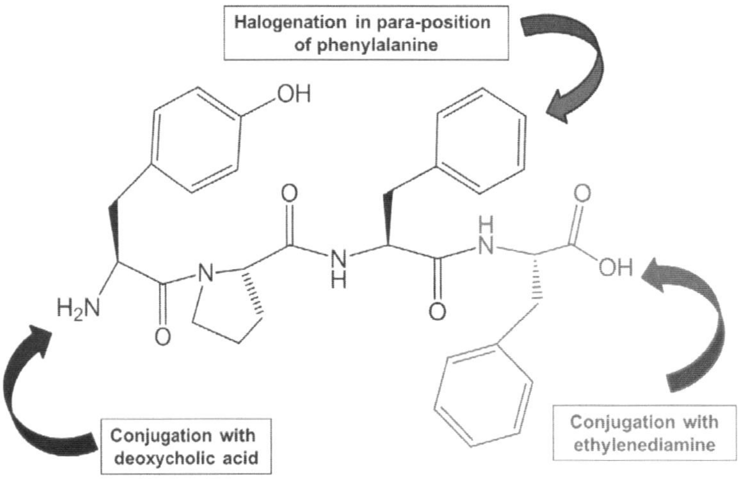

Figure 1: Modifications in chemical structure of endomorphin-2

The four endomorphin-2 analogues we used, showed potent and concentration-dependent inhibition of CYP3A4 in concentrations of 6,25 to 100 μM. The observed effect of two of them (1 and 2) was more distinct. The other two analogues (3 and 4), have shown slightly less pronounced inhibition of CYP3A4 (Figure 2). For comparison was used ketoconazole, classical inhibitor of CYP3A4, in concentration of 10 μM (90 % inhibition of CYP3A4).

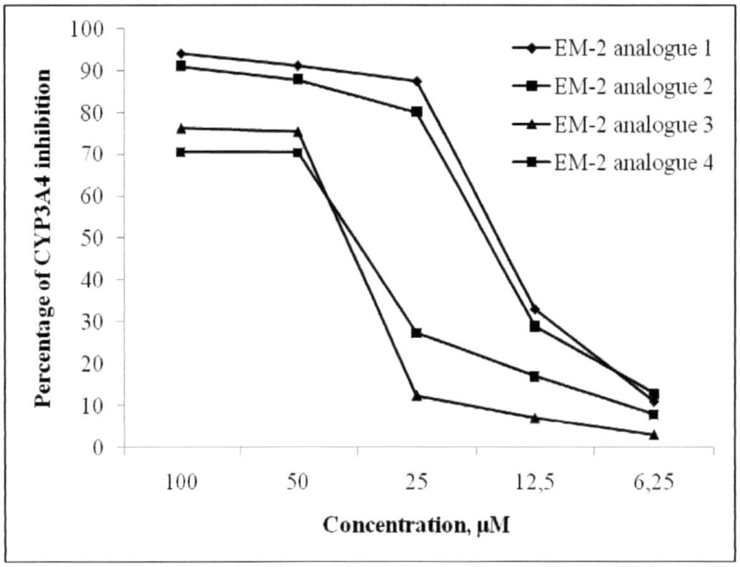

Figure 2: Inhibition of CYP3A4 by the endomorphin-2 analogues

After obtaining the results and analyzing them, we calculated the following IC_{50} values for each analogue (Table 1).

Table 1: IC50 values of EM-2 analogues

EM-2 analogues	***IC_{50} values with 95% CI***
Tyr-Pro-Phe(pF)-Phe-1,2-ethylenediamine	15,00 μM (9,89-22,73) p<0,05
Tyr-Pro-Phe-Phe-1,2-ethylenediamine	16,42 μM (10,28-26,25) p<0,05
Deoxycholic-Tyr-Pro-Phe(pF)-Phe-OH	31,37 μM (20,02-54,07) p<0,05
Deoxycholic-Tyr-Pro-Phe(pCl)-Phe-OH	29,04 μM (21,33-32,96) p<0,05

Conclusion

All four endomorphine-2 analogues showed potent inhibition of CYP3A4 isoenzyme, which should be considered if they are used in clinical practice for possible drug-drug interactions.

Acknowledgments

This work was supported by Project 16021"Investigation of drug interactions on biotransformation level", Fund Science, at the Medical University of Varna/Bulgaria.

References

1. Zadina, JE, Nilges, MR, Morgenweck, J, Zhang, X, Hackler, L, Fasold, MB Neuropharmacology, 2016, 105, 215-227.

2. Georgiev, K., Dzimbova, T., Iliev, I., Detcheva, R., Pajpanova, T. Comptes rendus de l'Academie bulgare des Sciences. 2014, 67, 1281-1286.

3. Georgiev, K., Dzimbova, T., Iliev, I., Detcheva, R., Pajpanova, T. In: Naydenova E, Danalev D, Pajpanova T (eds) Peptides 2014. Proceedings of the 33rd European Peptide Symposium, Sofia, Bulgaria,156-157 ISBN 978-619-90427-1-7.

Injectable peptide hydrogels for controlled drug release

Morgane Mannes[1], Charlotte Martin[1], Maria Dumitrascuta[2], Yannick Van Wanseele[3], Sophie Hernot[4], Ann Van Eeckhaut[3], Annemieke Madder[5], Richard Hoogenboom[6], Mariana Spetea[2], Steven Ballet[1]

[1]Research Group of Organic Chemistry, Vrije Universiteit Brussel, Belgium
[2]Department of Pharmaceutical Chemistry, University of Innsbruck, Austria
[3]Department of Pharmaceutical Chemistry and Drug Analysis, Vrije Universiteit Brussel, Belgium
[4]In Vivo Cellular and Molecular Imaging, Vrije Universiteit Brussel, Belgium
[5]Organic and Biomimetic Chemistry Research Group, Ghent University, Krijgslaan, 281, 9000 Ghent, Belgium
[6]Supramolecular Chemistry Group, Ghent University, Krijgslaan 281, 9000 Ghent, Belgium

https://doi.org/10.17952/35EPS.2018.286

Chronic pain remains one of the main challenges in human medicine at the beginning of the third millennium, with ca. 20-30% people worldwide suffering from chronic pain.[1] One of the major medical priorities is to provide effective pain control and sustained analgesia. Here, the focus was placed on the use of peptide-based hydrogels as a controlled drug release platform. Hydrogels are three-dimensional cross-linked networks able to retain large amounts of water. The porous structure of the gel allows the encapsulation of active compounds and their slow release from the hydrogel network in order to maintain a therapeutic concentration of the drug in the systemic circulation over an extended period (Figure 1).

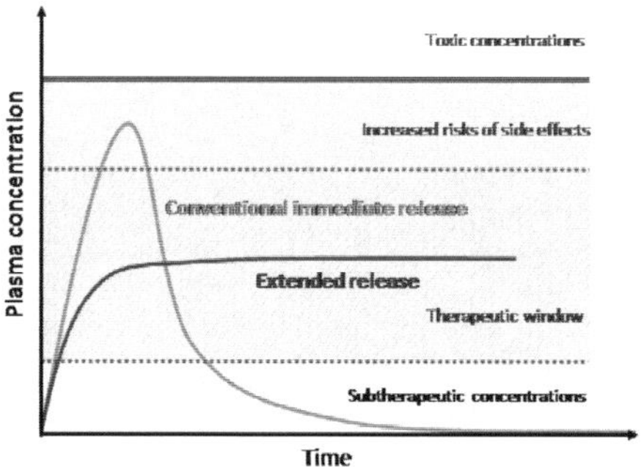

Figure 1: Schematic representation of plasma concentrations in function of time following administration of immediate-release and extended-release formulations.[2]

In case of pain therapy based on opioids, oscillating drug blood levels are avoided, reducing the frequency of administration and drug dosage, and resulting in less "clock-watching" and a better night time pain control.[2] The extended drug release, herein provided by the hydrogels, present a reduced risk of toxicity and side effects. Additionally, due to encapsulation into the matrix, the use of hydrogels protects the drug from premature *in vivo* degradation, increasing the drug's half-life.

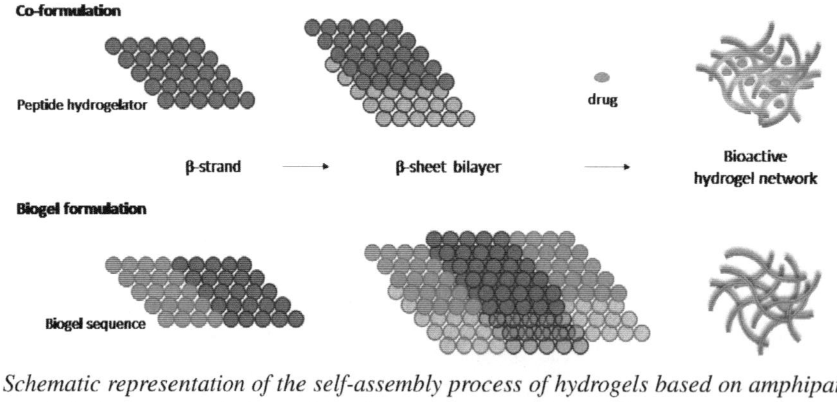

Figure 2: Schematic representation of the self-assembly process of hydrogels based on amphipathic peptide sequences. Both the 'co-formulation' and 'biogel' strategies are depicted.[5]

The focus was placed on amphipathic peptide-based hydrogels, because of their biocompatibility and biodegradability. The peptidic hydrogelator sequences are build-up of alternating hydrophobic and hydrophilic amino acids and have the tendency to self-assemble in aqueous media to form a hydrogel network.[3]

In this work, the peptide hydrogelators are short, but still tunable, sequences, enabling a plethora of possible modifications in view of the optimization of gelation and drug release properties. The injectability of these hydrogels allows an administration through parenteral delivery, including intramuscular and subcutaneous injection.[4]

In order to provide sustained analgesia with adequate efficacy, the focus was placed on the development of improved opioid-based therapies. To obtain the desired analgesic activity with reduced side effects over extended periods of time, two different strategies were exploited, as depicted in Figure 2. Either the opioid pharmacophore is encapsulated or co-formulated into the hydrogel network, or the peptidic opioid is part of the hydrogel sequence, giving rise to an analgesic gel, herein called biogel.[5] The peptide opioid pharmacophores were designed based on endogenous opioid peptides, and were linked with previously developed lead hydrogelators.[6,7]

The hydrogel-encapsulated peptide opioids and the biogel conjugates were tested *in vivo*, to determine their extended antinociceptive effect (Figure 3). The biogels, were compared to their corresponding parent opioid sequences co-formulated with a hydrogelator. A similar profile and strength were obtained for both strategies, even though the biogel required higher doses to achieve an equianalgesic effect. This variation in dosage is explained by a difference in release mechanism of the drug. While for the co-formulation the drug is released by diffusion through the hydrogel network, for the biogel an enzymatic degradation of the biogel is necessary to release the opioid pharmacophore and induce the observed analgesic effect.[5]

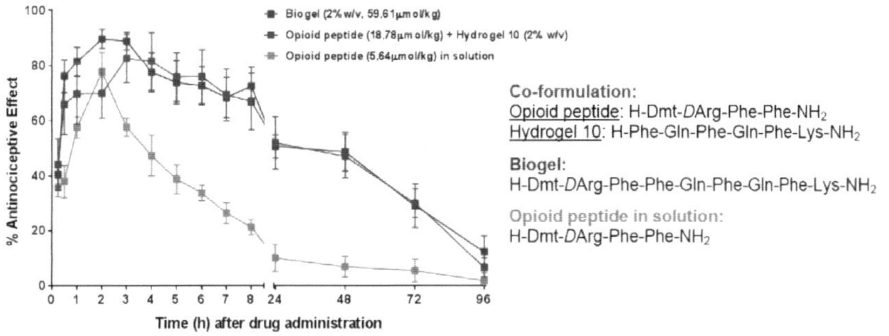

Figure 3: Comparison of antinociceptive activities of the opioid peptide co-formulated with Hydrogel 10 or conjugated to Hydrogel 10, resulting in the 'biogel', in the tail-flick test after s.c. administration in mice. [5]

We demonstrated that by using both strategies an extended antinociceptive effect up to 72 to 96 h was induced after subcutaneous injection in mice.[5] These promising results open up a research avenue for improved chronic pain therapies, but it is expected that the presented formulations are also more broadly applicable for other biologically active peptide therapeutics.

References

[1] Kapur, B. M.; Lala, P. K.; Shaw, J. L., Pharmacogenetics of chronic pain management. Clinical Biochemistry 2014, 47 (13), 1169-1187.

[2] Martin, C.; De Baerdemaeker, A.; Poelaert, J.; Madder, A.; Hoogenboom, R.; Ballet, S., Controlled-release of opioids for improved pain management. Materials Today 2016, 19 (9), 491-502.

[3] Dasgupta, A.; Mondal, J. H.; Das, D., Peptide hydrogels. RSC Advances 2013, 3 (24), 9117-9149.

[4] Bibian, M.; Mangelschots, J.; Gardiner, J.; Waddington, L.; Acevedo, M.M.D.; De Geest, B.G.; Van Mele, B.; Madder, A.; Hoogenboom, R.; Ballet, S., Rational design of a hexapeptide hydrogelator for controlled-release drug delivery. Journal of Materials Chemistry B 2015, 3 (5), 759-765.

[5] Martin, C.; Dumitrascuta, M.; Mannes, M.; Lantero, A.; Bucher, D.; Walker, K.; Van Wanseele, Y.; Oyen,

E.; Hernot, S.; Van Eeckhaut, A.; Madder, A.; Hoogenboom, R.; Spetea, M.; Ballet, S., Journal of Medicinal Chemistry 2018 DOI: 10.1021/acs.jmedchem.8b01282.

[6] Martin, C.; Oyen, E.; Mangelschots, J.; Bibian, M.; Ben Haddou, T.; Andrade, J.; Gardiner, J.; Van Mele, B.; Madder, A.; Hoogenboom, R.; Spetea, M.; Ballet, S., Injectable peptide hydrogels for controlled-release of opioids. MedChemComm 2016, 7, 542-549.

[7] Martin, C.; Oyen, E.; Van Wanseele, Y.; Haddou, T. B.; Schmidhammer, H.; Andrade, J.; Waddington, L.; Van Eeckhaut, A.; Van Mele, B.; Gardiner, J., Injectable peptide-based hydrogel formulations for the extended *in vivo* release of opioids. Materials Today Chemistry 2017, 3, 49-59.

Trichogin analogs for retinal drug delivery

Marta De Zotti[1], Ivan Guryanov[2], Tatiana Tennikova[2], Madhushree Bhattacharya[3], Arto Urtti[3]

[1]University of Padova, Italy
[2]St. Petersburg State University, Russia
[3]University of Helsinki, Finland

https://doi.org/10.17952/35EPS.2018.289

In ophthalmology retinal drug therapy is one of the major challenges, because it is hampered by ineffective and/or short-acting drug delivery to the targets [1]. Often these targets are located in intracellular compartments and there is a need of drug delivery systems able to transfer pharmaceutical cargo into retinal cells [2,3]. Many retinal therapeutics are based on proteins and oligonucleotides with poor cell permeability. The development of cell penetrating carriers of farmaceutical cargos can noticeably enhance their therapeutic effect. From this point of view membrane-active trichogin analogs are very promising candidates as components of retinal drug delivery systems [4]. The main disadvantage of thichogin as a cell penetrating peptide (CPP) is its low solubility in aqueous media. In order to increase water solubility, we introduced lysine or 4-aminopiperidine-4-carboxylic acid (Api) residues into the trichogin sequence at different positions (Table 1).

Table 1: Structures of trichogin analogs.

Peptide	Structure*
Tricogin	*n*Oct-UGLUGGLUGI-Lol
T-1	*n*Oct-UGLUKKLUGI-Lol
T-2	*n*Oct-UGLUKKLUGIL-NH-$(CH_2)_2$-NH_2
T-3	*n*Oct-UGLUGKLUGI-Lol
T-4	*n*Oct-UGLUKULUGI-Lol
T-5	*n*Oct-UGLUKULUGI- NH-$(CH_2)_2$-NH-FITC
AP-corto	*n*Oct-UGLUKKL-Lol
Corto-1	*n*Oct-UGLUKKLL- NH-$(CH_2)_2$-NH_2
Corto-2	*n*Oct-UKKLUGIL- NH-$(CH_2)_2$-NH_2
T-6	*n*Oct-Toac-GLUGGL-Api-I-Lol

*nOct = n-octanoyl; U = Aib; Lol = leucinol

To have a possibility for the attachment of a drug some analogs were modified with C-terminal amino group. The influence of the length of the peptide on the membrane activity and cell viability was studied by removal C- or N-terminal parts (peptides Corto-1 and Corto-2, respectively). For further study of peptide structure and mechanism of cell penetration the analog containing nitroxide spin-labeled amino acid Toac (2,2,6,6-tetramethyl-piperidine-1-oxy-4-amino-4-carboxylic acid) was prepared as well [5]. All analogs were synthesized by manual SPPS using either a Rink-amide or a 2-chlorotrityl resin preloaded with Lol or a 1,2-diaminoethane moiety with following modification with FITC label (peptide T-5). NMR conformational analysis of FITC labeled peptide showed no structural changes upon modification and native helical conformation was proved to be maintained (Figure 1).

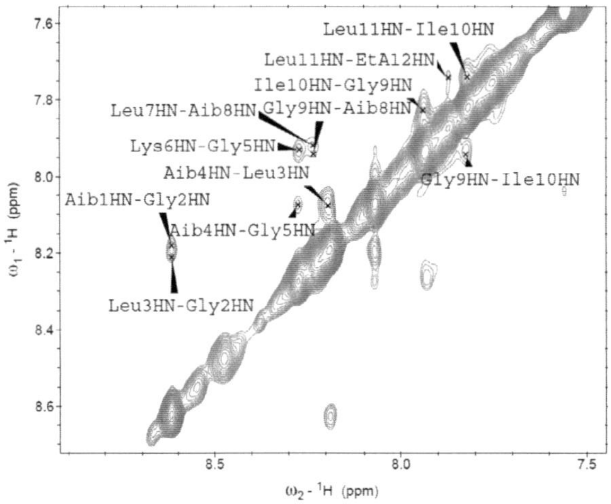

Figure 1: Amide portion of the Noesy spectrum of T-5 in SDS-d_{25} H_2O/D_2O 9:1 (2 mM, 600 MHz, 303K). Cross-peaks between corresponding amino acids are evidenced.

Cytotoxicity of peptide-carriers was evaluated for retinal pigment epithelium (ARPE-19) cells. The peptides were dissolved in a small amount of DMSO and then diluted with cell medium to the desired concentration. The cells were treated with peptides for 5 h and cell viability was evaluated by MTT cytotoxicity assay. Poly L-lysine (PLL) treated and untreated cells were used as positive and negative control, respectively (Figure 2).

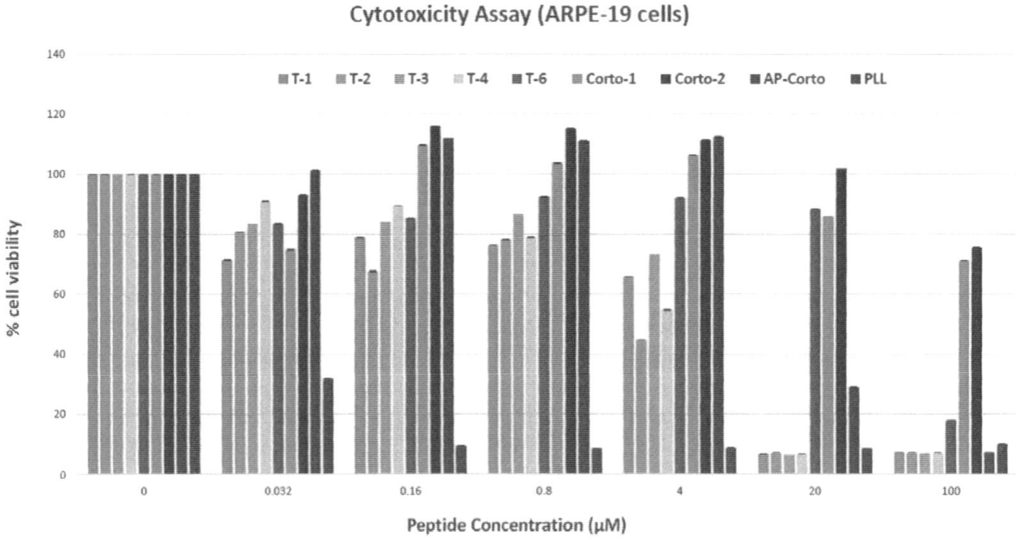

Figure 2: Cytotoxicity of trichogin analogs in ARPE-19 cells. The data are normalized based on the viability of untreated cell and are represented as means ± SD (n = 3).

Almost all the peptides showed very weak cytotoxicity at the concentration up to 0.8 μM with cell viability more than 80%. Thus, for the uptake studies 1 μM peptide concentration was used. The cell internalization experiments showed that more than 90% of the cell population has taken up the FITC labeled peptide (Figure 3).

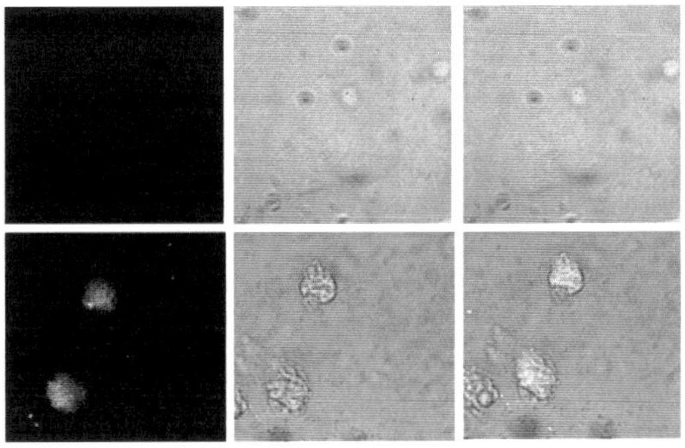

Figure 3: Confocal microscopy images of untreated cells (upper row) and treated with T-5 peptide (lower row). From left to right: fluorescence, transmission and merged images.

In summary, water soluble trichogin analogs described herein showed to be promising membrane permeating molecules with very efficient cell internalization that can be used for drug delivery for the treatment of various diseases of the eye.

Acknowledgements

The work was supported by Megagrant N° 14.W03.31.0025

References

[1] Del Amo, E. M. et alProg. Retin. Eye Res. 2017, 57, 134-185.

[2] Kamaleddin, M. A. Nanomedicine: NBM, 2017, 13, 1459-1472.

[3] Bhattacharya, M. *et al.* J. Control. Release. 2017, 251, 37-48.

[4] De Zotti, M. *et al.* Amino Acids. 2012 Oct;43(4):1761-1777.

[5] Bartucci, R. *et al.* Biophys. J. 2008, 94, 2698-2705.

Proceedings of the 35th European Peptide Symposium
Patrick B. Timmons, Chandralal M. Hewage, Michal Lebl (Editors)
European Peptide Society & PSP, 2018

Hyaluronan based scaffold for cardiomyocyte adhesion

Sergej Karel[1], Jana Sogorkova[1], Lucie Marholdova[1], Martin Flegel[2], Vladimír Velebný[1]

[1]Contipro a.s., Czechia
[2]University of Chemistry and Technology Prague, Czechia

https://doi.org/10.17952/35EPS.2018.292

Introduction

Hyaluronic acid (HA) is known as a linear polysaccharide consisting of β-(1-4)-d-sodium glucuronate-β-(1-3)-N-acetyl-d-glucosamine units. Hyaluronan derivatives are often used in biomedical applications as substrates for tissue engineering [1]. Peptide with active sequence YIGSR enabling adhesion of cardiomyocytes was chosen for this study.

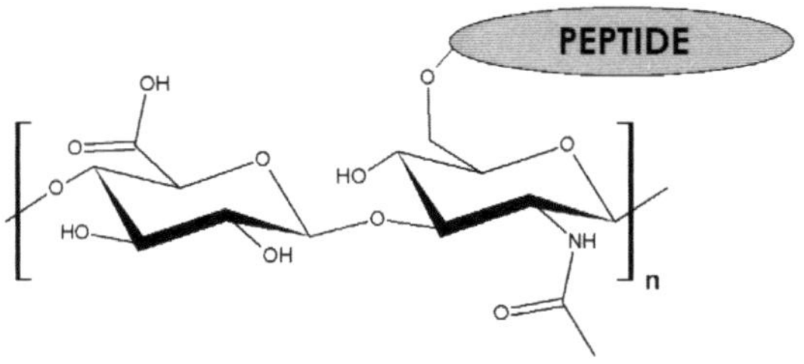

Methods and Results

Non-woven textile fabrication

The HA non-woven fabrics were formed using wet spinning technology that had been applied for hyaluronic acid (Mw 1 MDa, Contipro a.s., Czech Republic) [2]. Non-woven fabrics with basis weight in the range of 15 and 60 g.m^{-2} were prepared from staple fibers with typical diameter of 1 μm.

Figure 1: Non-woven textile (1×1 cm) – SEM image showing inner structure of non-woven textile (left) and macroscopic view of the same scaffold (right).

Solid Phase Peptide Synthesis using HA carrier

Couplings were performed in DMF where 3 molar excess of Fmoc amino acid and DIC/OxymaPure as coupling agents were employed for 2 h. N-terminal Fmoc group was removed with 20% piperidine in DMF. HA carrier bearing Fmoc protected peptide was used for the determination of fabric substitution with the aid of common Fmoc release test. Substitution of non-woven fabrics varied in the range 0.19 – 0.21 $mmol.g^{-1}$.

Approaches for synthesis or attachment of peptides to hyaluronan-based materials (fibers and textile scaffold) have been developed [3,4]. Fully protected peptides were assembled on Wang resin and after deprotection and splitting they were coupled to HA fibers. The structure of peptides was checked by amino acid analysis and HPLC.

Structure of applied peptides:

A H-Gly-Gly-Glu-Gly-Tyr-Gly-Glu-Gly-Tyr-Ile-Gly-Ser-Arg-Ahx-Ahx-Nle-OH

B H-Ile-Lys-Val-Ala-Val-Ahx-Ahx-Nle-OH

C H-Gly-Gly-Glu-Gly-Tyr-Gly-Glu-Gly-Ile-Lys-Val-Ala-Val-Ahx-Ahx-Nle-OH

D H-Ahx-Ahx-Nle-OH

Cardiomyocyte adhesion

Stem cells pre-differentiated to cardiomyocytes were stained by DiI (15 min, Ex_{max}/Em_{max} 514/570 nm) prior to seeding to UV-sterilized non-woven textile previously soaked in cultivation medium (10^5 cells/scaffold ca. 1,5 cm^2). The adhesion of cells was monitored during the 7-day long cultivation.

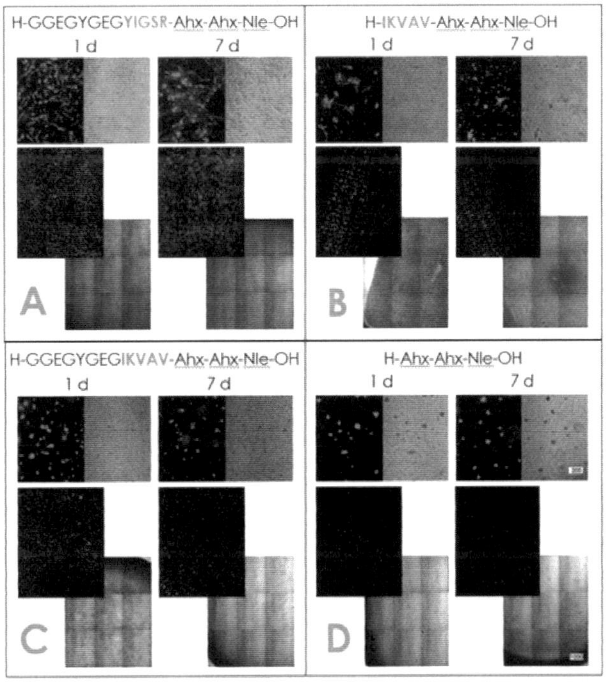

Figure 2: Stem cells pre-differentiated to cardiomyocyte and cultivated on non-woven textile with different conjugated peptides. Time points 1 day and 7 days are displayed. The scale bar unites are μm. Cells are stained by DiI (red).

Discussion

Scaffolds were stable under the cultivation conditions. Stem cells adhered uniformly on the non-woven HA biofunctionalized by YIGSR peptide, which should support the adhesion of cardiac myocytes (Fig. 2-A). Contrary, cardiomyocytes attached poorly and unequally, forming cell clumps on the textile modified by IKVAV peptide (Fig 2-B), which is not primarily used to enhance cardiomyocytes adhesion. Controls confirmed that the adhesion was not supported by the peptide used ahead of the sequence YIGSR (Fig. 2-C) as well as cells did not adhere on the textile with Ahx linkers used to anchor the adhesive motives to HA. These data indicate that the activity of YIGSR peptide was maintained, and its functional group was accessible for the cellular integrins enabling the focal adhesion. The negative results obtained with IKVAV peptide supported the selectivity and specificity for this process.

Conclusion

Preparation method employing the non-woven fabrics of HA as solid support has been studied with different short peptides specifically enabling adhesion of cardiomyocytes. Biocompatibility and biodegradability of the HA carrier in human body was considered as the main advantage. Such a kind of HA modification may improve the properties of wound healing materials because it enables to eliminate the need of its removal after surgery. Preliminary results and biological effects of the prepared peptide models confirmed the rightness of our approach.

References

1. Rosellini, E., et al., Protein/polysaccharide-based scaffolds mimicking native extracellular matrix for cardiac tissue engineering applications. J Biomed Mater Res A, 2018. 106(3): p. 769-781.

2. Burgert, L., et al., Method of Preparation of Polysaccharide Fibers, Wound Covers That Contain Them, Method of Manufacturing of Wound Covers, and Apparatus for Preparation of Polysaccharide Fibers, WO/2013/167098, 2013, Contipro a.s.

3. Karel, S., et al., Hyaluronan fibers as a carrier for Solid Phase Peptide Synthesis. J Pep Sci, 2016. 22: p. S65.

4. Karel, S., et al., Stabilization of hyaluronan-based materials by peptide conjugation and its use as a cell-seeded scaffold in tissue engineering. Carbohydrate Polymers, 2018. 201: p. 300-307.

5. Yu, J., et al., Electrospun PLGA fibers incorporated with functionalized biomolecules for cardiac tissue engineering. Tissue Eng Part A, 2014. 20(13-14): p. 1896-907.

Imaging evaluation of an *in vivo* long-acting Neuropeptide(NPY) analogue for multimodality breast tumor therapy

Ming-Hsin Li, Su-Jung Chen, Ming-Wei Chen, Chun-Fang Feng, Yuan-Ruei Huang, Sheng-Nan Lo, Cheng-Hui Chuang, Shih-Ying Lee, Chih-Hsien Chang

Institute of Nuclear Energy Research, Taiwan

https://doi.org/10.17952/35EPS.2018.295

Neuropeptide Y (NPY) is a 36-amino acid peptide and regulates in various physiological functions through its four receptors. Recently, NPY receptor- Y1R has been found to be overexpressed in breast carcinomas. Although the truncated NPY analog has high affinity with Y1R, it is rapidly metabolized, resulting in low tumor uptake. The aims of this study were to synthesize a long-circulated NPY analog and evaluate tumor image of the 4T1 animal model as a candidate for breast cancer therapy.

Materials and Methods

We synthesized INER-7218, which is a 16-amino acid NPY analog. Its structure consists of three parts-(1) chelate for radiolabeling(2) Long-acting linker (3)NPY analogue for high affinity with breast cancer. In the radiolabeling procedure, amount of INER-7218 was dissolved in buffer, followed by addition of (3-12mCi) In-111 or Lu-177, and incubated in high temperature. The labeling efficiency was determined by Radio-TLC. The Radio-HPLC are used for analyzing labeling purity. In animal study, tumor xenografts were performed in 6-wk-old female BALB/c mice by subcutaneous injection of $2^{*}\ 10^{6}$ 4T1 cells, and nanoSPECT/CT imaging was performed at 0.5 h to 48h after injecting of the ^{111}In- INER-7218

Results

We can get the high labeling efficiency (>90%) of Lu-177- INER-7218, and In-177- INER-7218 after reacting 15-30min. *In vivo* study, the nanoSPECT/CT image of the 4T1 animal model revealed that In-111-INER-7218 has with high tumor uptake value (ID%/g>10)and high tumor to muscle (T/M>9) ratio in 48 hours after tail vein injection

Conclusion

The result shows the INER-7218 is easy to radiolabel with Lu-177 and In-111. The two compounds are high labeling efficiency and high resolution of the image. From the *in vivo* study, we consider the INER-7218 as a long circulation radiopharmaceutical candidate for companion diagnostics (CDx). Furthermore, we apply such INER-7218 for the design of multimodality carriers. Peptide Receptor Radionuclide Therapy (PRRT) is a widely known molecular targeted therapy. PRRT is performed by using a small peptide which is combined with radionuclides.

INER-7218 is not only a good tracer but also a therapeutic drug by labeling with Lu177.

The capabilities of INER-7218 make it a potential drug for multimodality breast tumor therapy.

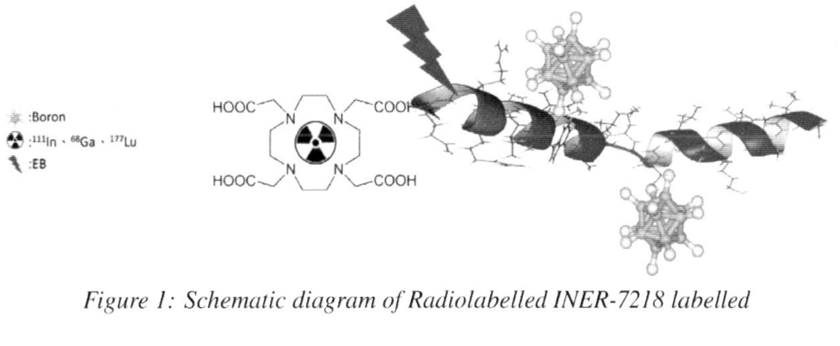

Figure 1: Schematic diagram of Radiolabelled INER-7218 labelled

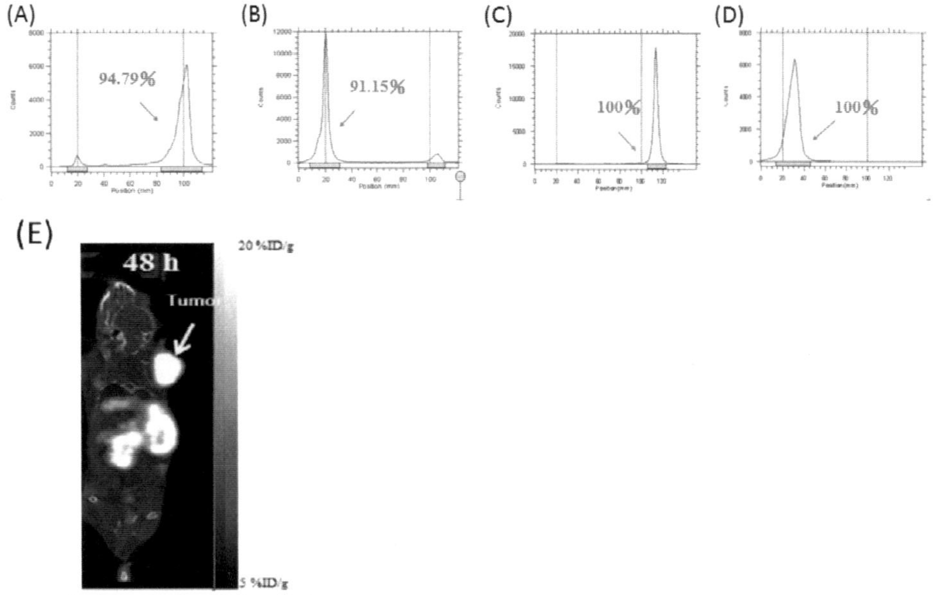

	Tumor (%ID/g)	T/M ratio	T/L ratio	T/K ratio
(E)	23.75	1.5	1.18	9.88

*Tumor/Muscle :(T/M)
Tumor/Liver : (T/L)
Tumor/Kidney : (T/K)

Figure 2: The labeling efficiency is analyzed by Radio-TLC. (A) Free In-111 at solvent front.(B) 111In-INER-7218 (Rf =0), the labeling efficacy is 91.15%. (C) Free Lu-177 at solvent front.(D) 177Lu-INER-7218 (Rf =0), the labeling efficacy is 100%.

NGR-DAU conjugates, a favorable tumor-homing motif with potential dual-targeting

Andrea Angelo Pierluigi Tripodi[1], Szilard Toth[2], Ivan Randelovic[3], Jozsef Tovari[3], Gábor Mezó[1]

[1]MTA-ELTE Research Group of Peptide Chemistry, Hungary
[2]Institute of Enzimology, Research Center for Natural Sciences, Hungarian Academy of Sciences, Hungary
[3]National Institute of Oncology, Dept. of Experimental Pharmacology, Hungary

https://doi.org/10.17952/35EPS.2018.297

Introduction

At present, cancer is one of the major health-related issues world-wide. Strategies to target and defeat tumor are limited due to the access to tumor, side effects and absence of selectivity to the tumor tissue. Small molecule drug conjugates containing the Asn-Gly-Arg (NGR) motif can act as tumor-homing compounds that are recognized by Aminopeptidase N (APN/CD13). CD13 is receiving an accurate attention due to its role in the progression of tumor vasculo-genesis and identifying it as a critical regulator of angiogenesis, generally expressed on the monolayer cell surface of malignant cells [1,2]. As established in the literature the Asn-Gly-Arg sequence is inclined to Asn deamidation bearing the formation of a five-membered succinimide ring intermediate, that under fast hydrolytic conditions generates isoDGR derivative which can bind to integrins [3-6]. According to literature data, one of the most stable and tumor selective cyclic NGR-peptides is c[KNGRE]-NH_2, in which the α-amino group of the N-terminal Lys is coupled to the γ-carboxyl group of the glutamic acid residue (head-to-side chain cycle) [7]. The preparation of a c[KNGRE] conjugate require a sophisticated synthetic route, for this reason the main goal of the present study was to investigate whether the exchange of the lysine in the cycle has any influence on the chemo-stability, selectivity and antitumor activity of the conjugates.

Results and Discussion

NGR cyclic peptides were prepared by SPPS on a Rink-Amide MBHA Resin, using Fmoc/tBu strategy. Daunomycin (Dau) as an anticancer agent was conjugated to an aminooxyacetic acid linker (oxime linkage) connected through an enzyme labile spacer (GFLG) *via* oxime linkage allowing an efficient drug release of active metabolite in lysosomes of cancer cells. We recently described that the Dau=Aoa-GFLGK(c[KNGRE]-GG-)-NH_2 conjugate (K) has a significant anti-tumor activity against both CD13 positive HT-1080 human fibrosarcoma and CD13 negative but integrin positive HT-29 human colon adenocarcinoma cells [8]. However, we have confirmed that the free ε-amino group of Lys in the cycle is irrelevant for the biological activity. Therefore, we developed novel cyclic NGR peptide – daunomycin conjugates in which Lys was replaced by different amino acids [9]. From these investigations the Nle conjugate (3) as the most efficient compound was selected for further studies. The cytostatic and cytotoxic effect of the novel cyclic NGR peptide Dau conjugates were evaluated *in vitro* on HT-1080 (human fibrosarcoma), HT-29 (human colon adenocarcinoma) and KS (Kaposi's sarcoma) cell lines (Table 2). The stability in cell culture medium (Table 1) and the lysosomal degradation in presence of rat liver homogenate were determined by LC-MS [9].

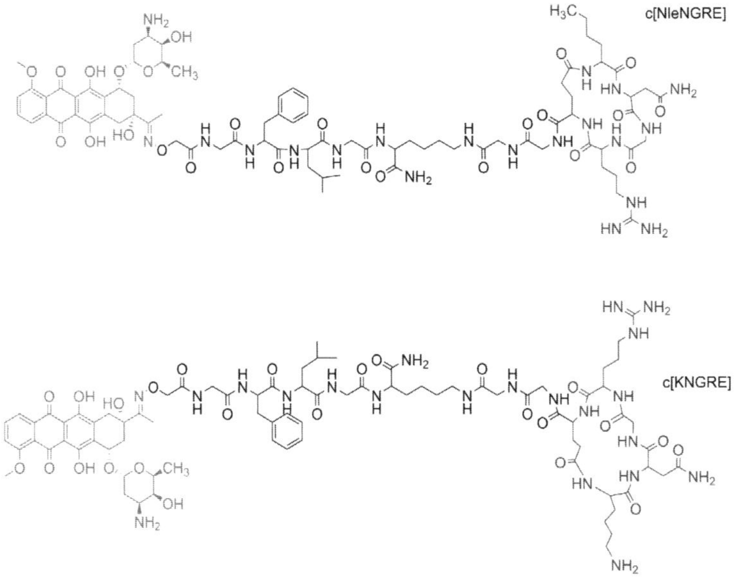

Figure 1

Table 1: Chemo-stability of cyclic NGR peptide - daunomycin conjugates

Ratio of Asn-/Asp-/*iso*Asp-derivatives (DMEM CM, 37 °C)							
Code	**AAA replacement**	**6 h**			**72 h**		
		NGR	**DGR**	***iso*DGR**	**NGR**	**DGR**	***iso*DGR**
1	Ala	96	0	4	58	11	33
2	Leu	93	0	7	54	11	35
3	Nle	93	1	6	58	09	33
4	Pro	73	14	13	19	46	35
5	Ser	93	0	7	56	12	31
K	Lys	100	0	0	100	0	0

The main cleavage site of the conjugates in lysosomes could be detected between Gly-Phe within the enzyme labile spacer resulting in the smallest active Dau containing metabolite Dau=Aoa-Gly-OH [9].

Detection of CD13 expression on tumor cell lines.

The surface expression of CD13 on KS (Kaposi's sarcoma) and on HT-1080 (human fibrosarcoma) was determined by immunocytochemistry with a FITC-conjugated anti-CD13 monoclonal murine antibody (OKM13). No fluorescence was detected in case of HT-29 (human colon adenocarcinoma) CD13 negative.

Table 2: In vitro cytostatic and cytotoxic effect IC50 (μM)

		HT1080 CD13+ 6h	**HT-29 CD13- 6h**	**KS CD13+ 6h**	**HT1080 CD13+ 72h**	**HT-29 CD13- 72h**	**KS CD13+ 72h**
3	**Dau=Aoa-GFLGK(c[NleNGRE]-GG-)-NH_2**	2.8	6.4	3.6	2.4	1.8	0.3
K	**Dau=Aoa-GFLGK(c[KNGRE]-GG-)-NH_2**	12.4	9.9	7.2	5.7	16.5	2.4
Dau	**Daunomycin**	0.004	0.03	0.04	0.002	0.001	0.001

In this experiment we measured the cytostatic effect (6 h treatment and further 66 h incubation after washing out the compounds) and the cytotoxic effect (72 h treatment), compounds were dissolved in serum containing (FBS+) RPMI 1640 medium. Non-treated cells in both conditions were used as controls.

Cell Uptake studies

As daunomycin present fluorescent properties, internalization of both peptides was also detected by flow cytometry.

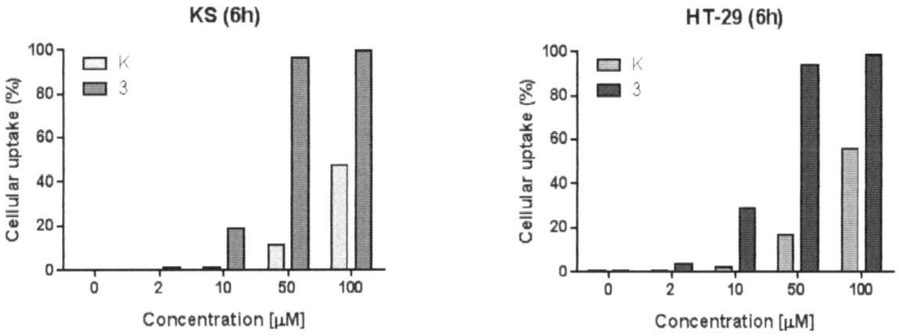

Figure 2

From this study we showed that the changes decrease the chemo-stability of the cyclic NGR moiety, resulting in the formation of isoAsp-derivatives in higher amount. Among the new cyclic NGR peptide - daunomycin conjugates the most effective compound was Dau=Aoa-GFLGK(c[Nle-NGRE]-GG)-NH_2. This conjugate (3) was especially active on Kaposi's sarcoma cell which highly expresses for CD13.

Acknowledgements

This project had received fundings from the European Union's Horizon 2020 research and Innovation program under the Marie Sklodowska-Curie grant agreement No 642004, and from the National Research, Development and Innovation Office (NKFIH K119552), Hungary.

References

[1] Curnis, F.; Cancer Res 2002, 62, 867-74.

[2] Corti A.; Blood. 2008 Oct 1;112(7):2628-35.

[3] Meinwald, Y. C.; *et al.* Int. J. Pept. Protein Res. 1986, 28, 79–84.

[4] Kirikoshi, R.; Manabe, N.; Takahashi, O. Int. J. Mol. Sci. 2017, 18, No. 429.

[5] Catak, S.; Monard, G.; Aviyente, V.; Ruiz-López, M. F. J. Phys. Chem. A 2008, 112, 8752–8761.

[6] Geiger, T.; *et al.* J. Biol. Chem. 1987, 262, 785–794.

[7] Negussie A. H.; J Control Release. 2010 April 19; 143(2): 265–273.

[8] Enyedi, K. N.; *et al.* PLoS One 2017, 12, e0178632.

[9] Tripodi A. A. P.; Beilstein J. Org. Chem. 2018, 14, 911-918.

Peptides specifically target bacteriophage MS2, filled with an apoptosis-inducing agent, to tumour

Ekaterina Kolesanova[1], Tatiana Bolshakova[2], Ekaterina Rybalkina[3], Marina Melnikova[1], Yulia Filatova[1], Alexey Chistov[1], Igor Sivov[4]

[1]Institute of Biomedical Chemistry, Russia
[2]N.F. Gamaleya National Research Centre of Epidemiology and Microbiology, Russia
[3]Russian Research Centre of Oncology, Russia
[4]"Biotechnologia" Ltd., Russia

https://doi.org/10.17952/35EPS.2018.300

Summary

We employed bacteriophage MS2 for the targeted delivery of an apoptosis-inducing agent into tumour tissue. The targeted delivery was provided by iRGD peptides, ligands of integrins presumably located on the surface of tumour tissue novel vasculature endotheliocytes. Peptides were synthesized and conjugated to MS2 capsid proteins. Tl^+ ions as TlNO3 were used as apoptosis-induced agents, which penetrated into phage particles and tightly bond to the phage RNA. Peptide-modified MS2 preparations filled with Tl^+ caused cell death in several tested solid tumour cell cultures and tumour mass loss in xenograft mice with inoculated with MCF7 and DA-MB-231 cells, while no effect was observed with control preparations (MS2 + peptide, MS2 + Tl^+). Overall concentration of TlNO3 in peptide-modified MS2 filled with Tl^+ was 500000 times lower than LD50, and therapeutic index of the phage-based preparation was about 15000. The results show the perspective of the use of bacteriophage MS2 filled with Tl^+ and targeted by iRGD peptides as a lead substance for solid tumour therapy.

Introduction

A target-delivering drug system (fig. 1) composed of: 1) a ligand - iRGD peptide (fig. 2), known to bind to $\alpha V\beta 3$ and $\alpha V\beta 5$ integrins that are expressed on endothelial cells of the newly formed tumour vasculature and certain tumour cells, and to activate an endocytic/exocytic transport pathway *via* the interaction with neuropilin-1 [1]; peptide ligands contained Gly3 spacers for their conjugation with phage particles; 2) a vehicle - bacteriophage MS2 with a single-stranded genomic RNA, replicating only in F-pilus-bearing enteric bacteria that are neither symbiotic nor pathogenic for human organisms. MS2 capsid contains pores permeable for metal ions (Tl^+). MS2 serves an impermeable vehicle for Tl^+, excluding its toxic action on tissues except tumour and its vasculature; 3) an active substance - Tl^+, a toxic metal ion known to cause cell apoptosis. Tl^+ tightly binds to viral RNA and hence does not leak from MS2 particles [2]. It is not removed from cancer cells *via* multidrug resistance-responsible systems [3].

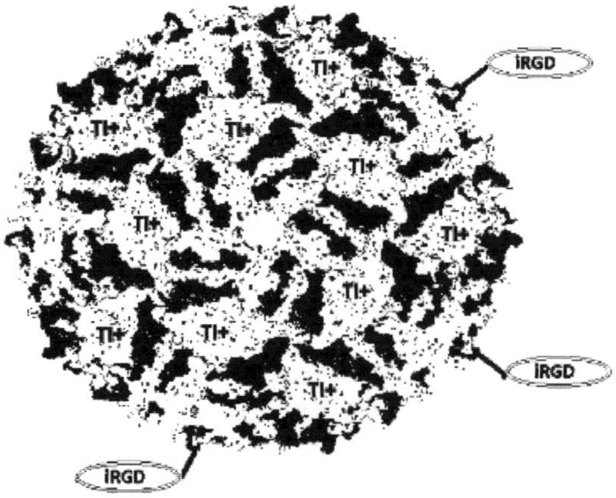

Figure 1: A schematic image of bacteriophage MS2 conjugated with iRGD peptide and filled with Tl+.

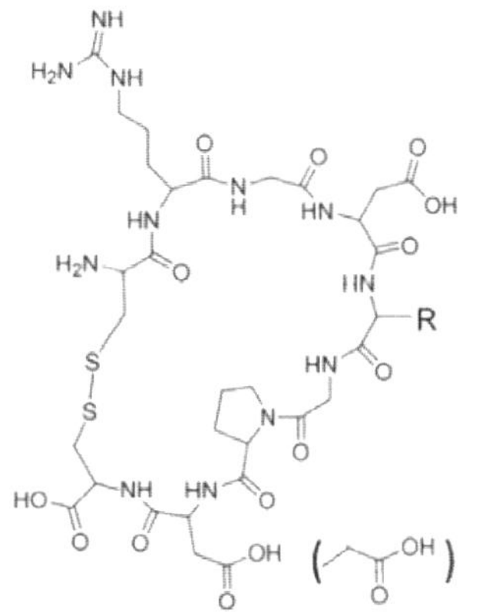

iRGD: c(CRGD[K,R]GP[D,E]C)

Figure 2: Structure of iRGD peptides

Experimental

iRGD peptides were prepared by automated SPPS (433A peptide synthesizer, Applied Biosystems) on 2-Cl-Trityl resin from Fmoc-amino acids *via* FastMoc procedure (activators - HBTU with Oxyma Pure); cysteine derivatives (Trt- and Acm-protected) were coupled with DIC/Oxyma Pure. Cyclization *via* formation of S-S bridges was achieved on totally protected peptides in solution by I_2 . After the cyclization the peptides were deprotected, purified by reverse-phase HPLC and characterized by LC-MS for the conformity of their structure and purity. Peptides were conjugated to MS2 phage particles by using a bifunctional reagent dimethyl adipimidate through protein and peptide free NH_2 groups [4]. Filling of peptide-MS2 particles with Tl^+ was achieved by incubating peptide-conjugated MS2 ($3.3x10^{10}$ pfu(plaque forming units)/mL) with 0.5 mkM $TlNO_3$ with further precipitation by polyethylene glycol 6000 and dialysis. Tl^+ dosage was $5x10^{-7}$ g-equiv per 1 pfu (determined fluorimetrically and mass-spectrometrically). Four different preparations of peptide-conjugated MS2 filled with Tl^+ containing four different iRGD peptides were mixed in about equimolar ratios for further testing. Testing of the drug delivery system was performed on cultivated MCF-7 (hormone-dependent breast cancer) and DA-MB-231 (hormone-independent breast cancer) cells and nude mice with inoculated MCF-7 and DA-MB-231 xenografts. Cell viability was tested in serum-free (enhanced phagocytosis) and serum-containing media by Evans Blue staining. Effect on animal xenografts was evaluated after 10 p.o. or i.p. injections of 10^8 pfu/kg/day by comparing tumour area in experimental and control animal groups. Peptide-MS2 without Tl^+ and MS2+ Tl^+ without peptides were used as controls.

Results and Discussion

iRGD peptides were prepared in 95% purity. Lys^8-containing iRDG peptides formed intermolecular rather than intramolecular S-S bridges by air oxidation of deprotected peptides and did not readily formed S-S bridges while on resin, and we could successfully cyclize these peptides only in a totally protected state. Arg^8-containing peptides were prepared by the same scheme as well. Incubation of peptide-conjugated MS2 particles in $TlNO_3$ solution at neutral pH resulted in the drop in Tl^+ concentration in the media up to 30 times compared to the initial one, thus representing the filling of the phage paricles with Tl^+ , which was tightly bound to the phage RNA. Free Tl^+ content of the preparation was <10%, while Tl^+ content inside phage particles was about 5x10-11 g-equiv./pfu. An equivalent of 10^8 pfu peptide-modified MS2 filled with Tl^+ added to cell cultures resulted in 80-85% both MCF7 and DA-MB-231 cell death in enhanced phagocytosis conditions, with much lower apoptosis in serum-containin gmedia. No cell death was observed in control experiments. When peptide-modified MS2 filled with Tl^+ was injected to MCF7- and DA-MB-231-xenograft nude mice, it caused tumour mass loss up to 2.5 times compared to control (fig. 3).

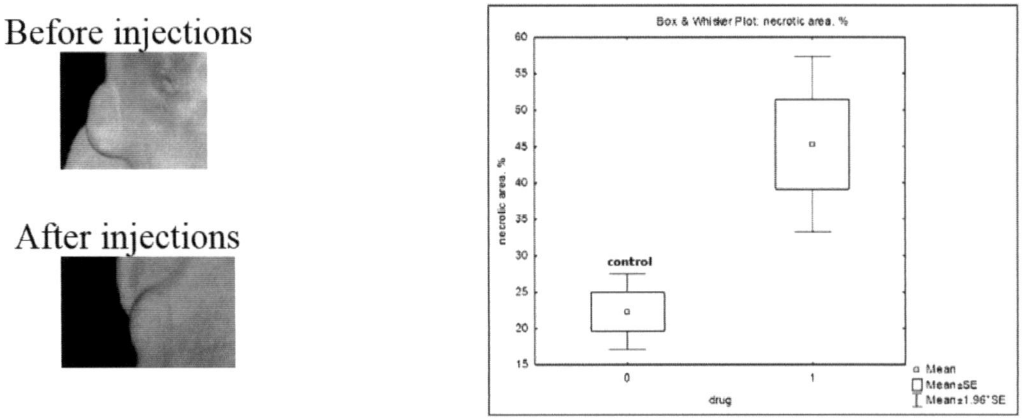

Figure 3: Tumour necrosis areas visualized on photographs and determined by ScanScope CS2

LD_{50} for iRGD-MS2-Tl was estimated as 1.5x10 ^{8} pfu/kg in Wistar-Kyoto rats. Overall concentration of $TlNO_3$ in peptide-modifiedMS2 filled with Tl^+ was 500000 times lower than $TlNO_3$ LD_{50}, and therapeutic index of the phage-based preparation was about 15000. The results show the perspective of the use of bacteriophage MS2 filled with Tl^+ and containing a target delivery system as a lead substance for solid tumour therapy. No Tl^+ leakage was observed during iRGD-MS2-Tl storage. Hence iRGD-MS2-Tl was recommended for preclinical investigations as an anticancer drug for the treatment of breast cancer.

Acknowledgments

The research was supported by the State contract with the Russian Ministry of Education and Science No. 14.N08.11.0188 (27.11.2017) (unique project ID 171771053913577100100100030807219241). Peptides were synthesized with the use of the equipment of "Human Proteome" Core Facility of the Institute of Biomedical Chemistry (Moscow, Russia) supported by Ministry of Education and Science of the Russian Federation (unique project ID RMEFI62117X0017). Animal experiments were performed at the N.N. Petrov Research Oncology Centre of the Russian Ministry of Healthcare.

The results are covered by the Russian Federation Patent No. 2599462, PCT WO 2017052419, and US patent claim 15/757,285 of 02.03.2018.

References

1. Tessalu T. *et al.* / Proc. Natl. Acad. Sci. USA 106 (2009) 16157-16162. doi: 10.1073/pnas.0908201106.

2. Zasukhina G.D. *et al.* /Mutat. Res. 124 (1983) 163-173

3. Korotkov S. M., *et al.* /J. Biochem. Mol. Toxicol. 28 (2014) pp. 149-156. doi: 10.1002/jbt.21547.

4. van Regenmortel M.H.V., Muller S. Synthetic peptides as antigens. Elsevier, 1999, pp. 88-90.

Antimicrobial activities of chimera peptides composed of human neutrophil peptide 1 (HNP-1) truncated analogues and bovine lactoferrampin

Natalia Ptaszyńska[1], Katarzyna Gucwa[2], Anna Łęgowska[1], Dawid Dębowski[1], Agata Gitlin-Domagalska[1], Jan Lica[2], Mateusz Heldt[2], Dorota Martynow[2], Mateusz Olszewski[2], Sławomir Milewski[2], Krzysztof Rolka[1]

[1]Univesity of Gdańsk, Faculty of Chemistry, Departmentof Molecular Biochemistry, Poland
[2]Gdańsk University of Technology, Faculty of Chemistry, Departmentof Pharmaceutical Technology and Biochemistry, Poland

https://doi.org/10.17952/35EPS.2018.303

Introduction

Many pathogenic bacteria and fungi strains have become resistant to the most of commercially available antibiotics as the result of their common and excessive use in medical practice. One of the most promising strategy for searching for new compounds with antimicrobial activity is bioconjugation, covalent or non-covalent connection, of at least two chemical molecules of compounds with different activity that in the end display a different mechanism of action than the parent compounds.

The main aim of this work was to obtain a new class of compounds with antimicrobial activity. These compounds are peptidic hybrids composed of two peptides with antimicrobial activity. First peptide is a modified fragment of human neutrophil peptide 1 (HNP-1), a member of α-defensins [1]. This peptide derived from the C-terminal fragment of HNP-1, comprised of its modified (Cys^{19} $\rightarrow$Cys(Acm), Gly^{23} $\rightarrow$ 2Abz, Cys^{29} $\rightarrow$Ser) middle 15 - 29 fragment. It was reported that this peptide displayed antimicrobial activity against a broad spectrum of pathogens, the closest to that of full length HNP-1. Second peptide is bovine lactoferrampin (LFampB) which is the fragment 268 - 284 of bovine lactoferrin [2]. This peptide also exhibits broad antimicrobial action against several gram-positive and gram-negative bacteria as well as candidacidal activity [3,4]. We decided to design and synthesize series of peptidic hybrids composed of 2Abz^{23}S^{29} analogues (named here as HNP) and bovine LFampB. Primary structures of synthesized chimeras and their constituent peptides are shown in Fig. 1. Two different covalent linkers were used: isopeptide bond, considered as non-cleavable in biological systems and redox-sensitive disulfide bridge. The first chimera (CH8) is composed of LFampB, which is coupled through the isopeptide bond formed by its C-terminal carboxyl group to the reference peptide HNP by the ε-amino group of the C-terminal Lys. In the second chimera (CH9), Cys(Acm) residue of HNP was replaced by l-α-aminobutyric acid (Abu), which is often used to substitute sulfur containing amino acids (Met or Cys). In the third chimera (CH10), constituent peptides were coupled by the disulfide bridge formed by the thiol groups of Cys^5 of HNP and the Cys residue attached at the C-terminus of LFampB, respectively. In order to determine the cellular uptake of CH10, its fluorescently-labelled analogue (CH10-CF) and constituent peptides (HNP[Cys^5]-CF and CF-LFampB-C) were also synthesized.

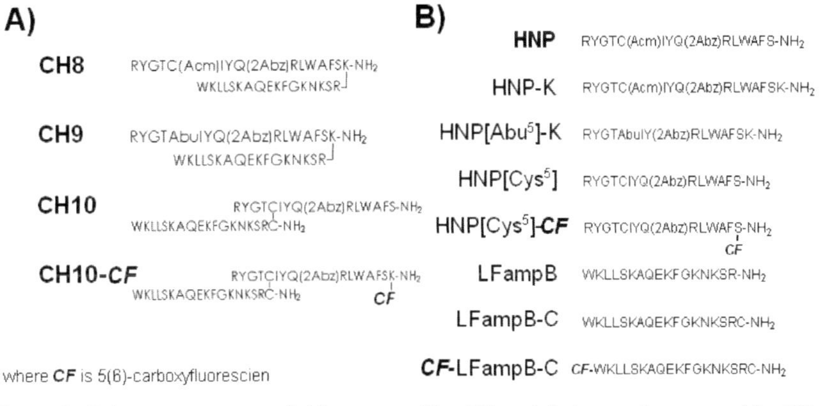

Figure 1: Primary structures of chimera peptides (A) and their constituent peptides (B).

Results and Discussion

All peptides were screened against the wide range of microorganisms, including Gram-negative, Gram-positive bacteria and fungi from *Candida* species. Broth microdilution method for MIC determination was used as a standard procedure. As a result, we obtained three peptidic hybrids that effectively inhibited the growth of selected bacterial strains, but none of the chimeric peptides was able to inhibit fungi growth. Under experimental conditions, they demonstrated significantly higher antimicrobial activity than constituent peptides. Chimeras containing the isopeptide bond as a linker (CH8 and CH9) were active against both Gram-positive and Gram-negative bacteria, whereas the one with a disulfide bridge (CH10) was specifically active against Gram-positive bacteria. CH8 and CH9 exhibited the most potent activity against all experimental strains, being the most effective against *S. epidermidis* and *P. aeruginosa*. Activity of CH10 against Gram positive bacteria was similar to that of CH8 and CH9 but this compound was practically inactive against Gram-negative strains. These results clearly show that the type of a linker used to connect AMPs determines the antibacterial activity of chimera peptides. Presence of the isopeptide bond in chimeras CH8 and CH9 resulted in their remarkable antimicrobial activity against both Gram-positive and Gram-negative bacteria, whereas connection of the two peptide chains by a disulfide bridge afforded CH10, specifically active against Gram-positive bacteria. The fact that the equimolar mixtures of constituent peptides were practically inactive, clearly indicates that antibacterial activity of chimera peptides results from the presence of a covalent linkage between the constituent peptides. The observed specificity correlated well with a significantly higher content of the helical structure determined in an environment that mimics a biological membrane in case of the chimera CH10 in compare to remaining two chimeras. The antimicrobial activity of the equimolar mixtures of studied peptides was lower than that of the individual peptides and chimeras. This observation indicates a lack of the synergistic effect of constituent peptides and importance of the covalent linkage between them for the enhancement of antibacterial activity.

In this study we also reported that antimicrobial activity depends strongly on cell penetration. Using carboxyfluorescein labeled chimera (CH10-CF) and its constituents we checked weather compounds undergo cellular uptake. Microscopic observations showed that CH10-CF as well as its constituent LFampB-C penetrate well into *S. epidermidis* cells (Fig. 2). In contrast, HNP[Cys^5] with the lowest activities neither penetrates the cell nor interact with cell membrane. In reference to antimicrobial drugs widely used in clinic, cytotoxicity of tested chimera peptides against human cells is low and rather limited to cancer cells which possess low negative charge (in contrast to no-cancer which are neutral or slightly positive [5]). Indeed, selective toxicity of chimera peptides against pathogenic bacteria is also associated with negative charge of their cell surface. We also showed that generation of ROS in human cells is limited what can be perceived as an advantage from the point of view of eventual toxicity to the host. Nevertheless, small amount of ROS generated may be advantageous in combating pathogenic bacteria. Type of the linker has a significant impact on the proteolytic stability of studied compounds which is an important issue to consider especially in case of peptides. Taking into consideration both antimicrobial activity and selective cytotoxicity not affecting normal cells and also other presented results, we claim that such chimera peptides are very promising lead structures for the development of antimicrobial drugs.

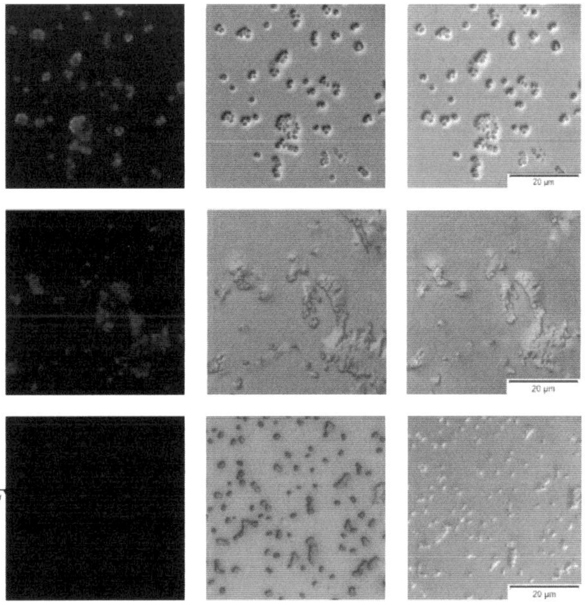

Figure 2: Fluorescence microscope images of CH10-CF and its constituents (CF-LFampB-C and HNP[Cys5]-CF) uptake by S. epidermidis. Cells were grown to the mid-log phase and exposed to the action of the compounds at concentration 200 μm for 30 min.

Acknowledgements

This work was supported by the National Science Centre (NCN) under grant No UMO-2016/21/B/ST5/00101.

References

[1]Lundy F.T., Nelson J., Lockhart D., Greer B., Harriott P., Marley J. Mol Immunol45, 190-193 (2008), DOI: 10.1016/j.molimm.2007.04.018.

[2]. 90, 245-251 (2012), DOI: 10.1139/o2012-018 .

[3]van der Kraan M.I.A., Groenink J., Nazmi K., Veerman E.C.I., Bolscher J.G.M., NieuwAmerongen A.V. Peptides, 25, 77–183 (2004), DOI: .

[4] Haney E.J., Nazmi K., Lau F., Bolscher J.G.M., Vogel H.J. Biochimie 91, 141-154 (2009), DOI:

[5] Chen B., Le W., Wang Y., Li Z., Wang D., Ren L., Lin L., Cui S., Hu J. J., Hu Y., *et al.* Theranostics 6, 1887–1898 (2016), DOI: 10.7150/thno.16358.

Design, chemical synthesis and evaluation of antimicrobial activity peptide conjugates of lactoferricin analogues and antibiotics

Natalia Ptaszyńska[1], Katarzyna Olkiewicz[1], Joanna Okońska[1], Anna Łęgowska[1], Katarzyna Gucwa[2], Agata Gitlin-Domagalska[1], Dawid Dębowski[1], Sławomir Milewski[2], Krzysztof Rolka[1]

[1]Univesity of Gdańsk, Faculty of Chemistry, Departmentof Molecular Biochemistry, Poland
[2]Gdańsk University of Technology, Faculty of Chemistry, Departmentof Drug Technology and Biochemistry, Poland

https://doi.org/10.17952/35EPS.2018.306

Introduction

Despite the increasing need for antibiotics to fight infectious diseases, fewer new antibiotics are available on the market. Modification or combination of existing antibiotics to improve their efficacy is one of a promising strategy. In this broad field, peptide - drug conjugates linked by non-cleavable or intracellular cleavable structures have evolved as highly promising agents.

Here we report synthesis and biological investigations of a series of peptide conjugates composed of modified bovine lactoferricin (LFcinB) truncated analogues and three antibiotics, ciprofloxacin (CIP), levofloxacin (LVX) and fluconazole (FLC) commonly used in medical practice. The first two belong to the class of fluoroquinolone displaying broad spectrum of antimicrobial activity against both Gram-positive and Gram-negative bacteria. They inhibit DNA gyrase and topoisomerases. FLC, similarly to other azole drugs, inhibits the lanosterol 14α-demethylase and decreases the level of ergosterol required for the integrity of cell membrane [1]. LFcinB is an antimicrobial peptide isolated in 1992 by Bellamy *et al.* [2] from bovine milk, representing 17–41 fragment of bovine, multifunctional iron-binding glycoprotein, lactoferrin (LF). LFcinB displays a broad spectrum of antibacterial and antifungal activity and also cooperative effect with azole types of antifungal agents [3]. Our intention was to design a new class of antimicrobial compounds. Chemical structures of constituent compounds and peptide conjugates are shown in Fig. 1.

Results and Discussion

A series of eight new peptide conjugates containing three above mentioned antibiotics was obtained. Four different approaches were applied to couple a peptide and the antimicrobial agent. We developed an optimized conditions for coupling of LVX *via* its carboxylic group, CIP *via* its carboxylic or amino group to peptide chain using solid phase peptide synthesis. In the last step of synthesis, both drugs were attached through their carboxyl groups to the N-terminal amino group of peptidyl-resin, whereby LVX-LFcinB(2-11)-NH_2 (I) and CIP-LFcinB(2-11)-NH_2 (II) were obtained. CIP was also coupled to the peptide by the method utilized for the synthesis of peptoids, named submonomeric [4]. In this case, the covalent bond was formed in reaction of the secondary amino group of CIP and bromoacetylated peptidyl-resin, yielding CIP-CH2-CO-LFcinB(2-11)-NH_2 (III). In the next two conjugate, CIP-Cys-S-S-LFcinB(2-11)-NH_2 (IV) and CIP-Cys-S-S-Nle-LFcinB-NH_2 (V) intermolecular disulfide bridge was used as a linker between peptide and CIP, formed by the thiol group of peptide Cys present in position 3 and thiol group of Cys attached to CIP amino group. In both conjugates (IV and V) a disulfide bridge was formed in solution phase with the help of S-(3-nitro-2-pyridylsulfanyl) (Npys) group introduced on Cys attached to CIP. In the case of fluconazole-based conjugates VI – VIII contained LFcinB(2-11)-NH_2, Nle-LFcinB and Nle-LFcinB-NH_2, respectively, a "click chemistry" method was used to attach this antifungal agent derivative to a peptidic component. FLC-based conjugates attached through their N-termini to the nitrogen atom of FLC and methylene-carbonyl moiety was used as a linker. This required synthesis of the FLC precursor which was obtained by the method described by Pore *et al.* [5].

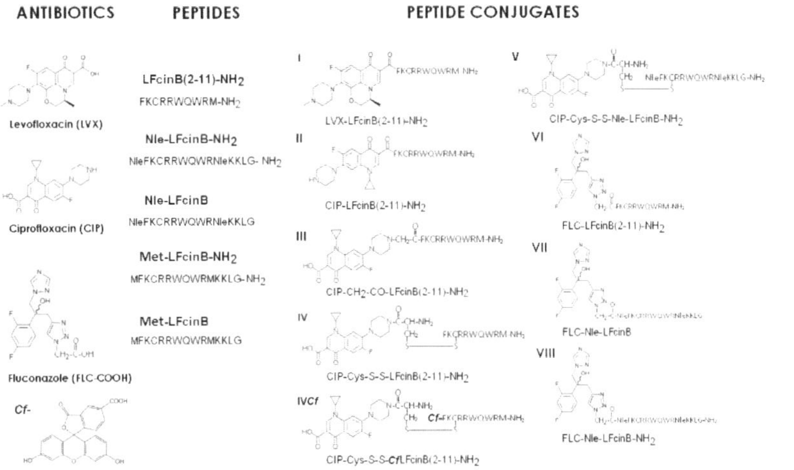

Figure 1: Chemical structures of peptide conjugates, constituent peptides and antibiotics.

All synthesised conjugates and their constituent compounds were investigated for fungicidal activity against five fungal stains *C. albicans* ATCC 10231, *C. krusei* DSM 6128, *C. parapsilosis* DSM 5784, *C. glabrata* DSM 11226 and *C. tropicalis* CZD 519 and for antibacterial activity against six bacterial strains, four Gram-positive (*S. aureus* ATCC 25923, *S. epidermidis* ATCC 12228, *B. subtilis* ATCC 6633, *B. cereus* PCM 2003), two Gram-negative (E.coli ATCC 25922, *P. aeruginosa*ATCC 27857). Constituent peptides displayed weak inhibitory activity against tested strains. The synthesized conjugates, except the conjugate (VII), displayed stronger antifungal activity than the constituent peptides, and in the most cases, than CIP and LVX. The most active appeared to be conjugate (I) composed of LVX and the shortest peptide LFcinB(2-11)-NH_2 which inhibited the growth of *C. krusei* and C. parapsilosis. The determined values of MIC_{50} were 12.5 and 1.56 μg/mL, respectively. All constituent peptides exhibited antibacterial activity, especially against Gram-positive bacteria. The lowest MIC_{50} and MIC90 values were determined for LFc inB(2-11)-NH_2 (3.13, 6.25 μg/ml) and Nle-LFcinB-NH_2 (3.13 μg/ml) against S. epidermidis. These two peptides were also active against *S. aureus* and *E. coli* . Fluorescently labeled CIP-Cys-S-S-Cf-LFcinB(2-11)-NH $_2$ (IVCf) as well as its constituent Cf-LFcinB(2-11)-NH_2 are able to penetrate both *C. albicans* and S. epidermidis. However, despite rapid and effective internalization they do not show antifungal activity.

Cytotoxicity of conjugates, constituent peptide and drugs (LVX, CIP, FLC) was assessed against four human cell lines: non-cancer cell line of embryonic kidney (HEK 293) and three cancer cell lines: non-small cell lung cancer (A549), breast cancer (BT-20) and acute myeloblastic leukemia (HL-60). Cytotoxicity assays on cancer and non-cancer cell lines showed that the proposed conjugates induce a relatively low cytotoxic effect in reference to antibiotics widely used in clinic (except FCL which was least toxic). Non-cancer cell line was the most resistant to the action of the tested compounds, where cell growth was virtually not affected up to the concentration of 250 μM.

ROS generation by conjugates and their constituents was tested on non-cancer cell-line HEK 293 and cancer cell line HL-60. In neither of tested conjugates the amount of ROS generated was higher than 25 %. In contrast CIP and LVX caused production of ROS in more than 50 % of the cells. The differences were more pronounced in case of cancer cells where the observed amount of ROS generated under influence of conjugates was comparable to the negative control (not treated with any compound)or slightly elevated, while in the case of antibiotics the amount of ROS was on the level of positive control (H2O2) or even higher.

Within this study, eight conjugates of antimicrobials, CIP, LVX and FLC and the LFcinB peptide were synthesized and evaluated for their antimicrobial activity. Three FLC-based conjugates, to our knowledge, are the first examples of peptide conjugates of this antifungal agent reported in literature. Summarizing the results presenting above we would like to stress that some of the obtained conjugates showed promising antibacterial activity, especially against Gram-positive bacteria, within which *S. epidermidis* was the most sensitive. The most active ones were conjugates containing CIP attached to the peptide by the redox-sensitive disulfide bridge. They displayed not only higher efficacy than constituent peptides, but also broader spectrum of activity, inhibiting the growth of other strains. The obtained results indicate an important role of a linker, especially a disulfide bridge, used to couple drug and peptide. In all three conjugates (III-V), CIP was attached through its secondary amino group, but the highest antimicrobial activity showed those with disulfide bridge. The primary novelty of this work is the activity of CIP and LVX based conjugates against *Candida* yeast. Studies of CIP/LVX with peptide combinations against *Candida* yeast did not indicate the possibility of synergistic effects which also indicate an important role of a linker, especially a disulfide bridge, used to couple drug and peptide. Taking all above into consideration, also low or moderate side effects (toxicity and ROS generation) of studied conjugates, we find that proposed conjugates are promising led structures for development of new antibacterial drugs.

Acknowledgements

This work was supported by theNational Science Centre (NCN) under grant No UMO-2016/21/B/ST5/00101.

References

[1] Kabir M.A., Ahmad Z., ISRN Preventive Medicine, https://doi.org/10.5402/2013/763628 (2013).

[2] Bellamy W., Takase M., Yamauchi K., Wakabayashi H., Kawase K., Tomita M. Biochim. Biophys. Acta 1121, 130–136 (1992)

[3] Wakabayashi H., Abe S., Teraguchi S., Hayasawa H., Yamaguchi H. Antimicrob Agents Chemother, 42, 1587–1591 (1998), DOI:~10.1128/AAC.42.7.1587.

[4] Ł ęgowska A., Dębowski D., Łukajtis R., Wysocka M., Czaplewski C., Lesner A., Rolka K. Bioorg. Med. Chem. 18, 8188–8193 (2010)

[5] Pore V. S., Aher N. G., Kumar M., Shukla P. K. Tetrahedron62, 11178–11186 (2006)

Peptide conjugates of transportan10 with antimicrobial and antifungal antibiotics

Katarzyna Olkiewicz[1], Anna Łęgowska[1], Natalia Ptaszyńska[1], Agata Gitlin-Domagalska[1], Dawid Dębowski[1], Joanna Okońska[1], Dorota Martynow[2], Marcin Serocki[2], Sławomir Milewski[2], Krzysztof Rolka[1]

[1]Department of Molecular Biochemistry, Faculty of Chemistry, University of Gdańsk, Poland
[2]Department of Pharmaceutical Technology and Biochemistry, Faculty of Chemistry, Gdańsk University of Technology, Poland

https://doi.org/10.17952/35EPS.2018.309

Introduction

Conjugates, in which two molecules differing by chemical and physical properties are connected, are a promising group of compounds. Conjugates consisting of a cell penetrating peptide (CPP) and a molecule with therapeutic properties are among most frequently studied compounds. CPPs can deliver to eukaryotic and prokaryotic cells a vast range of different biologically active compounds, such as proteins, nucleic acids, oligonucleotides, liposomes, nanoparticles, peptides, PNA, and low-molecular chemotherapeutic agents (e.g. antibiotics). Transportan(TP), reported by Langel group, is a chimeric peptide composed of the first 12 amino acid residues of neuropeptide galanin and 14 amino acid residues-long wasp venom peptide, mastoparan, connected *via* a lysine residue. A short variant, named TP10, with deletion of the N-terminal hexapeptide, retains the efficient cell penetration property of the parent compound with significantly less potential side effects [1].

In this paper we report synthesis and biological studies of peptide conjugates composed of TP10 and antibiotics: levofloxacin, ciprofloxacin and fluconazole. Three different linkers (methylene carbonyl, amide and disulfide bridge) were used to connect both components. TP10 may not only help to transport the antibiotic across the microbial cell membranes, but having intrinsic antimicrobial activity, it can potentiate or even change the molecular mechanism of action of the conjugated antibiotic. In order to determine the cellular uptake of TP10 and its conjugate with ciprofloxacin, their fluorescently-labeled analogues were also synthesized (TP10(F)-NH_2 and CIP-TP10(F)-NH_2). The chemical formulas of antibiotics and synthesized compounds are shown in Fig. 1.

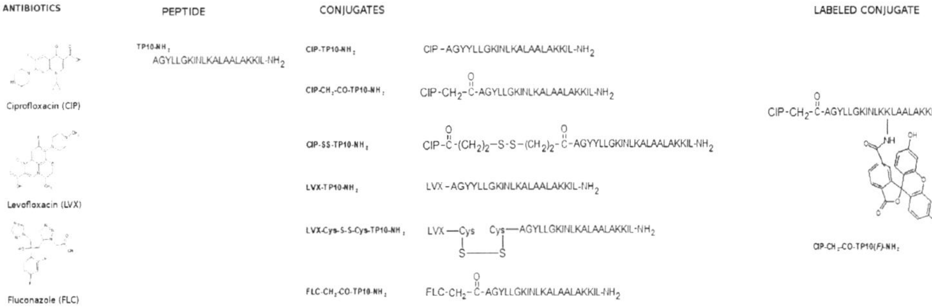

Figure 1: Chemical structures of antibiotics, TP10 and peptide conjugates.

Results and Discussion

All conjugates were obtained using solid phase synthesis and standard 9-fluorenylmethoxycarbonyl (Fmoc) chemistry. In CIP-TP10-NH_2 and LVX-TP10-NH_2 amide bond was formed between carboxyl group of CIP or LVX and peptide α-amino group. In CIP-CH2-CO-TP10-NH_2, methylene carbonyl linker was applied to connect antibiotic through its amine to the N-terminus of TP10. Intermolecular disulfide bridge, spanning both components in CIP-SS-TP10-NH_2 and LVX-Cys-S-S-Cys-TP10-NH_2, was formed using Lomant's reagent – dithio-bis(succinimidyl propionate) or between two Cys residues.

FLC-CH2-CO-TP10-NH_2 was synthesized *via* modified Huisgen 1,3-dipolar cycloaddition, known also as click reaction. This required synthesis of the FLC precursor (1 in Fig. 2.) containing alkyne group, which was obtained by the method described by Pore *et al.* [2], and then a click reaction of 2-(2,4-difluorophenyl)-1-(1H-1,2,4-triazol-1-yl)pent-4-yn-2-ol with a peptide containing azidoacetyl moiety attached to α-amino group was performed. In order to determine the cellular uptake, fluorescently-labeled CIP-CH2-CO-TP10(F)-NH_2 and TP10(F)-NH_2 were also synthesized. In both compounds, 5(6)-carboxyfluorescein was attached to the ε-amino group of peptide's Lys that replaced Leu in position 12.

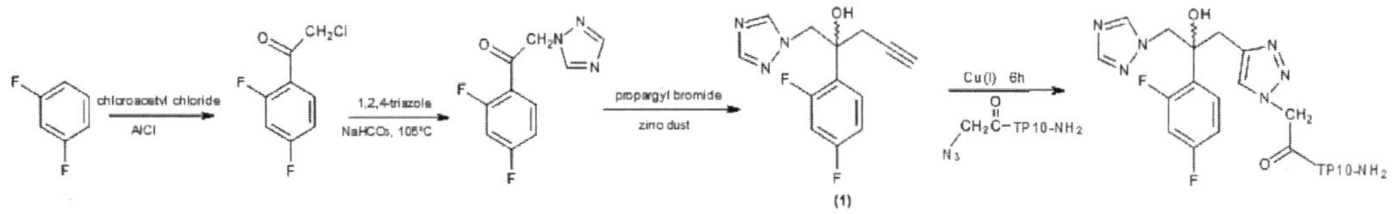

Figure 2: Synthesis of FLC-CH2-CO-TP10-NH_2.

All conjugates and their parent compounds were investigated for fungicidal activity against four fungal strains: *C. albicans* ATCC 10231, *C. albicans* SC 5314, *C. krusei* DSM 6128, and *C. glabrata* DSM 11226, and antibacterial activity against five bacterial strains, three Gram-positive: *S. aureus* ATCC 25923, *S. aureus* ATCC 29213, *S. epidermidis* ATCC 12228, and two Gram-negative (E.coli ATCC 25922, *P. aeruginosa*ATCC 27853). TP10-NH_2, levofloxacin and ciprofloxacin did not exhibit antifungal activity at concentrations up to 100 μg/ml; however, two conjugates with ciprofloxacin CIP-TP10-NH_2 and CIP-SS-TP10-NH_2 were active against both *C. albicans* strains (MIC_{50} values determined for CIP-TP10-NH_2 were 62 and 30 μg/mL, respectively and for CIP-SS-TP10-NH_2 MIC_{50} were 24 and 34 μg/mL, respectively). Unfortunately, FLC-CH2-CO-TP10-NH_2 was not active towards tested fungal strains. TP10 did not display antibacterial activity at concentrations up to 200 μg/ ml, whereas its two conjugates with ciprofloxacin CIP-TP10-NH_2 and CIP-SS-TP10-NH_2 and one with levofloxacin LVX-Cys-S-S-Cys-TP10-NH $_2$ inhibited growth of all studied bacteria (Table 1). Interestingly, fluconazole conjugate was also found active towards bacteria. To assess the cytotoxic effect of TP10, ciprofloxacin and its two conjugates, we estimated IC_{50} values (using MTT assay) for two human cell lines: liver cancer cells (HepG2) and cell line of embryonic kidney (HEK 293), as well as a cell line isolated from kidney of 3-4 weeks old male pig (LLC-PK1). The conjugates were slightly more cytotoxic than their constituents; however, they induced a relatively low cytotoxic effect in comparison with antibiotics widely used in clinic. Microscopic observations showed that TP10(F)-NH_2 as well as its conjugate CIP-CH2-CO-TP10(F)-NH_2 penetrate well into *C. albicans* SC 5314, and *C. albicans* ATCC 10231.

Table 1: MIC values determined for selected bacterial strains

Compound	MIC [µg/ml]									
	Gram (+)						Gram (-)			
	S. aureus ATCC 25923		*S. aureus* ATCC 29213		*S. epidermidis* ATCC 12228		*E. coli* ATCC 25922		*P. aeruginosa* ATCC 27853	
	MIC_{90}	MIC_{50}	MIC_{90}	MIC_{50}	MIC_{90}	MIC_{50}	MIC_{90}	MIC_{50}	MIC_{90}	MIC_{50}
Ciprofloxacin (CIP)	<0.3	<0.3	<0.3	<0.3	0.9	0.2	<0.3	<0.3	<0.3	<0.3
Levofloxacin (LVX)	<0.3	<0.3	<0.3	<0.3	2.2	1.5	<0.3	<0.3	1.5	1.1
Fluconazole (FLC)	>250	>250	>250	>250	>250	>250	>250	>250	>250	>250
TP10	>300	257	>300	231	>300	215	246	121	>300	284
CIP-CH_2CO-TP10	295	132	290	198	215	102	15	9	256	115
CIP-TP10	4	4	2	2	4	4	4	4	8	8
CIP-SS-TP10	1.1	0.7	1.3	0.8	1.6	0.2	<0.3	<0.3	1.1	0.8
LVX-TP10	>>300	199	>>300	289	274	211	74	28	164	140
LVX-Cys-SS-Cys-TP10	62	62	16	16	8	8	8	8	62	31
FLC-TP10	16	16	8	8	8	8	8	8	62	62

In conclusion, we obtained six new conjugates composed of antibiotic (ciprofloxacin, levofloxacin or fluconazole) and TP10-NH_2. Two ciprofloxacin conjugates CIP-TP10-NH_2 and CIP-SS-TP10-NH_2 exhibited fungicidal activity against *C. albicans* 10231 and *C. albicans* SC 5314, even though their constituents were inactive. These conjugates displayed also antibacterial activity, whereas TP10-NH_2 did not inhibit bacterial growth. It is worth noting that the activity of conjugate CIP-SS-TP10-NH_2 was the same as of ciprofloxacin.

Acknowledgements

This work was supported by the National Science Centre (NCN) under grant No UMO-2016/21/B/ST5/00101.

References

[1] M. Pooga, M. Hallbrink, M. Zorko, Ü. Langel, FASEB J., 1998, 12, 67–77;

[2] Pore V.S., Aher N.G., Kumarb M., Shukla P. K., Tetrahedron, 2006, 62, 11178-11186; doi.

Host cell targeting with peptide candidates derived from a HSV-1 entry protein

Szilvia Bősze [1], Ferenc Zsila [2], Zsuzsanna Majer [3], Ferenc Hudecz [3], Katalin Uray [1]

[1]MTA-ELTE Research Group of Peptide Chemistry, Hungary
[2]MTA-TTK, Institute of Materials and Environmental Chemistry, Research Centre for Natural Sciences, Hungary
[3]Departmentof Organic Chemistry, Eötvös L. University, Hungary

https://doi.org/10.17952/35EPS.2018.312

Introduction

Herpes Simplex viruses (HSV-1 and HSV-2 of *Alphaherpesvirinae*) have unique entry mechanism into the host cells using their gD glycoprotein as well as gB, gC and gH/gL, as usual in case of other herpes viruses. gD glycoprotein selectively binds nectin-1 adhesion protein on the surface of the host's neurons, keratinocytes or epithelial cells followed by major conformational changes of the gD glycoprotein [1]. Based on the literature [2] and knowing the 3-dimensional structure of the HSV-1 gD and of HSV-1 gD – nectin-1 [3] receptor complex, we have chosen 20mer HSV-1 gD fragments represented by synthetic peptides for cellular uptake studies.

Methods

Overlapping 20-mer peptide amides of the HSV-1 gD 239-280 region were prepared on either SynPhaseTM lanterns (Mimotopes) or Rink-amide-MBHA resin with Fmoc/tBu strategy with DIC/HOBt chemistry. Methionine residues were substituted by norleucine to avoid oxidation. The peptides were acetylated for ECD measurements or labelled with 5(6)-carboxyfluorescein (Cf) for cellular uptake studies, and then cleaved from the solid support with TFA and appropriate scavengers, purified by RP-HPLC and characterised by ESI-MS. Carboxyfluorescein labelled peptides were characterised by fluorescence spectroscopy as well on different pH values.

Cellular uptake of Cf-HSV peptides into host cell model [4-6], SH-SY5Y human neuroblastoma cells [7-9] was characterised by the enhanced intracellular fluorescence measured by BD LSR II flow cytometer (Coherent Sapphire laser; λ = 488 nm, 22 mW) (incubation time: 3h; channel FITC LP505; BP 530/30), data were analysed with FACSDiva 5.0 software. All measurements were performed in duplicates, and the Cf-positive (Cf+) live cells with standard error of the mean was presented (Figure 1). Uptake of HSV 253-272 at 10 μM was visualised in an Olympus CKX41 microscope.

Electronic circular dichroism (ECD) spectroscopy measurements were performed in the far-UV region on selected peptides with Jasco J-715 / Jasco J-810* spectropolarimeters, at 0.1 cm / 0.02* cm path length in quartz cuvette; in 2,2,2-trifluoroethanol (TFE); TFE-water 1:1 (v/v); water solvents, in ~0.14 mg/mL / 0.5-0.7 mg/mL* peptide concentration.

Results

Significant differences have been observed between the cellular uptake of individual peptides. Peptides containing the 253-268 region of the HSV-1 gD glycoprotein (HSV-249-268 and 253-272) were observable in more than 50 % of the cells in 10 μM concentration. Parallel with flow cytometry measurements, to visualise cell morphology after Cf-peptide treatment, microscopic images of trypsinised, washed and resuspended cells were also captured. Fluorescent signal was concentrated on the cytoplasm (data not shown).

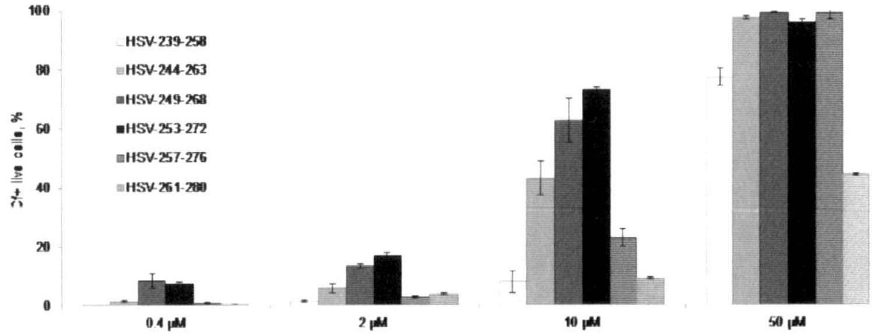

Figure 1: Cellular uptake profile of Cf-labeled HSV-1 gD peptides into SH-SY5Y neuroblastoma cells

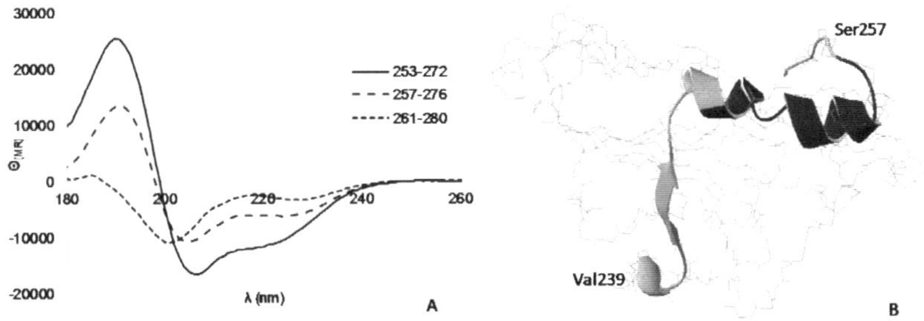

Figure 2: A) ECD spectra of acetylated HSV-1 gD peptides in TFE, Jasco J-810 spectropolarimeter. B) Localization of the studied region on the HSV-1 gD in ribbon structure (PDB id: 3SKU), the helical region 253-268 is represented in black ribbon.

The ECD spectral patterns clearly indicate that most of the peptides are stabilised in helical conformation which is the consequence of the helix promoting effect of TFE [10]. The spectroscopic data are in good agreement with the cell internalisation profile. The most efficient cellular uptake was observed for HSV-249-268 and 253-272 (Figure 2A), which exhibit pronounced helical folding in the membrane mimicking solvent TFE. Conversely, sequences associated with little helical character even in pure TFE such as HSV-239-258, 257-276 and 261-280 (Figure 2A) showed decreasing internalisation rates.

It can be concluded that the disorder-to-helix conformational transition of HSV peptides is a decisive structural feature for their cell entry mediated by biomembrane interactions. Taken together, our studies demonstrated that nectin-1 binding sequences of the HSV-1 gD glycoprotein can internalise into human SH-SY5Y neuroblastoma cells with high efficiency. The most potent peptides are corresponding to sequences on the FG-loop of HSV-1 gD, within that, a helical structure (Figure 2B). Therefore, these peptides might be promising candidates for the development of conjugates for antiviral drug delivery into HSV infected cells.

Acknowledgements

Supported by the Hungarian Scientific Research Fund (OTKA K104275 and 104385), MedInProt, the Foundation for Hungarian Peptide and Protein Research and Ministry of Human Capacities (1783-3/2018/FEKUTSRA T).

References

[1] Connolly SA, et al.: Nat Rev Microbiol (2011) 9: 369–381; doi:10.1038/nrmicro2548.

[2] Lazear E, Whitbeck JC, Zuo Y, Carfí A, Cohen GH, Eisenberg RJ, Krummenacher C.: Virology (2014) 448: 185–195; doi:10.1016/j.virol.2013.10.019

[3] Di Giovine P, Settembre EC, Bhargava AK, Luftig MA, Lou H, Cohen GH, Eisenberg RJ, Krummenacher C, Carfi A: PLoS Pathog (2011) 7(9):e1002277; doi:10.1371/journal.ppat.1002277

[4] Smith G: Annu Rev Microbiol. (2012) 66: 153–176; doi:10.1146/annurev-micro-092611-150051.

[5] Shipley MM, Mangold CA, Kuny CV, Szpara ML.: J Virol. (2017) 91(23) pii: e00958–17; doi:10.1128/JVI.00958-17.

[6] Yin K, Baillie GJ, Vetter I.: Mol Pain. (2016) 12: 1–17; doi:10.1177/1744806916646111

[7] Biedler JL, Helson L, Spengler BA: Cancer Res. (1973) 33: 2643–2652.

[8] Biedler JL, Roffler-Tarlov S, Schachner M, Freedman LS.: Cancer Res. (1978) 38(11 Pt 1):3751–3757.

[9] Kovalevich J, Langford D: Methods Mol Biol. (2013) 1078: 9–21; doi: 10.1007/978-1-62703-640-5_2.

[10] Buck M.: Q Rev Biophys (1998) 31: 297–355; doi:10.1017/s003358359800345x

Proceedings of the 35th European Peptide Symposium
Patrick B. Timmons, Chandralal M. Hewage, Michal Lebl (Editors)
European Peptide Society & PSP, 2018

Biochemical-activity studies of NGR-peptide-drug conjugates for targeted tumour therapy

Kata Nóra Enyedi, Andrea Angelo Pierluigi Tripodi, Gábor Mező

MTA-ELTE Research Group of Peptide Chemistry, Hungary

https://doi.org/10.17952/35EPS.2018.315

Introduction

Chemotherapy has a great importance in the treatment of cancer patients. However, the vast majority of chemotherapeutic drugs used today can cause serious side effects. Targeted tumour therapy may offer an opportunity to overcome this problem by conjugating chemotherapeutic drugs to tumour specific homing peptides. NGR (Asn-Gly-Arg) peptides received particular interest when phage display libraries were used to identify non-RGD integrin binding motifs. Among the non-RGD peptides the NGR motif was the most frequent one that showed integrin binding properties.[1] However, it has been found that peptides containing the NGR motif, are not integrin ligands, but specifically recognize Aminopeptidase N (CD13) receptor isoforms that are overexpressed in tumour vasculature and on some tumour cells.[2] In addition, these sequences prone to spontaneous decomposition through a succinimide-ring formation, forming isoAsp and Asp derivatives which is strongly influenced by the structure. The resulting isoaspartyl (isoDGR) derivatives are recognized by the members of RGD-integrin family that are essential for tumour metastasis.[3] Hence NGR peptides are of high interest, due to their simultaneous and selective targeting of CD13 and RGD-integrin receptors, therefore potential applications in dual drug targeting.

Results and Discussion

Based on the literature and our earlier work, we have chosen seven small cyclic NGR peptides as targeting molecules: c[KNGRE]-NH_2, c[NleNGRE]-NH_2, c[GNSarRGK], Ac-c[CNGRC]-NH_2 and H-c[CNGRC]-NH_2, and also the thioether bond containing c[CH2-CO-KNGRC]-NH_2 and c[CH2-CO-NGRC]-NH_2.[4,5,6] Using these small cyclic NGR derivatives as homing moieties, we designed, synthesised and characterized novel cyclic NGR peptide-drug conjugates with a well-known chemotherapeutic agent, daunomycin (Dau). Since sequences of certain peptides allowed different conjugation sites for Dau, where it was possible, not only the C-terminal elongated, but also branched variants were prepared. Because the efficient internalization of the target receptors is questionable, therefore an extracellular MMP-2 enzyme labile spacer (GPLGVRG) was incorporated between the homing peptide and drug molecule to ensure selective drug release. It is believed that the released drug or its active metabolite can enter the cells by diffusion afterwards, that make them suitable for targeted tumour therapy. The following conjugates were prepared (compound 1-9).

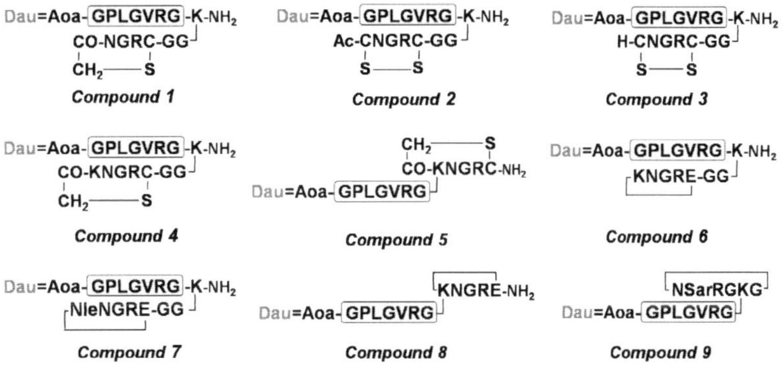

Figure 1

Linear peptides were synthesized on Rink Amide MBHA resin by Fmoc chemistry. To avoid side reactions the aminooxyacetyl group was blocked with isopropyliden protection. The cyclization through thioether bond formation was carried out in TRIS buffer (0.1 M, pH: 8.1), followed by deprotection of the aminooxyacetyl group with 1 M methoxylamine in NH4OAc buffer (0.2 M, pH: 5.0). The cyclization of the KNGRE, NleNGRE and GNSarRGK containing peptides were done in solution, through in situ active ester formation. Finally, the daunomycin was conjugated *via* oxime bond formation in NH4OAc-buffer in all cases.

The cytotoxicity effect of the novel cyclic NGR peptide-Dau conjugates were examined *in vitro* on CD13 positive HT1080 (human fibrosarcoma) and KS (Kaposi's sarcoma) and, as a CD13 negative control, on HT-29 (human colon adenocarcinoma)cell lines using MTT assay. All the three cell lines express RGD integrins and MMPs.

We theorised that the metabolite generated from the MMP cleavable spacer enters the cells *via* diffusion. This is supportedby the fact, that both the spacer (Dau=Aoa-GPLGVRG-OH) and metabolite (Dau=Aoa-GPLG-OH) show high activity on all cell lines. Also, a clear difference could be detected in conjugates with different conjugation sites: conjugates built through the side chain of Lys (5,8) have better anti-tumour activity and the longer treatment time resulted in a better CD13 selectivity, compared to the conjugates which were elongated through their C-terminal (4,6). Moreover, it was reported, that N-acetylation of CNGRC would impair the biological activity [7]. However, in our case (Compound2 and 3) acetylation highly enhanced the *in vitro* effect on all cell lines.

Nevertheless, more metabolism and cell uptake studies are needed and are in progress to determine the exact mechanism of action of the conjugates.

Table 1

Stability		Compounds	24 h IC_{50} (μM)			72 h IC_{50} (μM)		
24 h	72 h							
(Asn/*iso*Asp/Asp)			KS (CD13+)	HT1080 (CD13+)	HT-29 (CD13-)	KS (CD13+)	HT1080 (CD13+)	HT-29 (CD13-)
0/0/100	0/0/100	1	>>50	>>50	>>50	>>50	37.8 ± 0.7	>>50
0/0/100	0/0/100	2	2.9 ± 1.2	2.7 ± 1.0	2.9 ± 0.9	3.5 ± 1.3	2.7 ± 1.1	5.0 ± 1.2
0/0/100	0/0/100	3	22.5 ± 0.5	14.0 ± 0.2	7.5 ± 0.7	26.8 ± 0.1	14.0 ± 0.2	15.8 ± 0.4
0/100/0	0/100/0	4	11.3 ± 0.3	10.5 ± 0.3	12.6 ± 0.4	14.0 ± 0.3	7.9 ± 0.8	28.7 ± 0.4
12/0/88	0/0/100	5	7.3 ± 0.1	5.7 ± 0.7	11.2 ± 0.5	11.8 ± 0.2	6.4 ± 0.9	17.6 ± 0.8
100/0/0	100/0/0	6	12.5 ± 0.9	13.1 ± 0.1	16.2 ± 0.9	18.9 ± 0.1	10.7 ± 0.2	16.3 ± 0.6
84/0/16	58/0/42	7	>>50	>>50	>>50	>>50	>>50	>>50
75/0/15	52/0/48	8	3.4 ± 1.7	6.9 ±1.7	5.8 ± 0.8	2.6 ± 1.5	4.7 ± 1.9	16.5 ± 2.3
100/0/0	100/0/0	9	23.2 ± 0.6	6.5 ± 1.1	31.0 ± 0.7	14.0 ± 0.6	3.4 ± 0.9	>>50
-	-	Daunomycin	0.1 ± 0.1	0.2 ± 0.1	0.1 ± 0.1	0.1 ± 0.1	0.1 ± 0.1	0.1 ± 0.1
-	-	Dau=Aoa-GPLGVRG-OH	4.1 ± 1.0	2.6 ± 0.8	8.9 ± 0.6	3.6 ± 1.0	10.5 ± 1.2	12.0 ± 1.1
-	-	Dau=GPLG-OH	1.5 ± 0.3	1.1 ± 0.7	0.9 ± 0.1	1.2 ± 0.5	2.0 ± 1.0	2.0 ±0.7

Acknowledgements

The research was founded by NKFIH K119552 and Horizon 2020 research and innovation program under the Marie Sklodowska-Curie grant (No 642004, Magicbullet).

References

[1] Pasqualini R *et al.* Cancer Res. 2000; 60: 722-727.

[2] Curnis F. *et al.* Cancer Res. 2002; 62: 867-874.

[3] Curnis F *et al.* J Biol Chem. 2010; 285: 9114–9123. (DOI: 10.1074/jbc.M109.044297)

[4] Enyedi KN *et al.* J. Med. Chem. 2015, 58, 1806-1817. (DOI: 10.1021/jm501630j)

[5] Tripodi AAP *et al.* Beilstein J Org Chem. 2018; 14: 911–918. (DOI: 10.3762/bjoc.14.78)

[6] Corti A *et al.* Adv Funct Mater. 2017; 27: 1701245. (DOI: 10.1002/adfm.201701245)

Synthesis and in vitro biological effect of GnRH-protoporphyrin IX conjugates

Lilla Pethó[1], József Murányi[2], Nassima Kram[1], Gyöngyi Bökönyi[2], Gabriella Csík[3], Gábor Mező[1]

[1]MTA-ELTE Research Group of Peptide Chemistry, Eötvös Loránd University, Faculty of Science, Institute of Chemistry, Hungary
[2]MTA-SE Pathobiochemistry Research Group, Hungary
[3]SE Department of Biophysics and Radiation Biology, Hungary

https://doi.org/10.17952/35EPS.2018.317

Photodynamic therapy (PDT) combines non-toxic components, a photosensitizer (PS), light and oxygen. If the photosensitizer is activated to its excited states by irradiation with visible light in the presence of molecular oxygen, electron and energy transfers can produce reactive oxygen species (ROS) in the tissue. These ROS react rapidly with the biomolecules leading to cell death *via* apoptosis or necrosis. In most cases a singlet oxygen (^1O) is generated that has a short intracellular life time (3 μs) and a very small intracellular diffusion distance that makes PDT highly selective. Unfortunately, photosensitizers accumulate in healthy tissues too, causing severe side effects like prolonged skin and eye photosensitivity. Therefore, the conjugation of the PS to a carrier peptide that itself has anti-tumor effect can increase the efficacy and selectivity of the treatment.

Protoporphyrin IX (PpIX) is a second-generation PS, it is non-toxic without irradiation, but can efficiently absorb light in the visible area. 635 nm irradiation is usually used to reach deeper (3mm) penetration. PpIX has two carboxyl groups that are suitable for conjugation to targeting moieties.[1]

Gonadotropin-releasing hormone (GnRH) receptors are overexpressed on several tumor cells, e.g. on tumors of the reproductive organs or on oral and laryngeal cancer cells, which makes this receptor a proper target for targeted tumor therapy.[2] GnRH-I is a decapeptide (<EHWSYGLRPG-NH_2, where <E is pyroglutamic acid) synthesized and released in the hypothalamus that plays a central role in the vertebrate reproduction by regulating gonadal activity.[3] Several different isoforms were isolated from different species, like chicken GnRH-II (<EHWSHGWYPG-NH_2) that is expressed also in human mainly in the kidneys, bone marrow and prostate, and is found to be a neuromodulator that stimulates sexual behavior[4] or GnRH-III (<EHWSHDWKPG-NH_2) that was originally isolated from sea lamprey.

GnRH-III binds to both type I and type II GnRH receptors and inhibits proliferation of different cancer cells while having insignificant hormonal activity.[5]

The sequences show that the N- and C- terminal parts are conserved, but the amino acids 5-8 can be changed without significant loss of efficacy.[4] In the sequence of GnRH-I and GnRH-II, the glycine in position 6 can be replaced by D-lysine, which serves as conjugation site, increases enzymatic stability and enhances the agonistic effect too.[6] Rahimipour *et al.* have already conjugated PpIX to GnRH-I[6DLys], and the selective receptor mediated phototoxicity could be demonstrated on T3-1 pituitary gonadotrope cell line.[7]

In most isoform the serine in position 4 can be replaced by a butyric acid modified lysine (Lys(Bu)) since this change increases the receptor binding affinity and the stability of the molecule against enzymes.[8]

In this work GnRH-protoporphyrin IX conjugates were synthesized using various GnRH analogues. All peptides were synthesized manually using solid phase peptide synthesis according to standard Fmoc/tBu strategy. GnRH-I and GnRH-II analogues modified with butyrylated lysin (Lys(Bu)) in position 4 were synthesized with using Fmoc-Lys(Dde)-OH. After finishing the protected decapeptides, Dde group of ^{4}Lys was removed on the resin and the acylation was performed by using butyric anhydride and DIPEA. The GnRH-III analogue was synthesized with aspartic acid methyl ester (Asp(OMe)) in position 6 since the free carboxylic acid could have interfered with the PpIX conjugation. The methyl ester group is labile under basic conditions, so in this case Fmoc-Lys(Mtt)-OH was incorporated into the peptide sequence instead of the Dde protected lysine and the butyrylation was performed after the Mtt removal.

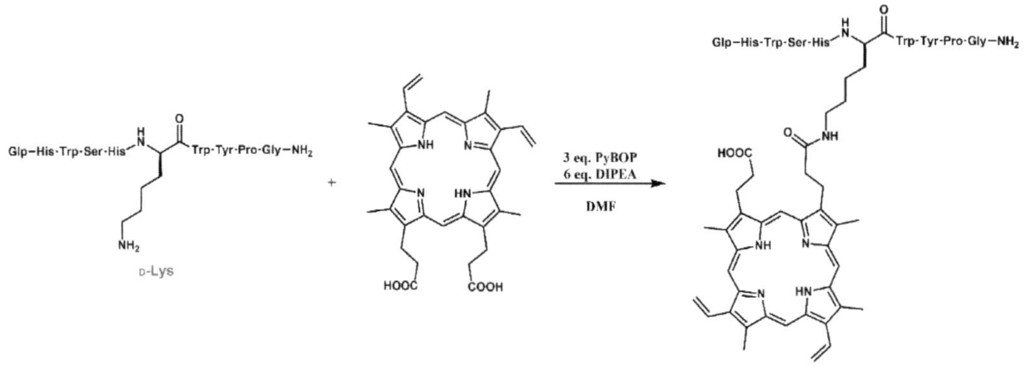

Figure 1: Synthesis of the GnRH-I - PpIX conjugate via amide bond formation

The purified peptides were conjugated to PpIX *via* an amide bond in solution phase using PyBOP (Figure 1).

The *in vitro* biological assays were performed on Detroit-562 human pharynx carcinoma cells that highly express GnRH receptors.[9] The cells were co-incubated with the conjugates, then after wash-out the cells were irradiated with 635 nm. MTT assay was performed after 72h.

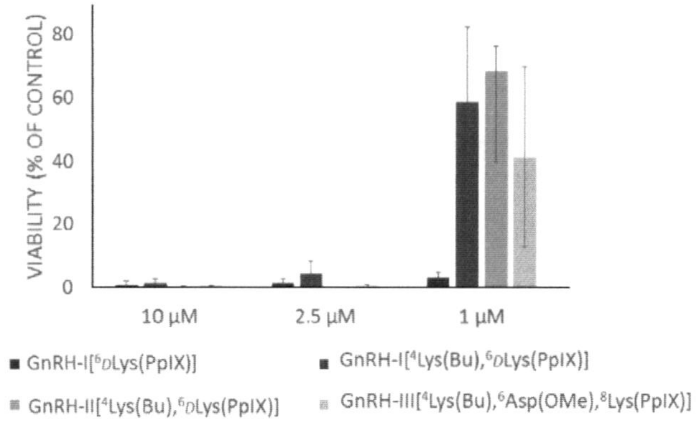

Figure 2: In vitro cytotoxicity of the GnRH-PpIX conjugates

We have investigated the compounds with and without irradiation and we found that the compounds alone, without irradiation, were not toxic in the assays (>10 μM). We could also show that a short irradiation time (10 min) is enough for the tests, since the compounds showed the same *in vitro* effect with 30 min than with 10 min irradiation. The shorter time is more beneficial for the further *in vivo* treatments. Furthermore, we found that all GnRH-Pp conjugates showed excellent anti-tumor effect at low concentrations (~1 μM), the best conjugate was GnRH-I[^{6}D-Lys(PpIX)] (Figure 2).

Acknowledgements

This project has received funding from the National Research, Development and Innovation Office (NKFIH K119552, NVKP_16-1-2016-0036), Hungary.

References

1. N. Malatesti, I. Munitic, I. Jurak, Biophys Rev. 2017 9. 149-168. doi: 10.1007/s12551-017-0257-7

2. A. Nagy, A.V. Schally, Biol. Reprod. 2005 73(5). 851-859. doi: 10.1095/biolreprod.105.043489

3. G. Mező, M. Manea, I. Szabó, B. Vincze, M. Kovács, Curr. Med. Chem. 2008 15. 2366-2379.
doi: 10.2174/092986708785909157

4. R.P. Millar, Anim. Reprod. Sci. 2005 88(1-2). 5-28. doi: 10.1016/j.anireprosci.2005.05.032

5. G. Mező, M. Manea, Expert Opin. Ther. Patents 2009 19(12). 1771-1785. doi: 10.1517/13543770903410237

6. T. Janáky et al., PNAS 1992 89(3). 972-976. doi: 10.1073/pnas.89.3.972

7. S. Rahimipour, N. Ben-Aroya, K. Ziv, A. Chen, M. Fridkin, Y. Koch, J. Med. Chem. 2003 46(19). 3965-3974.
doi: 10.1021/jm020535y

8. R. Hegedüs et al., Eur. J. Med. Chem. 2012 56. 155-165. doi: 10.1016/j.ejmech.2012.08.014

9. J. Murányi et al., J. Pept. Sci. 2016 22. 552-560. doi: 10.1002/psc.2904

Proceedings of the 35th European Peptide Symposium
Patrick B. Timmons, Chandralal M. Hewage, Michal Lebl (Editors)
320 *European Peptide Society & PSP, 2018*

Structure-activity relationship of HER2 receptor targeting peptide and its derivatives in targeted tumor therapy

Afrodité Emese Adorján[1], Szilvia Bósze[2], Ildikó Szabó[2], Gábor Mező[1]

[1]Eötvös Loránd University, Hungary
[2]MTA-ELTE Research Group of Peptide Chemistry, Hungary

https://doi.org/10.17952/35EPS.2018.320

Introduction

In normal cells, epidermal growth factor receptor 2 (HER2/ErbB2) plays a vital role in various cellular processes and the expression level of HER2 remains stable. When overexpression of HER2 occurs, it can disrupt the dynamic balance of many cellular processes and lead to uncontrolled tumor growth [1].

Breast cancer is one of the leading causes of death worldwide. In 20% of all cases, HER2 is found on breast cancer tissues. The overexpression of HER2 leads to an aggressive course of disease. Recently, targeted tumor therapy became a promising research area based on the increased selectivity of antitumor drugs attached to a targeting molecule, which binds to tumor specific antigens/receptors. Since HER2 might be a good target to prevent tumor growth, current researches are directed to discover HER2 targeting moieties for drug delivery. Trastuzumab-MCC-DM1 (Kadcyla®), an antibody–drug conjugate (ADC) is on the market now [2]. In spite of the benefits of ADCs (e.g. long half-life in circulation, high selectivity), they also have some disadvantages (e.g. low tissue penetration, high costs). Therefore, small molecule–drug conjugates (SMDCs) based on peptides as targeting moieties might have advantages over ADCs [3]. Appropriate homing peptides can be selected by phage display or molecular dynamic (MD) simulations. In our research, KCCYSL peptide selected by phage display technique [4], as well as GYYNPT peptide that binds selectively to HER2 according to MD modelling [5] were chosen. The *in vitro* cellular uptake profile of the peptides and their derivatives was studied by HER2+ cancer cells.

Results

The N-terminally 5(6)-carboxyfluorescein (CF) labelled KCCYSL-NH_2 peptide and its derivatives (one or both of the cysteines were replaced by Ser or Ala) were synthesized by SPPS using Fmoc/tBu strategy. Furthermore, CF-GYYNPT-NH_2 was prepared (Table 1.) for cellular uptake studies. The *in vitro* cellular uptake was measured by flow cytometry on MDA-MB-453 HER2+ cancer cells (medium expression level). The results indicate that CF-KCCYSL-NH_2 and its modified analogs were only moderately taken up by the cells, while CF-GYYNPT-NH_2 did not internalize. The internalization profile of KCCYSL was not significantly influenced by the amino acid replacements in the peptide sequence (Figure 1A).

Table 1: Characteristics of the CF-labelled peptides developed for HER2 receptor recognition

Compounds	HPLC (R_t / min)	MW_{calc} $[M+H]^+$	MW_{meas} $[M+H]^+$
CF-KAAYSL-NH_2	21.25	1007.8	1008.4
CF-KSCYSL-NH_2	21.27	1056.7	1056.4
CF-GYYNPT-NH_2	21.47	1070.8	1070.4
CF-KAAYSLGYYNPT	21.37	1704.5	1704.4
CF-KSCYSLGYYNPT	21.28	1751.9	1752.0
CF-TAKLYPGYANYS-NH_2	21.10	1704.5	1704.0
CF-YSLGYYNPT-NH_2	22.62	1433.8	1433.7

HPLC: Knauer analytical RP-HPLC (Bad Homburg, Germany); Nucleosil 100-5 C18 (250 mm x 4.6 mm; 5 μm, 100Å); eluents: 0.1% TFA/water (A), 0.1% TFA in water-CH3CN (20:80, v/v) (B) linear gradient elution (0 min 2% - 30 min 90%); flow rate 1 mL/min; detection λ = 220 nm.
MS: ESI-MS (Bruker Esquire 3000+ ion trap; Bruker Daltonics, Bremen, Germany); acquisition in 50-2000 m/z range, 4μL/min 0.1% acetic acid in water-CH3CN (50:50, v/v)

Peptides KAAYSLGYYNPT and KSCYSLGYYNPT were designed and synthesized using a combination of two different sequences to increase the biological activity. In contrast to the CF-labelled hexapeptides, the fluorescent signal intensity of CF-KAAYSLGYYNPT , CF-KSCYSLGYYNPT was the highest following the treatment. This observation was in accordance with the microscopic images and fluorescent signal mainly was detected on the cell surface, and only a small amount of signal concentrated on the cytoplasm (Figure 1C). The cell surface localization was also confirmed by trypsinization that resulted in significantly lower fluorescence intensity (Figure 1B).

To identify receptor specificity, a scrambled variant of KAAYSLGYYNPT (CF-TAKLYPGYANYS-NH_2) was also prepared and no cellular uptake was detected. In addition, the usage of an N-terminally shortened homing peptide (CF-YSLGYYNPT -NH_2) did not result in binding. These observations suggest that the KCCYSL sequence based peptides provide the receptor binding capacity, but the GYYNPT sequence that does not recognize the HER2 receptor alone, can increase the binding affinity in the combined version.

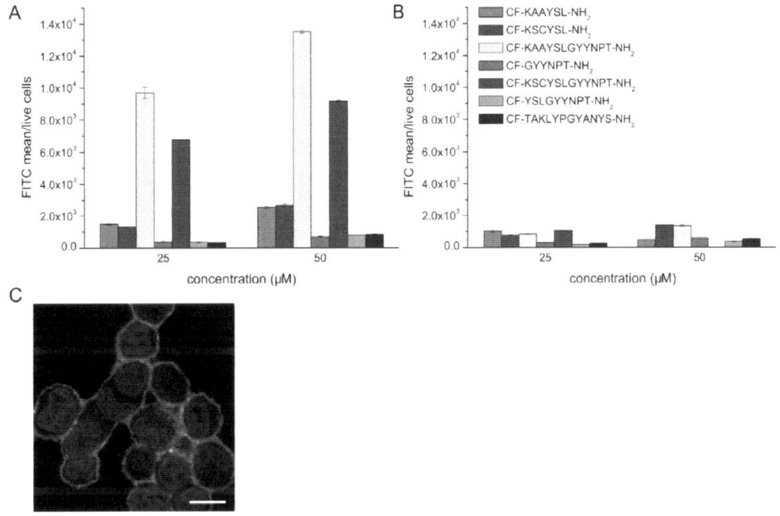

Figure 1: Cellular uptake of CF-labelled peptides by MDA-MB-453 HER2+ breast cancer cells. Uptake was measured after 3 h treatment (A) without trypsinization and (B) followed by 5 min trypsinization by flow cytometry. (C) Cellular uptake of peptide CF-KAAYSLGYYNPT-NH_2 (green) by MDA-MB-453 cells (1 h incubation, concentration: 25 µM) was detected by confocal microscopy. Nuclei are stained with DAPI (blue). The scale bar represents 10 µm.

PEP-FOLD3.5 prediction method was applied for the study of structure – activity relationship [6,7]. The data indicate that the combined peptides might have helical structure, while the shortened and scrambled ones are disordered (Figure 2). This observation is in accordance with the structural features of an engineered HER2 affibody containing three helical parts [8]. To confirm this structure, further CD measurements are in progress.

Figure 2: Predicted structure of peptides by PEP-FOLD3.5 prediction method.

Altogether, the results suggest that the combined peptide is suitable for high affinity receptor binding, but the internalization of the receptor conjugate complex is low. Therefore, these homing peptides may be applied for tumor diagnostic (e.g. PET) or selective delivery of radiotracers with therapeutic activity. In addition, the application of extracellular enzyme (elastase, MMPs) cleavable spacers between the homing peptide and an antitumor agent might be a good choice for the development of conjugates as drug delivery systems.

This project was founded by the National Research, Development and Innovation Office (NKFIH K119552, NVKP_16-1-2016-0036), MedInProt, Hungary and Ministry of Human Capacities (1783-3/2018/FEKUTSRA T).

References

[1] Ménard S.,et al.;Oncogene 22: 6570-6578 (2003) doi: 10.1038/sj.onc.1206779

[2] Lewis Philips GD., et al.; Cancer Res. 68: 9280-9290 (2008) doi: 10.1158/0008-5472.CAN-08-1776

[3] Altmann K-H.; Chimica 72: 154-155 (2018) doi:10.2533/chimia.2018.154

[4] Karasseva NG., et al.; J. Prot. Chem. 21: 287-296 (2002) doi.org/10.1023/A:101974950

[5] Geng L., et al.; Theranostics 6: 1261-1273 (2016) doi: 10.7150/thno.14302

[6] Alland C., et al.; Nucleic Acid Res 33(Web Server issue): W44-49 (2005) DOI: 10.1093/nar/gki477

[7] Néron B., et al.; Bioinformatics 25: 3005-3011 (2009) doi: 10.1093/bioinformatics/btp493

[8] Eigenbrot C., et al.; Proc. Nat. Acad. Sci. USA 107: 15039-15044 (2010) doi: 10.1073/pnas.1005025107

Proceedings of the 35th European Peptide Symposium
Patrick B. Timmons, Chandralal M. Hewage, Michal Lebl (Editors)
European Peptide Society & PSP, 2018

The power of peptides: Cell-penetrating and signal peptide sequences for precision targeting of probes in imaging and sensing in live cells

Aisling Byrne, Christopher Burke, Tia Keyes, Aaron Martin

Dublin City University, Ireland

https://doi.org/10.17952/35EPS.2018.323

Cell-penetrating peptides (CPPs) are short peptide sequences, typically less than 30 amino acids long that are effective in carrying appended impenetrable cargo across the plasma membrane. The most commonly studied CPPs include Transactivator of transcription(TAT), Penetratin, a domain from Antennapedia protein and polyarginines, which are cationic, and are believed to drive transport through endocytosis. Their primary application to date has been to transport therapeutic agents across the cell membrane in clinical applications and have also increasingly been applied to driving luminescent probes into cells for imaging applications.

The main advantage of using CPPs in cell imaging and sensing is that they promote and facilitate cellular uptake of molecules and can do so at low concentrations,without damage to the membrane and with minimal cytotoxicity. A drawback is that endocytosis is the transport mechanism, this can lead to isolation of the cargo in endosomes rather than released into the cytoplasm or organelle of interest. We have exploited the following CPPs based on sequences found in proteins in organelle membranes. In order to direct our Ru(II) probes for imaging and sensing to specific organelles in live cells we have also exploited signal peptide sequences such as the mitochondrial penetrating peptide (MPP) sequence FrFKFrFK (r = d-arginine),[1] nuclear localising signal (NLS) sequence VQRKRQKLMP which is derived from the NF-κB transcription factor,[2] and the endoplasmic reticulum (ER) localising sequence RQIKIWFQNRRMKWKK which is derived from the Drosophila transcription factor Antennepedia.[3] All Ru(II) parent complexes, i.e. without a peptide present, are membrane impermeable without the use of detergents or organic solvent. An advantage of Ruthenium polypyridyl complexes is their synthetic flexibility wherein the ligand can be altered in order to create the required response, for example, 1,10-phenanthroline(phen) or 4,7 diphen1,10-phenanthroline containing complexes are suitable for O2/ROS sensitivity, dipyridophenazine (dppz) promotes DNA binding/ intercalation and causes metal complex emission to switch off in water but switch on when bound and tetraazaphenathrene(tap) for phototoxic activity induced by photoinduced electron transfer to DNA. By combining our Ru(II) probes with CPP sequences we have created a library of probes capable of high precision targeting for imaging, including super-resolution STED imaging, and sensing in live cells.

Measurement of O_2 and ROS in the mitochondria of live cells

By combining an oxygen sensitive Ruthenium probe with a mitochondrial penetrating peptide (MPP) sequence FrFKFrFK (Fig 1A), delivery of the O_2 probe selectively to the mitochondria was achieved. Uptake is quick, with $[(Ru(bpy)_2phen\text{-}Ar)_2\text{-}MPP]^{7+}$ (70 μM) localising in the mitochondria after 2 h. Using Antimycin A, a mitochondrial uncoupler, we were able to increase the O_2 levels in the mitochondria, and measure the concentrations based on the luminescent lifetime using Luminescent Lifetime Imaging Microscopy (LLIM). The lifetime data was collected every 10 minutes following Antimycin A treatment and a decrease in lifetime was observed, consistent with rising oxygen levels as mitochondrial metabolism decreases, and O_2 quenches the emission of the complex. The emission lifetime decrease was observed to fall below the value expected for O_2 saturation, indicating that ROS was produced at the mitochondria which was more effective in quenching the complex luminescence.[4]

Targeting and Imaging mtDNA in live cells

Using the same MPP sequence (Fig 1A) but changing the ruthenium probe to target DNA using the dppz 'light-switch' ligand results in a mitochondrial penetrating complex $[Ru(dppz\text{-}FrFKFrFK)]^{5+}$, that was capable of binding and emitting from mtDNA. Figure 1A (i) shows $[Ru(dppz\text{-}FrFKFrFK)]^{5+}$ (green) colocalised with MitoTracker Deep Red (red) (ii). Along the mitochondrial strands, bright punctate spots in yellow indicate binding of the probe to mtDNA (iii). As the dppz 'light-switch' ligand only emits when protected from water, we can attribute the bright emission to interactions with mtDNA. The luminescence lifetime data was also consistent with DNA binding, and protein or RNA binding could be excluded on this basis.[5]

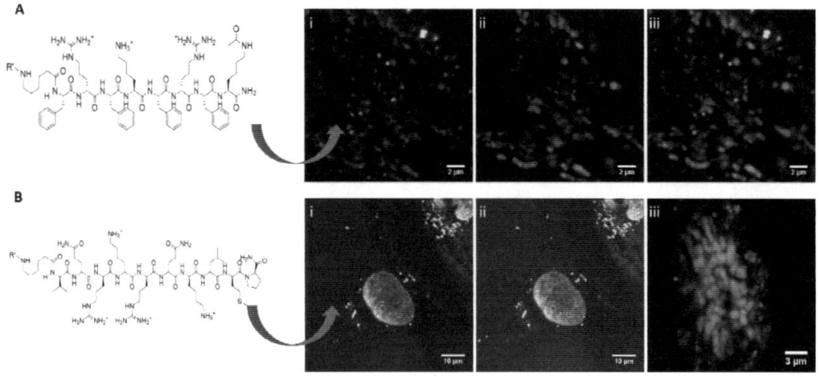

Figure 1: Structure of the mitochondrial penetrating peptide (MPP) sequence FrFKFrFK (A). [Ru(dppz-FrFKFrFK)]5+in the mitochondria of a live HeLa cell in green (i), the commercial probe MitoTracker Deep Red (ii), and their colocalisation in orange (iii). Structure of nuclear localising sequence VQRKRQKLMP (B).[Ru(dppz)(bpy)(bpy-Ar-NLS)]6+ in the nucleus of a live HeLa cell (i), colocalised with DAPI (ii), and it bound to nuclear chromosomes (iii).

STED of Nuclear DNA and chromosomes during mitosis

Using the DNA light-switch ligand we conjugated the nuclear localising sequence (NLS) VQRKRQKLMP (Fig 1B) to this complex resulting in $[Ru(dppz)(bpy)(bpy\text{-}Ar\text{-}NLS)]^{6+}$. Our DNA targeting probe successfully crosses the cell membrane, penetrates the nuclear envelope (Fig 1B (ii)) and colocalises with DAPI (Fig 1B (ii)). Figure 1B (iii) shows that $[Ru(dppz)(bpy)(bpy\text{-}Ar\text{-}NLS)]^{6+}$ binds to nuclear chromosomes.[6] After directing $[Ru(dppz)(bpy)(bpy\text{-}Ar\text{-}NLS)]^{6+}$ to the nucleus, we applied Stimulated Emission Depletion (STED), a super resolution imaging technique. The image contrast of the chromosomes is clearer and more resolved after applying the 660 nm STED depletion laser, compared to confocal, and the groove detail of the sister chromatids can be resolved.[6]

STED of Endoplasmic Reticulum structures

To target the endoplasmic reticulum (ER) we conjugated the endoplasmic reticulum directing signal peptide sequence RQIKIWFQNRRMKWKK to a Ru(II) complex which resulted in $[(Ru(bpy)_2\text{-phen-Ar)-ER}]^{9+}$. $[(Ru(bpy)_2\text{-phen-Ar)-ER}]^{9+}$ (70 μM) successfully crosses the cell membrane, and localizes in the ER, which was confirmed using ER-Tracker Blue. In the confocal image, the ER appear as small, punctate spots. However, upon using the STED 660 nm depletion laser, the tubule structure of the smooth ER becomes clearly resolved. This is reflected in the FWHM of the ER, where it decreased from 532.44 nm to 226.66 nm.[6] Overall, combining CPP or signal peptides with Ruthenium polypyridyl complexes is a powerful approach to imaging cells with super-resolution and can enable real time insights into metabolic changes at discrete organelles.

References

1 Ragin *et al.* Chem Biol. 2002, 9 ,943-948.

2 Horton*et al.* Chemistry & Biology. 2008, 15, 375-382. DOI 10.1016/j.chembiol.2008.03.015

3 Pietersz *et al.* Vaccine. 2001, 19, 1397-1405.

4 Martin *et al.* J. Am. Chem. Soc., 2014, 136, 15300−15309. DOI: 10.1021/ja508043q

5 Burke *et al.* Angew.Chem. Int.Ed., 2018, 57,2420 -12424. DOI: 10.1002/anie.201806002

6 Byrne *et al.* Chem. Sci., 2016, 7, 6551. DOI: 10.1039/c6sc02588a

Development of a new tetrafunctional hybrid to target cancer cells

Maya Georgieva[1], Androniki Kostagianni[2], Eirinaios Vrettos[2], Maria Chatziathanasiadou[2], Andreas Tzakos[2], Tamara Pajpanova[1]

[1]Institute of Molecular Biology Roumen Tsanev, Bulgarian Academy of Sciences, Bulgaria
[2]University of Ioannina, Department of Chemistry, Section of Organic Chemistry and Biochemistry, Greece

https://doi.org/10.17952/35EPS.2018.325

Introduction

Peptide sequences AVPI- and RGD- are interesting subject of targeted drug design in recent years. AVPI (Ala-Val-Pro-Ile) motif is the functionally essential sequence of the N-terminus of pro-apoptotic regulator Smac (Second mitochondria derived activator of caspases). It is shown that different AVPI- and Smac-mimetics could help proceeding of extrinsic and intrinsic apoptotic pathways by inhibition of IAP proteins (Inhibitor of Apoptosis Proteins) [1, 2]. An inverse correlation between Smac and IAP protein levels in a wide range of cancer cells has been reported [2]. RGD (Arg-Gly-Asp) is known for its targeting potential to cancerous cells by binding to molecules over-expressed on their cell membranes [3].

Combining of several motifs with different functional activities into single molecule is an advantageous strategy in drug development in recent decades. Herein, we designed a novel tetra-functional peptide hybrid containing Smac peptide fragment along with RGD targeting unit, a unit for conjugation of drugs/dyes and canavanine-tail in order to lower down the apoptotic thresholds and trigger apoptosis in cancer cells.

Results and Discussion

In order to enhance the anti-proliferative potential of AVPI motif and its cell membrane permeability we designed the following peptide:

We modified the $A^1V^2P^3I^4$ sequence by incorporating L-hydroxyproline instead of proline at the 3rd position. Then lysine residue was attached as a site for further conjugations regarding the data that free N-terminal NH_2 group is needed for more effective functional binding of AVPI sequence. RGD delivering motif was also attached. Then a tail of three canavanines was incorporated at the C-terminus of the peptide (Fig. 1). Cell permeability predictions pointed us that at least three arginine residues in a row are needed for good CPP index of the designed peptide. Instead of arginine we used L-canavanine, the δ-oxa analogue of L-arginine. Canavanine is a non-proteinogenic arginine analogue that shows cell viability inhibitory effect on transformed cells [4.5].

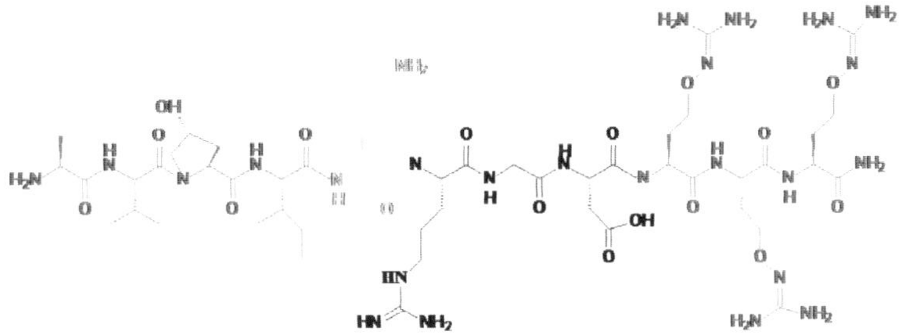

Figure 1: Tetra-functional peptide

The peptide we synthesized by manual standard Fmoc solid-phase protocol. Commercially available Rink-Amide resin (0.59mmol/g substitution)was used. The coupling of each amino acid was performed in the presence of 3 mol excess of: Fmoc-amino acid, HOBt, DIC, in a solution of DMF:DCM (1:1). The successful coupling and the following Fmoc-deprotection were monitored by Kaiser test. Fmoc-groups were removed with solution of 20% Piperidine in DMF with HOBt (0.1M final concentration). Cleavage step from the resin and side chain protecting groups was accomplished by treatment with 10 ml of TFA/dH2O/TIS cocktail (95% :2 .5% : 2.5%).

Crude peptide was purified by semi-preparative HPLC (Fig.2) and characterized by LC/MS.

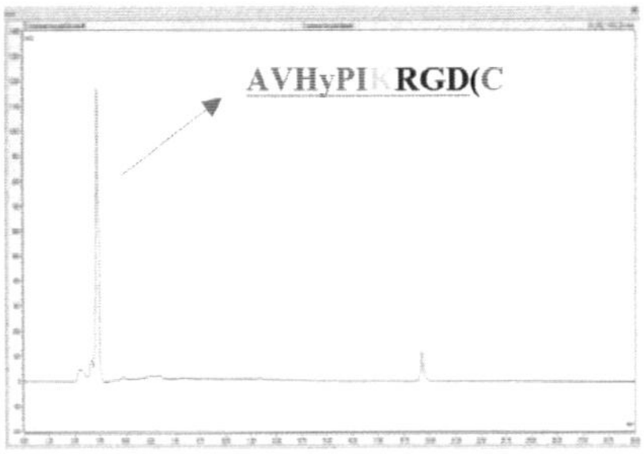

Figure 2: HPLC chromatogram of the crude peptide; Conditions: Gradient: 10% - 100% ACN + 0.1% TFA; Time: 30 min; Detection: 214 nm

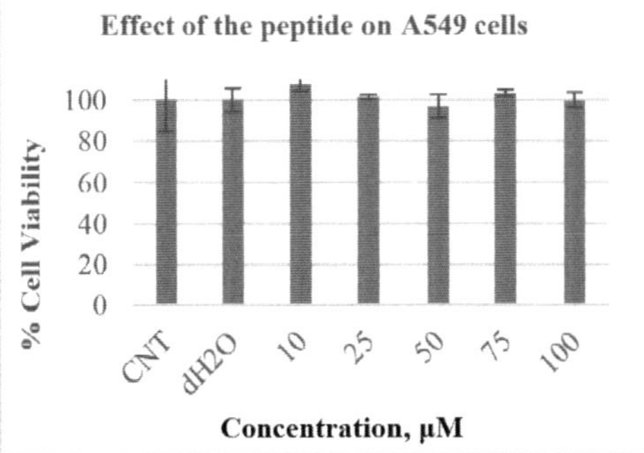

Figure 3: Effect of the peptide on A549 cells after 72h treatment. Each concentration was performed in triplicates.

The anti-proliferative activity of the pure peptide was tested over A549 lung cancer cell line by MTT assay. We found that that the peptide had no effect in the tested concentrations, when it was used as single agents (Fig. 3). Still that is in concert with the literature data describing AVPI-mimetics also as sensitizing agents to different used in practice anticancer drugs [6]. In summary, for first time it was synthesized an AVPI-mimetic with three-canavanine tail used as penetrating vector. The penetration potential of the peptide is under examination.

Acknowledgments

The study was supported by National Science Fund, Ministry of Education and Science, Bulgaria; Project: DN19/17, 20.12.2017; EPS Mobility Fellowship 2018.

References

1. Chen, D., and Huerta S. J. Anti-Cancer Drugs 20, 646-658 (2009).
2.. Martinez-Ruiz, G., *et al.* J Experimental and Clinical Cancer Research 27:48 (2008), .
3. Boohaker, R.J. et al., Curr Med Chem 19 (22), 3794–3804 (2012).
4. Bence, A. and Crooks P, J Enzyme Inhibition and Medical Chemistry 18, 383-394 (2003).
5. Park, H., *et al.* J. Life Science 19, 1529-1537 (2009).
6. Greer, R., *et al.* Cancer Res. 71 (24), 7640–7648 (2011).

Homology modelling and evaluation of the cannabinoid receptor type 2

Fatima Sapundzhi, Tatyana Dzimbova, Nevena Pencheva, Peter Milanov

South-West University "Neofit Rilski", Bulgaria

https://doi.org/10.17952/35EPS.2018.327

Introduction

The human cannabinoid receptor type 2 (CB2) is linked to a variety of immune functional events which are found primarily in brain and tissues of immune and hematopoietic systems like spleen, tonsil, and thymus. In the absence of crystal structure of CB2 receptor, 3D homology patterns with different templates are built [1,2].

The aims of the present study are: (1) to choose within the recently published crystallographic structures templates for homology modelling of the CB2 receptor; (2) to evaluate the models with different computational tools; (3) to precise the most reliable model on the basis of the correlation between data from molecular docking and the values of the biological activity of the cannabinoid ligands.

Materials and Methods

A different software is used in the present work in order to perform computational studies. The protein sequence of the human CB2 receptor was obtained from UNIPROT (Accession number - P34972). Homology modelling studies were carried out using Swiss-Model [3]. Ligand preparation was done with Avogadro. Docking studies were performed by using GOLD 5.2 (Genetic Optimization for Ligand Docking) [4]. Molegro Molecular Viewer was used for generating figures (CLC Bio. Qiagon Inc). In order to find relationship between sets of data derived from *in vitro* assay and docking results, we tried to predict it with the help of Pearson's correlation, using GraphPad Prism 3.0.

Table 1: The values of the inhibitory constant (Ki) of the cannabinoid ligands with known affinity, used in the docking studies [5].

Ligands	CB1 Affinity (K_i)	Type
Anandamide	78 nM	Endogenous
N-Arachidonoyl dopamine	-	Endogenous
2-Arachidonoylglycerol	-	Endogenous
2-Arachidonyl glyceryl ether	21 nM	Endogenous
Δ-9-Tetrahydrocannabinol	10 nM	Phytogenic
EGCG(Epigallocatechin Gallate)	33.6 μM	Phytogenic
Yangonin	0.72 μM	Phytogenic
UR-144	150 nM	Synthetic

The human CB2 receptor protein sequence was obtained from Swiss-Prot website (P34972) and aligned with the sequence of the crystal structure of the human CB1 receptor in complex with agonist AM115421U19 (PDBid:5xra.1) with transmembrane(TM) sequence identity = 46.78%, resolution = 2.8 Å using the PAM-250 matrix, which aligns the sequence based on the conservation of charged, bulky 13 aliphatic or aromatic residues (https://swissmodel.expasy.org/). Alignment was refined manually by taking the 7 TM regions and structurally conserved regions into consideration. The 7 TM regions were identified based on the conservation of residues in each putative TM region.

Table 2: The values of scoring functions in GOLD 5.2 for performed docking between the model of the CB2 receptor obtained by homology modelling and affinity constant Ki of the known cannabinoid ligands.

Ligands	**ASP Score**	**ChemScore**	**ChemPLP**	**GoldScore**
Δ-9-Tetrahydrocannabinol	13,99	37,30	23,33	45,24
Yangonin	14,57	30,06	18,90	40,57
UR-144	19,84	46,48	23,69	33,71
EGCG(Epigallocatechin Gallate)	17,06	30,34	0,89	37,70
Anandamide	18,03	56,40	20,03	61,24
2-Arachidonoylglycerol	21,29	54,27	18,03	53,70
N-Arachidonoyl dopamine	20,20	64,42	25,60	42,64

The binding site was defined as residues within 10 Å radius of Cys257. Docking was performed with the chosen model and 7 ligands (Table 2) [6]. There is a significant correlation between the results of the docking (ChemScore function) and the biological activity of the ligands (Table 3, Fig.1), (Pearson R = -0.945). This correlation describes the relationship between the biological activity of the compounds and the docking results.

Table 3: The values of Pearson's coefficient for the correlation between GOLD scoring functions for model of the CB2 receptor obtained by homology modelling and affinity constant Ki of the cannabinoid ligands.

ASP Score	ChemPLP	ChemScore	GoldScore
R=-0,1343 (p_{value}=0,7740)	R=-0,5118 (p_{value}=0,2403)	R=-0,9456 (p_{value}=0,0013)	R=-0,3426 (p_{value}=0,4519)

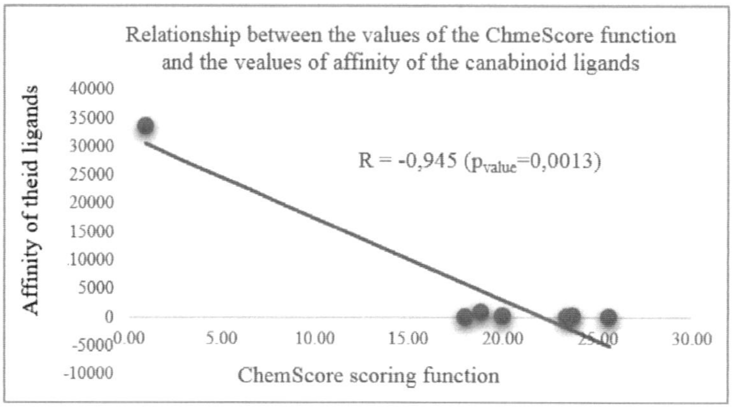

Figure 1: Pearson's correlation between the values of Affinity of cannabinoid ligands and the values of ChemScore scoring function from molecular docking.

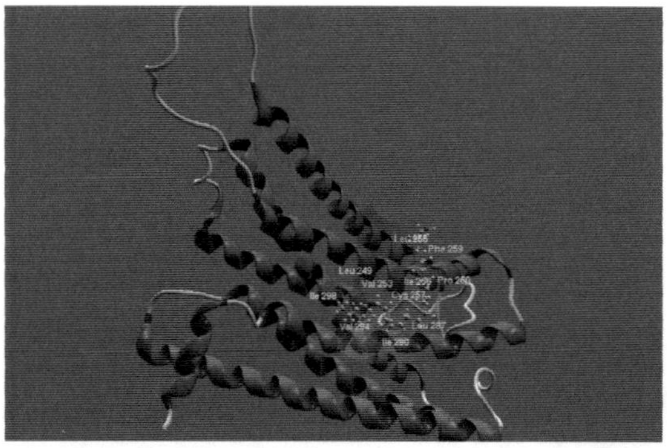

Figure 2: Graphical representation of the docking between the cannabinoid ligand Anandamid and the ChemScore scoring function for model of CB2 receptor obtained by homology modelling.

New model of the CB2 receptor was generated and evaluated by different approaches. Newly generated model by homology modelling could be used further for in silico experiments and along with the correlation found between *in vitro* and docking experiments will give possibility for faster and more correct design of selective and effective ligands for CB receptors [7-10].

References

1. Mackie K. Journal of Neuroendocrinology, 20 (1), 10-14, (2008)

2. Begg M., *et al.* Pharmacology & Therapeutics 106, 133 – 145, (2005)

3. Waterhouse A., *et al.* Nucleic Acids Res. 46(W1), W296-W303 (2018).

4. Verdonk M., et al., Proteins, 52, 609-623, (2003)

5. Pertwee R., Tocris Reviews,16, 1-16, (2006)

6. Pei Y., *et al.* Chemistry & biology, 15 (11), 1207–1219, (2008)

7. Sapundzhi F., Dzimbova T *et al.* Journal of Peptide Science, 20, S294-S295, (2014).

8. Sapundzhi F., *et al.* Bulgarian Chemical Communications, 49 (4), 768, (2017).

9. Sapundzhi F., *et al.* Bulgarian Chemical Communications, 49 (Special issue E), 23 – 30, (2017).

10. Dzimbova T., *et al.* Journal of Peptide Science, 18, S84, (2012).

Author Index

European Peptide Society
c/o Professor Paula Gomes
Rua do Campo Alegre, s/n
PT-4169-007 Porto
Portugal

ISBN 978-1-5108-8036-8

9 781510 880368

CURRAN ASSOCIATES INC.
proceedings.com